T0402181

Plant Genome Diversity Volume 1

Jonathan F. Wendel
Editor-in-chief

Johann Greilhuber · Jaroslav Doležel ·
Ilia J. Leitch
Editors

Plant Genome Diversity Volume 1

Plant Genomes, Their Residents, and Their Evolutionary Dynamics

 Springer

Editor-in-chief
Jonathan F. Wendel
Department of Botany
Iowa State University
Ames, Iowa
USA

Editors

Johann Greilhuber
Department of Systematic and Evolutionary Botany
Faculty Center Botany
University of Vienna
Vienna
Austria

Jaroslav Doležel
Institute of Experimental Botany ASCR
Centre of the Region Hana for
Biotechnological and Agricultural
Research
Olomouc
Czech Republic

Ilia J. Leitch
Jodrell Laboratory
Royal Botanic Gardens, Kew
Richmond, Surrey
United Kingdom

ISBN 978-3-7091-1129-1   ISBN 978-3-7091-1130-7 (eBook)
DOI 10.1007/978-3-7091-1130-7
Springer Wien Heidelberg New York Dordrecht London

Library of Congress Control Number: 2012935228

Printed on acid-free paper

Springer is part of Springer Science+Business Media (www.springer.com)

Preface

Ever since Darwin, biologists have been interested in understanding the intricacies of natural variation patterns and the evolutionary forces that shape this diversity. Although these dual objectives have long been central goals in evolutionary biology, they recently have assumed a new prominence arising from the development of breathtaking new genomic technologies and their application to natural systems. These technological leaps have invigorated the already thriving discipline variously referred to as *molecular evolution*, or *genome evolution*, or sometimes *evolutionary genomics*, empowering it with a vastly expanded insight into the diversity of genomes and their evolutionary dynamics. As a result, excitement in the discipline has never been higher. Notwithstanding this vibrancy and the attendant meteoric rise in the number of published papers and professional journals devoted to molecular evolution, remarkably few books on the topic are available. To be sure, there are a number of classic texts, written more than a decade ago, just as the genomics revolution was ramping up, and a larger suite of hybrid books variously combining aspects of bioinformatics, genome evolution, population genetics, and methods of phylogenetic inference. To date, however, there exists no single modern treatment of what we have learned about the diversity and evolution of plant genomes and their various genomic residents.

It was to fill this void that the present project was initiated. Inspired by the seemingly ever-expanding pace of insights into the evolution of plant genomes, and motivated by the desire to provide for students and researchers a single point of entry into a burgeoning literature, we invited leading authorities in plant molecular evolution to participate in a project aimed at providing a comprehensive (but not encyclopedic) yet accessible introduction to the current state of the art in the field. This is accomplished here in a total of 16 chapters that collectively cover the discipline. Although these are arranged in a logical progression and are interconnected, each chapter also serves as a stand-alone introduction and review, thus providing a text that may flexibly be used by advanced undergraduate students, graduate students, and professionals in many fields in the plant sciences and beyond.

The volume appropriately begins (Flagel and Blackman, Chap. 1) with a review of the immense insights that have been gleaned from plant genome sequencing projects, as well as a prospective view of both the promises and challenges that lie ahead. This is followed by two complementary chapters on the primary constituents of plant genomes, namely, transposable elements (TEs); the first of these (Kejnovsky et al., Chap. 2) focuses on the diversity of TEs, their genomic ecology, and their role in genome size evolution, whereas the second (Slotkin et al., Chap. 3) reviews the remarkable role TEs play in genetic and epigenetic regulation, and as evolutionary fodder for the origin of novel genes and for chromosomal evolution. Perhaps the most obvious features of chromosomes are centromeres and telomeres, for which our knowledge regarding structure and evolution have been dramatically increased by genomic technologies, as reviewed by Hirsch and Jiang (centromeres, Chap. 4) and Siomos and Riha (telomeres, Chap. 5).

Having described the major structural features and organization of plant genomes, we turn our attention to smaller genomic residents, including small RNAs, for which Lee et al. (Chap. 6) present a synopsis of the diversity, regulatory roles, and evolution of the different classes of small RNAs. This is followed by a chapter on genic evolution, with a special focus on rate variation within and among lineages and the utility of this information for timing divergence events (Burleigh, Chap. 7), and on the detection and significance of conserved non-coding DNA (Subramaniam and Freeling, Chap. 8). Mowers et al. (Chap. 9) and Wolf (Chap. 10) offer timely reviews of the structure and evolutionary dynamics of plant mitochondrial and plastid genomes, respectively.

One of the key emergent realizations of the genomics era has been that plant genomes are replete with evidence of historical and ongoing duplications, large and small. Barker et al. (Chap. 11) review the processes that generate duplications as well as their longer term evolutionary outcomes, whereas Nieto-Feliner and Rossello (Chap. 12) present an update on a curious non-Mendelian consequence of sequence multiplicity, namely, sequence homogenization via one or more means of "concerted evolution". Paterson et al. (Chap. 13) further describe the consequences of genome duplication and divergence on longer-term colinearity and synteny relationships among divergent lineages. A final consequence of genome divergence, variation in base composition, is considered by Šmarda and Bureš (Chap. 14), who provide an overview of the phenomenon and its possible causative forces. We close with two chapters devoted to the vibrant new frontier of plant epigenomics, one (Zhang, Chap. 15) describing the epigenetic landscape in plants and the various forms of chromatin modification, and the second (Richards et al., Chap. 16) devoted to the evolutionary signification of epigenetic variation.

We are living in a tremendously exciting time to be a biologist, perhaps one that in the future will be thought of as having been a "golden era", replete with technological and conceptual breakthroughs. We hope that you find this volume evocative in this sense, as stimulating to read as it was to produce, and inspiring in the promise of its content.

Of course there are many people to thank for bringing this project to fruition. First and foremost are the many authors, who are experts in their field and hence are very busy people. Yet they willingly and generously set aside the time to imagine and create their contributions. To them I offer my sincere appreciation. I also offer thanks to many of my professional colleagues, who reviewed drafts of the manuscripts and offered numerous insights and suggestions for improvement. It should go without saying that this book would not have been possible without the vision of the publisher, Springer-Verlag, who recognized that this volume would fill an important and presently largely vacant niche, and for moving the project along expeditiously toward completion. Finally, I need to express my delighted gratitude to my co-editors of this soon-to-be two volume set, Ilia Leitch, Jaroslav Doležel, and especially Editor-in-Chief Johann Greilhuber, for their many modes of assistance throughout project conception and execution.

Ames, Iowa (United States)   Jonathan Wendel

Contents

Contributions

A.J. Alverson Department of Biology, Indiana University, Bloomington, IN, USA, andy.alverson@gmail.com

M.S. Barker Department of Ecology & Evolutionary Biology, University of Arizona, Tucson, USA, msbarker@email.arizona.edu

G.J. Baute Department of Botany, University of British Columbia, Vancouver, Canada, gregbaute@gmail.com

Benjamin K. Blackman Department of Biology, Duke University, Durham, NC, USA, bkb7@duke.edu

O. Bossdorf Institute of Plant Sciences, University of Bern, Bern, Switzerland

Petr Bureš Faculty of Science, Department of Botany and Zoology, Masaryk University, Brno, Czech Republic, bures@sci.muni.cz

J.G. Burleigh Department of Biology, University of Florida, Gainesville, FL, USA, gburleigh@ufl.edu

Cédric Feschotte Department of Biology, University of Texas, Arlington, TX, USA, cedric@uta.edu

Lex E. Flagel Department of Biology, Duke University, Durham, NC, USA, lex.flagel@duke.edu

M. Freeling Department of Plant and Microbial Biology, University of California, Berkeley, CA, USA, freeling@berkeley.edu

Jennifer S. Hawkins Department of Biology, West Virginia University, Morgantown, WV, USA, jhawkins@uga.edu

C.D. Hirsch Department of Horticulture, University of Wisconsin-Madison, Madison, WI, USA, cdhirsch@wisc.edu

J. Jiang Department of Horticulture, University of Wisconsin-Madison, Madison, WI, USA, jjiang1@wisc.edu

N. Jiang Department of Horticulture, Michigan State University, Michigan, USA, jiangn@msu.edu

Eduard Kejnovsky Laboratory of Plant Developmental Genetics, Institute of Biophysics, ASCR, Brno, Czech Republic, kejnovsk@ibp.cz

T.-F. Lee Delaware Biotechnology Institute, University of Delaware, Newark, DE, USA, tzuufen@dbi.udel.edu

T. H. Lee   Plant Genome Mapping Laboratory, University of Georgia, Athens, GA, USA, alfalfa@gmail.com

P. Li   Delaware Biotechnology Institute, University of Delaware, Newark, DE, USA, li@dbi.udel.edu

S.-L. Liu   Department of Botany, University of British Columbia, Vancouver, Canada, shaolunliu@gmail.com

B. C. Meyers   Delaware Biotechnology Institute, University of Delaware, Newark, DE, USA, meyers@dbi.udel.edu

Jeffrey P. Mower   Center for Plant Science Innovation, E128 Beadle Center, University of Nebraska Lincoln, Lincoln, NE, USA, jmower2@unl.edu

Gonzalo Nieto-Feliner   Real Jardin Botnico, CSIC, Madrid, Spain, nieto@rjb.csic.es

S. Nuthikattu   Plant Cellular and Molecular Biology, The Ohio State University, 570 Aronoff Laboratory, Columbus, OH, USA, nuthikattu.1@osu.edu

Andrew H. Paterson   Plant Genome Mapping Laboratory, University of Georgia, Athens, GA, USA, paterson@dogwood.botany.uga.edu

C. L. Richards   Department of Integrative Biology, University of South Florida, Tampa, FL, USA, christinalrichards@gmail.com

K. Riha   GMI – Gregor Mendel Institute of Molecular Plant Biology GmbH, Vienna, Austria, karel.riha@gmi.oeaw.ac.at

J. A. Rossello   Jardín Botnico, Universidad de Valencia, Valencia and Marimurtra Botanical Garden, Carl Faust Foundation, Blanes, Spain, rossello@uv.es

M. Siomos   Department of Biochemistry and Cell Biology, University of Vienna, Vienna, Austria, maria.siomos@univie.ac.at

D. B. Sloan   Department of Biology, University of Virginia, Charlottesville, VA, USA, dbs4a@virginia.edu

K. Slotkin   Plant Cellular and Molecular Biology, The Ohio State University, 570 Aronoff Laboratory, Columbus, OH, USA, slotkin.2@osu.edu

P. Smarda   Department of Botany and Zoology, Masaryk University, Brno, Czech Republic, smardap@sci.muni.cz

S. Subramaniam   Department of Plant and Microbial Biology, University of California, Berkeley, Berkeley, CA, USA, sshabari@gmail.com

H Tang   Plant Genome Mapping Laboratory, University of Georgia, Athens, GA, USA, bao@uga.edu

K. J. F. Verhoeven   Department of Terrestrial Ecology, Netherlands Institute of Ecology (NIOO-KNAW), Wageningen, The Netherlands

X. Wang   Plant Genome Mapping Laboratory, University of Georgia, Athens, GA, USA, wangxy@uga.edu

P. G. Wolf   College of Science, Department of Biology, Utah State University, Logan, UT, USA, paul.wolf@usu.edu

X. Zhang   Department of Plant Biology, University of Georgia, Athens, GA, USA, xiaoyu@plantbio.uga.edu

The First Ten Years of Plant Genome Sequencing and Prospects for the Next Decade

1

Lex E. Flagel and Benjamin K. Blackman

Contents

1.1   Introduction

The genome of *Arabidopsis thaliana*, the first completed plant genome sequence, was published in December of 2000 (The Arabidopsis Genome Initiative 2000). This event marked the beginning of the plant genomics era. Over the next 10 years, there has been swift and striking progress in the field of plant genomics. An outpouring of effort coupled with technological advances has made it possible to sequence, assemble, and analyze the genomes of many additional plant species, including several genomes far larger and more complex than *Arabidopsis*. These changes have ushered in comparative plant genomics, the study of relationships between genomes of different species. Comparative genomic analysis has proven particularly enlightening in revealing recent and ancient events that have impacted the structure and contents of plant genomes; and this field is poised to grow rapidly with the advent of high throughput "Next-Generation" sequencing technologies.

In this chapter we revisit some of the breakthroughs and insights that have emerged from the first decade of plant genomics. We summarize our current understanding of plant genomes, based on more than a dozen published sequences. This is followed by a discussion of the opportunities and challenges that we anticipate over the next decade. We have organized these topics in three chronological phases. In the first phase we focus the on *A. thaliana* genome, which on its own yielded a plethora of discoveries and set the stage for further plant genome sequencing. In the second phase we look beyond *Arabidopsis* to the next wave of plant genomes and the lessons learned from comparing them to *A. thaliana* and to one another. Finally we look forward, into a future where the cost of sequencing a plant genome may be orders of magnitude cheaper than it is today, and we consider the rewards and challenges that this new technological capability will bring.

L.E. Flagel (✉)
Department of Biology, Duke University, 90338, Durham, NC 27708, USA
e-mail: lex.flagel@duke.edu

J.F. Wendel et al. (eds.), *Plant Genome Diversity Volume 1*,
DOI 10.1007/978-3-7091-1130-7_1, © Springer-Verlag Wien 2012

1.2 *Arabidopsis*: The Beginning of the Plant Genomics Era

Scientific studies of *A. thaliana* began in its native Europe in the early 1900s. Seminal research was conducted by Friedrich Laibach and colleagues who developed the first mutants and promoted *A. thaliana* as a tractable species for experimentation (Somerville and Koornneef 2002; Koornneef and Meinke 2010). *Arabidopsis thaliana* possesses a suite of life history traits that make it a convenient study system, including prodigious seed production, small stature, modest growth requirements, cross- and self-pollination compatibility, and a short life cycle. Interest grew steadily among other European researchers who were attracted by the convenience of working with *Arabidopsis* and by stimulating early experiments that demonstrated the power of this new system (Meinke et al. 1998; Somerville and Koornneef 2002; Koornneef and Meinke 2010). By the late 1970s *Arabidopsis* research communities were well-established in the United States and elsewhere, and by the early 1990s—following a string of major discoveries—*A. thaliana* had cemented itself as *the* model organism for many forms of basic plant research (Somerville and Koornneef 2002; Koornneef and Meinke 2010). By this time *Arabidopsis* researchers had developed a rapid genetic transformation system (Feldmann and Marks 1987), and work was well underway toward characterizing thousands of mutants and assessing the natural variation of the species. Simultaneous advances in DNA sequencing technology made whole genome sequencing a reasonable proposition. With an active community and an enviable collection of research tools now in place, it became clear that *A. thaliana*—with its diminutive 125 million base pair genome—was the top plant candidate for whole genome sequencing.

The Arabidopsis Genome Initiative (AGI) was formed in 1996 (Bevan 1997), putting in place the funding, organizational, and intellectual apparatus needed to sequence and analyze the genome. The AGI elected to sequence the accession "Columbia", a fecund inbred line that had been used as the wild type strain in many studies dating back to the 1960s (Koornneef and Meinke 2010). AGI also chose the genome sequencing strategy, the so-called BAC-by-BAC method, which involves recapitulating the linear order of the genome in a series of large genomic fragment clones. Initial reports of the second and fourth chromosome assemblies (Lin et al. 1999; Mayer et al. 1999) were followed by reports of the remaining three chromosomes and a detailed analysis of the complete genome in December of 2000 (The Arabidopsis Genome Initiative 2000; the publications of first, third and fifth chromosomes referenced therein). At the time of publication *A. thaliana* was the third multicellular eukaryote to

have its entire genome sequenced, following the nematode (*Caenorhabditis elegans*) and fruit fly (*Drosophila melanogaster*).

The genome assembly consisted of approximately 115.4 million nucleotides. The estimated genome size of *A. thaliana* is approximately 125 million base pairs, however, indicating that approximately 10 million base pairs remained unassembled (The Arabidopsis Genome Initiative 2000). A substantial portion of this unassembled sequence likely originates from highly repetitive areas such as centromeres, and consequently the finished assembly is arranged in ten linear molecules, each representing a chromosome arm from one of *A. thaliana*'s five chromosomes. Now, even a full decade later, much of this unassembled sequence is still unaccounted for, and may remain so for some time to come. Despite this limitation, the assembled portion of the *Arabidopsis* genome would prove to be invaluable to the plant research community.

It is difficult to overstate the novelty and transformative impact of the AGI's findings. Their analyses offered numerous insights, many of which may now seem almost self-evident given the rapid pace of plant genomics research. With the benefit of hindsight, we must consider the weight of these discoveries and basic descriptions at a time when there was little knowledge regarding plant genome architecture. For example, the basic genome organization offered interesting perspectives. Prior to the availability of the *Arabidopsis* genome, cytogeneticists had determined that plant genomes had two well-defined domains, heterochromatin and euchromatin. Heterochromatin consists of densely packed DNA, whereas euchromatin is loosely packed, and both are easily distinguished by chromosome staining techniques. This cytogenetic work suggested that genes would largely be found in euchromatin, because regulatory machinery could access them within the loosely packed structure. In *Arabidopsis*, genes were largely found in regions between the centromere and telomere that closely matched cytogenetically identified euchromatin (Lin et al. 1999; Mayer et al. 1999; The Arabidopsis Genome Initiative 2000). Moreover, transposable elements showed the opposite pattern, being most abundant near the heterochromatic centromeres and rare in gene-rich euchromatin. Despite being an anticipated result, this finding confirmed the general layout of a plant genome and provided a link between the cytogenetic and physical landscapes of the genome. Only a decade ago the aforementioned plant chromosomal organization was an unconfirmed hypothesis; the *Arabidopsis* genome provided the first concrete support.

Additional major insights came from cataloging gene content. As previously mentioned, *A. thaliana* has a small genome, in fact one of the smaller genomes known in plants. It was initially reported that approximately 25,500 genes reside within this relatively small genome (The Arabidopsis

Genome Initiative 2000). Since that time, this number has been revised slightly upward to approximately 27,400 (TAIR 9 genome release; www.arabidopsis.org), reflecting improvements in gene discovery and annotation. Nonetheless, given its small genome size, the *A. thaliana* genome is densely populated with genes, having approximately one gene every 4.5 kilobases (kb) on average (The Arabidopsis Genome Initiative 2000). A surprising amount of insight derives from this simple finding. For one, the total number of genes in *A. thaliana* is greater than in some eukaryotes (*D. melanogaster* and *Saccharomyces cerevisiae*, for example), and comparable to others (for example, *C. elegans* and several mammals). Plants are sessile and must adapt to their local environment, while at the same time maintaining all the necessary equipment needed to generate their own energy and synthesize scores of complex biomolecules. In

light of these demands, it is impressive that *A. thaliana* accomplishes these feats with such a modest number of genes. A second implication of this finding is that plants with much larger genomes than *A. thaliana* must either have many more genes or considerably lower gene density. As we know now, the answer appears to be primarily the latter alternative, as no diploid plant genome sequenced to date has more than approximately 60,000 genes (Velasco et al. 2010) despite haploid genome sizes more than an order of magnitude greater than *A. thaliana* (Fig. 1.1).

Beyond simply counting genes, the initial *Arabidopsis* genome analysis focused on the functional characteristics and evolutionary relationships of these genes. By comparing the gene content of *A. thaliana* to other major lineages available at the time (bacteria: *Haemophilus influenzae*, fungi: *S. cerevisiae*, animals: *D. melanogaster* and *C. elegans*), the

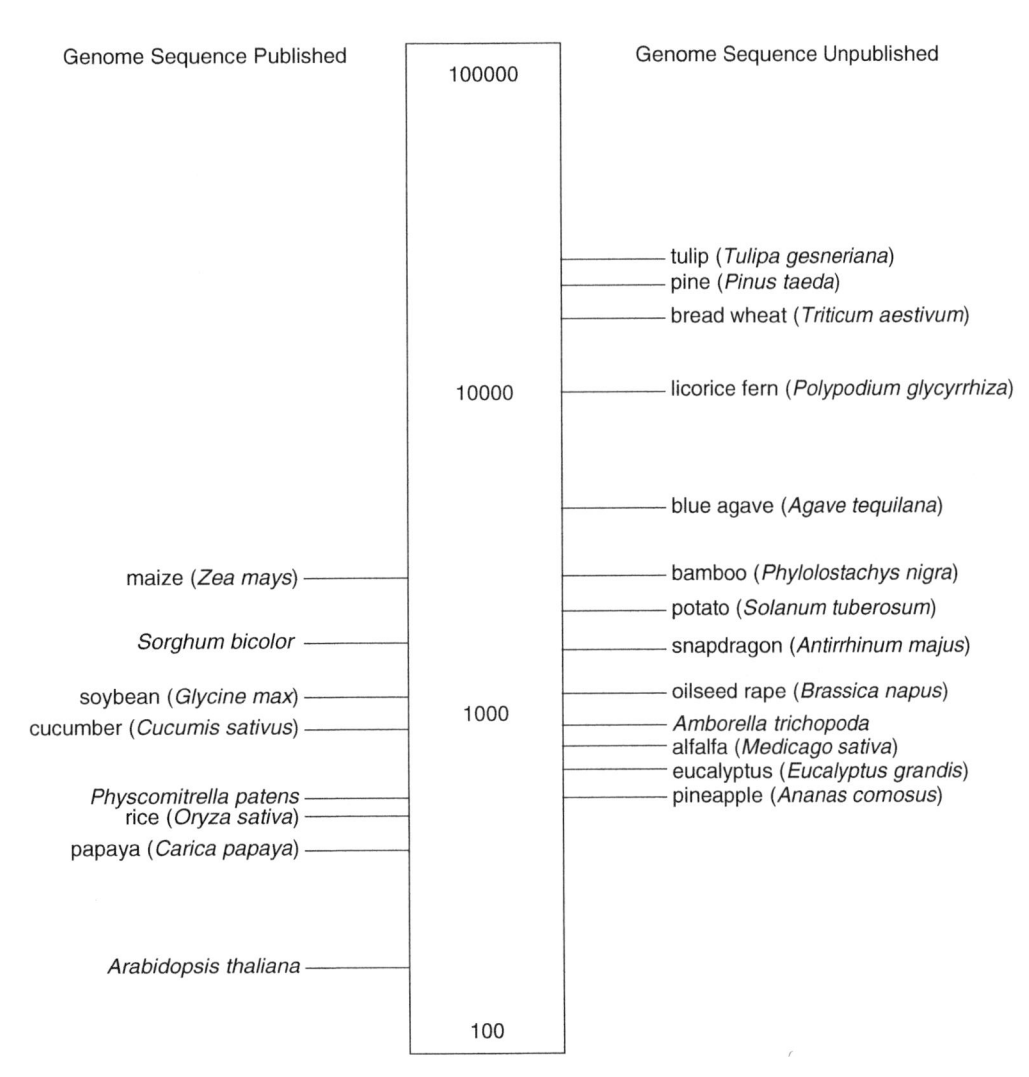

Fig. 1.1 Comparison of plant genome sizes for selected species, including some with (*left side*) and some without (*right side*) sequenced genomes. All plant genome size estimates were taken from the Kew Plant C-values Database (Bennett and Leitch 2010), and are plotted in millions of base pairs (Mbp) on a logarithmic scale

AGI concluded that about 150 major protein families were unique to *A. thaliana* (The Arabidopsis Genome Initiative 2000). At first blush this seems surprisingly low. The divergence of the plant lineage from an ancestral eukaryote is estimated to have occurred 1.5 billion years ago (Yoon et al. 2004), leaving a very long time period for plants to invent new kinds of genes. Two factors likely explain this result. First, there are a large number of basic processes conserved among most organisms—signaling, transcription, genomic maintenance, and metabolic functions are good examples—and these processes involve conserved genes and gene families that encode the core metabolic functions of a living cell. For example, a collection of orthologous eukaryotic proteins show that *A. thaliana* shares 3,285 gene families with at least one other anciently-derived lineage, including the animals *C. elegans*, human, and *D. melanogaster*, the fungi *Schizosaccharomyces pombe* and *S. cerevisiae*, and a basal eukaryote *Encephalitozoon cuniculi* (accessed at NCBI's KOG server: www.ncbi.nlm.nih.gov/COG/). A second explanation for the relative rarity of gene families unique to *A. thaliana* is ascertainment bias. It is easier to annotate known gene families in a newly assembled genome than it is to discover unknown genes. In any case, despite more than a billion years of independent evolution and despite all of the outward differences between plants and other eukaryotes, the *A. thaliana* genome demonstrates that these lineages have much in common at the genetic level.

Another major finding revealed by examining the gene families in the *A. thaliana* genome was their incredible size and redundancy. For example, about 41% of *A. thaliana* genes belong to a gene family with five or more members (The Arabidopsis Genome Initiative 2000), a value considerably higher than what is observed in animal species, where singleton genes are abundant (Lockton and Gaut 2005). Since the publication of the genome sequence, reverse genetic analysis of *A. thaliana* gene families has become something of a cottage industry, with research groups around the world using tools such as T-DNA insertion mutant collections, cDNA and tiling microarrays, and genetic transformation to explore the functions of numerous gene families. Amazingly, some of the largest gene families were scarcely detected before the *A. thaliana* genome sequence became available. One example, the pentatricopeptide repeat genes (PPRs)—including approximately 450 genes in *A. thaliana* and believed to play a role in RNA processing in the mitochondria and chloroplasts—were virtually unknown to plant geneticists prior to a careful analysis of the genome (Aubourg et al. 2000; Small and Peeters 2000; Schmitz-Linneweber and Small 2008). Since their discovery, knockouts of some members of the PPR gene family have been shown to cause disparate and severe mutant phenotypes (e.g., Lurin et al. 2004; Cushing et al. 2005), indicating that they did not escape notice for lack of meaningful function. Another example are the S1 self-incompatibility-like

proteins (Ride et al. 1999), which account for approximately 80 genes in the *Arabidopsis* genome and are predicted to be involved in signaling via protein–protein interactions. Despite their large sizes, these gene families "came out of the blue" following whole genome sequencing. These findings speak to the importance of whole genome sequencing in gene discovery and also underscore the enormous role that the *Arabidopsis* genome played in this regard.

The expansion of *A. thaliana* gene families can be largely attributed to duplicated chromosomal segments, many of which are predicted to have arisen via whole genome duplication (polyploidy) (The Arabidopsis Genome Initiative 2000; Vision et al. 2000; Tang et al. 2008a, b). The initial analysis of the *A. thaliana* genome identified 24 large duplications, which in total make up approximately 60% of the genome (The Arabidopsis Genome Initiative 2000). Contemporary *A. thaliana* is a diploid; thus these polyploidy events, apparent from their enduring gene duplications, must have occurred in a distant ancestor (a condition termed *paleopolyploidy*), with the corollary that the genome has since mutationally eroded back to diploidy. The AGI concluded that the duplications were likely the result of an ancient tetraploidy event, as they found evidence for only two pairs of duplicates. As we will see in the next section, comparative genomics has greatly enriched our understanding of paleopolyploidy in plants, and we now have convincing evidence that *Arabidopsis* has experienced multiple rounds of whole genome duplication including an ancient hexaploidy (Ming et al. 2008; Tang et al. 2008a, b).

Beyond sequencing and assembling the genome of the Columbia accession, AGI also partially shotgun sequenced another *A. thaliana* accession, Landsberg *erecta* (The Arabidopsis Genome Initiative 2000). With both genome sequences in hand, these accessions were compared to assess basic parameters of diversity and polymorphism within the species. AGI found that Columbia and Landsberg *erecta* have a single nucleotide polymorphism (SNP) every 3.3 kb, on average, in addition to approximately 14,500 insertion/deletion (indel) polymorphisms scattered throughout the genome. These polymorphisms were found within genes and in intergenic regions. After combining SNPs and indels, the AGI found that 7% of exons had a polymorphism, including numerous indels and amino acid changing substitutions, providing a sizable number of coding region differences that distinguish the two accessions. Finally, they found evidence for more significant structural changes, including transposition events and gene rearrangements. These data were the first assessment of genome-wide polymorphism in a plant species, and ushered in the era of plant *population genomics* (population genetics extended to a genome-wide scope), now a major thrust of plant genome research.

One final virtue of the *Arabidopsis* genome project that merits attention is the path it blazed for future plant genome

Table 1.1 Published plant genomes

	Species	Citation
1	*Arabidopsis thaliana*	The Arabidopsis Genome Initiative (2000)
2	Rice (*Oryza* spp.)	Goff et al. (2002) and Yu et al. (2002)
3	Poplar (*Populus trichocarpa*)	Tuskan et al. (2006)
4	Grape (*Vitis vinifera*)	Jaillon et al. (2007)
5	Papaya (*Carica papaya*)	Ming et al. (2008)
6	*Physcomitrella patens*	Rensing et al. (2008)
7	Maize (*Zea mays*)	Schnable et al. (2009)
8	Sorghum (*Sorghum bicolor*)	Paterson et al. (2009)
9	Apple (*Malus* × *domestica*)	Velasco et al. (2010)
10	*Brachypodium distachyon*	International Brachypodium Initiative (2010)
11	Castor bean (*Ricinus communis*)	Chan et al. (2010)
12	Cucumber (*Cucumis sativus*)	Huang et al. (2009)
13	Soybean (*Glycine max*)	Schmutz et al. (2010)
14	Cacao (*Theobroma cacao*)	Argout et al. (2011)
15	Strawberry (*Fragaria vesca*)	Shulaev et al. (2011)

sequencing efforts in terms of its community structure and public accessibility. At the time of publication there was little precedent for funding, sequencing, and data access for genome sequencing projects. In the late 1990s and early 2000s many differing ideologies and funding models were in circulation, ranging from full public funding with open data access to private funding and data access fees. The *Arabidopsis* genome project presented a model of collaboration that has guided many subsequent plant genome initiatives. It was organized and paid for through a large, multinational collaboration between academic, governmental, and corporate interests, each sharing data and contributing expertise. The genome sequence and annotation were released to the public without access fees. Furthermore, the *Arabidopsis* community organized a web portal called The *Arabidopsis* Information Database (TAIR; www.arabidopsis.org) shortly following publication. This site continues to serve as a primary hub for *Arabidopsis* information, allowing access to sequence data, a genome browser, gene annotations, and ordering information for *Arabidopsis* germplasm. By fostering interaction, sharing, and the foresight to freely distribute genomic information through a centralized hub, the AGI can be credited with putting forth an effective model that has been built upon by many subsequent plant genome sequencing projects.

1.3 The Next Series of Plant Genomes and the Beginning of Comparative Plant Genomics

Arabidopsis opened the floodgates for plant genome sequencing. To our knowledge, the primary descriptions of 15 plant genomes have been published at the time of this writing. Table 1.1 lists these species along with their year of publication and citation, while Fig. 1.2 shows their hypothesized phylogenetic relationships. Beyond these 15 plant species, there are dozens more with genome projects underway, at stages ranging from early planning to near-completion. Indeed, 5 of the 15 published plant genomes in Table 1.1 were released in 2010, and we can expect 2011 and beyond to be equally productive given the large number of ongoing sequencing projects.

As more plant genomes become available, the possibility emerges for functional and evolutionary comparisons among species. Many interesting developments have emerged from this comparative approach to plant genomics. In this section we highlight three key findings—conserved gene order, ancestral polyploidy, and conserved gene content—each with relevance to the future of plant genome sequencing. For in-depth coverage of these topics, as well as other topics in comparative plant genomics, see other chapters in this volume.

Conserved gene order within and between various plant lineages is a major revelation that has emerged over the last several decades. The first hints came from comparing marker order among genetic maps of related species in clades such as the grasses and the Solanaceae (Moore et al. 1995; Gale and Devos 1998). These comparative mapping efforts showed that marker order was often conserved, though in many cases the species being compared no longer shared orthologous chromosomes because of rearrangements. Thus, even over significant evolutionary timescales—encompassing chromosome structural modifications—an ancestral gene order is found between many contemporary plant species. This shared gene order is termed *synteny*, and it is useful to establish as it helps assign orthology and paralogy, the first step in describing evolutionary relationships between genomic regions. Comparative mapping, however, has a limited resolution that

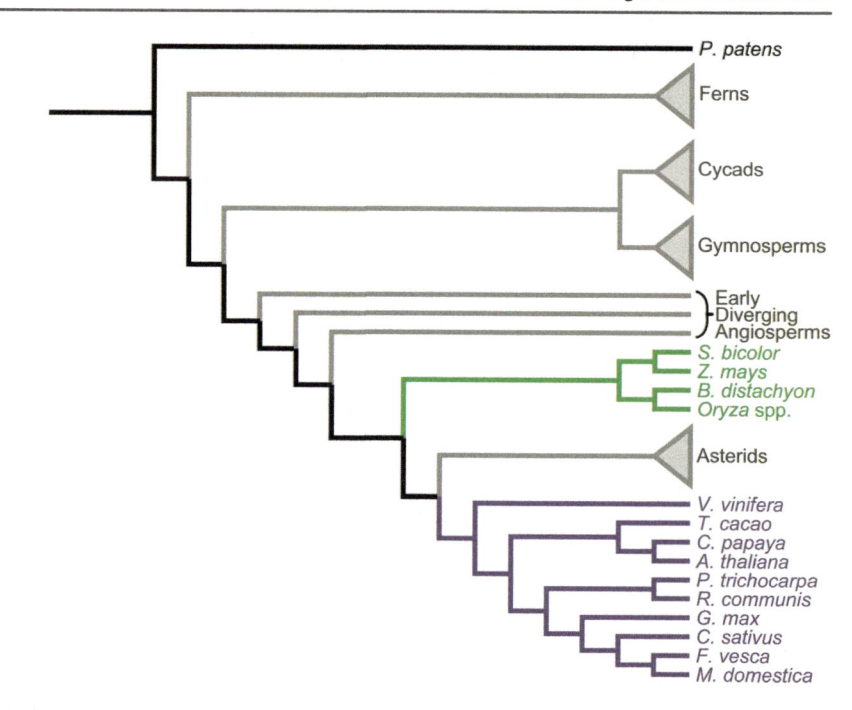

Fig. 1.2 Phylogeny of sequenced land plant genomes. The moss *P. patens* is listed in *black*; all monocots are in *green*; and rosids are in *purple*. Several important lineages lacking a published genome sequence are listed in *grey*. Phylogenetic relationships follow the AGP III system (The Angiosperm Phylogeny Group 2009)

does not permit assessment of the extent to which gene order is conserved at the nucleotide level. Fortunately, following a proliferation of plant genome sequences, we now have a wealth of fine-scale synteny analyses (see the chapter by Paterson et al. (2012, this volume)). The first genomic search for synteny came in the wake of the publication of the second plant genome sequence, rice (*Oryza sativa* spp.) (Goff et al. 2002; Yu et al. 2002), which could be compared to *A. thaliana*. Rice and *Arabidopsis* diverged approximately 200 million years ago—a considerable length of time for the accumulation of chromosomal gains, losses, and rearrangements—and some pre-genomic investigations predicted that little synteny would remain (Devos et al. 1999). This prediction was corroborated by the rice genome sequence, where it was determined that only modest stretches of synteny remained between rice and *A. thaliana* (Goff et al. 2002); the longest stretch contained 119 *Arabidopsis* proteins. Interestingly, among the identified syntenic segments, rice frequently showed a one-to-many relationship with *A. thaliana*, indicating that whole genome duplications in *Arabidopsis* occurred subsequent to its divergence from a common ancestor with rice (Goff et al. 2002). This limited synteny cast doubt over the feasibility of constructing a robust ancestral gene order among the monocots (rice) and eudicots (*Arabidopsis*) and contrasted with previous findings, as noted above, of strong evidence for synteny at closer phylogenetic distances. With the release of several genome sequences more closely related to *Arabidopsis* and rice (Table 1.1 and Fig. 1.2), intra-rosid and intra-monocot searches for synteny could be performed. Within each clade, these comparisons were typically updated with each new genome release, increasing in turn

the power and sophistication of the analyses. Looking back, it is apparent that every additional genome sequenced has sharpened our still slightly hazy picture of gene order and its evolutionary dynamics across plants. We need not elaborate this story here, as this has been well covered in the chapter by Paterson et al. (2012, this volume). We will instead jump ahead to highlight pertinent aspects of contemporary knowledge of synteny within the rosids.

At present the rosids—featuring ten published genomes (Fig. 1.2)—offer the most complete view of genomic synteny. The evolutionary history stored within these genomes is a convoluted one. This should not be surprising, in that the rosids include approximately 75,000 species in 17 diverse orders and are found in great abundance in virtually all parts of the world. Moreover, rosid chromosome counts and genome sizes range over at least an order of magnitude (Goldblatt and Johnson 1979; Bennett and Leitch 2010), and concomitant with this structural divergence, gene order also has been greatly disrupted. *Arabidopsis thaliana* in particular has had a wonderfully convoluted genomic history. It is the product of at least three detectable rounds of polyploidy, the last two occurring after its relatively recent divergence from papaya (*Carica papaya*) (Tang et al. 2008a), yet despite multiple genome duplication events, it also has suffered enormous genomic downsizing, both in terms of DNA content and chromosome number. The apparent conflict between this duplication rich history and modest contemporary genome size can only be remedied by invoking massive amounts of genomic loss. Remarkably, *Arabidopsis* still shares tracts of clear synteny with its relatives, and with substantial cutting and pasting its gene order can be projected onto the gene order of other rosid species (Jaillon et al. 2007;

Ming et al. 2008; Tang et al. 2008a). Like *A. thaliana*, poplar (*Populus trichocarpa*) also appears to have a recent history of whole genome duplication, with a single polyploidy event occurring within the salicoid lineage (Tuskan et al. 2006). In contrast to *Arabidopsis* and poplar, the genomes of grape (*Vitis vinifera*) and papaya have had a more static duplication history, showing only a widely shared hexaploidy event embedded at a deep and poorly resolved position within the angiosperm phylogeny (Ming et al. 2008; Tang et al. 2008a, b). Together these findings highlight several intriguing facts. First, there is a predicted hexaploidy event deep within the eudicots, and this hexaploidy may be correlated with the rapid expansion and radiation of these species (De Bodt et al. 2005). Second, through processes that are still poorly understood, some polyploid plant genomes (e.g., *Arabidopsis* and poplar) break down over time and return to diploidy (Wolfe 2001). Finally, some species—like grape and papaya—appear to have been unduplicated for an extensive period of time. These findings bring to light fundamental mysteries about synteny and plant genome structure. For example, how important was polyploidy in driving plant diversification? Also, why do some lineages appear to have dynamic and punctuated genome content evolution while other lineages appear to have maintained a fairly static genomic architecture for millions of years? Answers to these vexing questions are likely to emerge from multiple avenues of investigation, including a continuation and extension of comparative investigations of plant genomes.

Plant researchers have long sought to understand the genetics underlying many of the biological features that are specific to plants. The age of comparative plant genomics has cast a bright light on this area by making it possible to catalog and compare diverse plant gene repertoires. As mentioned earlier, many genes and gene families perform essential tasks and are conserved among living organisms. This is also true within plants; both generic and plant-specific gene families are broadly conserved. Notably, gene structure is also broadly conserved due to low rates of intron gain and loss. For instance, only approximately 5% of genes differ in intron content between *A. thaliana* and rice (Roy and Penny 2007), and exon–intron structure is also highly conserved among poplar, soybean, and grape (Schmutz et al. 2010). However, because of gene duplication and loss, the *size* of conserved genes families varies widely between plant species (Velasco et al. 2007; Rensing et al. 2008). These observations have led some to suggest that amplification and reduction in gene family size could be a more important contributor to evolution and diversification in plants than is the production of novel genes (Flagel and Wendel 2009). In light of this viewpoint, understanding the evolutionary history of plant gene family expansion and contraction can be seen as a tool for elucidating functional evolution.

Early diverging plant lineages—such as the green algae, bryophytes, and lycophytes—offer an enlightening perspective on the dynamic nature of plant gene family size. When compared to angiosperms, these early diverging lineages reveal extensive gene family conservation. For example, the moss *Physcomitrella patens*, which diverged from the angiosperms approximately 450 million years ago, shares 5,809 gene families with *Arabidopsis*, a tally that accounts for approximately 69% of the genes in the *A. thaliana* genome (data extracted from Phytozome version 5.0; www.phytozome.org). In Fig. 1.3 we compare gene family size between some early diverging plants, monocots, and rosids. These comparisons reveal several interesting trends. The green algae *Chlamydomonas reinhardtii* and the lycophyte *Selaginella moellendorffii*—both early diverging plant lineages—generally have fewer genes per shared gene family when compared to angiosperm species (Fig. 1.3). *Physcomitrella patens*, on the other hand, has many conserved gene families that are larger than those found in the angiosperms (Fig. 1.3), which is consistent with the fact that *P. patens* has approximately the same number of predicted genes as *Arabidopsis* (Rensing et al. 2008), while *C. reinhardtii* and *S. moellendorffii* have fewer. Finally, within the angiosperms the grasses rice and maize (*Zea mays*)—which both have a large number of genes (Goff et al. 2002; Yu et al. 2002; Schnable et al. 2009)—also tend to have larger conserved gene families when compared to the rosids *Arabidopsis* and papaya. Taken as a whole these patterns indicate that increases in gene content often result from gene family expansion, rather than novel gene creation.

By looking at the types of genes that have expanded or contracted in each lineage, we may be able to gain insights into the selective pressures plant genomes face. For example, following recent genome duplications in *A. thaliana* and maize lineages there is strong evidence for biased retention of transcription factors (Blanc and Wolfe 2004; Seoighe and Gehring 2004; Schnable et al. 2009). Moreover, in *A. thaliana*, genes involved in basic enzymatic processes and DNA repair appear to have been preferentially *lost* following duplication (Blanc and Wolfe 2004; Seoighe and Gehring 2004). Notably, different conclusions were reached regarding the functional categories preferentially retained in duplicate subsequent to polyploidy events in the Compositae (Barker et al. 2008). Here, transcription factors are underrepresented and genes associated with structural or cellular organization are overrepresented. Finally, a comparison of *A. thaliana*, poplar, rice, and moss revealed that when paralogs arise by tandem duplication rather than polyploidy, genes responsive to abiotic and biotic environmental stimuli are preferentially preserved over evolutionary time (Hanada et al. 2008).

These observations, derived from distantly related species, give an indication that not all duplicated genes have the

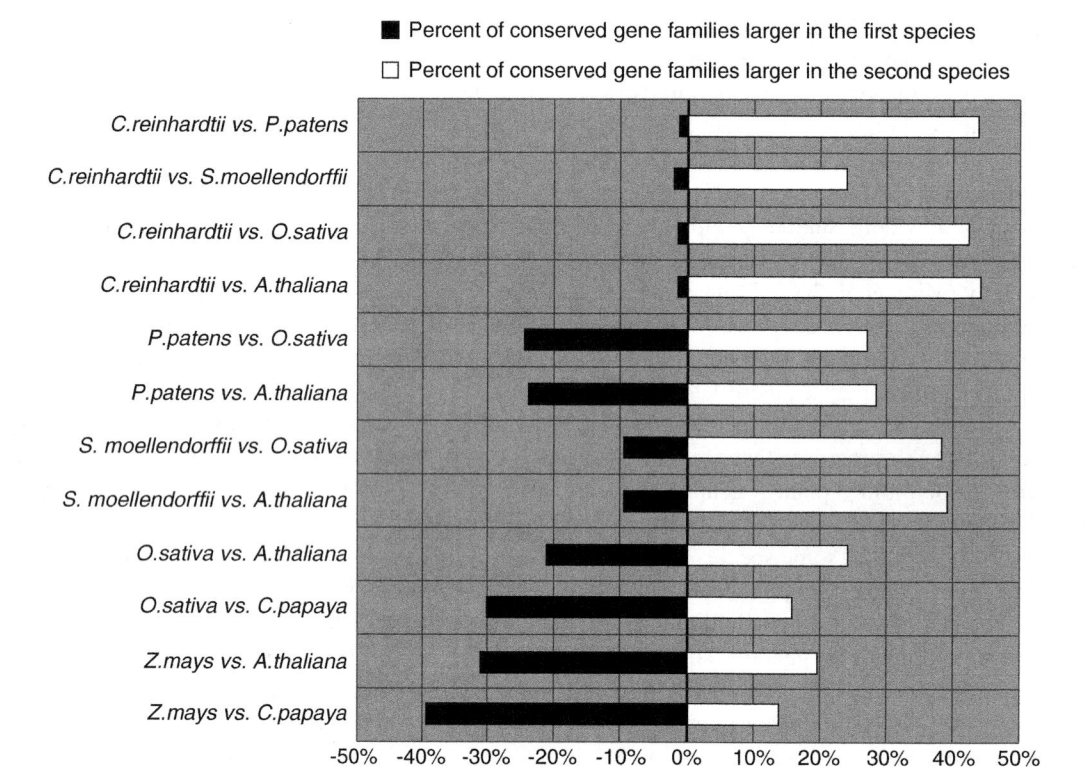

Fig. 1.3 Gene family size comparison between early diverging plant species and angiosperms. Early diverging species include the green alga *Chlamydomonas reinhardtii*, the bryophyte *Physcomitrella patens*, and the lycophyte *Selaginella moellendorffii*. Angiosperms are represented by the monocots *Oryza sativa* and *Zea mays*, and the eudicots *Arabidopsis thaliana* and *Carica papaya*. Select pairwise comparisons of these species are presented, showing the percent of conserved gene families with more genes in the first species (*black*) versus the second species (*white*). Families of the same size in both species are omitted from this tally. All data extracted from Phytozome version 5.0 gene family database (www.phytozome.org)

same probability of retention following duplication. One emerging perspective is that gene family expansion and contraction proceeds in a non-random fashion, and can have secondary effects on interacting genes (Paterson et al. 2010). These patterns may emerge because plants are constantly subjected to new biotic and abiotic stresses, and retaining genes that recognize and react to these agents is adaptively favorable, though neutral processes such as subfunctionalization (the division of ancestral functions among duplicated genes) may also contribute. Because nearly all genes and gene families are embedded within genetic networks—be they regulatory or metabolic—their expansion and contraction can have effects on their interacting partners. Again, looking at the most recent genome duplication in *Arabidopsis* lineage, there is evidence that certain suites of interacting genes are preferentially retained, while their paralogous counterparts are preferentially lost (Thomas et al. 2006). Thus, there is a need to consider gene family expansion and contraction in a network-informed context. Over evolutionary time-scales, these processes can greatly alter gene content, giving rise to highly skewed gene family sizes, and ultimately shaping the entire composition of the genome.

The availability of many plant genome sequences has given rise to an era of comparative plant genomics. By comparing the structure and contents of various species in a phylogenetic context we are beginning to discover the evolutionary forces that have shaped modern plant genomes. This work has shown that plants possess a conserved gene repertoire, yet each species experiences significant gains and losses of genetic material. This occurs in the form of recent and ancient genome duplication, as well as lineage-specific gene family expansion and contraction. Future work will continue to shed light on these unique aspects of genomic evolution in plants.

1.4 Moving Forward: The Second Decade of Plant Genomics

After considering the great strides plant genomics has made in the last decade, we now take the opportunity to imagine the potential triumphs and challenges that lie ahead. But, before we delve into these topics, it is worth considering the rationale for continued plant genome sequencing. As documented in the previous section, full genome sequences are available or in development for several plant research models and many of the world's major crop species. Notwithstanding these many accomplishments, existing genome descriptions remain in some respects incomplete. That is, every genome assembly contains missing or inaccurate information, and this incompleteness is exacerbated by the fact that all available genome sequences, even *A. thaliana*, contain thousands of annotated genes that lack an experimentally determined function. Thus, it could be argued that we have already surpassed our capacity to study the genomic information we have in hand, or to put it another way, the first decade of plant genomics has created an information reservoir sufficient to fuel decades of follow-up research. Based on this, an argument could be made that new resources might be more optimally diverted toward a more complete analysis of existing genomes. In contrast to this perspective, here we present the case that continued investment in plant genome sequencing is warranted. Our primary reasoning is based on the recognition that genome sequencing is both a tool for new discovery and a means of achieving functional insight. In this section we discuss the power of genome sequencing in both of these capacities, demonstrating how genome sequencing can be used to reveal the functions of known genes, in addition to enabling new and unexpected discoveries.

Genome sequencing itself is merely a starting point in the analysis of the molecular and cellular function of genes. Immediately after release, many useful tools can be developed to open up an array of new research questions. Classic examples include sequence-based marker development, gene-targeted mutational forward genetics, and microarray platform design for gene expression and copy number variation studies (e.g. Schmid et al. 2005; Borevitz et al. 2007; Springer et al. 2009). In addition, sophisticated tools can be developed to access small RNAs and probe epigenetic modifications (e.g. Zhang et al. 2006; Lister et al. 2009). All of these tools have consistently led to new discoveries, and they will continue to be a primary impetus for further genome sequencing. Moreover, the benefits of plant genome sequencing continue to emerge with the addition of new technologies. For example, with a high-quality reference genome, additional individuals can be cost effectively *resequenced* (generating and aligning short Next-Generation sequence reads from a new genotype to a reference genome to produce a novel draft genome sequence). Genomic information from resequenced individuals can in turn be used understand genetic diversity within species. We are now beginning to see this concept in action and once again *Arabidopsis* has taken the lead with the 1001 Genomes Project, which aims to sequence and disseminate genome-wide diversity data from 1001 *Arabidopsis* accessions (Ossowski et al. 2008). Today, a researcher interested in an *Arabidopsis* gene can scan polymorphism data at the 1001

Genomes Project website (www.1001genomes.org), and identify accessions with interesting polymorphisms without ever entering a laboratory. As expected, this resource is quickly replacing the time consuming process of cloning and sequencing a gene of interest from various *Arabidopsis* accessions. A major lesson has been learned from the initial genome sequencing projects: one genome representing an entire species is simply not sufficient for many research questions. More powerful forms of analysis, utilizing principles from population genetics, can be performed when multiple genome sequences are available for a species. Along with the continued development of classical genomic tools, projects similar to the *Arabidopsis* 1001 Genomes Project have been initiated for a few major plant research models and crops (Huang et al. 2010; Lai et al. 2010; Lam et al. 2010). With the steadily falling price of DNA sequencing, we expect this to be an area of great growth in the next decade.

Additional information can be extracted from genomic data once it is put in an evolutionary context because genomic features can be analyzed in a phylogenetic manner. Consequently, the accumulation of new genomes enriches the value of existing genome sequences. Notably, plants have possibly the best-resolved phylogeny of any major eukaryotic lineage (Savolainen and Chase 2003), which provides a powerful framework for evolutionary comparison. The first available plant genome sequences came from distantly related species, making them only appropriate for broad evolutionary comparisons. As discussed in the previous section and in additional chapters throughout this volume, these broad comparisons have proved fruitful. The next phase of comparative genomics will likely harness the evolutionary information found in closely related species, perhaps at the relatively shallow taxonomic depths of populations or genera. These analyses will benefit greatly from the strong evolutionary signal that is maintained at relatively close phylogenetic proximity. With multiple genome sequences available at the genus level we may someday be able to study genome evolution with sufficient precision to ascertain the exact spectrum and perhaps sequence of molecular evolutionary events that have shaped chromosome organization or gene family composition. As an example of the potential of fine-scale evolutionary analysis, various monocot genomes were utilized to demonstrate a history of chromosome fusions within the *Brachypodium distachyon* genome (International Brachypodium Initiative 2010). This example shows that some plant lineages are approaching the critical mass of fully sequenced species needed to make precise evolutionary comparisons (notably the Brassicaceae and Poaceae; Fig. 1.2). In the future, expanded plant genome sequencing efforts will only continue to add more power to this approach, further enabling our ability to discover new aspects of plant genome evolution. Similarly, combining new computational tools with the evolutionary signal from additional plant genomes of intermediate phylogenetic distance will substantially enhance our power to functionally annotate the coding and regulatory portions of the genome (Wang et al. 2009; Picot et al. 2010). For many species we are, or will soon be able, to annotate and analyzed their genomes in a phylogenetic framework. In the future this approach will dramatically benefit the quality and utility of plant genome sequences.

1.5   Challenges That Lie Ahead

Though the plant genomics community has made impressive progress in the decade following the release of the *Arabidopsis* genome, many critical plant species remain to be sequenced (some can be found in Figs. 1.1, 1.2). Among the species without a sequenced genome are important crop plants, including, for example, coffee (*Coffea arabica*), common bean (*Phaseolus vulgaris*), cotton (*Gossypium hirsutum*), potato (*Solanum tuberosum*), and sugarcane (*Saccharum* spp.). Other species lacking genome sequences are important research models such as snapdragon (*Antirrhinum majus*) and tobacco (*Nicotiana tabaccum*; both a crop and research model). Finally, there are large clades in the angiosperm tree of life that have yet to be included in the tabulation of sequenced genomes, including early diverging angiosperms, asterids, and several anciently diverged lineages such as the ferns, cycads, and gymnosperms (Fig. 1.2; note some of the aforementioned species have ongoing genome sequencing projects). Most striking among these might be the asterids, one of the two major eudicot clades, which at present have no representative members with a published genome sequence (Fig. 1.2). The asterids include over 60,000 species and contain numerous crop species, including coffee, potato, sunflower (*Helianthus annuus*), and olive (*Olea europaea*), to name just a few. Interestingly, the other comparably large and diverse eudicot clade, the rosids, makes up the majority of published plant genome sequences (Fig. 1.2). This contrast will likely soon end, as draft asterid genome sequences are in the finishing stages for the monkey flower (*Mimulus guttatus*) and tomato (*Solanum lycopersicum*) (pers. comm. T.J. Vision). Over the next decade many of the pockets of uninvestigated plant diversity will likely be explored, because there are important species in these clades and because understanding the full biodiversity of plant life is a major goal of many research organizations around the world.

Possibly the greatest future challenge in plant genome sequencing will be tackling the massive genomes found among some of the world's most important plants. Figure 1.1 displays the genome sizes of a number of published plant genome sequences, compared to a range of unpublished plant species selected for their economic, cultural, or research significance. Clearly we have only completed genome sequencing projects at the bottom of the range of plant genome sizes. Some exceptionally important plants have very large genome sizes, and sequencing these genomes will be a great challenge. As an example, upland cotton (*Gossypium hirsutum*), bread wheat (*Triticum aestivum*), and pines (species in the genus *Pinus*) have genome sizes that are approximately 21, 108, and 138 times the size of *Arabidopsis*, respectively (Bennett and Leitch 2010). Many essential human activities are reliant on these species, and it seems imperative that at some point we will endeavor to sequence their genomes. What we might find in the complex genomes of these species will be quite exciting. We know that, in part, upland cotton and bread wheat have large genome sizes because they are polyploids (cotton is an allotetraploid while bread wheat is an allohexaploid). To our knowledge, no recent polyploid plant genome sequence has been published (though some recent paleopolyploids have, such as maize and soybean (Schnable et al. 2009; Schmutz et al. 2010)). The duplicate nature of polyploid genomes creates several hurdles for accurate assembly. For example, polyploids with divergent genomes (allopolyploids, such as cotton and bread wheat) add complexity as both co-resident genomes will need to be differentiated and assembled separately. Difficulties aside, many critically important plant species are polyploid, creating a strong motivation to surmount these challenges. Notably, despite their massive genome size, most pines are thought to be diploid, and what makes their genome so large is a fascinating mystery. The pine genome could be beset with enormous numbers of transposable elements or replete with segmental duplications. At present, producing a quality pine genome would be a tremendous undertaking, both technically and monetarily, but if the next decade witnesses similar technological improvements and cost reductions as the last, the feasibility of completing a pine genome sequence could be just on the horizon (Neale and Kremer 2011).

The discussion of massive and complex plant genomes raises another interesting question: is it necessary to have a nearly complete genome assembly, or could great advances be stimulated by much less expensive and less technically demanding low-coverage sequencing scans? The answer, of course, depends on how the research community intends to use the information. Many of the first genome sequences originated from model organisms, and the research communities backing these species often required a robust and detailed understanding of genomic structure and content. On the other hand, a low-coverage and incomplete assembly may be all that is needed to aid some forms of plant improvement, like marker-based selection. In the next decade, plant genome sequencing endeavors will move beyond the most utilized model species and the smallest genome sizes, and we may see a concomitant adjustment in the level of genomic detail that is sufficient for the research goals of these communities. Some movement in this direction is already evident from the great abundance of plant expressed sequence tag (EST) projects. EST sequencing is simply the capture and sequencing of mRNA transcripts, an effective way to isolate the expressed portion of the genome. Surely, EST sequencing—like low-coverage genome sequencing—has major limitations; however, for some research questions these low cost alternatives may provide sufficient data. Nevertheless, we anticipate a future where most research communities will seek whole plant genome sequences, because this resource facilitates the greatest range of genomic tools and analyses and offers insights that cannot be matched by low-coverage alternatives. Indeed, in many cases, low-coverage sequences are likely to be used as a stepping-stone toward the eventual goal of producing a complete genome sequence.

Current and future plant genome sequencing projects are greatly aided by new sequencing technologies, which have caused a precipitous drop in the price per base pair of DNA sequence (Huang et al. 2009; Argout et al. 2011; Shulaev et al. 2011). In a just a matter of weeks, a small group of researchers can now produce billions of nucleotides of sequence for only thousands of US dollars. Though the price of sequencing has fallen dramatically, *assembling* a new plant genome is far from routine or easy. Assembly still requires significant infrastructure, technological proficiency, and effort. The assembly issue has become even thornier with current Next-Generation sequencing technologies, primarily because these platforms produce short reads (generally between 30 and 500 base pairs in length). For most plant genomes it is unlikely that reads of these lengths will assemble into large contiguous pieces, instead producing a highly fragmented assembly that offers more limited utility. In large part this fragmentation occurs because many plant repeats are significantly longer than these short read lengths and thus cannot be traversed, effectively terminating any non-arbitrary elongation from the point of the repeat onward (Pop and Salzberg 2008). A greater depth of sequence coverage—the area in which Next-Generation technologies excel—cannot alleviate this problem. Instead, alternative library construction and sequencing techniques, often acquired at a far greater cost, must be used to bridge these large repeats. Thus, despite the impressive technological gains of the last few years, producing a high-quality genome remains an expensive and laborious task. One obvious way to overcome this limitation is the development of sequencing technologies that can produce longer reads (for example $>1,000$ base pairs, approximately the current ceiling for Sanger sequencing technology) in a rapid and cost effective manner. Exceptionally long reads ($>10,000$ base pairs),

could potentially be an even larger advance, as they are likely to span many plant repeats and consequently greatly reduce the complexity of genome assembly. Single-molecule sequencing (Eid et al. 2009), is emerging as a potential front-runner in this area. Single-molecule sequencers that produce read lengths comparable to Sanger sequencing are beginning to appear, though it is too early to predict where the technology might be headed. It does, however, seem likely that over the next decade DNA sequencing devices with tremendous throughput and exceptionally long read lengths will be developed. This will prove to be a breakthrough for plant genome sequencing efforts. It will precipitate new strategies for genome sequencing and assembly, and likely open up the door for plant genome sequencing in species that have largely been neglected by funding agencies.

Finally, to better utilize future plant genome sequences, researchers will need to integrate data between research groups. At present, a research community typically conducts in-house genome assembly and annotation, and stores this information in a species-specific database. So far this model has served each community well, but has also created some impediments to multi-species comparative analyses. The plant community would benefit greatly from unified data annotation terminology and distribution models. For example, in some species probable pseudo-genes are left in the final gene catalog, while in other species they are systematically excluded. If one wishes to compare genes between these species, one must first perform a tedious filtering task. Fortunately, genome database experts have undertaken some of this data integration. A few good examples of well-curated, multi-species, comparative plant genome databases include Phytozome (www.phytozome.org), Plant Genome Database (www.plantgdb.org), and Plant Genome Duplication Database (http://chibba.pgml.uga.edu/duplication/). The research community behind each newly sequenced species will want to rapidly integrate their new sequence into a framework with existing plant genome sequences for comparative purposes. This creates an incentive to adopt a common framework. We anticipate that this motivation will stimulate researchers to gravitate toward packaging their data in a standardized way, which will ultimately benefit the entire research community.

Conclusions

Tremendous resources have been devoted to plant genome sequencing. This outlay has thus far rewarded us with 15 published sequences. Today—one decade into the plant genomics era—comparative plant genomics is beginning to realize its promise, revealing the nature of the evolutionary forces that have shaped the structure and contents of modern plant genomes. In the future, with the advent of improved sequencing technologies, comparative plant genomics will continue to be a major a source of new insights. Existing plant genomes have also greatly benefited plant research as a platform for gaining functional insights. Work along these lines will continue to flourish, including greater within-species genome resequencing, which will prove to be a key method for elucidating the genetic consequences of population level processes. There is cause for continued optimism about future developments in these areas, particularly as we surmount the challenges inherent in sequencing and assembling massive and redundant plant genomes. Similarly, we anticipate continued progress in integrating genomic data from disparate sources, creating new opportunities for comparative discovery. Because of this, we anticipate that the second decade of plant genomics will surpass the first in terms of scientific breakthroughs and advances to our knowledge of plant diversity.

References

Argout X, Salse J, Aury J-M, Guiltinan MJ, Droc G, Gouzy J, Allegre M, Chaparro C, Legavre T, Maximova SN, Abrouk M, Murat F, Fouet O, Poulain J, Ruiz M, Roguet Y, Rodier-Goud M, Barbosa-Neto JF, Sabot F, Kudrna D, Ammiraju JSS, Schuster SC, Carlson JE, Sallet E, Schiex T, Dievart A, Kramer M, Gelley L, Shi Z, Berard A, Viot C, Boccara M, Risterucci AM, Guignon V, Sabau X, Axtell MJ, Ma Z, Zhang Y, Brown S, Bourge M, Golser W, Song X, Clement D, Rivallan R, Tahi M, Akaza JM, Pitollat B, Gramacho K, D'Hont A, Brunel D, Infante D, Kebe I, Costet P, Wing R, McCombie WR, Guiderdoni E, Quetier F, Panaud O, Wincker P, Bocs S, Lanaud C (2011) The genome of *Theobroma cacao*. Nat Genet 43:101–108

Aubourg S, Boudet N, Kreis M, Lecharny A (2000) In *Arabidopsis thaliana*, 1% of the genome codes for a novel protein family unique to plants. Plant Mol Biol 42:603–613

Barker MS, Kane NC, Matvienko M, Kozik A, Michelmore RW, Knapp SJ, Rieseberg LH (2008) Multiple paleopolyploidizations during the evolution of the Compositae reveal parallel patterns of duplicate gene retention after millions of years. Mol Biol Evol 25:2445–2455

Bennett M, Leitch I (2010) Plant DNA C-values database (release 5.0). http://www.kew.org/cvalues/

Bevan M (1997) Objective: the complete sequence of a plant genome. Plant Cell 9:476–478

Blanc G, Wolfe KH (2004) Functional divergence of duplicated genes formed by polyploidy during *Arabidopsis* evolution. Plant Cell 16:1679–1691

Borevitz JO, Hazen SP, Michael TP, Morris GP, Baxter IR, Hu TT, Chen H, Werner JD, Nordborg M, Salt DE, Kay SA, Chory J, Weigel D, Jones JDG, Ecker JR (2007) Genome-wide patterns of single-feature polymorphism in *Arabidopsis thaliana*. Proc Natl Acad Sci USA 104:12057–12062

Chan AP, Crabtree J, Zhao Q, Lorenzi H, Orvis J, Puiu D, Melake-Berhan A, Jones KM, Redman J, Chen G, Cahoon EB, Gedil M, Stanke M, Haas BJ, Wortman JR, Fraser-Liggett CM, Ravel J, Rabinowicz PD (2010) Draft genome sequence of the oilseed species *Ricinus communis*. Nat Biotechnol 28:951–956

Cushing DA, Forsthoefel NR, Gestaut DR, Vernon DM (2005) *Arabidopsis emb175* and other *ppr* knockout mutants reveal essential roles for pentatricopeptide repeat (PPR) proteins in plant embryogenesis. Planta 221:424–436

De Bodt S, Maere S, Van de Peer Y (2005) Genome duplication and the origin of angiosperms. Trends Ecol Evol 20:591–597

Devos KM, Beales J, Nagamura Y, Sasaki T (1999) *Arabidopsis*-rice: will collinearity allow gene prediction across the eudicot–monocot divide? Genome Res 9:825–829

Eid J, Fehr A, Gray J, Luong K, Lyle J, Otto G, Peluso P, Rank D, Baybayan P, Bettman B, Bibillo A, Bjornson K, Chaudhuri B, Christians F, Cicero R, Clark S, Dalal R, Dewinter A, Dixon J, Foquet M, Gaertner A, Hardenbol P, Heiner C, Hester K, Holden D, Kearns G, Kong X, Kuse R, Lacroix Y, Lin S, Lundquist P, Ma C, Marks P, Maxham M, Murphy D, Park I, Pham T, Phillips M, Roy J, Sebra R, Shen G, Sorenson J, Tomaney A, Travers K, Trulson M, Vieceli J, Wegener J, Wu D, Yang A, Zaccarin D, Zhao P, Zhong F, Korlach J, Turner S (2009) Real-time DNA sequencing from single polymerase molecules. Science 323:133–138

Feldmann KA, Marks MD (1987) *Agrobacterium*-mediated transformation of germinating seeds of *Arabidopsis thaliana*: a non-tissue culture approach. Mol Gen Genet 208:1–9

Flagel LE, Wendel JF (2009) Gene duplication and evolutionary novelty in plants. New Phytol 183:557–564

Gale MD, Devos KM (1998) Plant comparative genetics after 10 years. Science 282:656–659

Goff SA, Ricke D, Lan T-H, Presting G, Wang R, Dunn M, Glazebrook J, Sessions A, Oeller P, Varma H, Hadley D, Hutchison D, Martin C, Katagiri F, Lange BM, Moughamer T, Xia Y, Budworth P, Zhong J, Miguel T, Paszkowski U, Zhang S, Colbert M, Sun W-l, Chen L, Cooper B, Park S, Wood TC, Mao L, Quail P, Wing R, Dean R, Yu Y, Zharkikh A, Shen R, Sahasrabudhe S, Thomas A, Cannings R, Gutin A, Pruss D, Reid J, Tavtigian S, Mitchell J, Eldredge G, Scholl T, Miller RM, Bhatnagar S, Adey N, Rubano T, Tusneem N, Robinson R, Feldhaus J, Macalma T, Oliphant A, Briggs S (2002) A draft sequence of the rice genome (*Oryza sativa* L. ssp. *japonica*). Science 296:92–100

Goldblatt P, Johnson D (1979) Index to plant chromosome numbers. Missouri Botanical Garden, St. Louis

Hanada K, Zou C, Lehti-Shiu MD, Shinozaki K, Shiu S-H (2008) Importance of lineage-specific expansion of plant tandem duplicates in the adaptive response to environmental stimuli. Plant Physiol 148:993–1003

Huang S, Li R, Zhang Z, Li L, Gu X, Fan W, Lucas WJ, Wang X, Xie B, Ni P, Ren Y, Zhu H, Li J, Lin K, Jin W, Fei Z, Li G, Staub J, Kilian A, van der Vossen EAG, Wu Y, Guo J, He J, Jia Z, Ren Y, Tian G, Lu Y, Ruan J, Qian W, Wang M, Huang Q, Li B, Xuan Z, Cao J, Asan WuZ, Zhang J, Cai Q, Bai Y, Zhao B, Han Y, Li Y, Li X, Wang S, Shi Q, Liu S, Cho WK, Kim J-Y, Xu Y, Heller-Uszynska K, Miao H, Cheng Z, Zhang S, Wu J, Yang Y, Kang H, Li M, Liang H, Ren X, Shi Z, Wen M, Jian M, Yang H, Zhang G, Yang Z, Chen R, Liu S, Li J, Ma L, Liu H, Zhou Y, Zhao J, Fang X, Li G, Fang L, Li Y, Liu D, Zheng H, Zhang Y, Qin N, Li Z, Yang G, Yang S, Bolund L, Kristiansen K, Zheng H, Li S, Zhang X, Yang H, Wang J, Sun R, Zhang B, Jiang S, Wang J, Du Y, Li S (2009) The genome of the cucumber, *Cucumis sativus* L. Nat Genet 41:1275–1281

Huang X, Wei X, Sang T, Zhao Q, Feng Q, Zhao Y, Li C, Zhu C, Lu T, Zhang Z, Li M, Fan D, Guo Y, Wang A, Wang L, Deng L, Li W, Lu Y, Weng Q, Liu K, Huang T, Zhou T, Jing Y, Li W, Lin Z, Buckler ES, Qian Q, Zhang Q-F, Li J, Han B (2010) Genome-wide association studies of 14 agronomic traits in rice landraces. Nat Genet 42:961–967

International Brachypodium Initiative (2010) Genome sequencing and analysis of the model grass *Brachypodium distachyon*. Nature 463:763–768

Jaillon O, Aury JM, Noel B, Policriti A, Clepet C, Casagrande A, Choisne N, Aubourg S, Vitulo N, Jubin C, Vezzi A, Legeai F, Hugueney P, Dasilva C, Horner D, Mica E, Jublot D, Poulain J, Bruyere C, Billault A, Segurens B, Gouyvenoux M, Ugarte E, Cattonaro F, Anthouard V, Vico V, Del Fabbro C, Alaux M, Di Gaspero G, Dumas V, Felice N, Paillard S, Juman I, Moroldo M, Scalabrin S, Canaguier A, Le Clainche I, Malacrida G, Durand E, Pesole G, Laucou V, Chatelet P, Merdinoglu D, Delledonne M, Pezzotti M, Lecharny A, Scarpelli C, Artiguenave F, Pe ME, Valle G, Morgante M, Caboche M, Adam-Blondon AF, Weissenbach J, Quetier F, Wincker P (2007) The grapevine genome sequence suggests ancestral hexaploidization in major angiosperm phyla. Nature 449:463–467

Koornneef M, Meinke D (2010) The development of *Arabidopsis* as a model plant. Plant J 61:909–921

Lai J, Li R, Xu X, Jin W, Xu M, Zhao H, Xiang Z, Song W, Ying K, Zhang M, Jiao Y, Ni P, Zhang J, Li D, Guo X, Ye K, Jian M, Wang B, Zheng H, Liang H, Zhang X, Wang S, Chen S, Li J, Fu Y, Springer NM, Yang H, Wang J, Dai J, Schnable PS, Wang J (2010) Genome-wide patterns of genetic variation among elite maize inbred lines. Nat Genet 42:1027–1030

Lam H-M, Xu X, Liu X, Chen W, Yang G, Wong F-L, Li M-W, He W, Qin N, Wang B, Li J, Jian M, Wang J, Shao G, Wang J, Sun SS-M, Zhang G (2010) Resequencing of 31 wild and cultivated soybean genomes identifies patterns of genetic diversity and selection. Nat Genet 42:1053–1059

Lin X, Kaul S, Rounsley S, Shea TP, Benito M-I, Town CD, Fujii CY, Mason T, Bowman CL, Barnstead M, Feldblyum TV, Buell CR, Ketchum KA, Lee J, Ronning CM, Koo HL, Moffat KS, Cronin LA, Shen M, Pai G, Van Aken S, Umayam L, Tallon LJ, Gill JE, Adams MD, Carrera AJ, Creasy TH, Goodman HM, Somerville CR, Copenhaver GP, Preuss D, Nierman WC, White O, Eisen JA, Salzberg SL, Fraser CM, Venter JC (1999) Sequence and analysis of chromosome 2 of the plant *Arabidopsis thaliana*. Nature 402:761–768

Lister R, Pelizzola M, Dowen RH, Hawkins RD, Hon G, Tonti-Filippini J, Nery JR, Lee L, Ye Z, Ngo Q-M, Edsall L, Antosiewicz-Bourget J, Stewart R, Ruotti V, Millar AH, Thomson JA, Ren B, Ecker JR (2009) Human DNA methylomes at base resolution show widespread epigenomic differences. Nature 462:315–322

Lockton S, Gaut BS (2005) Plant conserved non-coding sequences and paralogue evolution. Trends Genet 21:60–65

Lurin C, Andres C, Aubourg S, Bellaoui M, Bitton F, Bruyere C, Caboche M, Debast C, Gualberto J, Hoffmann B, Lecharny A, Le Ret M, Martin-Magniette M-L, Mireau H, Peeters N, Renou J-P, Szurek B, Taconnat L, Small I (2004) Genome-wide analysis of *Arabidopsis* pentatricopeptide repeat proteins reveals their essential role in organelle biogenesis. Plant Cell 16:2089–2103

Mayer K, Schuller C, Wambutt R, Murphy G, Volckaert G, Pohl T, Dusterhoft A, Stiekema W, Entian KD, Terryn N, Harris B, Ansorge W, Brandt P, Grivell L, Rieger M, Weichselgartner M, de Simone V, Obermaier B, Mache R, Muller M, Kreis M, Delseny M, Puigdomenech P, Watson M, Schmidtheini T, Reichert B, Portatelle D, Perez-Alonso M, Boutry M, Bancroft I, Vos P, Hoheisel J, Zimmermann W, Wedler H, Ridley P, Langham SA, McCullagh B, Bilham L, Robben J, Van der Schueren J, Grymonprez B, Chuang YJ, Vandenbussche F, Braeken M, Weltjens I, Voet M, Bastiaens I, Aert R, Defoor E, Weitzenegger T, Bothe G, Ramsperger U, Hilbert H, Braun M, Holzer E, Brandt A, Peters S, van Staveren M, Dirkse W, Mooijman P, Lankhorst RK, Rose M, Hauf J, Kotter P, Berneiser S, Hempel S, Feldpausch M, Lamberth S, Van den Daele H, De Keyser A, Buysshaert C, Gielen J, Villarroel R, De Clercq R, Van Montagu M, Rogers J, Cronin A, Quail M, Bray-Allen S, Clark L, Doggett J, Hall S, Kay M, Lennard N, McLay K, Mayes R, Pettett A, Rajandream MA, Lyne M, Benes

V, Rechmann S, Borkova D, Blocker H, Scharfe M, Grimm M, Lohnert TH, Dose S, de Haan M, Maarse A, Schafer M, Muller-Auer S, Gabel C, Fuchs M, Fartmann B, Granderath K, Dauner D, Herzl A, Neumann S, Argiriou A, Vitale D, Liguori R, Piravandi E, Massenet O, Quigley F, Clabauld G, Mundlein A, Felber R, Schnabl S, Hiller R, Schmidt W, Lecharny A, Aubourg S, Chefdor F, Cooke R, Berger C, Montfort A, Casacuberta E, Gibbons T, Weber N, Vandenbol M, Bargues M, Terol J, Torres A, Perez-Perez A, Purnelle B, Bent E, Johnson S, Tacon D, Jesse T, Heijnen L, Schwarz S, Scholler P, Heber S, Francs P, Bielke C, Frishman D, Haase D, Lemcke K, Mewes HW, Stocker S, Zaccaria P, Bevan M, Wilson RK, de la Bastide M, Habermann K, Parnell L, Dedhia N, Gnoj L, Schutz K, Huang E, Spiegel L, Sehkon M, Murray J, Sheet P, Cordes M, Abu-Threideh J, Stoneking T, Kalicki J, Graves T, Harmon G, Edwards J, Latreille P, Courtney L, Cloud J, Abbott A, Scott K, Johnson D, Minx P, Bentley D, Fulton B, Miller N, Greco T, Kemp K, Kramer J, Fulton L, Mardis E, Dante M, Pepin K, Hillier L, Nelson J, Spieth J, Ryan E, Andrews S, Geisel C, Layman D, Du H, Ali J, Berghoff A, Jones K, Drone K, Cotton M, Joshu C, Antonoiu B, Zidanic M, Strong C, Sun H, Lamar B, Yordan C, Ma P, Zhong J, Preston R, Vil D, Shekher M, Matero A, Shah R, Swaby IK, O'Shaughnessy A, Rodriguez M, Hoffman J, Till S, Granat S, Shohdy N, Hasegawa A, Hameed A, Lodhi M, Johnson A, Chen E, Marra M, Martienssen R, McCombie WR (1999) Sequence and analysis of chromosome 4 of the plant *Arabidopsis thaliana*. Nature 402:769–777

Meinke DW, Cherry JM, Dean C, Rounsley SD, Koornneef M (1998) *Arabidopsis thaliana*: a model plant for genome analysis. Science 282:662–682

Ming R, Hou S, Feng Y, Yu Q, Dionne-Laporte A, Saw JH, Senin P, Wang W, Ly BV, Lewis KLT, Salzberg SL, Feng L, Jones MR, Skelton RL, Murray JE, Chen C, Qian W, Shen J, Du P, Eustice M, Tong E, Tang H, Lyons E, Paull RE, Michael TP, Wall K, Rice DW, Albert H, Wang M-L, Zhu YJ, Schatz M, Nagarajan N, Acob RA, Guan P, Blas A, Wai CM, Ackerman CM, Ren Y, Liu C, Wang J, Wang J, Na J-K, Shakirov EV, Haas B, Thimmapuram J, Nelson D, Wang X, Bowers JE, Gschwend AR, Delcher AL, Singh R, Suzuki JY, Tripathi S, Neupane K, Wei H, Irikura B, Paidi M, Jiang N, Zhang W, Presting G, Windsor A, Navajas-Perez R, Torres MJ, Feltus FA, Porter B, Li Y, Burroughs AM, Luo M-C, Liu L, Christopher DA, Mount SM, Moore PH, Sugimura T, Jiang J, Schuler MA, Friedman V, Mitchell-Olds T, Shippen DE, dePamphilis CW, Palmer JD, Freeling M, Paterson AH, Gonsalves D, Wang L, Alam M (2008) The draft genome of the transgenic tropical fruit tree papaya (*Carica papaya* Linnaeus). Nature 452:991–996

Moore G, Devos KM, Wang Z, Gale MD (1995) Cereal genome evolution: grasses, line up and form a circle. Curr Biol 5:737–739

Neale DB, Kremer A (2011) Forest tree genomics: growing resources and applications. Nat Rev Genet 12:111–122

Ossowski S, Schneeberger K, Clark RM, Lanz C, Warthmann N, Weigel D (2008) Sequencing of natural strains of *Arabidopsis thaliana* with short reads. Genome Res 18:2024–2033

Paterson AH, Bowers JE, Bruggmann R, Dubchak I, Grimwood J, Gundlach H, Haberer G, Hellsten U, Mitros T, Poliakov A, Schmutz J, Spannagl M, Tang H, Wang X, Wicker T, Bharti AK, Chapman J, Feltus FA, Gowik U, Grigoriev IV, Lyons E, Maher CA, Martis M, Narechania A, Otillar RP, Penning BW, Salamov AA, Wang Y, Zhang L, Carpita NC, Freeling M, Gingle AR, Hash CT, Keller B, Klein P, Kresovich S, McCann MC, Ming R, Peterson DG, Mehboob ur R, Ware D, Westhoff P, Mayer KFX, Messing J, Rokhsar DS (2009) The *Sorghum bicolor* genome and the diversification of grasses. Nature 457:551–556

Paterson AH, Freeling M, Tang H, Wang X (2010) Insights from the comparison of plant genome sequences. Annu Rev Plant Biol 61:349–372

Paterson A, Wang X, Tang H, Lee TH (2012) Synteny and genomic rearrangements. In: Wendel JF (ed) Plant genome diversity, vol 1, Plant genomes, their residents, and their evolutionary dynamics. Springer-Verlag, Wien, New York

Picot E, Krusche P, Tiskin A, Carré I, Ott S (2010) Evolutionary analysis of regulatory sequences (EARS) in plants. Plant J 64:165–176

Pop M, Salzberg SL (2008) Bioinformatics challenges of new sequencing technology. Trends Genet 24:142–149

Rensing SA, Lang D, Zimmer AD, Terry A, Salamov A, Shapiro H, Nishiyama T, Perroud P-F, Lindquist EA, Kamisugi Y, Tanahashi T, Sakakibara K, Fujita T, Oishi K, Shin IT, Kuroki Y, Toyoda A, Suzuki Y, Hashimoto S-i, Yamaguchi K, Sugano S, Kohara Y, Fujiyama A, Anterola A, Aoki S, Ashton N, Barbazuk WB, Barker E, Bennetzen JL, Blankenship R, Cho SH, Dutcher SK, Estelle M, Fawcett JA, Gundlach H, Hanada K, Heyl A, Hicks KA, Hughes J, Lohr M, Mayer K, Melkozernov A, Murata T, Nelson DR, Pils B, Prigge M, Reiss B, Renner T, Rombauts S, Rushton PJ, Sanderfoot A, Schween G, Shiu S-H, Stueber K, Theodoulou FL, Tu H, Van de Peer Y, Verrier PJ, Waters E, Wood A, Yang L, Cove D, Cuming AC, Hasebe M, Lucas S, Mishler BD, Reski R, Grigoriev IV, Quatrano RS, Boore JL (2008) The *Physcomitrella* genome reveals evolutionary insights into the conquest of land by plants. Science 319:64–69

Ride JP, Davies EM, Franklin FCH, Marshall DF (1999) Analysis of *Arabidopsis* genome sequence reveals a large new gene family in plants. Plant Mol Biol 39:927–932

Roy SW, Penny D (2007) Patterns of intron loss and gain in plants: intron loss-dominated evolution and genome-wide comparison of *O. sativa* and *A. thaliana*. Mol Biol Evol 24:171–181

Savolainen V, Chase MW (2003) A decade of progress in plant molecular phylogenetics. Trends Genet 19:717–724

Schmid M, Davison TS, Henz SR, Pape UJ, Demar M, Vingron M, Scholkopf B, Weigel D, Lohmann JU (2005) A gene expression map of *Arabidopsis thaliana* development. Nat Genet 37:501–506

Schmitz-Linneweber C, Small I (2008) Pentatricopeptide repeat proteins: a socket set for organelle gene expression. Trends Plant Sci 13:663–670

Schmutz J, Cannon SB, Schlueter J, Ma J, Mitros T, Nelson W, Hyten DL, Song Q, Thelen JJ, Cheng J, Xu D, Hellsten U, May GD, Yu Y, Sakurai T, Umezawa T, Bhattacharyya MK, Sandhu D, Valliyodan B, Lindquist E, Peto M, Grant D, Shu S, Goodstein D, Barry K, Futrell-Griggs M, Abernathy B, Du J, Tian Z, Zhu L, Gill N, Joshi T, Libault M, Sethuraman A, Zhang X-C, Shinozaki K, Nguyen HT, Wing RA, Cregan P, Specht J, Grimwood J, Rokhsar D, Stacey G, Shoemaker RC, Jackson SA (2010) Genome sequence of the palaeopolyploid soybean. Nature 463:178–183

Schnable PS, Ware D, Fulton RS, Stein JC, Wei F, Pasternak S, Liang C, Zhang J, Fulton L, Graves TA, Minx P, Reily AD, Courtney L, Kruchowski SS, Tomlinson C, Strong C, Delehaunty K, Fronick C, Courtney B, Rock SM, Belter E, Du F, Kim K, Abbott RM, Cotton M, Levy A, Marchetto P, Ochoa K, Jackson SM, Gillam B, Chen W, Yan L, Higginbotham J, Cardenas M, Waligorski J, Applebaum E, Phelps L, Falcone J, Kanchi K, Thane T, Scimone A, Thane N, Henke J, Wang T, Ruppert J, Shah N, Rotter K, Hodges J, Ingenthron E, Cordes M, Kohlberg S, Sgro J, Delgado B, Mead K, Chinwalla A, Leonard S, Crouse K, Collura K, Kudrna D, Currie J, He R, Angelova A, Rajasekar S, Mueller T, Lomeli R, Scara G, Ko A, Delaney K, Wissotski M, Lopez G, Campos D, Braidotti M, Ashley E, Golser W, Kim H, Lee S, Lin J, Dujmic Z, Kim W, Talag J, Zuccolo A, Fan C, Sebastian A, Kramer M, Spiegel L, Nascimento L, Zutavern T, Miller B, Ambroise C, Muller S, Spooner W, Narechania A, Ren L, Wei S, Kumari S, Faga B, Levy MJ, McMahan L, Van Buren P, Vaughn MW, Ying K, Yeh C-T, Emrich SJ, Jia Y, Kalyanaraman A, Hsia A-P, Barbazuk WB, Baucom RS, Brutnell TP, Carpita NC, Chaparro C, Chia J-M,

Deragon J-M, Estill JC, Fu Y, Jeddeloh JA, Han Y, Lee H, Li P, Lisch DR, Liu S, Liu Z, Nagel DH, McCann MC, SanMiguel P, Myers AM, Nettleton D, Nguyen J, Penning BW, Ponnala L, Schneider KL, Schwartz DC, Sharma A, Soderlund C, Springer NM, Sun Q, Wang H, Waterman M, Westerman R, Wolfgruber TK, Yang L, Yu Y, Zhang L, Zhou S, Zhu Q, Bennetzen JL, Dawe RK, Jiang J, Jiang N, Presting GG, Wessler SR, Aluru S, Martienssen RA, Clifton SW, McCombie WR, Wing RA, Wilson RK (2009) The B73 maize genome: complexity, diversity, and dynamics. Science 326:1112–1115

Seoighe C, Gehring C (2004) Genome duplication led to highly selective expansion of the *Arabidopsis thaliana* proteome. Trends Genet 20:461–464

Shulaev V, Sargent DJ, Crowhurst RN, Mockler TC, Folkerts O, Delcher AL, Jaiswal P, Mockaitis K, Liston A, Mane SP, Burns P, Davis TM, Slovin JP, Bassil N, Hellens RP, Evans C, Harkins T, Kodira C, Desany B, Crasta OR, Jensen RV, Allan AC, Michael TP, Setubal JC, Celton J-M, Rees DJG, Williams KP, Holt SH, Rojas JJR, Chatterjee M, Liu B, Silva H, Meisel L, Adato A, Filichkin SA, Troggio M, Viola R, Ashman T-L, Wang H, Dharmawardhana P, Elser J, Raja R, Priest HD, Bryant DW, Fox SE, Givan SA, Wilhelm LJ, Naithani S, Christoffels A, Salama DY, Carter J, Girona EL, Zdepski A, Wang W, Kerstetter RA, Schwab W, Korban SS, Davik J, Monfort A, Denoyes-Rothan B, Arus P, Mittler R, Flinn B, Aharoni A, Bennetzen JL, Salzberg SL, Dickerman AW, Velasco R, Borodovsky M, Veilleux RE, Folta KM (2011) The genome of woodland strawberry (*Fragaria vesca*). Nat Genet 43:109–116

Small ID, Peeters N (2000) The PPR motif—a TPR-related motif prevalent in plant organellar proteins. Trends Biochem Sci 25:45–47

Somerville C, Koornneef M (2002) A fortunate choice: the history of *Arabidopsis* as a model plant. Nat Rev Genet 3:883–889

Springer NM, Ying K, Fu Y, Ji T, Yeh C-T, Jia Y, Wu W, Richmond T, Kitzman J, Rosenbaum H, Iniguez AL, Barbazuk WB, Jeddeloh JA, Nettleton D, Schnable PS (2009) Maize inbreds exhibit high levels of copy number variation (CNV) and presence/absence variation (PAV) in genome content. PLoS Genet 5:e1000734

Tang H, Bowers JE, Wang X, Ming R, Alam M, Paterson AH (2008a) Synteny and collinearity in plant genomes. Science 320:486–488

Tang H, Wang X, Bowers JE, Ming R, Alam M, Paterson AH (2008b) Unraveling ancient hexaploidy through multiply-aligned angiosperm gene maps. Genome Res 18:1944–1954

The Angiosperm Phylogeny Group (2009) An update of the Angiosperm Phylogeny Group classification for the orders and families of flowering plants: APG III. Bot J Linn Soc 161:105–121

The Arabidopsis Genome Initiative (2000) Analysis of the genome sequence of the flowering plant *Arabidopsis thaliana*. Nature 408:796–815

Thomas BC, Pedersen B, Freeling M (2006) Following tetraploidy in an *Arabidopsis* ancestor, genes were removed preferentially from one homeolog leaving clusters enriched in dose-sensitive genes. Genome Res 16:934–946

Tuskan GA, DiFazio S, Jansson S, Bohlmann J, Grigoriev I, Hellsten U, Putnam N, Ralph S, Rombauts S, Salamov A, Schein J, Sterck L, Aerts A, Bhalerao RR, Bhalerao RP, Blaudez D, Boerjan W, Brun A, Brunner A, Busov V, Campbell M, Carlson J, Chalot M, Chapman J, Chen G-L, Cooper D, Coutinho PM, Couturier J, Covert S, Cronk Q, Cunningham R, Davis J, Degroeve S, Déjardin A, dePamphilis C, Detter J, Dirks B, Dubchak I, Duplessis S, Ehlting J, Ellis B, Gendler K, Goodstein D, Gribskov M, Grimwood J, Groover A, Gunter L, Hamberger B, Heinze B, Helariutta Y, Henrissat B, Holligan D, Holt R, Huang W, Islam-Faridi N, Jones S, Jones-Rhoades M, Jorgensen R, Joshi C, Kangasjärvi J, Karlsson J, Kelleher C, Kirkpatrick R, Kirst M, Kohler A, Kalluri U, Larimer F, Leebens-Mack J, Leplé J-C, Locascio P, Lou Y, Lucas S, Martin F, Montanini B, Napoli C, Nelson DR, Nelson C, Nieminen K,

Nilsson O, Pereda V, Peter G, Philippe R, Pilate G, Poliakov A, Razumovskaya J, Richardson P, Rinaldi C, Ritland K, Rouzé P, Ryaboy D, Schmutz J, Schrader J, Segerman B, Shin H, Siddiqui A, Sterky F, Terry A, Tsai C-J, Uberbacher E, Unneberg P, Vahala J, Wall K, Wessler S, Yang G, Yin T, Douglas C, Marra M, Sandberg G, Van de Peer Y, Rokhsar D (2006) The genome of black cottonwood, *Populus trichocarpa* (Torr. & Gray). Science 313:1596–1604

Velasco R, Zharkikh A, Troggio M, Cartwright DA, Cestaro A, Pruss D, Pindo M, FitzGerald LM, Vezzulli S, Reid J, Malacarne G, Iliev D, Coppola G, Wardell B, Micheletti D, Macalma T, Facci M, Mitchell JT, Perazzolli M, Eldredge G, Gatto P, Oyzerski R, Moretto M, Gutin N, Stefanini M, Chen Y, Segala C, Davenport C, Demattè L, Mraz A, Battilana J, Stormo K, Costa F, Tao Q, Si-Ammour A, Harkins T, Lackey A, Perbost C, Taillon B, Stella A, Solovyev V, Fawcett JA, Sterck L, Vandepoele K, Grando SM, Toppo S, Moser C, Lanchbury J, Bogden R, Skolnick M, Sgaramella V, Bhatnagar SK, Fontana P, Gutin A, Van de Peer Y, Salamini F, Viola R (2007) A high quality draft consensus sequence of the genome of a heterozygous grapevine variety. PLoS One 2: e1326

Velasco R, Zharkikh A, Affourtit J, Dhingra A, Cestaro A, Kalyanaraman A, Fontana P, Bhatnagar SK, Troggio M, Pruss D, Salvi S, Pindo M, Baldi P, Castelletti S, Cavaiuolo M, Coppola G, Costa F, Cova V, Dal Ri A, Goremykin V, Komjanc M, Longhi S, Magnago P, Malacarne G, Malnoy M, Micheletti D, Moretto M, Perazzolli M, Si-Ammour A, Vezzulli S, Zini E, Eldredge G, Fitzgerald LM, Gutin N, Lanchbury J, Macalma T, Mitchell JT, Reid J, Wardell B, Kodira C, Chen Z, Desany B, Niazi F, Palmer M, Koepke T, Jiwan D, Schaeffer S, Krishnan V, Wu C, Chu VT, King ST, Vick J, Tao Q, Mraz A, Stormo A, Stormo K, Bogden R, Ederle D, Stella A, Vecchietti A, Kater MM, Masiero S, Lasserre P, Lespinasse Y, Allan AC, Bus V, Chagne D, Crowhurst RN, Gleave AP, Lavezzo E, Fawcett JA, Proost S, Rouze P, Sterck L, Toppo S, Lazzari B, Hellens RP, Durel C-E, Gutin A, Bumgarner RE, Gardiner SE, Skolnick M, Egholm M, Van de Peer Y, Salamini F, Viola R (2010) The genome of the domesticated apple (*Malus × domestica* Borkh.). Nat Genet 42:833–839

Vision TJ, Brown DG, Tanksley SD (2000) The origins of genomic duplications in *Arabidopsis*. Science 290:2114–2117

Wang X, Haberer G, Mayer K (2009) Discovery of cis-elements between sorghum and rice using co-expression and evolutionary conservation. BMC Genomics 10:284

Wolfe KH (2001) Yesterday's polyploids and the mystery of diploidization. Nat Rev Genet 2:333–341

Yoon HS, Hackett JD, Ciniglia C, Pinto G, Bhattacharya D (2004) A molecular timeline for the origin of photosynthetic eukaryotes. Mol Biol Evol 21:809–818

Yu J, Hu S, Wang J, Wong GK-S, Li S, Liu B, Deng Y, Dai L, Zhou Y, Zhang X, Cao M, Liu J, Sun J, Tang J, Chen Y, Huang X, Lin W, Ye C, Tong W, Cong L, Geng J, Han Y, Li L, Li W, Hu G, Huang X, Li W, Li J, Liu Z, Li L, Liu J, Qi Q, Liu J, Li L, Li T, Wang X, Lu H, Wu T, Zhu M, Ni P, Han H, Dong W, Ren X, Feng X, Cui P, Li X, Wang H, Xu X, Zhai W, Xu Z, Zhang J, He S, Zhang J, Xu J, Zhang K, Zheng X, Dong J, Zeng W, Tao L, Ye J, Tan J, Ren X, Chen X, He J, Liu D, Tian W, Tian C, Xia H, Bao Q, Li G, Gao H, Cao T, Wang J, Zhao W, Li P, Chen W, Wang X, Zhang Y, Hu J, Wang J, Liu S, Yang J, Zhang G, Xiong Y, Li Z, Mao L, Zhou C, Zhu Z, Chen R, Hao B, Zheng W, Chen S, Guo W, Li G, Liu S, Tao M, Wang J, Zhu L, Yuan L, Yang H (2002) A draft sequence of the rice genome (*Oryza sativa* L. ssp. *indica*). Science 296:79–92

Zhang X, Yazaki J, Sundaresan A, Cokus S, Chan SWL, Chen H, Henderson IR, Shinn P, Pellegrini M, Jacobsen SE, Ecker JR (2006) Genome-wide high-resolution mapping and functional analysis of DNA methylation in *Arabidopsis*. Cell 126:1189–1201

Plant Transposable Elements: Biology and Evolution 2

Eduard Kejnovsky, Jennifer S. Hawkins, and Cédric Feschotte

Contents

E. Kejnovsky (✉)
Institute of Biophysics, ASCR, Kralovopolska 135, 612 00 Brno, Czech Republic
e-mail: kejnovsk@ibp.cz

2.1 Introduction

Beginning with the pioneering work in the 30s and 40s of Barbara McClintock, R.A. Brink, Rollins Emerson, Marcus Rhoades, and other prominent maize geneticists, transposable elements (TEs) have come to occupy a central position in the study of plant genomes. Not only did McClintock's discovery of the *Activator/Dissociation* (*Ac/Ds*) system of maize change forever our appreciation of the dynamic nature of chromosomes, her seminal characterization of the regulatory influence of 'controlling elements' (such as *Ac/Ds* and later the *Enhancer/Suppressor-Mutator* (*En/Spm*) system) on adjacent gene expression paved the way for decades of exciting research on the control, both genetic and epigenetic, of gene regulation in plants and other eukaryotes.

It took four decades after McClintock's groundbreaking discoveries and the rise of recombinant DNA technology for the first TEs to be cloned and sequenced in the 1980s. One of the surprises from these early molecular studies was the striking similarity in structure, genetic organization, and even sometimes nucleotide sequence, among the first TEs characterized in maize, snapdragon, *Drosophila* and bacteria (Green 1980; Fedoroff et al. 1983; Levis et al. 1984; Saedler et al. 1984). At that time, and over the next two decades, the biology of TEs was assessed primarily on the basis of the mutations they engendered. Myriad mutant alleles caused by insertions and/or rearrangements of transposons were collected by geneticists in the field, the greenhouse and the fly room, and meticulously analyzed at the molecular level in the lab. Although this era furnished many crucial insights regarding the mechanistic underpinnings and mutagenic capabilities of transposition (for review, Berg and Howe 1989), it yielded little information regarding the abundance and diversity of TEs, much less the long-term evolutionary impact of TE activity.

The advent of large-scale DNA sequencing over the last two decades, combined with advances in functional

J.F. Wendel et al. (eds.), *Plant Genome Diversity Volume 1*,
DOI 10.1007/978-3-7091-1130-7_2, © Springer-Verlag Wien 2012

genomics and bioinformatics, has transformed the study of TE biology. This "genomics revolution" has resulted in a greater understanding of the many ways that TEs influence the function and evolution of genes and genomes, and consequently, their host organisms. In particular the genomics era has revealed that, although only a tiny fraction of TEs are transpositionally active, most eukaryotic genomes, and especially plant genomes, are packed with a plethora of seemingly dormant or inactivated TE families (Feschotte et al. 2002). Given the inherent mutagenic potential of active transposition, it should come as no surprise that the majority of these TEs are either defective, fossilized copies or potentially active copies that are restrained by host silencing systems; however, active transposition, as evidenced by instances of mutagenic (yet potentially evolutionarily significant) insertions, has been demonstrated. For example, TEs have been shown to silence or alter expression of genes adjacent to insertion sites, become integrated into functional genes as newly acquired exons (exapted), acquire host gene sequences and insert them into new genomic locations, contribute to chromosomal rearrangements via recombination, epigenetically alter regional methylation patterns, and provide template sequences for RNA interference (Feschotte et al. 2002; Bennetzen 2005; Morgante et al. 2007; Weil and Martienssen 2008; and see Slotkin et al. 2012, this volume). This diverse functional impact of TEs, and their intrinsic contribution to genomic plasticity, suggests that these elements play a major role in molecular diversification and, ultimately, species divergence.

In this chapter, we provide the reader with the fundamentals of TE biology, with an emphasis on plant elements. We begin with an overview of TE classification and transposition mechanisms, followed by an examination of the extensive variability in both inter- and intra-specific TE content across diverse plant taxa. Finally, we explore some of the general principles characterizing and influencing the genomic distribution, activity and evolution of TEs.

2.2 Transposable Element Classification

TEs can be broadly defined as DNA segments capable of chromosomal movement, either via replicative or conservative (cut-and-paste) mechanisms (discussed in more detail below). The TE classification system that we present here is similar to the one proposed by Wicker et al. (2007) and to the one implemented in Repbase, the most popular database of repetitive DNA sequences (http://www.girinst.org/). At the highest level, eukaryotic TEs comprise two major classes, and each class can be divided into subclasses based on their mechanism of chromosomal integration, which is reflective of the protein-coding capabilities and organizational structure of each class and subclass of elements (Figs. 2.1, 2.2).

Class I elements, also known as retrotransposons, transpose via an RNA intermediate, which must be reverse transcribed prior to integration into the genome, while Class II elements transpose via a DNA intermediate (Finnegan 1989). Transposition of both classes of elements may result in a heritable increase in genomic copy number; hence, individual TE types are found in multiple copies (often referred to as a TE family) and comprise the majority of the repetitive fraction of eukaryotic genomes (e.g. Adams et al. 2000; The Arabidopsis Genome Initiative 2000; Lander et al. 2001; International Rice Genome Sequencing Project 2005). TEs have been found in virtually every organism studied to date (with few exceptions, such as *Plasmodium falciparum* and other Apicomplexa), although significant qualitative and quantitative variation abounds, even among closely related organisms (see below for a comparison among selected plant species).

The genomes of plants are packed with many and diverse TEs, and continue to serve as excellent models to yield some of the most significant advances in the field of transposon biology. The vast majority of repetitive DNA in the nuclear genomes of plants is derived from the proliferation of TEs, most often Class I RNA elements (Fig. 2.1) (e.g. SanMiguel et al. 1996; Vicient et al. 1999; Hawkins et al. 2006; Neumann et al. 2006; Vitte and Bennetzen 2006). Two major subclasses of Class I elements have been identified in plants: (1) Long terminal repeat (LTR) retrotransposons, whose reverse-transcription and subsequent integration as double-stranded DNA is mediated by an element-encoded reverse transcriptase and integrase, respectively, (2) non-LTR retrotransposons (sometimes called retroposons), which include long and short interspersed elements (LINEs and SINEs) and use target-primed reverse transcription, a mechanism coupling reverse transcription and integration. DIRS-like elements (named after *Dictyostelium* intermediate repeat sequence) represent a third subclass of retrotransposons integrated through an element-encoded tyrosine recombinase. They are relatively common in animals and fungi, but have yet to be found in flowering plants. Class II elements have been identified in every plant genome that has been thoroughly examined, and these can be divided in two major subclasses: (1) classic 'cut-and-paste' DNA transposons, characterized by terminal inverted repeats (TIRs), which are excised and reintegrated as double-stranded DNA by the action of an element-encoded transposase and (2) *Helitrons*, or rolling-circle transposons, which most likely transpose via a replicative mechanism involving a single-stranded DNA intermediate and which encode recombinase with Replicator initiator motif (Rep) and DNA Helicase domains (Fig. 2.1).

In plants, Class I elements (particularly LTR retrotransposons) make up the largest fraction of the TE complement (SanMiguel et al. 1996, 1998; Vicient et al. 1999;

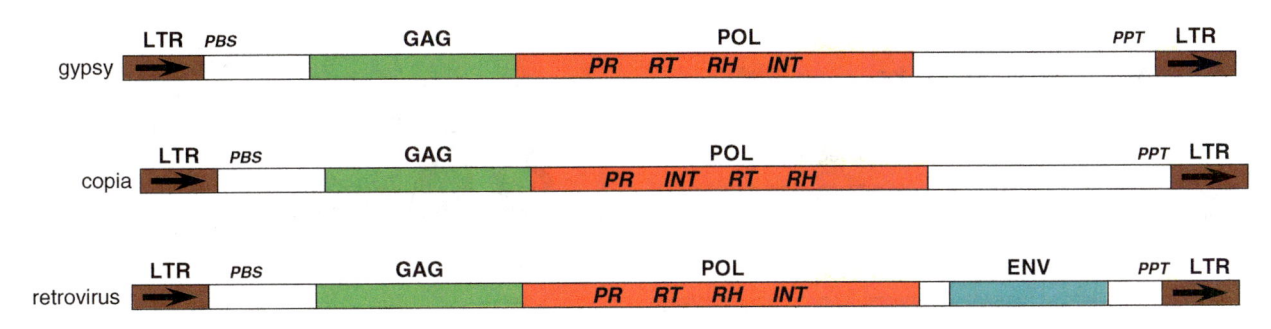

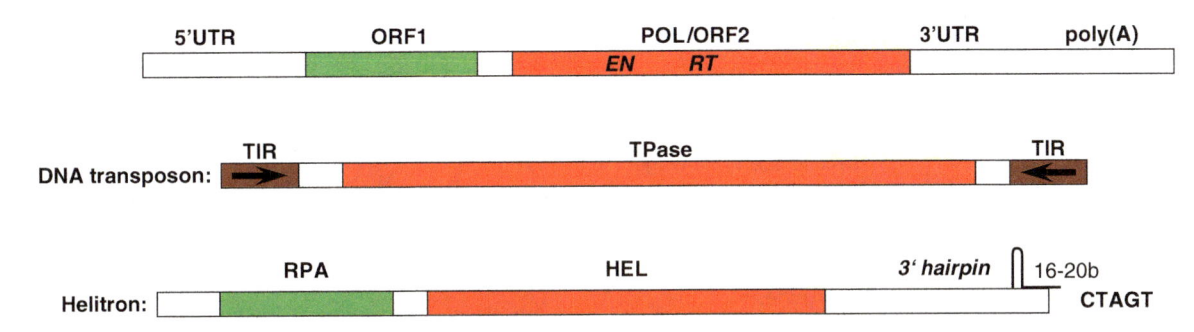

Fig. 2.1 Structure of main types of transposable elements. GAG and POL genes of LTR retrotransposons, ORF1 of non-LTR retrotransposons, transposase (TPase) of DNA transposons and replicative protein A (RPA) and helicase (HEL) of Helitrons are marked. Long terminal repeats (LTRs), primer-binding site (PBS) and polypurine tract (PPT) of LTR retrotransposons, 5′ UTR, 3′ UTR and poly(A) of non-LTR retrotransposons, terminal inverted repeats (TIR) of DNA transposons and 3′ hairpin of Helitrons are labeled. LTR retrotransposons are exemplified by *gypsy*, *copia* and retrovirus superfamilies. Protease (PR), reverse transcriptase (RT), RNaseH (RH), integrase (INT) and endonuclease (EN) domains are marked

Hawkins et al. 2006; Neumann et al. 2006; Vitte and Bennetzen 2006). The LTRs flanking a retrotransposon can range from just a few hundred base pairs to as much as 6 kb, and usually begin with 5′-TG-3′ and end with 5′-CA-3′. The LTR retrotransposons typically contain GAG and POL protein coding ORFs, which encode several enzymes (reverse transcriptase – RT; protease – PR; RNaseH – RH; integrase – INT) responsible for reverse transcription and integration of daughter sequences into new chromosomal locations. Two major superfamilies of LTR retrotransposons are found in plants, *gypsy*-like and *copia*-like (also known as *Metaviridae* and *Pseudoviridae*, respectively). Both types of LTR retrotransposons contain the same protein coding domains, but these are arranged in a different order. Their ancient origin is evidenced by the fact that they form deeply diverged monophyletic clades in phylogenetic analyses of reverse transcriptases (Eickbush and Malik 2002; Havecker et al. 2004). Non-LTR retrotransposons (LINEs and SINEs) are, as their name indicates, not flanked by LTRs, but complete LINEs can reach several thousand base pairs in length, contain coding sequences responsible for transposition, and often display a stretch of adenines or a simple sequence repeat at their 3′ end (Figs. 2.1, 2.2c).

Class II DNA elements are found in most eukaryotes, and despite their conservative transposition mechanism, have been capable of attaining relatively high copy numbers in some plants (see Sect. 2.3.1, Feschotte and Pritham 2007). Class II elements encode the machinery to facilitate their own transposition, usually in the form of a transposase (TPase) encoded by a single gene. "Cut-and-paste" transposition is associated with Subclass 1 DNA transposons, and occurs via TPase binding to the terminal inverted repeats (TIRs) of the element (Fig. 2.1), followed by excision and reintegration of the transposon at a new chromosomal location (Craig et al. 2002). The transposition mechanism of *Helitrons* has not been investigated in functional detail, but these elements are believed to employ a mechanism where only one DNA strand is cut, displaced and which serves as a template for replication of the element at a new locus (Kapitonov and Jurka 2007).

Both Class I and Class II TEs may be further divided into autonomous or non-autonomous elements dependent upon their ability to encode the enzymatic machinery responsible for movement. Non-autonomous elements may still be mobilized *in trans* if they retain the capacity to be recognized by the enzymes encoded by autonomous

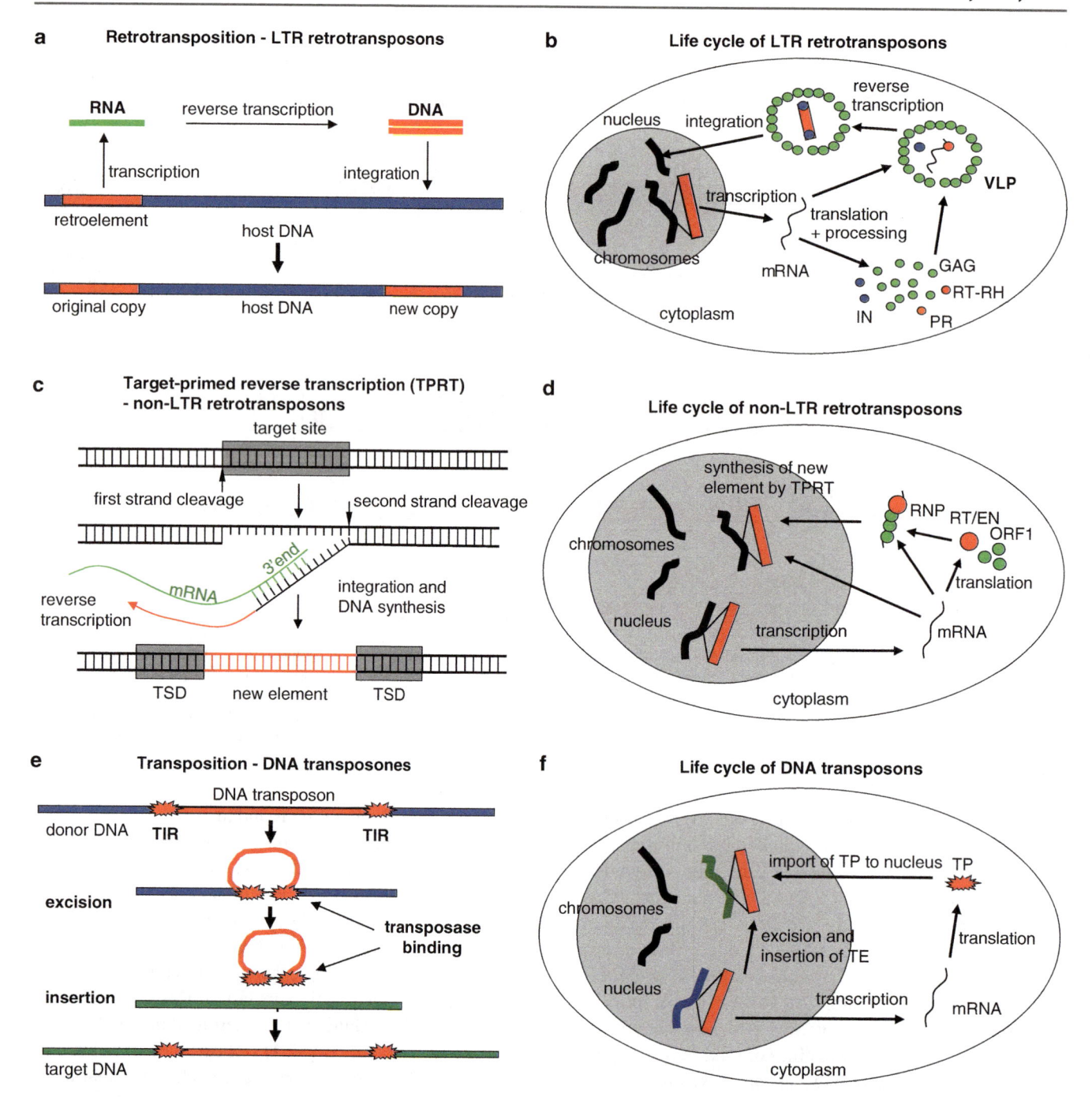

Fig. 2.2 Transpositional mechanism of main types of transposable elements. (**a**) Schematic retrotransposition of LTR retrotransposons and (**b**) their life cycle in the cell has a "copy and paste" character. Target-primed reverse transcription of non-LTR retrotransposons where cDNA is synthesized *in situ* (**c**) and the life cycle of non-LTR retrotransposons in the cell (**d**). Transposition of DNA transposons using "cut and paste" mode (**e**) and their life cycle in the cell (**f**). GAG gene, reverse transcriptase (RT), endonuclease (EN), integrase (INT), protease (PR) domains, transposase (TP), terminal inverted repeat (TIR), target-site duplication (TSD), ribonucleoparticle (RNP) and virus-like particle (VLP) are marked

elements located elsewhere in the genome. Although this concept was initially described for classic, two-component DNA transposon systems, such as *Ac/Ds* in maize, it seems that virtually all types of TEs may include both autonomous elements and non-autonomous counterparts that are movable *in trans* (Feschotte et al. 2002; Wicker

et al. 2007). Non-autonomous Class I elements in plants include SINEs (short interspersed elements, Deragon and Zhang 2006), TRIMs (terminal repeat retrotransposons in miniature, Witte et al. 2001) and LARDs (large retrotransposon derivates, Kalendar et al. 2004). MITEs (miniature inverted-repeat elements, Bureau and Wessler 1992) represent the most abundant type of non-autonomous DNA transposon in plant genomes thus far examined.

Non-autonomous elements may originate in a variety of ways. Most commonly, they derive from autonomous copies that have suffered mutations (substitutions or insertions/ deletions) disabling their coding capabilities. For example, most *Ds* elements are directly derived from *Ac* by internal deletions (Yan et al. 1999). Note, however, that autonomous and non-autonomous elements need not share extensive sequence similarity to form a functional pair. Indeed, the original *Ds1* element from maize, which is recognized and mobilized by the *Ac*-encoded transposase, shares only the outermost 11 nucleotides of its TIRs (terminal inverted repeats) with *Ac* (Kunze and Starlinger 1989). Likewise, many families of high-copy number MITEs are not always directly related to autonomous elements present in the same genome. Nonetheless, there is evidence that some MITEs, such as *Stowaway*, can be mobilized with high efficiency by distantly related autonomous transposons (*mariner*-like elements in the case of *Stowaway*, Yang et al. 2009). With respect to the origin of such 'orphan' MITE families, it remains possible that their progenitors are direct derivatives of autonomous elements that did not reach fixation or are no longer recognizable in the genome (Feschotte et al. 2003). Alternatively, some may have arisen 'de novo', by juxtaposition of sequences that were fortuitously recognized by transposition enzymes produced in *trans*. This scenario has been documented at least once in *Drosophila* (Tsubota and Huong 1991), but to our knowledge, never in plants.

SINEs represent another atypical category of non-autonomous elements that derive from non-coding genes transcribed by RNA polymerase III (pol III), most commonly tRNA genes (Deragon and Zhang 2006). The simplest SINE families are equivalent to amplified tRNA retrogene families, which apparently result from accidental *trans*-recognition by the enzymatic machinery of autonomous LINEs. The use of an internal promoter (retained after retroposition) coupled to the short length and perhaps also the cellular localization of pol III transcripts may explain the recurrent amplification of tRNA genes by retroposition. More complex SINEs are formed either by multimerization, duplication and/or fusion with the 3′ terminus of a LINE (Deragon and Zhang 2006). Such chimeric SINEs may become highly efficient at hijacking the machinery of their partner LINEs. Perhaps the best-known SINE is the *Alu* element of primates, which is present in over a million copies per haploid human genome (Lander et al. 2001). SINEs have been identified in a wide range of plant species and individual families may attain several thousand copies (Deragon and Zhang 2006), but due to their short size they tend to make up a relatively small fraction of the repetitive DNA content of plant genomes (Fig. 2.5).

2.3 Transposable Elements Biology: Intrinsic Factors of Transposon Proliferation

Although the total quantitative amount of TEs varies tremendously among (and possibly within) plant species, every genome analyzed so far has been found to harbor representatives of both Class I and Class II TEs (Figs. 2.4, 2.5). As mentioned above, however, the relative qualitative contribution of the two classes and their subclasses to the total TE population varies substantially among species. For example, LTR-retrotransposons predominate in the genomes of cotton and maize (Hawkins et al. 2006; Vitte and Bennetzen 2006), but less so in the genomes of rice or *Lotus japonicus*, where DNA transposons are as (or more) successful than other TE types, as measured by copy numbers (Holligan et al. 2006). Additionally, there are differences in the chromosomal distribution of Class I and Class II elements in the genome (e.g. Peterson-Burch et al. 2004; International Rice Genome Sequencing Project 2005; Baucom et al. 2009). These variations correspond, in part, to the disparate histories of TE invasion experienced by different plant lineages, in addition to how an organism copes with these invasions, which is greatly influenced by host biology, as discussed in Sect. 2.4. In the present section, we examine how the biology and properties of the TEs themselves may lead to significant variation in TE composition among species, and possibly, within the genome.

2.3.1 Mechanisms of Transposition

The mechanism by which a TE family is amplified may determine, in part, their pattern of proliferation and diversification in the genome. Part of the proliferative success of Class I retrotransposons in many taxonomic groups (particularly plants) is ensured by their replicative mode of transposition, where in principle, a small number of 'master' copies can produce hundreds or thousands of 'daughter' copies during a single amplification event (Fig. 2.2a, b). Evidence of such transpositional "bursts" comes from phylogenetically informed analyses in both rice and *Gossypium*, where comparative sequence analyses within various TE families indicates waves of TE accumulation surrounded by periods of relative quiescence (Piegu et al. 2006; Hawkins et al. 2008, 2009). Additionally, diversification can be accomplished via "template-switching" during reverse transcription, first described in retroviruses (Pathak and Hu 1997), in which two different RNA molecules co-localized in a virus-like particle combine, leading to a new, chimeric element. Recent data point to this mechanism as an important force driving the evolution of maize LTR-retrotransposons (Sharma et al. 2008). Template-switching may also occur during the transposition of non-LTR retrotransposons (Garcia-Perez et al. 2007), and this mechanism may explain the chimeric structure and modular evolution of SINEs (Deragon and Zhang 2006). Exchange of sequences is also possible at the DNA level and may promote the diversification of DNA transposons, including Helitrons (Yang and Bennetzen 2009), providing a mechanism for the acquisition of host gene fragments by various plant TEs (Bureau et al. 1994; Jiang et al. 2004; see also Chap. 2). In fact, template-switching or other forms of inter-element recombination may be viewed as a primitive form of sex, promoting the genetic diversification of TEs.

Class II DNA transposons are mobilized by a cut-and-paste mechanism where the element is excised from one locus and re-inserted elsewhere in the genome (Fig. 2.2e, f). This process, by itself, does not result in an increase in copy number, as the element is not replicated; however, increases in Class II element copy number can occur through two known mechanisms, both of which are dependent upon host cellular activities. First, upon excision of a Class II element, the consequential double-stranded DNA break can be repaired by homologous recombination using the transposon copy located on the homologous chromosome as a template (Engels et al. 1990), or alternatively, the sister chromatid if excision takes place during S phase. DNA replication offers a second opportunity for duplication: when a transposon jumps ahead of a replication fork, from a post- to a pre-replicated region, it can effectively be replicated twice (Ros and Kunze 2001). Nevertheless, each of these mechanisms produces a net gain of only one copy per transposition event.

In spite of the conservative nature of cut-and-paste transposition, it is clear that DNA transposons can amplify to very high copy numbers (up to several thousands per family), as documented by the explosive bursts of MITEs in many angiosperms (for a spectacular example of MITE amplification in 'real-time', see Naito et al. 2009). How MITEs could achieve such high copy number has remained a mystery for nearly two decades, but some important clues have surfaced recently, thanks to the study of actively transposing MITE families discovered in the rice genome. It is now established that MITEs rely on a transposase encoded by larger, autonomous elements (Feschotte and Mouchès 2000; Zhang et al. 2001; Feschotte et al. 2003). Furthermore, the data point to a typical cut-and-paste mechanism involving excision and re-insertion similar to that of other eukaryotic DNA transposons (Petersen and Seberg 2000; Nakazaki et al.

2003; Yang et al. 2006, 2009). One key to the mystery of MITE amplification seems to lie in the complexity of their interactions with transposases, as revealed by functional studies of *Stowaway* MITEs and their partner *Osmar* transposases in rice. First, a single source of *Osmar* transposase is capable of interaction and mobilization of a diversity of *Stowaways* having different origins, even in the absence of extensive sequence similarity between *Osmar* and *Stowaway* termini (Feschotte et al. 2005; Yang et al. 2009). Second, some *Stowaway* elements possess the inherent ability to excise at higher efficiency in response to *Osmar* transposase than other substrates, including the cognate *Osmar* element providing the source of transposase. Sequence-swapping experiments indicate that the excision hyperactivity of the MITE stems from a combination of properties, including short size, the absence of *cis*-elements present in the autonomous *Osmar* element that repress transposition, and conversely the presence of *cis*-elements in the MITE internal sequence that enhance transposition (Yang et al. 2009). Thus, multiple, overlaying rampant amplification of MITEs in plant genomes.

2.3.2 Targeting Strategies

2.3.2.1 Transposable Elements Occupy Different Genomic Niches

Genomes can be partitioned into a variety of "chromosomal niches" that are colonized by various repetitive sequences (Kidwell and Lisch 2001). In particular, constitutive heterochromatic regions of the plant genome, such as pericentromeric regions, knobs, and subtelomeres, represent chromosomal niches heavily occupied by LTR-retrotransposons (Miller et al. 1998; Lippman et al. 2004; Kejnovsky et al. 2006b). By contrast, most DNA transposons, and MITEs especially, are found at higher density in euchromatic regions where they often reside within or in close proximity to genes (Bureau and Wessler 1992; International Rice Genome Sequencing Project 2005).

To account for the chromosomal distribution of TEs, one must consider the action of several, non-mutually exclusive forces acting at the time of insertion and often long after insertion. Some of these forces are inherent to the transposition machinery of the elements that confer insertion preference for certain chromosomal or sequence features. Also, natural selection will favor the fixation of beneficial insertions and the elimination of deleterious ones from the population. Finally, an array of indirect forces may act more gradually, influencing the decay of the elements or their removal by deletion or recombination (for review, Pritham 2009). These latter forces include rates of substitution, deletion and recombination, which can vary dramatically along chromosomes (e.g. low recombination in peri-centromeric regions) and also among species. It is often difficult to discern the relative importance of these many forces on TE accumulation differentially over time. The effect of insertion preference tends to be more apparent for younger TE insertions, while recombination and deletional processes become more significant as TEs become older and accumulate in the genome.

With respect to TE insertion preference targeting of TEs into specific chromosomal locations, such as heterochromatin where they likely have less deleterious effects, represents a mechanism minimizing the negative impact of TEs on the host. Thus different targeting strategies are likely to evolve among TEs to occupy diverse genomic niches and thereby contribute to their evolutionary persistence. The biased TE populations of two diverged yeast species, *S. cerevisiae* and *S. pombe*, provide an extreme example. In these streamlined genomes, only a handful of LTR retrotransposon families co-exist, and remarkably, all have adopted different targeting strategies. Ty1 and Ty3 of *S. cerevisiae* preferentially insert upstream of tRNA genes and other units transcribed by Pol III (Ji et al. 1993), while Ty5 targets the silent chromatin located in subtelomeric regions and around the mating loci (Zou et al. 1996). In *S. pombe*, Tf elements preferentially insert upstream of Pol II-transcribed genes (Bowen et al. 2003).

In plants, there is evidence that the accumulation of *Arabidopsis* LTR retrotransposons in pericentromeric regions and other highly heterochromatic chromosomal compartments is the result of both active targeting and selective retention over time (Pereira 2004; Peterson-Burch et al. 2004). Comparison of the age and chromosomal distribution of TEs in *Arabidopsis* indicate that *copia*-like elements are integrated fairly randomly into the genome while *gypsy*-like elements preferentially insert into the pericentromeric heterochromatin (Pereira 2004). In maize, high-copy-number LTR retrotransposon families are found to primarily accumulate in gene-poor regions, while LINEs, SINEs, and low-copy-number LTR retrotransposons show biased insertion in gene-rich regions (Baucom et al. 2009). The accumulation of LTR elements in heterochromatic regions is also evident in rice, while conversely, MITEs and most other DNA transposons are found in higher density close to or within protein-coding genes (International Rice Genome Sequencing Project 2005). An examination of a large number of *de novo* insertions of *Mutator* DNA elements in rice and maize (Dietrich et al. 2002; Liu et al. 2009; Jiang et al. 2011) and *mPing* MITEs in *Arabidopsis* and rice (Yang et al. 2007; Naito et al. 2009) demonstrate that these transposons actively target genes, with a preference for insertion in their 5′ upstream region. A preference for insertion within or near the same family of elements (self-preference) was also observed for *Tourist* MITEs in maize and rice (Jiang and Wessler 2001) and for *Helitrons* in maize (Yang and Bennetzen 2009).

The molecular mechanisms underlying the targeting of plant TEs remain poorly understood, but may involve the recognition of specific DNA motifs (Zhang et al. 2001), their position relative to the nucleosome (Jiang and Wessler 2001), or the epigenetic state of the insertion sites (Brady et al. 2008). Indeed, emerging evidence suggests that mobile elements may, in some cases, possess the inherent capacity to target their integration toward particular chromatin domains. For example, chromoviruses are *gypsy*-like retrotransposons that contain 40–50 amino acid "chromodomains" at the C-terminus of their integrase (Kordis 2005; Novikova 2009). These chromodomains are thought to direct integration into heterochromatic regions via interaction with methylated histone residues, thereby facilitating targeted insertion into relatively gene-poor regions (Gao et al. 2008).

2.3.2.2 The Special Relationship of Plant Retrotransposons with Centromeres

In spite of their conserved function, centromeres are highly dynamic at the sequence level (see Hirsch and Jiang 2012, this volume). The major components of plant centromeres are large arrays of tandem repeat, called satellite DNA (Jiang et al. 2003), as exemplified by rice CentO (Cheng et al. 2002) and maize CentC satellites (Ananiev et al. 1998). In most plants examined, centromeric satellites are intermingled with a particular group of *gypsy*-like elements called centromeric retrotransposons (*CRs*). *CRs* were originally found in many grass species, such as *CRM* in maize (Zhong et al. 2002; Nagaki et al. 2003), *CRR* and *RIRE7* in rice (Kumekawa et al. 2001; Cheng et al. 2002; Nagaki et al. 2005), *CEREBA* in barley (Presting et al. 1998), *CRW* in wheat (Liu et al. 2008), *CRS* in sugarcane (Nagaki and Murata 2005), and *Bilby* in rye (Francki 2001). However *CRs* are not restricted to grasses, as they were recently discovered in *Arabidopsis*, soybean (Du et al. 2010) and many other eudicot species (Neumann et al. 2011). These findings suggest that *CRs* colonized centromeres (or pericentromeres) before the divergence of monocots and eudicots and have been stable components of angiosperm genomes ever since (Du et al. 2010). Consistent with this scenario, phylogenetic analyses revealed that *CRs* represent a deeply rooted, monophyletic clade of *gypsy*-like elements (Gorinsek et al. 2004; Kordis 2005; Neumann et al. 2011). There are exceptions, nonetheless, such as in *Oryza brachyantha*, where *CentO* satellites and *CRR* retrotransposons have disappeared from functional centromeres and were subsequently replaced by *FRetro3*, a retrotransposon belonging to a different lineage of *gypsy*-like elements (Gao et al. 2009).

The ancient origin, vertical persistence and relatively high level of sequence conservation across species set *CRs* apart from other plant LTR retrotransposons, and these characteristics prompted several investigators to hypothesize that *CRs* could have been co-opted for a cellular function (Zhong et al. 2002). One possibility is that that they provide an abundant source of promoters for the transcription of satellite repeats, which may be important for the establishment of centromere identity and/or chromosome segregation (May et al. 2005). Indeed, plant centromeric satellites and *CRs* themselves are often transcribed (Topp et al. 2004; Neumann et al. 2007), and there is evidence that the transcripts of rice *CRR* elements are partially processed into small RNAs through the RNA interference (RNAi) pathway (Neumann et al. 2007). It is tempting to speculate that *CR*-derived small RNAs are implicated in the formation and/or maintenance of centromeric chromatin (Neumann et al. 2007), akin to the mechanism underlying the formation of pericentromeric heterochromatin in fission yeast which are also initiated by transcription and RNAi-dependent processing of repetitive elements (Volpe et al. 2002; Grewal and Jia 2007). Furthermore maize *CRM* DNA and, surprisingly, *CRM*-derived transcripts, both interact with the centromeric histone CENH3 (Zhong et al. 2002; Topp et al. 2004) and at least one subfamily of *CRM* elements (*CRM2*) exhibit tightly phased positioning on CENH3-containing nucleosomes (Gent et al. 2011). Together these data point at a functional association of *CRs* with centromeric chromatin, although further experiments are needed to clarify the role of *CRs* in plant centromere biology.

2.4 Influence of Host Biology on Transposable Element Proliferation

Factors acting at the level of the host and affecting the likelihood of fixation of TE insertions, subsequent decay (via nucleotide substitutions or indels), or their physical removal (via large deletions and other recombination events) may have a significant influence on shaping TE content over time. We highlight here three of these forces, effective population size, sexual reproduction and recombination rate that are likely to have prominent effects on TE persistence and accumulation. We also describe the role of recombination in shaping TE proliferation and distribution on sex chromosomes.

2.4.1 Effective Population Size

For long-term persistence any TE family must, on average, give rise to at least one daughter element for each element inactivated by mutation or eliminated by deletion. Strongly deleterious TE insertions are eliminated by selection

(Le Rouzic et al. 2007), and the efficiency of selection is proportional to the host effective population size. The persistence of TEs in species with large effective populations, like in most unicellular species, is rare (Wagner 2006), especially in the absence of sex or horizontal transfer. The critical effective population size above which eukaryotic populations appear to be immune to retrotransposon proliferation is suggested to be $\sim 7 \times 10^7$, whereas for DNA transposons it is $\sim 2 \times 10^7$ (Lynch and Conery 2003). Additionally, total genome size is inversely correlated with long-term effective population size because TEs form a significant part of the genomes of multicellular eukaryotes having generally smaller effective population sizes (Lynch and Conery 2003). Long-term effective population size reduction then probably enabled increases in genome sizes as well as organism sizes. In plants, it was suggested that species with small population sizes should purge TE insertions less efficiently and hence accrue DNA more rapidly (Lockton et al. 2008); however, a recent study of 205 species of seed plants determined no relationship between effective population size and genome size, suggesting that effective population size is not an especially significant factor in the relative level of proliferation and persistence of TEs in plants (Whitney et al. 2010).

2.4.2 Breeding System

Almost three decades ago Hickey (1982) suggested that the potential for TE proliferation is related to the rate of outcrossing in a given host species. Population genetics and mathematical modeling predict that obligatory out-crossing species should contain a larger number of and more active TEs than self-fertilizing or facultative sexual species, while at the other end of the spectrum obligate asexuals and uniparental organelle genomes should rapid purge active TEs and essentially be free of selfish genetic elements, unless they have recently re-entered by horizontal transfer (Hickey 1982; Bestor 1999; Schön and Martens 2000). Although these theoretical arguments were grounded in arguments of population genetics, they have proven difficult to test empirically (but see Zeyl et al. 1996; Arkhipova and Meselson 2000; Schaack et al. 2010a,b).

Plants, which include closely related selfing and outcrossing species, offer a valuable system to investigate these questions because the genetics of selfing species resemble that of asexuals. Consistent with this theory, the outcrossing *Arabidopsis lyrata* displays higher transposition frequency, stronger selection against new TE insertions, and faster removal of insertions by ectopic recombination than in the selfing *A. thaliana* (Wright et al. 2001, 2003; Lockton and Gaut 2010; Hollister and Gaut 2007). Perhaps consequently, the diversity of TEs is greater in *A. lyrata* than in

A. thaliana (Lockton et al. 2008). However the difference in breeding system may be only partially or indirectly causative of these patterns. As discussed above, demographic history, such as population bottlenecks, has the power to explain most of these variations and to exert a substantial influence on TE dynamics (Lockton et al. 2008; Tenaillon et al. 2010).

2.4.3 Recombination Rates Shape the Chromosomal Distribution of Transposable Elements

The chromosomal distribution of TEs is influenced by many factors, such as local variation in recombination rates or gene density (as reviewed above). Genomic regions with no or low recombination are represented by most of the Y chromosome, B chromosomes, or (peri)centromeres. In particular, the non-recombining Y chromosome is subject to a suite of processes leading to the accumulation of deleterious mutations (Charlesworth and Charlesworth 2000). These processes include (1) Muller's ratchet, (2) genetic hitchhiking and (3) background selection (reviewed in Bachtrog 2006). There are two predictable consequence of these processes, namely, the degeneration of genes and the accumulation of selfish genetic elements, including TEs (Charlesworth et al. 1994). Two models have been evoked to explain the accumulation of TEs in gene-poor regions with low or no recombination, such as the Y chromosome or the peri-centromeric regions of chromosomes (reviewed in Dolgin and Charlesworth 2008). In the "insertion model", there is weaker selection against TEs in gene-poor regions due to the decreased possibility of deleterious insertions, resulting in higher TE abundance in these regions. The "ectopic recombination model" postulates that TEs accumulate in regions of low recombination because ectopic recombination between copies, which is a powerful deletional force, is less frequent in these regions than in regions with high recombination rate (Langley et al. 1988).

The accumulation of TEs in non-recombining regions has been observed empirically on the Y chromosome of humans (Erlandsson et al. 2000; Skaletsky et al. 2003), *Drosophila melanogaster* (Pimpinelli et al. 1995), as well as on the neo-Y chromosome of *Drosophila miranda* (Steinemann and Steinemann 1992; Bachtrog 2003). In plants, however, the relationship of recombination rate to TE distribution is not clear. As noted above, in *A. thaliana* and many other angiosperms examined, TEs tend to accumulate in pericentromeric regions. TE distribution in *A. thaliana*, however, does not correlate with recombination rate, but is negatively correlated with gene density (Wright et al. 2003). Shorter LTR retrotransposons and their fragments accumulate in regions with higher recombination rates, indicating that both recombination and gene density can

influence the rate and pattern of TE elimination (Swigonová et al. 2005; Tian et al. 2009).

Some dioecious plants possess sex chromosomes that often are in the early stages of evolution (compared to the more ancient mammalian sex chromosomes), where the Y chromosomes have expanded, rather than contracted, compared to X (Vyskot and Hobza 2004). It is often assumed, but not yet demonstrated, that the increased size of plant Y chromosomes results from the accumulation of TEs in non-recombining regions. Consistent with this idea, various types of repetitive DNA are specific to or enriched on the Y chromosome of several plant species, e.g., RAYS tandem repeats in *Rumex acetosa* (Shibata et al. 1999), LINE elements in *Cannabis sativa* (Sakamoto et al. 2000) and *copia*-like elements in *Marchantia polymorpha* (Okada et al. 2001). In papaya, which possesses the youngest studied plant Y chromosome, the male-specific region of this nascent Y chromosome is associated with a high density of various DNA repeats (Liu et al. 2004). In *Silene latifolia*, the most popular dioecious plant model (Kejnovsky and Vyskot 2010), the Y chromosome is strikingly enlarged (Fig. 2.3a). This is due in part to an accumulation of *copia*-like elements but also to chloroplast DNA insertions and to an expansion of tandem repeats (Hobza et al. 2006; Kejnovsky et al. 2006a; Cermak et al. 2008; Kubat et al. 2008). However, not all TEs of *S. latifolia* accumulate on the Y chromosome. For example, *Ogre*-like *gypsy* elements (Fig. 2.3b) are abundant on all chromosomes but virtually absent on the non-recombining parts of the Y chromosome. Several mechanisms might account for this unexpected distribution, including female-specific transposition activity or specific

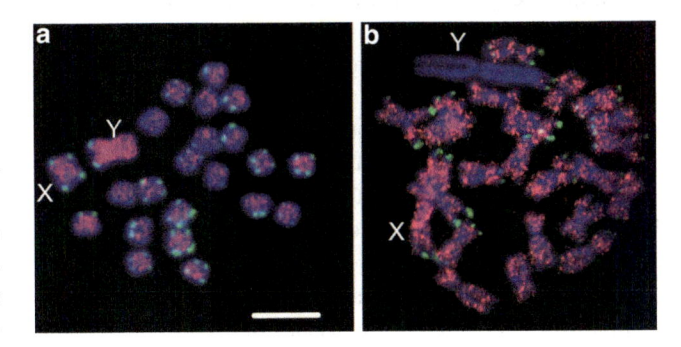

Fig. 2.3 Examples of TE localization on sex chromosomes using FISH in model dioecious plant *Silene latifolia* (*white campion*), species with heteromorphic sex chromosomes. Accumulation of *Copia* elements (in *red*) on the Y chromosome (**a**) in contrast with the *Ogre-like* gypsy retrotransposon (in *red*) that colonizes only recombining parts of genome (**b**). *Ogre-like* elements are ubiquitously distributed on all autosomes and the X chromosome but on the Y chromosome occupy only short pseudoautosomal region while are absent in the large non-recombining parts (**b**). The tandem repeat X-43.1 labels most subtelomeres but on the Y chromosomes only its q-arm (*green signals*). Chromosomes are counterstained by DAPI (*blue*). The X and Y chromosomes are indicated. Bar represents 10 μm. Reproduced by courtesy of Cytogenetic and Genome Research

targeting of recombining regions of genome (Cermak et al. 2008; Kejnovsky et al. 2009a).

The massive aggregation of TEs in early stages of Y chromosome evolution suggests that TEs themselves may be involved in the degeneration of genes located on the Y chromosome, through insertional disruption, rearrangements or post-insertional effects on gene expression (Marais et al. 2008). However, it is still unknown whether the accumulation of retrotransposons causes gene degeneration or whether these elements accumulate on the Y chromosome only after the erosion of most gene content (Steinemann and Steinemann 2005). The comparison of sex chromosomes at different stages of evolution, which should be possible in plants, may provide an opportunity to address this question and evaluate the generality of the models and processes shaping sex chromosomes in plants and animals.

2.5 Transposable Elements and Genome Size Evolution

2.5.1 The C-Value Paradox and Plant TE Composition

The "C-value paradox", a term derived to describe the lack of correlation between morphological complexity and total nuclear DNA content, was resolved in part by the discovery that eukaryotic genomes harbor large and dynamic populations of repetitive sequences, primarily transposable elements. Over the past few decades, numerous studies (summarized in Table 2.1) have described the total TE contribution to genome size and compositional diversity of TEs among and within various plant genomes. These studies have converged upon the conclusion that often the greatest fraction of plant genomes are composed of TEs, particularly in those plants with greater total nuclear content (Zhang and Wessler 2004; Hawkins et al. 2006; Vitte and Bennetzen 2006; Wicker and Keller 2007; Sweredoski et al. 2008; Wicker et al. 2009) (Fig. 2.4). The total TE copy number in plant genomes ranges widely, from as little as a few hundred in those with smaller genome sizes, such as *Arabidopsis*, to hundreds of thousands in their larger genome counterparts (e.g. maize, *Triticum*, *Hordeum*). Notably, this positive correlation between genome size and TE copy number generally holds across a broad range of eukaryotes (Bennett and Leitch 2005).

Copy number of a particular TE family or subfamily is a reflection of its relative success in terms of amplification and subsequent retention in the genome. Comparisons of TE composition across a wide range of plant species suggests that, although the same general TE types are found in all plants, the relative proportions contributed by various classes and subclasses can differ dramatically (Fig. 2.5).

Table 2.1 Genome size and proportion of TEs in plant species

Species	Genome size (Mbp)	Proportion TE (%)	Reference
Arabidopsis thaliana	120	14	The Arabidopsis Genome Initiative (2000)
Fragaria vesca	240	23	Shulaev et al. (2011)
Cucumis sativus	243	24	Huang et al. (2009)
Carica papaya	372	52	Ming et al. (2008)
Medicago truncatula	375	10	Wang and Liu (2008)
Oryza sativa	420	35	Paterson et al. (2009)
Theobroma cacao	430	26	Argout et al. (2011)
Lotus japonicus	470	19	Holligan et al. (2006)
Populus trichocarpa	485	42	Tuskan et al. (2006)
Vitis vinifera	487	17	French-Italian Public Consortium for Grapevine Genome Characterization (2007)
Brassica oleracea	600	20	Qiu et al. (2009)
Sorghum bicolor	740	62	Paterson et al. (2009)
Gossypium raimondii	880	54	Hawkins et al. (2006)
Glycine max	1,100	59	Schmutz et al. (2010)
Zea mays	2,045	76	Paterson et al. (2009)
Hordeum vulgare	5,439	80	Wicker et al. (2009)
Triticum aestivum	16,979	80	Bennett and Smith (1976)
Pinus taeda	21,516	80	Kovach et al. (2010)

Generally speaking, the genomes of eudicots contain fewer transposable elements relative to that of monocots, which have experienced recent and rampant LTR retrotransposon activity (SanMiguel et al. 1998; Vitte and Bennetzen 2006). LTR retrotransposon turnover in monocots appears to be extremely rapid, with both gains and losses of TE sequences occurring over as little as a few million years (see Sect. 2.5.2; Ma et al. 2004). In striking contrast, gymnosperm LTR retrotransposons are distinguished by their high level of decay and significant degree of divergence from angiosperm LTR retrotransposons, indicative of their ancient origin and subsequent long-term retention (Kovach et al. 2010). Additionally, these types of qualitative differences are not necessarily restricted to comparisons among major plant lineages. For example, significant differences in TE content have been observed among species within the genus *Gossypium*, where *gypsy*-like retrotransposons comprise the majority of the TE fraction

Fig. 2.4 Positive correlation between genome size and TE amount in selected plant species. *Arabidopsis thaliana* (At), *Fragaria vesca* (Fv), *Cucumis sativus* (Cs), *Carica papaya* (Cp), *Medicago truncatula* (Mt), *Oryza sativa* (Os), *Theobroma cacao* (Tc), *Lotus japonicus* (Lj), *Populus trichocarpa* (Pt), *Vitis vinifera* (Vv), *Brassica oleracea* (Bo), *Sorghum bicolor* (Sb), *Gossypium raimondii* (Gr), *Glycine max* (Gm), *Zea mays* (Zm) and *Hordeum vulgare* (Hv)

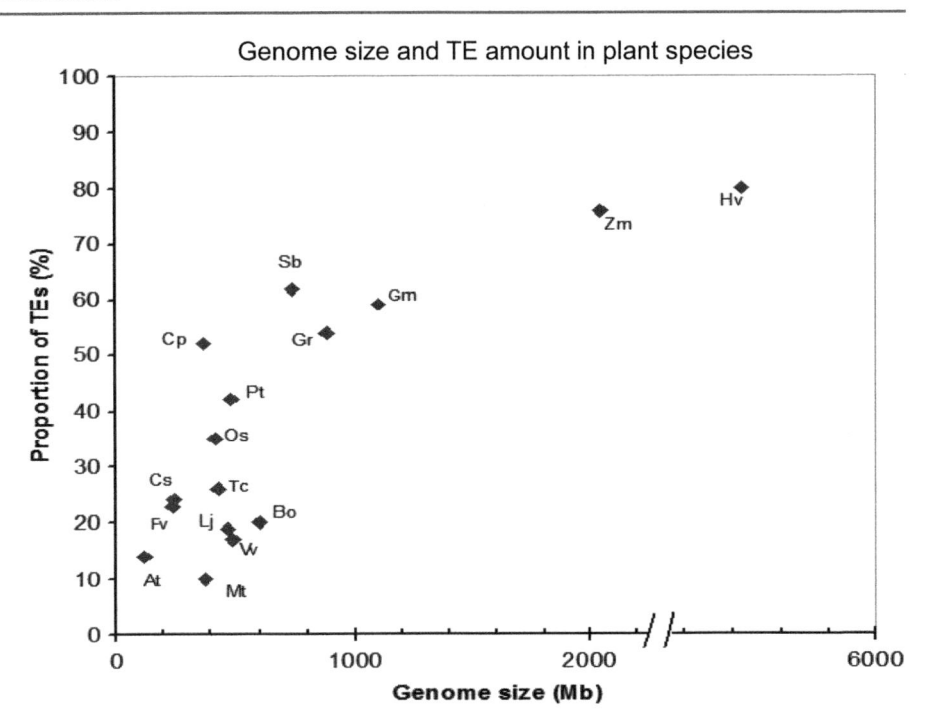

in species with larger genomes, while *copia*-like elements dominate the TE fraction in species with smaller genomes (Hawkins et al. 2006).

Sequence diversity among closely related species or even [individuals] within a TE family may also differ substantially, and the extent of divergence is often a direct reflection of a particular TE family's age. TE families that have undergone relatively recent proliferation are usually absent among closely related species, while families that have undergone amplification at more distant time points can be shared among closely related organisms as a product of their shared evolutionary history. An example of the former was demonstrated via comparisons of TE families among wheat and barley, where families that were highly abundant in one species were virtually absent in their close relative (Wicker et al. 2009). The rate at which new families appear (through either vertical or horizontal transfer), as well as the rate at which older TE families decay (by nucleotide mutation leading to sequence erosion or sequence removal via deletion) often differ substantially between species, and will be discussed further in the next section. These forces act to mold genome structure and composition not only at higher levels of taxonomic divergence, but often even at the species level.

2.5.2 Variable TE Insertion and Deletion Rates as a Driving Force in Plant Genome Size Evolution

As outlined above, large-scale amplification of transposable elements can lead to extraordinarily high copy numbers within plant genomes, often over short evolutionary timescales (Bennetzen 2005). One of the best-known examples in plants comes from maize, where repeated bursts of retrotransposon amplification over the past 6 million years have been responsible for generating approximately half of the modern maize genome (SanMiguel et al. 1998; Walbot and Petrov 2001). Similarly, a three-fold increase in the genome size of diploid members of *Gossypium* is due to the accumulation of LTR retrotransposons over the past 5–10 Myr (Hawkins et al. 2006). In *Oryza australiensis*, three LTR retrotransposon families proliferated during the last 3 million years leading to a two-fold increase in genome size compared to that of *Oryza sativa* (Piegu et al. 2006). The two- to threefold higher copy number of TEs in *Arabidopsis lyrata* compared to *A. thaliana* correlate and may be attributed to the higher expression of TEs in *A. lyrata*, apparently caused by less efficient TE silencing in this species (Hollister et al. 2011). These studies, in addition to several other plant genome surveys, clearly demonstrate that amplification of TEs, together with persistent rounds of genome doubling via polyploidization, are the primary mechanisms responsible for genome size expansion and variation in plants (Vitte and Bennetzen 2006; Kejnovsky et al. 2009b). These examples specifically implicate LTR retrotransposons as the agents most often responsible for massive TE-mediated increases in plant genome size. In contrast, comparative analyses of *A. thaliana* and *Brassica oleracea* indicate that several families of DNA transposons have amplified to high copy number in the lineage of *B. oleracea*, and that this activity has contributed to genome expansion in this lineage (Zhang and Wessler 2004), suggesting that DNA transposons may also play a significant role in shaping genome size in plants.

a Relative contribution of main groups of TEs to genome coverge in various plants (species arrayed according to genome size)

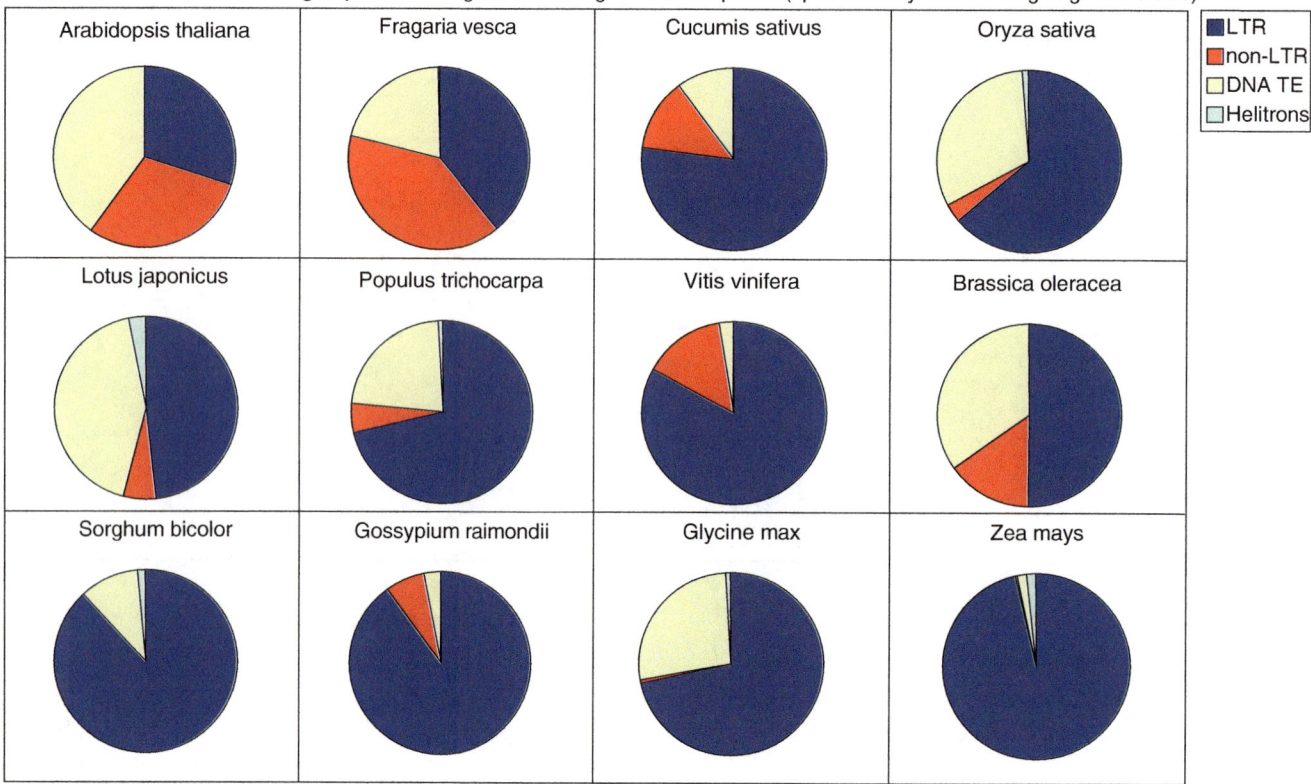

b Relative coby numbers of main groups of TEs in various plant genomes (species arrayed according to genome size)

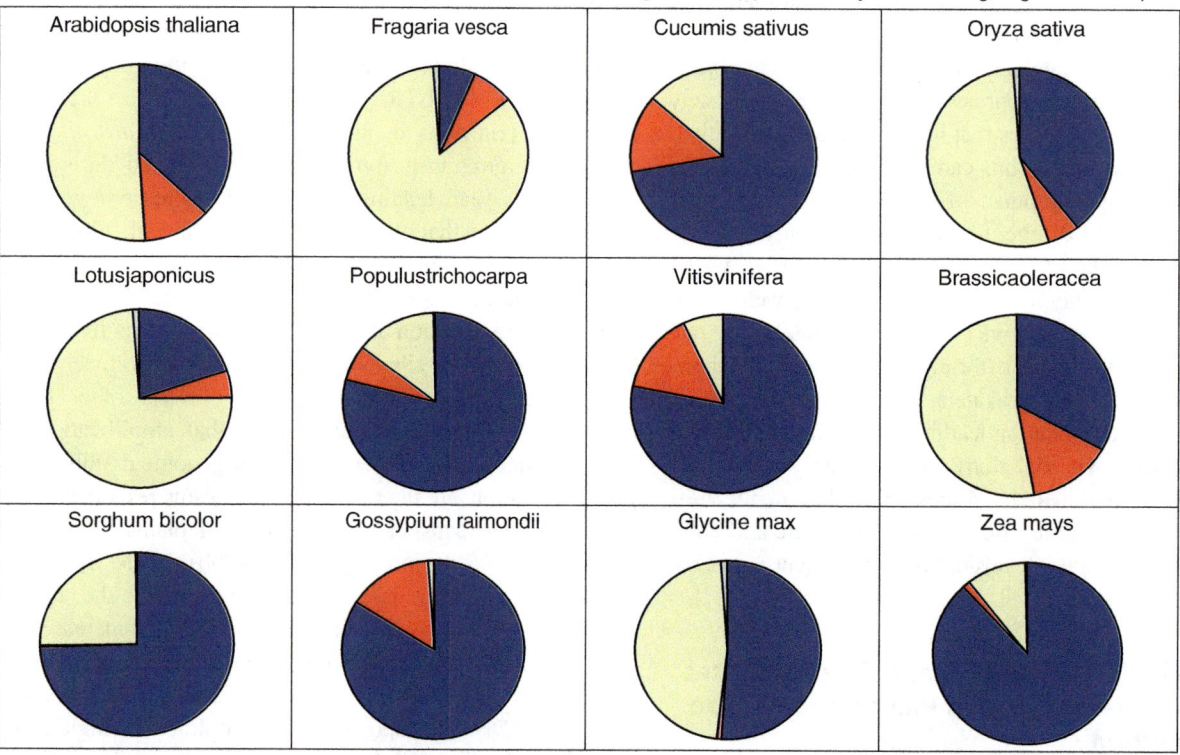

Fig. 2.5 Relative contributions of main groups of transposable elements in 12 plant genomes calculated by either genome coverage (**a**) or TE copy numbers (**b**). LTR retrotransposons (*dark blue*), non-LTR retrotransposons (*red*), DNA transposons (*yellow*) and Helitrons (*white blue*). It is evident that the contribution of LTR retrotransposons to the genome coverage increases with increasing genome size (**a**). DNA transposons are successful in their amplifications in both small and large genomes (**b**) but their contribution to the genome coverage is because their small length not so evident in large genomes (**a**)

The discovery that plant genomes expand via TE amplification combined with a paucity of information regarding mechanisms that might counteract this process and lead to DNA removal raised the question whether the process is a "one-way ticket to genomic obesity" (Bennetzen and Kellogg 1997). Analyses of plant genome size variation across a wide taxonomic range and within a phylogenetic framework show that several species with small genomes are embedded within clades of species characterized by much larger genomes, suggesting that genome downsizing can and does occur (Ma et al. 2004; Leitch et al. 2005; Hawkins et al. 2009). These observations have stimulated significant efforts to discover the genetic mechanisms responsible for DNA removal that might lead to substantial decreases in genome size (Vitte and Panaud 2003, 2005). Presently, two primary mechanisms of genome contraction have been proposed: intra-strand homologous recombination and illegitimate recombination.

Intra-strand homologous recombination, a form of ectopic recombination, is a process in which recombination occurs between non-allelic sequences of high sequence similarity. Such recombination may occur between two different TEs (inter-element recombination) or among highly similar sequences within a single TE, such as the long terminal repeats (LTRs), of the same retroelement (intra-element recombination) (Fig. 2.6). The latter mechanism is straightforward to detect and quantify as it results in the formation of solo LTRs. Since solo LTRs have been identified in virtually all species known to be colonized by LTR retrotransposons, this process appears to occur frequently and it is believed to play a major role in DNA removal in plant genomes. For example, the vast majority of BARE-1 elements in barley are represented by solo LTRs (14,000 full-length and 64,000 solo LTR), indicating massive amplification and subsequent removal of these elements in recent evolutionary history (Vicient et al. 1999). The ratio of full-length elements to solo LTRs is 1:1 in *Arabidopsis*, 2:3 in rice (Devos et al. 2002), 5:1 in maize (SanMiguel et al. 1996) and 1:7–11 in barley (Vicient et al. 1999), suggesting recent amplification of elements in maize as evidenced by the prevalence of intact to partial elements, and conversely, element removal by intra-strand homologous recombination in barley (prevalence of solo LTRs). Intra-strand homologous recombination has been coined a "partial return ticket from genomic obesity" (Vicient et al. 1999) as some portion of the nucleotide sequences involved in recombination (for example, one of the LTRs) is left behind, preventing complete deletion of extraneous DNA.

This type of recombination is expected to operate more strongly (1) to remove TE insertions in regions of high recombination, (2) on larger TE families, and (3) on longer copies of element (Petrov et al. 2003). Because longer elements increase the probability of ectopic recombination, longer TE copies persist in genomes for shorter periods than smaller elements, which is true not only for LTR retrotransposons, but also for Helitrons (Hollister and Gaut 2007). For the same reason, solo LTRs are preferentially formed by TEs with longer LTRs (Du et al. 2010) suggesting selection may occur for shorter LTRs in order to escape these deletions. The mechanisms suppressing genetic recombination may reduce the frequency of the formation of solo-LTRs as was demonstrated in pericentromeric regions of soybean (Du et al. 2010). Additional support for this hypothesis comes from *Oryza sativa*, where short elements accumulate in regions of high recombination while long elements accumulate in regions of low recombination (International

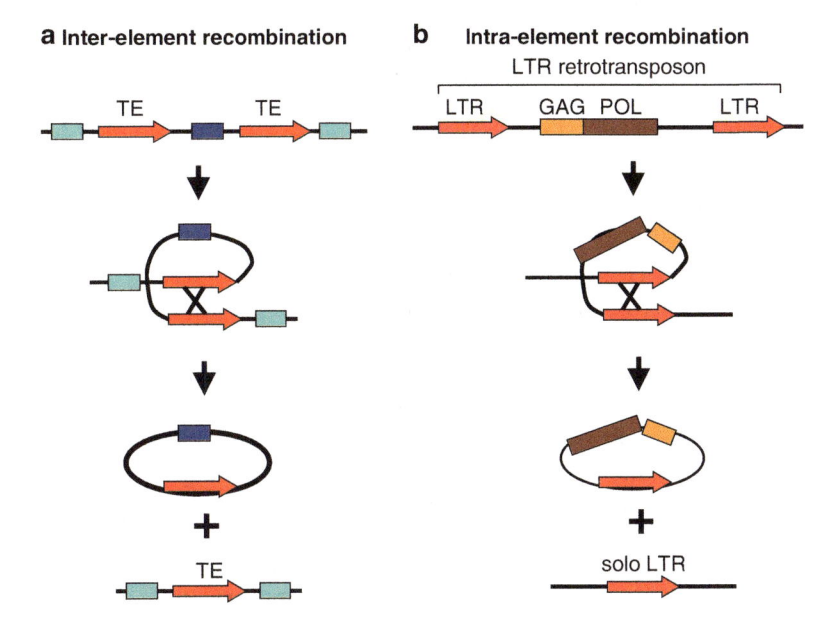

Fig. 2.6 Removal of transposable elements by homologous recombination. (**a**) Recombination between two transposable elements results in deletion of an in-between region. (**b**) Recombination between long terminal repeats (LTRs) of the same retrotransposon results in solo LTRs and deletion of internal region

Rice Genome Sequencing Project 2005). Because the accumulation of substitutions and indels reduce recombination frequency, opportunities for LTR–LTR recombination are rapidly lost in species with high rates of substitution and indels.

Illegitimate recombination is a RecA-independent form of recombination involving sequences of microhomology in which small deletions result due to non-homologous end-joining (NHEJ) or slip-strand mispairing. Sequences of microhomology may be as small as a few nucleotides, and the resulting deletions are often less than 10 bp in length, although they can be much larger. DNA loss leading to genome contraction in *Arabidopsis* and wheat is primarily attributed to illegitimate recombination (Devos et al. 2002; Wicker et al. 2003) but to both intra-strand homologous recombination and illegitimate recombination in rice (Ma et al. 2004). It is unclear at this time as to the evolutionary significance of DNA loss in shaping extant genome size; however, assuming removal rates at a high enough level to counteract genome expansion via TE proliferation, differences in the repair/recombination machinery of the host species might be a driving force in shaping extant genome size (Orel and Puchta 2003).

2.6 Closing

Upon her discovery of transposable elements in the 1950s, Barbara McClintock suggested that these sequences might operate to control gene expression and play a major role in evolution. This suggestion was remarkably prophetic, even though the concept it embodied had to survive several decades of misinterpretation of TEs as mere "junk" prior to emerging in the last decade or more as major players in the organization and function of plant genomes. Our increasing understanding of TE abundance, distribution, and behavior has revealed that the selfish nature of TEs is not incompatible with them playing a significant role in genome evolution at multiple levels, from genome-wide (total nuclear content, chromatin structure, recombination, RNAi, etc.) to local effects (chromosomal rearrangements, regulation of neighboring genes, co-option of individual TE sequences to form new genes, TE-mediated gene duplication, etc.) as summarized above and discussed in more detail in Chap. 3. Thus, after almost 60 years, Barbara McClintock's vision, considered radical at the time and dismissed by most, is receiving growing empirical support. Plant research continues to be at the forefront of TE biology, and the ongoing genomic revolution is promised to yield many more exciting discoveries in the years to come.

Acknowledgments Research on transposable elements in the authors' laboratories has been supported by the Grant Agency of the Czech Republic (grant P305/10/0930), grants No AV0Z50040507 and AV0Z50040702 from the Academy of Sciences of the Czech Republic.

References

Adams MD, Celniker SE, Holt RA, Evans CA, Gocayne JD, Amanatides PG, Scherer SE, Li PW, Hoskins RA, Galle RF et al (2000) The genome sequence of *Drosophila melanogaster*. Science 287:2185–2195

Ananiev EV, Phillips RL, Rines HW (1998) Chromosome-specific molecular organization of maize (*Zea mays* L.) centromeric regions. Proc Natl Acad Sci USA 95:13073–13078

Argout X, Salse J, Aury J-M, Guiltinan MJ, Droc G, Gouzy J, Allegre M, Chaparro T, Maximova SN, Abrouk M et al (2011) The genome of *Theobroma cacao*. Nat Genet 43:101–109

Arkhipova I, Meselson M (2000) Transposable elements in sexual and ancient asexual taxa. Proc Natl Acad Sci USA 97:14473–14477

Bachtrog D (2003) Accumulation of Spock and Worf, two novel non-LTR retrotransposons, on the neo-Y chromosome of *Drosophila miranda*. Mol Biol Evol 20:173–181

Bachtrog D (2006) A dynamic view of sex chromosome evolution. Curr Opin Genet Dev 16:578–585

Baucom RS, Estill JC, Chaparro C, Upshaw N, Jogi A, Deragon J, Westerman RP, Sanmiguel PJ, Bennetzen JL (2009) Exceptional diversity, non-random distribution, and rapid evolution of retroelements in the B73 maize genome. PLoS Genet 5:e1000732

Bennett MD, Leitch IJ (2005) Nuclear DNA amounts in angiosperms: progress, problems and prospects. Ann Bot 95:45–90

Bennett MD, Smith JB (1976) Nuclear DNA amounts in angiosperms. Philos Trans Roy Soc B 274:227–274

Bennetzen JL (2005) Transposable elements, gene creation and genome rearrangement in flowering plants. Curr Opin Genet Dev 15:621–627

Bennetzen JL, Kellogg EA (1997) Do plants have a one-way ticket to genomic obesity? Plant Cell 9:1509–1514

Berg DE, Howe MM (1989) Mobile DNA. American Society for Microbiology, Washington, DC

Bestor TH (1999) Sex brings transposons and genomes into conflict. Genetica 107:289–295

Bowen NJ, Jordan IK, Epstein JA, Wood V, Levin HL (2003) Retrotransposons and their recognition of pol II promoters: a comprehensive survey of the transposable elements from the complete genome sequence of *Schizosaccharomyces pombe*. Genome Res 13:1984–1997

Brady TL, Fuerst PG, Dick RA, Schmidt C, Voytas DF (2008) Retrotransposon target site selection by imitation of a cellular protein. Mol Cell Biol 28:1230–1239

Bureau TE, Wessler SR (1992) Tourist: a large family of small inverted repeat elements frequently associated with maize genes. Plant Cell 4:1283–1294

Bureau TE, White SE, Wessler SR (1994) Transduction of a cellular gene by a plant retroelement. Cell 77:479–480

Cermak T, Kubat Z, Hobza R, Koblizkova A, Widmer A, Macas J, Vyskot B, Kejnovsky E (2008) Survey of repetitive sequences in *Silene latifolia* with respect to their distribution on sex chromosomes. Chromosome Res 16:961–976

Charlesworth B, Charlesworth D (2000) The degeneration of Y chromosomes. Philos Trans Roy Soc B 355:1563–1572

Charlesworth B, Sniegowski P, Stephan W (1994) The evolutionary dynamics of repetitive DNA in eukaryotes. Nature 371:215–220

Cheng Z, Dong F, Langdon T, Ouyang S, Buell CR, Gu M, Blattner FR, Jiang J (2002) Functional rice centromeres are marked by a satellite repeat and a centromere-specific retrotransposon. Plant Cell 14:1691–1704

Craig NL, Craigie R, Gellert M, Lambowitz AM (2002) Mobile DNA II. ASM Press, Washington, DC

Deragon J, Zhang X (2006) Short interspersed elements (SINEs) in plants: origin, classification, and use as phylogenetic markers. Syst Biol 55:949–956

Devos KM, Brown JKM, Bennetzen JL (2002) Genome size reduction through illegitimate recombination counteracts genome expansion in *Arabidopsis*. Genome Res 12:1075–1079

Dietrich CR, Cui F, Packila ML, Li J, Ashlock DA, Nikolau BJ, Schnable PS (2002) Maize Mu transposons are targeted to the $5'$ untranslated region of the gl8 gene and sequences flanking Mu target-site duplications exhibit nonrandom nucleotide composition throughout the genome. Genetics 160:697–716

Dolgin ES, Charlesworth B (2008) The effects of recombination rate on the distribution and abundance of transposable elements. Genetics 178:2169–2177

Du J, Tian Z, Hans CS, Laten HM, Cannon SB, Jackson SA, Shoemaker RC, Ma J (2010) Evolutionary conservation, diversity and specificity of LTR-retrotransposons in flowering plants: insights from genome-wide analysis and multi-specific comparison. Plant J. http://www.ncbi.nlm.nih.gov/pubmed/20525006. Accessed 13 Sept 2010

Eickbush TH, Malik HS (2002) Origin and evolution of retrotransposons. In: Craig NL, Craigie R, Gellert M, Lambowitz AM (eds) Mobile DNA. ASM Press, Washington, DC, pp 1111–1146

Engels WR, Johnson-Schlitz DM, Eggleston WB, Sved J (1990) High-frequent P element loss in *Drosophila* is homolog dependent. Cell 10:515–525

Erlandsson R, Wilson JF, Pääbo S (2000) Sex chromosomal transposable element accumulation and male-driven substitutional evolution in humans. Mol Biol Evol 17:804–812

Fedoroff N, Wessler S, Shure M (1983) Isolation of the transposable maize controlling elements Ac and Ds. Cell 35:235–242

Feschotte C, Mouchès C (2000) Evidence that a family of miniature inverted-repeat transposable elements (MITEs) from the *Arabidopsis thaliana* genome has arisen from a pogo-like DNA transposon. Mol Biol Evol 17:730–737

Feschotte C, Pritham EJ (2007) DNA transposons and the evolution of eukaryotic genomes. Annu Rev Genet 41:331–368

Feschotte C, Jiang N, Wessler SR (2002) Plant transposable elements: where genetics meets genomics. Nat Rev Genet 3:329–341

Feschotte C, Swamy L, Wessler SR (2003) Genome-wide analysis of mariner-like transposable elements in rice reveals complex relationships with stowaway miniature inverted repeat transposable elements (MITEs). Genetics 163:747–758

Feschotte C, Osterlund MT, Peeler R, Wessler SR (2005) DNA-binding specificity of rice mariner-like transposases and interactions with Stowaway MITEs. Nucleic Acids Res 33:2153–2165

Finnegan DJ (1989) Eukaryotic transposable elements and genome evolution. Trends Genet 5:103–107

Francki MG (2001) Identification of Bilby, a diverged centromeric Ty1-copia retrotransposon family from cereal rye (*Secale cereale* L.). Genome 44:266–274

French-Italian public consortium for grapevine genome characterization (2007) The grapevine genome sequence suggests ancestral hexaploidization in major angiosperm phyla. Nature 449:463–468

Gao X, Hou Y, Ebina H, Levin HL, Voytas DF (2008) Chromodomains direct integration of retrotransposons to heterochromatin. Genome Res 18:359–369

Gao D, Gill N, Kim H, Walling JG, Zhang W, Fan C, Yu Y, Ma J, SanMiguel P, Jiang N et al (2009) A lineage-specific centromere retrotransposon in *Oryza brachyantha*. Plant J 60:820–831

Garcia-Perez JL, Doucet AJ, Bucheton A, Moran JV, Gilbert N (2007) Distinct mechanisms for trans-mediated mobilization of cellular RNAs by the LINE-1 reverse transcriptase. Genome Res 17:602–611

Gent JI, Schneider KL, Topp CN, Rodriguez C, Presting GG, Dawe RK (2011) Distinct influences of tandem repeats and retrotransposons on CENH3 nucleosome positioning. Epigenetics Chromatin 4:3

Gorinsek B, Gubensek F, Kordis D (2004) Evolutionary genomics of chromoviruses in eukaryotes. Mol Biol Evol 21:781–798

Green MM (1980) Transposable elements in *Drosophila* and other Diptera. Annu Rev Genet 14:109–120

Grewal SIS, Jia S (2007) Heterochromatin revisited. Nat Rev Genet 8:35–46

Havecker ER, Gao X, Voytas DF (2004) The diversity of LTR retrotransposons. Genome Biol 5:225

Hawkins JS, Kim H, Nason JD, Wing RA, Wendel JF (2006) Differential lineage-specific amplification of transposable elements is responsible for genome size variation in *Gossypium*. Genome Res 16:1252–1261

Hawkins JS, Hu G, Rapp RA, Grafenberg JL, Wendel JF (2008) Phylogenetic determination of the pace of transposable element proliferation in plants: copia and LINE-like elements in *Gossypium*. Genome 51:11–18

Hawkins JS, Proulx SR, Rapp RA, Wendel JF (2009) Rapid DNA loss as a counterbalance to genome expansion through retrotransposon proliferation in plants. Proc Natl Acad Sci USA 106:17811–17816

Hickey DA (1982) Selfish DNA: a sexually-transmitted nuclear parasite. Genetics 101:519–531

Hirsch C, Jiang J (2012) Centromeres: sequences, structure, and biology. In: Wendel JF (ed) Plant genome diversity, vol 1, Plant genomes, their residents, and their evolutionary dynamics. Springer, Wien, New York

Hobza R, Lengerova M, Svoboda J, Kubekova H, Kejnovsky E, Vyskot B (2006) An accumulation of tandem DNA repeats on the Y chromosome in *Silene latifolia* during early stages of sex chromosome evolution. Chromosoma 115:376–382

Holligan D, Zhang X, Jiang N, Pritham EJ, Wessler SR (2006) The transposable element landscape of the model legume *Lotus japonicus*. Genetics 174:2215–2228

Hollister JD, Gaut BS (2007) Population and evolutionary dynamics of Helitron transposable elements in *Arabidopsis thaliana*. Mol Biol Evol 24:2515–2524

Hollister JD, Smith LM, Guo Y-L, Ott F, Weigel D, Gaut BS (2011) Transposable elements and small RNAs contribute gene expression divergence between *Arabidopsis thaliana* and *Arabidopsis lyrata*. Proc Natl Acad Sci USA 108:2322–2327

Huang S, Li R, Zhang Z, Li L, Gu X, Fan W, Lucas WJ, Wang X, Xie B, Ni P et al (2009) The genome of the cucumber, *Cucumis sativus* L. Nat Genet 41:1275–1283

International Rice Genome Sequencing Project (2005) The map-based sequence of the rice genome. Nature 436:793–800

Ji H, Moore DP, Blomberg MA, Braiterman LT, Voytas DF, Natsoulis G, Boeke JD (1993) Hotspots for unselected Ty1 transposition events on yeast chromosome III are near tRNA genes and LTR sequences. Cell 73:1007–1018

Jiang N, Wessler SR (2001) Insertion preference of maize and rice miniature inverted repeat transposable elements as revealed by the analysis of nested elements. Plant Cell 13:2553–2564

Jiang J, Birchler JA, Parrott WA, Dawe RK (2003) A molecular view of plant centromeres. Trends Plant Sci 8:570–575

Jiang N, Bao Z, Zhang X, Eddy SR, Wessler SR (2004) Pack-MULE transposable elements mediate gene evolution in plants. Nature 431:569–573

Jiang N, Ferguson AA, Slotkin RK, Lisch D (2011) Pack-Mutator-like transposable elements (Pack-MULEs) induce directional modification of genes through biased insertion and DNA acquisition. Proc Natl Acad Sci USA 108:1537–1542

Kalendar R, Vicient CM, Peleg O, Anamthawat-Jonsson K, Bolshoy A, Schulman AH (2004) Large retrotransposon derivatives: abundant, conserved but nonautonomous retroelements of barley and related genomes. Genetics 166:1437–1450

Kapitonov VV, Jurka J (2007) Helitrons on a roll: eukaryotic rolling-circle transposons. Trends Genet 23:521–529

Kejnovsky E, Vyskot B (2010) Silene latifolia: the classical model to study heteromorphic sex chromosomes. Cytogenet Genome Res 129:250–262

Kejnovsky E, Kubat Z, Hobza R, Lengerova M, Sato S, Tabata S, Fukui K, Matsunaga S, Vyskot B (2006a) Accumulation of chloroplast DNA sequences on the Y chromosome of Silene latifolia. Genetica 128:167–175

Kejnovsky E, Kubat Z, Macas J, Hobza R, Mracek J, Vyskot B (2006b) Retand: a novel family of gypsy-like retrotransposons harboring an amplified tandem repeat. Mol Genet Genomics 276:254–263

Kejnovsky E, Hobza R, Cermak T, Kubat Z, Vyskot B (2009a) The role of repetitive DNA in structure and evolution of sex chromosomes in plants. Heredity 102:533–541

Kejnovsky E, Leitch IJ, Leitch AR (2009b) Contrasting evolutionary dynamics between angiosperm and mammalian genomes. Trends Ecol Evol 24:572–582

Kidwell MG, Lisch DR (2001) Perspective: transposable elements, parasitic DNA, and genome evolution. Evolution 55:24

Kordis D (2005) A genomic perspective on the chromodomain-containing retrotransposons: chromoviruses. Gene 347:161–173

Kovach A, Wegrzyn JL, Parra G, Holt C, Bruening GE, Loopstra CA, Hartigan J, Yandell M, Langley CH, Korf I et al (2010) The Pinus taeda genome is characterized by diverse and highly diverged repetitive sequences. BMC Genomics 11:420

Kubat Z, Hobza R, Vyskot B, Kejnovsky E (2008) Microsatellite accumulation on the Y chromosome in Silene latifolia. Genome 51:35356

Kumekawa N, Ohmido N, Fukui K, Ohtsubo E, Ohtsubo H (2001) A new gypsy-type retrotransposon, RIRE7: preferential insertion into the tandem repeat sequence TrsD in pericentromeric heterochromatin regions of rice chromosomes. Mol Genet Genomics 265:48488

Kunze R, Starlinger P (1989) The putative transposase of transposable element Ac from Zea mays L. interacts with subterminal sequences of Ac. EMBO J 8:3173–3185

Lander ES, Linton LM, Birren B, Nusbaum C, Zody MC, Baldwin J, Devon K, Dewar K, Doyle M, FitzHugh W et al (2001) Initial sequencing and analysis of the human genome. Nature 409:860–921

Langley CH, Montgomery E, Hudson R, Kaplan N, Charlesworth B (1988) On the role of unequal exchange in the containment of transposable element copy number. Genet Res 52:223–235

Le Rouzic A, Boutin TS, Capy P (2007) Long-term evolution of transposable elements. Proc Natl Acad Sci USA 104:19371–19380

Leitch IJ, Soltis DE, Soltis PS, Bennett MD (2005) Evolution of DNA amounts across land plants (embryophyta). Ann Bot 95:207–217

Levis R, O'Hare K, Rubin GM (1984) Effect of transposable element insertions on RNA encoded by the white gene of Drosophila. Cell 38:471–481

Lippman Z, Gendrel AV, Black M, Vaughn MW, Dedhia N, McCombie WR, Lavine K, Mittal V, May B, Kasschau KD et al (2004) Role of transposble elements in heterochromatin and epigenetic control. Nature 430:471–476

Liu Z, Moore PH, Ma H, Ackerman CM, Ragiba M, Yu Q, Pearl HM, Kim MS, Charlton JW, Stiles JI et al (2004) A primitive Y chromosome in papaya marks incipient sex chromosome evolution. Nature 427:348–352

Liu Z, Yue W, Li D, Wang RR, Kong X, Lu K, Wang G, Dong Y, Jin W, Zhang X (2008) Structure and dynamics of retrotransposons at wheat centromeres and pericentromeres. Chromosoma 117:445–456

Liu S, Yeh CT, Ji T, Ying K, Wu H, Tang HM, Fu Y, Nettleton D, Schnable PS (2009) Mu transposon insertion sites and meiotic recombination events co-localize with epigenetic marks for open chromatin across the maize genome. PLoS Genet 5:e1000733

Lockton S, Gaut BS (2010) The evolution of transposable elements in natural populations of self-fertilizing Arabidopsis thaliana and its outcrossing relative Arabidopsis lyrata. BMC Evol Biol 10:10

Lockton S, Ross-Ibarra J, Gaut BS (2008) Demography and weak selection drive patterns of transposable element diversity in natural populations of Arabidopsis lyrata. Proc Natl Acad Sci USA 105:13965–13970

Lynch M, Conery JS (2003) The origins of genome complexity. Science 302:1401–1404

Ma J, Devos KM, Bennetzen JL (2004) Analyses of LTR-retrotransposon structures reveal recent and rapid genomic DNA loss in rice. Genome Res 14:860–869

Marais GAB, Nicolas M, Bergero R, Chambrier P, Kejnovsky E, Monéger F, Hobza R, Widmer A, Charlesworth D (2008) Evidence for degeneration of the Y chromosome in the dioecious plant Silene latifolia. Curr Biol 18:545–549

May BP, Lippman ZB, Fang Y, Spector DL, Martienssen RA (2005) Differential regulation of strand-specific transcripts from Arabidopsis centromeric satellite repeats. PLoS Genet 1:e79

Miller JT, Dong F, Jackson SA, Song J, Jiang J (1998) Retrotransposon-related DNA sequences in the centromeres of grass chromosomes. Genetics 150:1615–1623

Ming R, Hou S, Feng Y, Yu Q, Dionne-Laporte A, Saw JH, Senin P, Wang W, Ly BV, Lewis KLT et al (2008) The draft genome of the transgenic tropical fruit tree papaya (Carica papaya Linnaeus). Nature 452:991–996

Morgante M, De Paoli E, Radovic S (2007) Transposable elements and the plant pan-genomes. Curr Opin Plant Biol 10:149–155

Nagaki K, Murata M (2005) Characterization of CENH3 and centromere-associated DNA sequences in sugarcane. Chromosome Res 13:195–203

Nagaki K, Song J, Stupar RM, Parokonny AS, Yuan Q, Ouyang S, Liu J, Hsiao J, Jones KM, Dawe RK et al (2003) Molecular and cytological analyses of large tracks of centromeric DNA reveal the structure and evolutionary dynamics of maize centromeres. Genetics 163:759–770

Nagaki K, Neumann P, Zhang D, Ouyang S, Buell CR, Cheng Z, Jiang J (2005) Structure, divergence, and distribution of the CRR centromeric retrotransposon family in rice. Mol Biol Evol 22:845–855

Naito K, Zhang F, Tsukiyama T, Saito H, Hancock CN, Richardson AO, Okumoto Y, Tanisaka T, Wessler SR (2009) Unexpected consequences of a sudden and massive transposon amplification on rice gene expression. Nature 461:1130–1134

Nakazaki T, Okumoto Y, Horibata A, Yamahira S, Teraishi M, Nishida H, Inoue H, Tanisaka T (2003) Mobilization of a transposon in the rice genome. Nature 421:170–172

Neumann P, Koblížková A, Navrátilová A, Macas J (2006) Significant expansion of Vicia pannonica genome size mediated by amplification of a single type of giant retroelement. Genetics 173:1047–1056

Neumann P, Yan H, Jiang J (2007) The centromeric retrotransposons of rice are transcribed and differentially processed by RNA interference. Genetics 176:749–761

Neumann P, Navrátilová A, Koblížková A, Kejnovský E, Hřibová E, Hobza R, Widmer A, Doležel J, Macas J (2011) Plant centromeric retrotransposons: a structural and cytogenetic perspective. Mob DNA 2:4

Novikova O (2009) Chromodomains and LTR retrotransposons in plants. Commun Integr Biol 2:158–162

Okada S, Sone T, Fujisawa M, Nakayama S, Takenaka M, Ishizaki K, Kono K, Shimizu-Ueda Y, Hanajiri T, Yamato KT et al (2001) The Y chromosome in the liverwort Marchantia polymorpha has accumulated unique repeat sequences harboring a male-specific gene. Proc Natl Acad Sci USA 98:9454–9459

Orel N, Puchta H (2003) Differences in the processing of DNA ends in *Arabidopsis thaliana* and tobacco: possible implications for genome evolution. Plant Mol Biol 51:523–531

Paterson AH, Bowers JE, Bruggmann R, Dubchak I, Grimwood J, Gundlach H, Haberer G, Hellsten U, Mitros T, Poliakov A et al (2009) The *Sorghum bicolor* genome and the diversification of grasses. Nature 457:551–556

Pathak VK, Hu W-S (1997) "Might as well jump!" Template switching by retroviral reverse transcriptase, defective genome formation, and recombination. Semin Virol 8:141–150

Pereira V (2004) Insertion bias and purifying selection of retrotransposons in the *Arabidopsis thaliana* genome. Genome Biol 5:R79

Petersen G, Seberg O (2000) Phylogenetic evidence for excision of Stowaway miniature inverted-repeat transposable elements in triticeae (Poaceae). Mol Biol Evol 17:1589–1596

Peterson-Burch BD, Nettleton D, Voytas DF (2004) Genomic neighborhoods for *Arabidopsis* retrotransposons: a role for targeted integration in the distribution of the Metaviridae. Genome Biol 5: R78

Petrov DA, Aminetzach YT, Davis JC, Bensasson D, Hirsh AE (2003) Size matters: non-LTR retrotransposable elements and ectopic recombination in *Drosophila*. Mol Biol Evol 20:880–892

Piegu B, Guyot R, Picault N, Roulin A, Saniyal A, Kim H, Collura K, Brar DS, Jackson S, Wing RA et al (2006) Doubling genome size without polyploidization: dynamics of retrotransposition-driven genomic expansions in *Oryza australiensis*, a wild relative of rice. Genome Res 16:1262–1269

Pimpinelli S, Berloco M, Fanti L, Dimitri P, Bonaccorsi S, Marchetti E, Caizzi R, Caggese C, Gatti M (1995) Transposable elements are stable structural components of *Drosophila melanogaster* heterochromatin. Proc Natl Acad Sci USA 92:3804–3808

Presting GG, Malysheva L, Fuchs J, Schubert I (1998) A Ty3/gypsy retrotransposon-like sequence localizes to the centromeric regions of cereal chromosomes. Plant J 16:721–728

Pritham EJ (2009) Transposable elements and factors influencing their success in eukaryotes. J Hered 100:648–655

Qiu D, Gao M, Li G, Quiros C (2009) Comparative sequence analysis for *Brassica oleracea* with similar sequences in *B. rapa* and *Arabidopsis thaliana*. Plant Cell Rep 28:649–661

Ros F, Kunze R (2001) Regulation of activator/dissociation transposition by replication and DNA methylation. Genetics 157:1723–1733

Saedler H, Bonas U, Gierl A, Harrison BJ, Klösgen RB, Krebbers E, Nevers P, Peterson PA, Schwarz-Sommer Z, Sommer H (1984) Transposable elements in *Antirrhinum majus* and *Zea mays*. Cold Spring Harb Symp Quant Biol 49:355–361

Sakamoto K, Ohmido N, Fukui K, Kamada H, Satoh S (2000) Site-specific accumulation of a LINE-like retrotransposon in a sex chromosome of the dioecious plant *Cannabis sativa*. Plant Mol Biol 44:723–732

SanMiguel P, Tikhonov A, Jin YK, Motchoulskaia N, Zakharov D, Melake-Berhan A, Springer PS, Edwards KJ, Lee M, Avramova Z et al (1996) Nested retrotransposons in the intergenic regions of the maize genome. Science 274:765–768

SanMiguel P, Gaut BS, Tikhonov A, Nakajima Y, Bennetzen JL (1998) The paleontology of intergene retrotransposons of maize. Nat Genet 20:43–45

Schaack S, Choi E, Lynch M, Pritham EJ (2010a) DNA transposons and the role of recombination in mutation accumulation in *Daphnia pulex*. Genome Biol 11:R46

Schaack S, Pritham EJ, Wolf A, Lynch M (2010b) DNA transposon dynamics in populations of *Daphnia pulex* with and without sex. Proc Biol Sci 7:2381–2387

Schmutz J, Cannon SB, Schlueter J, Ma J, Mitros T, Nelson W, Hyten DL, Song Q, Thelen JJ, Cheng J et al (2010) Genome sequence of the palaeopolyploid soybean. Nature 463:178–183

Schön I, Martens K (2000) Transposable elements and asexual reproduction. Trends Ecol Evol 15:287–288

Sharma A, Schneider KL, Presting GG (2008) Sustained retrotransposition is mediated by nucleotide deletions and interelement recombinations. Proc Natl Acad Sci USA 105:15470–15474

Shibata F, Hizume M, Kuroki Y (1999) Chromosome painting of Y chromosomes and isolation of a Y chromosome-specific repetitive sequence in the dioecious plant *Rumex acetosa*. Chromosoma 108:266–270

Shulaev V, Sargent DJ, Crowhurst RN, Mockler TC, Folkerts O, Delcher AL, Jaiswal P, Mockaitis K, Liston A, Mane SP et al (2011) The genome of woodland strawberry (*Fragaria vesca*). Nat Genet 43:109–118

Skaletsky H, Kuroda-Kawaguchi T, Minx PJ, Cordum HS, Hillier L, Brown LG, Repping S, Pyntikova T, Ali J, Bieri T et al (2003) The male-specific region of the human Y chromosome is a mosaic of discrete sequence classes. Nature 423:825–837

Slotkin R, Nuthikattu S, Jiang N (2012) The impact of transposable elements on gene and genome evolution. In: Wendel JF (ed) Plant genome diversity, vol 1, Plant genomes, their residents, and their evolutionary dynamics. Springer, Wien, New York

Steinemann M, Steinemann S (1992) Degenerating Y chromosome of *Drosophila miranda*: a trap for retrotransposons. Proc Natl Acad Sci USA 89:7591–7595

Steinemann S, Steinemann M (2005) Y chromosomes: born to be destroyed. Bioessays 27:1076–1083

Sweredoski M, DeRose-Wilson L, Gaut BS (2008) A comparative computational analysis of nonautonomous helitron elements between maize and rice. BMC Genomics 9:467

Swigonová Z, Bennetzen JL, Messing J (2005) Structure and evolution of the r/b chromosomal regions in rice, maize and sorghum. Genetics 169:891–906

Tenaillon MI, Hollister JD, Gaut BS (2010) A triptych of the evolution of plant transposable elements. Trends Plant Sci 15:471–478

The Arabidopsis Genome Initiative (2000) Analysis of the genome sequence of the flowering plant *Arabidopsis thaliana*. Nature 408:796–815

Tian Z, Rizzon C, Du J, Zhu L, Bennetzen JL, Jackson SA, Gaut BS, Ma J (2009) Do genetic recombination and gene density shape the pattern of DNA elimination in rice long terminal repeat retrotransposons? Genome Res 19:2221–2230

Topp CN, Zhong CX, Dawe RK (2004) Centromere-encoded RNAs are integral components of the maize kinetochore. Proc Natl Acad Sci USA 101:15986–15991

Tsubota SI, Huong DV (1991) Capture of flanking DNA by a P element in *Drosophila melanogaster*: creation of a transposable element. Proc Natl Acad Sci USA 88:693–697

Tuskan GA, DiFazio S, Jansson S, Bohlmann J, Grigoriev I, Hellsten U, Putnam N, Ralph S, Rombauts S, Salamov A et al (2006) The genome of black cottonwood, *Populus trichocarpa* (Torr. & Gray). Science 313:1596–1604

Vicient CM, Suoniemi A, Anamthawat-Jónsson K, Tanskanen J, Beharav A, Nevo E, Schulman AH (1999) Retrotransposon BARE-1 and its role in genome evolution in the genus hordeum. Plant Cell 11:1769–1784

Vitte C, Bennetzen JL (2006) Analysis of retrotransposon structural diversity uncovers properties and propensities in angiosperm genome evolution. Proc Natl Acad Sci USA 103:17638–17643

Vitte C, Panaud O (2003) Formation of solo-LTRs through unequal homologous recombination counterbalances amplifications of LTR retrotransposons in rice *Oryza sativa* L. Mol Biol Evol 20:528–540

Vitte C, Panaud O (2005) LTR retrotransposons and flowering plant genome size: emergence of the increase/decrease model. Cytogenet Genome Res 110:91–107

Volpe TA, Kidner C, Hall IM, Teng G, Grewal SIS, Martienssen RA (2002) Regulation of heterochromatic silencing and histone H3 lysine-9 methylation by RNAi. Science 297:1833–1837

Vyskot B, Hobza R (2004) Gender in plants: sex chromosomes are emerging from the fog. Trends Genet 20:432–438

Wagner A (2006) Periodic extinctions of transposable elements in bacterial lineages: evidence from intragenomic variation in multiple genomes. Mol Biol Evol 23:723–733

Walbot V, Petrov DA (2001) Gene galaxies in the maize genome. Proc Natl Acad Sci USA 98:8163–8164

Wang H, Liu J-S (2008) LTR retrotransposon landscape in *Medicago truncatula*: more rapid removal than in rice. BMC Genomics 9:382

Weil C, Martienssen R (2008) Epigenetic interactions between transposons and genes: lessons from plants. Curr Opin Genet Dev 18:188–192

Whitney KD, Baack EJ, Hamrick JL, Godt MJW, Barringer BC, Bennett MD, Eckert CG, Goodwillie C, Kalisz S, Leitch IJ et al (2010) A role for nonadaptive processes in plant genome size evolution? Evolution 64:2097–2109

Wicker T, Keller B (2007) Genome-wide comparative analysis of copia retrotransposons in Triticeae, rice, and *Arabidopsis* reveals conserved ancient evolutionary lineages and distinct dynamics of individual copia families. Genome Res 17:1072–1081

Wicker T, Yahiaoui N, Guyot R, Schlagenhauf E, Liu Z, Dubcovsky J, Keller B (2003) Rapid genome divergence at orthologous low molecular weight glutenin loci of the A and Am genomes of wheat. Plant Cell 15:1186–1197

Wicker T, Sabot F, Hua-Van A, Bennetzen JL, Capy P, Chalhoub B, Flavell A, Leroy P, Morgante M, Panaud O et al (2007) A unified classification system for eukaryotic transposable elements. Nat Rev Genet 8:973–982

Wicker T, Taudien S, Houben A, Keller B, Graner A, Platzer M, Stein N (2009) A whole-genome snapshot of 454 sequences exposes the composition of the barley genome and provides evidence for parallel evolution of genome size in wheat and barley. Plant J 59:712–722

Witte CP, Le QH, Bureau T, Kumar A (2001) Terminal-repeat retrotransposons in miniature (TRIM) are involved in restructuring plant genomes. Proc Natl Acad Sci USA 98:13778–13783

Wright SI, Le QH, Schoen DJ, Bureau TE (2001) Population dynamics of an Ac-like transposable element in self- and cross-pollinating arabidopsis. Genetics 158:1279–1288

Wright SI, Agrawal N, Bureau TE (2003) Effects of recombination rate and gene density on transposable element distributions in *Arabidopsis thaliana*. Genome Res 13:1897–1903

Yan X, Martínez-Férez IM, Kavchok S, Dooner HK (1999) Origination of Ds elements from Ac elements in maize: evidence for rare repair synthesis at the site of Ac excision. Genetics 152:1733–1740

Yang L, Bennetzen JL (2009) Distribution, diversity, evolution, and survival of Helitrons in the maize genome. Proc Natl Acad Sci USA 106:19922–19927

Yang G, Weil CF, Wessler SR (2006) A rice Tc1/mariner-like element transposes in yeast. Plant Cell 18:2469–2478

Yang G, Zhang F, Hancock CN, Wessler SR (2007) Transposition of the rice miniature inverted repeat transposable element mPing in *Arabidopsis thaliana*. Proc Natl Acad Sci USA 104:10962–10967

Yang G, Nagel DH, Feschotte C, Hancock CN, Wessler SR (2009) Tuned for transposition: molecular determinants underlying the hyperactivity of a stowaway MITE. Science 325:1391–1394

Zeyl C, Bell G, Green DM (1996) Sex and the spread of retrotransposon Ty3 in experimental populations of *Saccharomyces cerevisiae*. Genetics 143:1567–1577

Zhang X, Wessler SR (2004) Genome-wide comparative analysis of the transposable elements in the related species *Arabidopsis thaliana* and *Brassica oleracea*. Proc Natl Acad Sci USA 101:5589–5594

Zhang X, Feschotte C, Zhang Q, Jiang N, Eggleston WB, Wessler SR (2001) P instability factor: an active maize transposon system associated with the amplification of tourist-like MITEs and a new superfamily of transposases. Proc Natl Acad Sci USA 98:12572–12577

Zhong CX, Marshall JB, Topp C, Mroczek R, Kato A, Nagaki K, Birchler JA, Jiang J, Dawe RK (2002) Centromeric retroelements and satellites interact with maize kinetochore protein CENH3. Plant Cell 14:2825–2836

Zou S, Ke N, Kim JM, Voytas DF (1996) The *Saccharomyces* retrotransposon Ty5 integrates preferentially into regions of silent chromatin at the telomeres and mating loci. Genes Dev 10:634–645

The Impact of Transposable Elements on Gene and Genome Evolution

3

R. Keith Slotkin, Saivageethi Nuthikattu, and Ning Jiang

Contents

3.1   Introduction

Transposable elements (TEs) are fragments of DNA than can move, or transpose, from one location in the genome to another. Plant TEs are extremely powerful mutagens, inserting into genes and resulting in chromosomal inversions, duplications, and deletions. This feature of TEs has made them extremely useful as tools to gain genetic understanding of the plant genome. However, in addition to their well-described destructive roles as mutagens and molecular parasites on a generation-to-generation timescale, TE activity also has a creative side, constructing novel features in the genome on an evolutionary timescale. The ability of TEs to naturally alter the genome has made them excellent molecular mechanics, generating novel regulator regions, coding regions and chromosomal loci. Occasionally, some of this genomic tinkering generates useful products and is therefore selected for. This chapter explores the creative (rather than destructive) aspects of TE activity.

The identification of the constructive role that TEs play dates back to the original discovery of TEs themselves. Shortly after the discovery of mobile DNA, Barbara McClintock referred to TEs as controlling elements because of their ability to exert an effect on development (McClintock 1956). Expanding on these ideas, Britten and Davidson theorized that novel gene structure and function is regulated by the repetitive fraction of the genome, and results in a coordinated regulatory network responsible for cellular differentiation (Britten and Davidson 1969). Controversial at the time, it is now accepted that some TEs play a role as important regulators of neighboring gene expression, altering developmental patterns. In addition to the role of regulating the expression of single genes, recent data have demonstrated that TEs play a larger role regulating genome-wide patterns of expression by simultaneously bringing entire suites of genes under their control. TEs have also been shown to mediate the formation of new genes, as well as act as the building material of essential chromosomal

R.K. Slotkin (✉)
Department of Molecular Genetics, The Ohio State University, Columbus, OH, USA

Department of Molecular Genetics, The Ohio State University, 500 Aronoff Laboratory, 318 West 12th Avenue, Columbus, OH 43210, USA
e-mail: Slotkin.2@OSU.edu

J.F. Wendel et al. (eds.), *Plant Genome Diversity Volume 1*,
DOI 10.1007/978-3-7091-1130-7_3, © Springer-Verlag Wien 2012

structures such as centromeres, greatly altering genome size. This chapter focuses on the origin of new genomic regulation and features by TEs using recent examples in plant genomes. In addition to single gene examples, this chapter attempts to provide an understanding of the overall contribution that TEs are making to the regulation and function of the genome.

3.2 Gene Regulation By Transposable Elements

Plant biologists have understood at the molecular level for over 20 years that TEs can exert their effect on the genes and the regulatory regions they neighbor (Masson et al. 1987). This understanding has primarily come from the genetic dissection of the anthocyanin pathway in maize. The dozens of examples of this type of regulation in *cis* (where the TE is adjacent to the gene) can be roughly subdivided into two non-mutually exclusive mechanisms. First is the genetic alteration of a gene by structural polymorphism, usually produced by insertion of the TE and resulting in nucleotide modification of the promoter, coding region or distal enhancer of a gene. Second is the influence on a neighboring gene's expression that the TE may exert by recruiting repressive epigenetic regulation to the insertion locus. This second mechanism of regulation is not dependent on a new insertion or polymorphism, but rather responds to the epigenetic status of the TE family as a whole. In either case, the TE-based regulation may be artificially or naturally selected for, or be in a sufficiently small population to spread in frequency through genetic drift. In many instances of purifying selection, the TE will be eroded leaving a *cis* regulatory unit where a now unidentifiable TE once controlled gene expression (Fig. 3.1).

3.2.1 Regulation Through Genetic Means

The mechanism of transposition of DNA occurs differently for the various types of elements (see chapter by Kejnovsky et al. 2012, this volume). While some TE families favor insertion sites of heterochromatin and gene-poor regions of the genome, others favor gene-rich regions, and particularly the 5′ regions of genes. Upon insertion into the coding region of a gene, the TE will likely interrupt and destroy normal gene function and result in the generation of a mutant allele. The vast majority of these TE-mediated mutant alleles will be selected against or at least not selected for, and therefore these events are not the focus of this chapter. In addition to TEs that interrupt gene coding potential and function, many TEs insert into the 5′UTR or promoter of a gene. This may result not in the interruption of a protein's function, but

rather in an alternate regulatory pattern for that protein. These gain-of-function mutant alleles generated by TE insertion can have profound effects on plant development.

3.2.1.1 Altered Regulation By TE-Induced Rearrangements

TE insertion is an inherently imperfect process, with the TE relying on host machinery to reconcile any gaps or unligated DNA generated. Polymorphisms are produced upon insertion or excision of a TE and this may alter a neighboring gene's regulation. An excellent example of how an atypical transposition event can alter the regulation of a gene with consequences on development is the SUN gene of tomato. The wild ancestor of domesticated tomato has round fruit, while some varieties display elongated fruit. This change in morphology is the result of a complex and aberrant transposition event that rearranged the promoter region of the SUN locus, altering SUN expression in the developing fruit and resulting in an elongated fruit shape (Fig. 3.2a) (Xiao et al. 2008). The LTR retrotransposon that generated this rearrangement, *Rider*, used an alternate 3′ LTR to duplicate several genes from the original donor site of the element to the insertion site. The change in the SUN upstream region due to the transposition event resulted in a new regulatory pattern and in a trait that has subsequently been artificially selected. In tomato, the *Rider* family of LTR retrotransposons is high copy and is often associated with genic regions, and is either currently active or has been active in the recent past, affecting the regulation of at least two other genes (Jiang et al. 2009). Thus, having *Rider*, or any other TEs, in or near a gene's promoter will greatly increase the chances of producing a new mutant allele with a novel regulatory pattern. TE presence and activity in upstream regions of a gene often results in the production of an allelic series of variants, due to repeated insertion or excision. The production of such an allelic series is best studied when the regulatory region controls the expression of an easily assayed phenotype, such as plant pigmentation, which has been dissected for multiple maize genes (Kloeckener-Gruissem and Freeling 1995). A recent example has been the production of 35 deletion alleles in the p1 gene promoter by an *Ac* DNA transposon, reaching sizes measured in cM (Zhang and Peterson 2005).

3.2.1.2 TEs as Enhancer Elements

In addition to TEs rearranging the promoter or regulatory region of a gene, the TE itself can also carry regulatory information, exerting an effect on the neighboring gene. For example, the over-branching of the grape fruit meristem, termed reiterated reproductive meristem (RRM), occurs as a somatic variant due to the insertion of a hAT family DNA transposon in the promoter of the VvTFL1A regulatory gene

Fig. 3.1 Production of new gene regulatory regions by TEs. (**a**) A representation of a normal protein-coding gene under the control of typical gene regulation. *Boxes* in the promoter represent *cis*-regulatory promoter regions. (**b**) Insertion of a TE into the promoter of the gene. (**c**) The presence of the TE in the gene's promoter results in new layers of regulation and a new expression pattern. This additional regulation can include the overlapping layers of the normal tissue-specific expression of the TE promoter, as well as the repressive chromatin modifications targeted to the TE, which may result in epigenetic patterns of gene regulation. (**d**) Like any mutation, the new TE-regulated allele may undergo purifying selection. The ability of the TE to transpose, or even our ability to recognize the TE, may be degraded until just another *cis*-regulatory promoter element remains

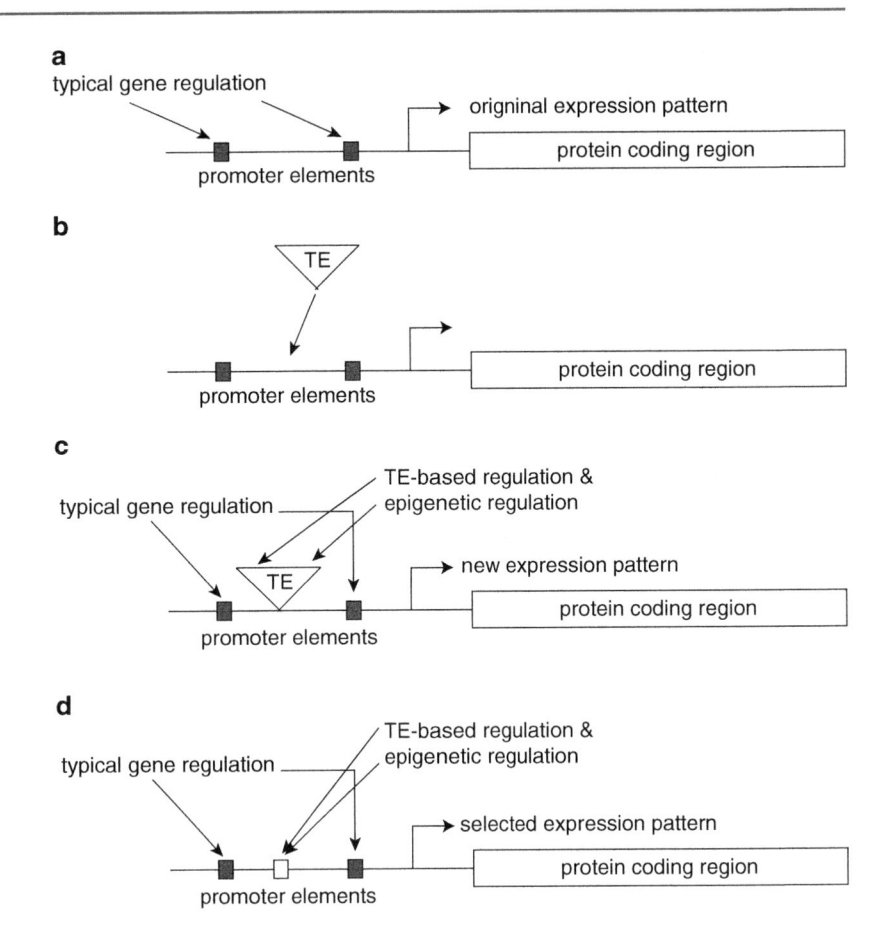

in the reproductive meristem (Fernandez et al. 2010). The insertion of the 5 kb TE in this gain-of-function allele results in over expression and a broader expression pattern of the VvTFL1 gene and the reiteration of branching in the floral meristem, without changing the transcriptional start site of the gene (Fig. 3.2b). Thus, the entire effect of the TE insertion is on the regulation of the promoter element, and the TE insertion acts as an enhancer of transcription. Opposed to acting as an enhancer element, TEs may also potentially act as insulator elements, blocking the association of a distal enhancer element and a gene's proximal promoter. Although this feature has not been described in plants, this function has been well studied using the *Gypsy* LTR retrotransposon of *Drosophila*, where the insulator element has been mapped down to a ~350 bp motif in the 5′UTR (Geyer and Corces 1992).

3.2.1.3 TEs as Promoters

TEs can themselves act as promoters to neighboring genes. Autonomous TEs, and many of their deletion derivatives, carry their own promoters used to drive the expression of the TE-encoded proteins. If a TE fragmentation event occurs, a native TE promoter could potentially drive the expression of whatever is downstream, be it TE region or gene. Even without a rearrangement, many TEs harbor cryptic outward

reading promoters, which have been defined for both the inward transcribing terminal inverted repeats (TIRs) of DNA transposons and the 3′ outward reading long terminal repeat (LTR) of LTR retrotransposons. In this respect a TE may drive the expression of the gene it inserts next to or in, not under the control of the gene's native promoter, but rather under control of the TE's promoter. One example is the *hcf*106 gene involved in the chloroplast electron transport pathway of maize. The insertion of a non-autonomous TE into the 5′UTR disrupts normal gene function, producing a loss-of-function allele (Martienssen et al. 1989). However, this null phenotype is dependent on the regulation of the TE itself. The non-autonomous TE in *hcf*106 responds to the presence of the transposase protein, produced in *trans* by a separate autonomous element. In the presence of the transposase protein, the non-autonomous TE disrupts *hcf*106. In this respect the transposase acts as a transcriptional repressor, however the same transposase protein is a known transcriptional activator in separate situations (Lin et al. 2007). In the absence of transposase protein, a cryptic outward reading promoter on the non-autonomous element is activated, transcribing the gene and restoring gene function of *hcf*106 (Fig. 3.2c) (Barkan and Martienssen 1991). These TE-suppressible alleles have been defined for multiple genes induced from several different TE families (Girard

Fig. 3.2 Phenotypes of TE-induced alleles and epialleles. (**a**) A complex rearrangement of the 5′ region of the SUN gene induced by a *Rider* LTR retrotransposon leads to over expression of SUN (*red arrow*) (Xiao et al. 2008). The increase in SUN expression produces an elongated fruit. (**b**) The reiterated reproductive meristem (RRM) phenotype of grape is caused by a hAT family TE insertion's effect on a downstream gene (Fernandez et al. 2010). (**c**) The *hcf106::Mu1* allele of maize responds to the activity state of the TE family. When autonomous *Mutator* TEs are expressed in *trans*, the HCF106 protein is not expressed, and white chloroplast-mutant sectors are produced on leaves. When the *Mutator* family of TEs is inactive, the TE TIRs are methylated and act as a cryptic promoter driving HCF106 expression. Photo from (Martienssen et al. 1990). (**d**) FWA expression is repressed in the plant body due to the methylation of a SINE TE located in the promoter. Removal of this methylation in mutants leads to ectopic FWA expression and a late flowering phenotype. (**e**) A TE downstream of a gene can nucleate heterochromatin and silence the neighboring gene. The insertion of a hAT TE and the subsequent spreading of DNA methylation result in the silencing of the CmWIP1 gene and the production of a female melon flower. (**f**) The epigenetic activation of a TE can lead to the silencing of the neighboring gene. In *Arabidopsis*, the *bonsai* mutant allele is formed when the DNA methylation is lost from a LINE TE and transcription reads into the downstream convergent BONSAI gene. This overlapping transcription leads to the spread of DNA methylation and heritable epigenetic silencing. Photo from (Saze and Kakutani 2007)

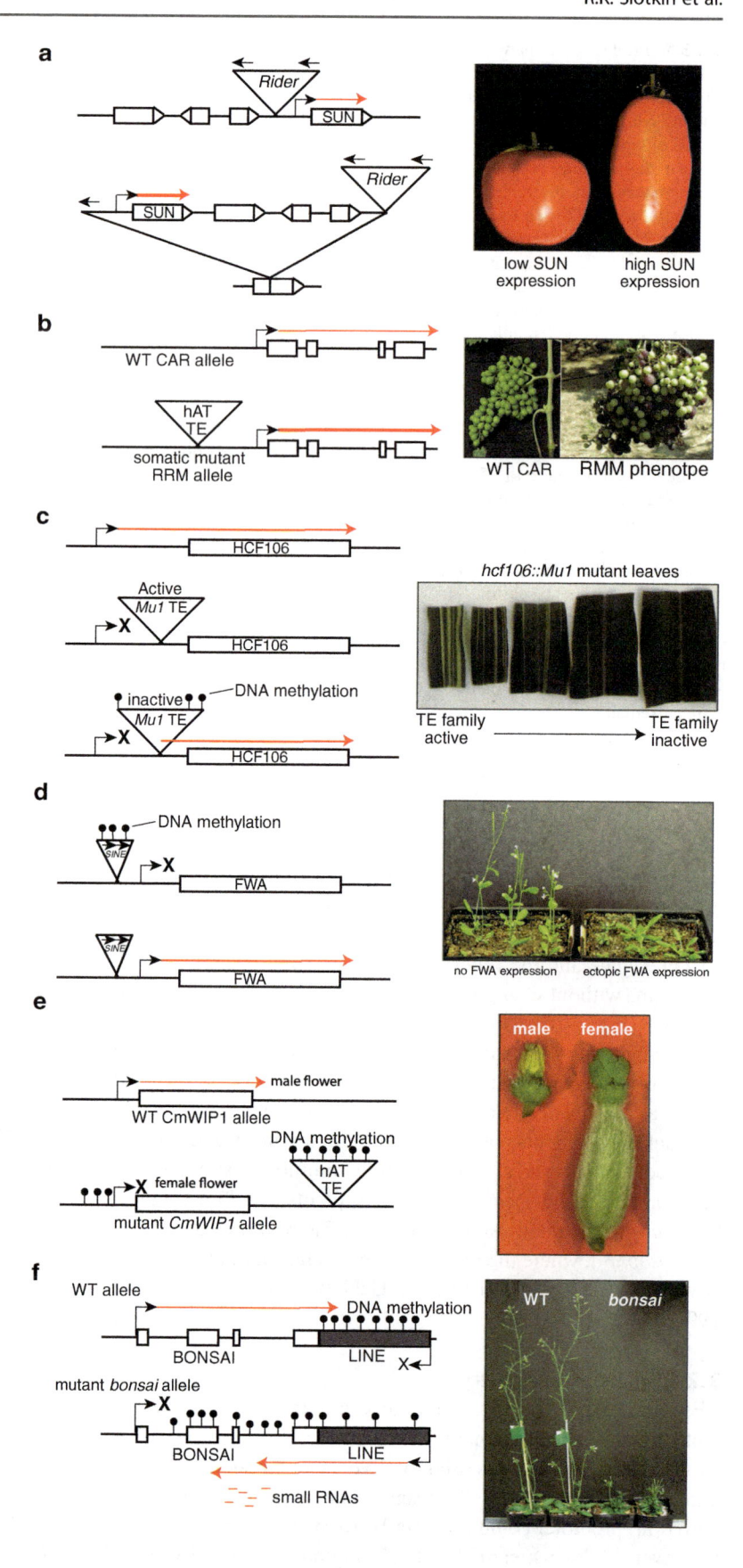

and Freeling 2000; Masson et al. 1987; McClintock 1971). In this way, the activity state of the TE family is acting as a master regulator, modifying the expression of genes adjacent to the insertion site of non-autonomous elements—truly defining McClintock's controlling elements.

3.2.2 Regulation Through Epigenetic Means: Epialleles

TEs are the major targets of gene silencing, which results in the flagging of the repetitive portion of the genome with repressive chromatin modifications such as cytosine DNA methylation and certain histone post-translational modifications (reviewed in Slotkin and Martienssen 2007). These chromatin modifications are targeted to the TEs by the action of small interfering RNAs, and once established, these modifications (particularly DNA methylation) are inherited from cell-to-cell and from generation-to-generation in plants. This epigenetic gene regulation ensures that the majority of TEs are heritably inactivated and are not sources of new mutations.

Repressive chromatin modifications are dynamic by nature. They have the ability to target similar sequences in the genome (guided by small RNAs), and once established, spread to neighboring DNA (guided by transcription). This, along with the fact that the activity of TEs is impossible to separate from their epigenetic regulation, provides TEs with the ability to act as sites of repressive chromatin nucleation. Once initiated, this silencing can influence or spread to adjacent genes, altering their regulation. Returning to the *hcf106* example from above, the TIR of the non-autonomous TE inserted into the *hcf106* 5'UTR responds to the presence or absence of the transposase protein via cytosine DNA methylation (Martienssen et al. 1990). When the transposase protein is absent, the TIR is hyper-methylated, and acts as an outward reading promoter (Fig. 3.2c), counter-intuitive to the regulation of inward-reading TE transcripts, which are generally suppressed by DNA methylation. The inability to separate TE activity from epigenetic regulation leads to an additional layer of epigenetic regulation for TE-neighboring genes. These genes are now subject to the same regulatory patterns as the TE, which has two interconnected layers: First the regulation of the TE (via TE promoters), and second by the targeting of TEs for epigenetic silencing. Importantly, once established, the silenced state of this TE-neighboring allele can be subject to patterns of epigenetic inheritance, remaining silenced from cell-to-cell or generation-to-generation. These epigenetically regulated TE-induced alleles are called *epialleles*.

Of all epialleles, the FWA homeodomain transcription factor gene of *Arabidopsis* is best studied. FWA expression is activated only in the central cell of the female gametophyte, and helps program the maternal contribution to the developing endosperm (Kinoshita et al. 2004). Upstream of the FWA gene is a tandem repeat of a SINE non-autonomous non-LTR retrotransposon that controls expression of FWA. This SINE repeat is normally densely covered with DNA methylation in the plant body. In the central cell, or in mutants that abolish DNA methylation (Soppe et al. 2000), the SINE repeat is demethylated, leading to expression of the FWA gene, and in the mutants a late flowering phenotype (Fig. 3.2d). The methylation of the FWA SINE element is well conserved among *Arabidopsis* ecotypes (Vaughn et al. 2007). Interestingly, a survey of the FWA upstream region in various closely related species found that the tandem repeat orientation of the SINE repeat is not necessary for repressing FWA expression in the plant body, but the presence of the SINE element itself must be the essential factor, proving that this TE is the regulatory agent (Fujimoto et al. 2008). Heritably stable epialleles of FWA can be isolated from DNA methylation mutants, in which a loss of DNA methylation of the SINE repeat mediates the ectopic expression of the FWA coding region (Soppe et al. 2000). Transformation of these unmethylated *fwa* epiallele plants with a naïve SINE repeat and FWA transgene results in *de novo* DNA methylation of the SINE repeat by the RNA-dependent DNA methylation (RdDM) pathway (Cao et al. 2003; Chan et al. 2004b). The SINE repeat matches siRNAs in the genome, potentially encoded by the FWA SINE repeat itself. Upon transformation, these small RNAs target the FWA SINE repeat via RdDM, and reestablishes proper heritable DNA methylation of the SINE repeat and expression pattern of FWA. The FWA example demonstrates how regulation of DNA methylation patterns can regulate a gene by targeting a neighboring TE. Over evolutionary time, this has led to a heritably stable expression state of a transcription factor (with major developmental consequences) by a now barely-recognizable TE acting as a *cis*-regulatory promoter element (Fig. 3.1).

Another example of an epiallele formed by targeting of a TE fragment by small RNAs is the FLC flowering time regulatory gene. The FLC gene in the *Arabidopsis* ecotype *Landsberg erecta* has a *Mutator* (*Mu*) family TE inserted into an intron. This TE insertion acts to nucleate repressive chromatin modifications, reducing the expression of this FLC allele and leading to the vernalization-independent early flowering of *Ler* (Liu et al. 2004). Similar to the FWA example, the FLC TE fragment is likely targeted by TE siRNAs produced in *trans* from another unidentified location elsewhere in the genome that matches the TE fragment. Thus, the origins of epialleles are likely in the repetitive nature of TEs, and in the ability of small RNAs to work in *trans*. In these examples a TE fragment is closely associated with a gene, rendering the gene subject to the induction of heterochromatic silencing that the older previously silent TEs elsewhere in the genome exert on the younger elements or new insertions. With respect to heterochromatin-nucleating TE insertions, one example that must be mentioned, although it is not understood in the detail that FWA and FLC are, is the sex determination in melon. Melon flowers are unisexual, and gynoecious flowers are controlled by the *g* locus. Martin et al. (2009) found that the change from male to female flowers is due to DNA methylation of a hAT family TE adjacent to the *g* locus gene *CmWIP1* (Martin et al. 2009). *CmWIP1* expression is reduced when hAT element DNA methylation spreads to the *CmWIP1* promoter resulting in feminization of the flower (Fig. 3.2e).

In addition to TEs acting as sites of heterochromatin nucleation from siRNAs produced in *trans*, epialleles can also be formed by RdDM-mediated from direct transcription of the TE. Such an example is the *bonsai* mutant allele of *Arabidopsis*. *bonsai* was first defined as a mutant phenotype that spontaneously arose in a *ddm1* mutant background, in which silencing of repetitive elements is lost (Lippman et al. 2004) and ectopic DNA methylation and silencing of some genes occurs (Jacobsen et al. 2000). The *bonsai* phenotype is due to the transcription of a LINE non-LTR retrotransposon ectopically spreading into the downstream convergently transcribed gene, Anaphase Promoting Complex (APC) 13, which becomes epigenetically silenced when the LINE is activated (Fig. 3.2f) (Saze and Kakutani 2007). Thus, in contrast to the FWA and FLC examples, even when the epigenetic control of TE activity is lost, this does not necessarily lead to activation of the neighboring gene. Rather, in this example the loss of TE methylation and a gain in TE activity is the driver of APC13 silencing. In each of these single gene examples (*FWA, FLC, CmWIP1* and *Bonsai*), the complex epigenetic regulation of a TE has exerted its regulation on the neighboring gene, producing new forms of gene regulation that at least with the examples of FWA and *CmWIP1*, have been selected for and retained due to the novel function the TE provided.

3.2.3 Genome Wide Influence

The examples given to this point represent an interesting few cases of TE-controlled genes that display the diversity of mechanisms of TE control of gene expression. Over the last decade, plant biologists have observed nearly every imaginable variation in mechanism of a TE influencing or controlling the expression of a neighboring gene. However, one major question remains: to what extent does this phenomenon occur genome-wide? In other words, what percentage of all genes are regulated by their neighboring TEs? It is important to take into consideration that not all of the TE-induced gene regulation as depicted in Fig. 3.2 will have such drastic phenotypes. Most of the TE-induced gene

regulation may be small quantitative changes in gene expression that do not manifest themselves as morphological phenotypes. In the past 3 years the data collected on this question of genome-wide influence have revealed that of the 16% of *Arabidopsis* genes with a TE insertion 500 bp upstream or downstream of the gene, only a relatively small number of genes per (*Arabidopsis thaliana*) genome are influenced by their neighboring TE (Hollister and Gaut 2009). However, these calculations of neighboring TE-influence on gene expression are likely underestimates due to only sampling some tissues for gene expression, and the rapid erosion of TE sequences in plant genomes. This number is higher in species other than *Arabidopsis thaliana*, which is a self-pollinating plant with a small genome and tight control of TE activity. In the primarily outcrossing *Arabidopsis lyrata* genome, which contains more TEs and displays higher TE activity, the number of genes regulated by TEs is substantially higher (Hollister et al. 2011). This data suggests that the evolutionary history of species likely factors into how efficiently that genome is able to control TE activity, which affects TE abundance and number of genes regulated by TEs.

Another reason that the number of genes regulated by TEs may be low is due to selection against TE and methylation in promoters of genes, and the repressive chromatin environment that they generate. Methylated TEs near genes show signs of negative selection, suggesting that there is an evolutionary trade-off between the silencing of TE, and the effect on neighboring gene expression (Hollister and Gaut 2009). Stated differently, the TEs must be targeted for methylation by the host organism to ensure the TE's heritable repression, but this will result in alteration of regulation for many TE-neighboring genes, and like most mutations, these alleles or TE fragments are selected against.

Modification of a genome by TEs likely does not occur gradually, but rather as interspersed proliferations of activity followed by long periods of inactivity (see below section on genome size expansion). TEs can quickly and massively modify the genome, and bring entire suites of genes under their control. In rice, the proliferation of just one TE family can now make many of their adjacent genes stress-inducible (Naito et al. 2009). In maize it is thought that the activity of *Mutator* family TEs drastically alters the expression of developing anthers, with as much as 25% of the transcriptome responding to the expression status of one single TE family (Skibbe et al. 2009). Although these examples show that many genes can be regulated by TEs, two examples stand out among the others as entire pathways or gene families driven by TE diversification and regulation.

3.2.3.1 A Potential Role for TEs in Plant Immunity
Plants are constantly attacked by pathogens, which can change by season, by location, and in addition evolve

rapidly. To counteract pathogen attack, plants use large number of resistance (R) genes to detect and eventually resist pathogens (Jones and Dangl 2006). Similar to the mammalian adaptive immune system, the plant immune system seems to have used TE-mediated rearrangement and diversification of R gene clusters as a mechanism to generate the raw diversity needed to consistently generate new R genes. R gene clusters are extremely TE rich (Song et al. 1997; Wei et al. 2002) and highly variable between species and even between individuals of the same species. Examples exist of TEs knocking-out R gene expression (Luck et al. 1998; Wang et al. 1998), TEs driving R gene expression from internal promoters (Hayashi and Yoshida 2009), as well as the production and regulation of R epialleles in the same *ddm1* epigenetic mutant background that affects FWA (Stokes and Richards 2002). Thus, R gene clusters may represent a portion of the genome that has selected for a higher TE density and activity because of the TE's ability to create R gene diversity.

In addition to the R gene encoding regions of plant genomes being enriched for TEs, effector regions of pathogen genomes rapidly change to remain effective pathogens. These effector regions of the pathogen genomes (such as powdery mildew and *Phytophthora*) are also rich in repeats and TEs, suggesting that TEs are used to generate diversity in both the plant (to defend against infection), as well as the pathogen genomes (to adapt to a specific host or switch hosts) (Raffaele et al. 2010; Spanu et al. 2010), highlighting an evolutionary arms race between pathogen and host that both use the same mechanism of sequence diversification.

3.2.3.2 The TE Control of Plant Imprinting

The final example of the genome-wide effect that TEs exert on their neighboring genes is regulation of imprinting of gene expression in the developing endosperm. As in the FWA example described above, maternal copies of imprinted genes are de-methylated and expressed in the fertilized endosperm, while copies of the same gene from the paternal parent remain methylated and silenced. This imprinted gene expression balance is necessary for proper endosperm development and seed viability. Gehring et al. (2009) found that the maternal imprinting (activation) of gene expression that regulates FWA is also responsible for regulating ~50 other genes (Gehring et al. 2009). In many of these cases, TEs adjacent to genes have been selected based on their ability to silence neighboring genes. Pared-down by evolution, these fragmented TEs now contribute as a *cis-* regulatory unit to the promoters of these genes (Fig. 3.1). By only expressing the maternal copy of the gene in the endosperm, the transcriptional activation of these imprinted genes are simply reflecting the control of the TE, which is naturally and specifically de-methylated and activated in the

central cell and endosperm (Gehring et al. 2009; Hsieh et al. 2009; Pillot et al. 2010). In this way, TEs have contributed the gene regulatory information required to evolve parent-of-origin specific regulation, and in a broader sense play a key role in gene regulation of the endosperm and seed viability.

3.3 Origin of New Genes

The finding that TEs are capable of duplicating genomic sequences is relatively recent, and the consequences of such duplication is still being revealed. Nevertheless, emerging evidence indicates that amplification of gene sequences by TEs may provide raw material for the evolution of new genes and result in the modification of existing genes. This section will focus on how novel genes may arise through the action of TEs.

3.3.1 From Autonomous to Non-autononmous Elements

As mentioned in the chapter by Kejnovsky et al. (the volume, 2012), TEs fall into two categories based on whether they encode proteins required for transposition. Autonomous elements encode transposase or reverse transcriptase and are able to transpose by themselves. In contrast, non-autonomous elements do not encode transposition-related proteins and rely on the cognate autonomous elements for their transposition. Some non-autonomous elements are direct deletion derivatives of their autonomous elements. For instance, *mPing*, an active MITE in rice, inherited all of its sequence from its autonomous element *Ping* (Fig. 3.3a) (Jiang et al. 2003b; Kikuchi et al. 2003; Nakazaki et al. 2003). In other cases, non-autonomous elements may carry sequences that are not present in their cognate autonomous elements. A classic example is the maize autonomous *Mutator* element (*MuDR*) and its non-autonomous elements (*Mu1–Mu8*) (Fig. 3.3b). *MuDR* harbors two open reading frames—*mudrA* (the transposase) and *mudrB*, both of which seem to be required for transposition (Lisch 2002). The non-autonomous *Mu* elements share ~220 bp TIRs with *MuDR*; however, each non-autonomous element has a distinct internal region with similarity to non-TE, genic sequence in maize (Lisch 2002). Thus, it seems that internal regions of non-autonomous elements can be formed by acquiring non-TE sequences, including genes. As described in the following sections, this seems to be a generic attribute for many types of TEs.

In addition to non-autonomous elements, some autonomous elements carry acquired gene fragments as well. A peptide domain, which might be derived from a small ubiquitin-like protease gene, is present in autonomous MULEs as well as some *Spm* elements in different plant

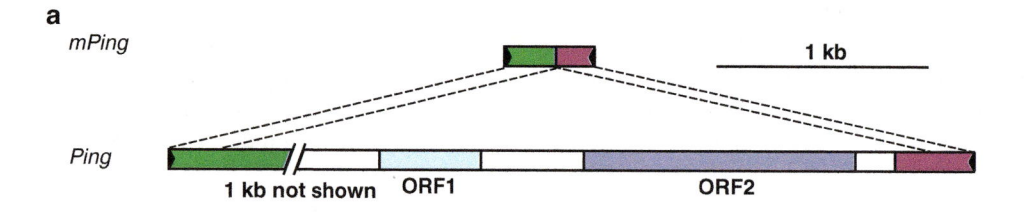

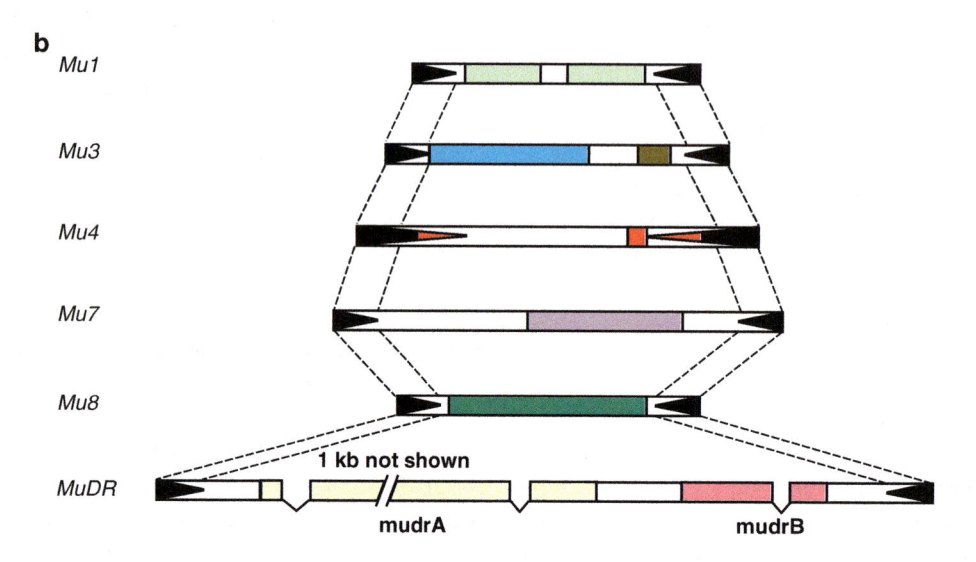

Fig. 3.3 The structural organization of autonomous and non-autonomous elements from two TE families. (**a**) The rice miniature inverted repeat transposable element (MITE) *mPing* and its autonomous element *Ping* (Jiang et al. 2003b). TIRs are depicted as *black arrowheads*. *Colored boxes* represent sub-terminal regions or open reading frames, and other sequences are depicted as *white boxes*. The homologous sequences between *Ping* and *mPing* are connected with *dashed lines*. (**b**) Maize autonomous *Mutator* element *MuDR* and its non-autonomous *Mu1–Mu8* (Lisch 2002). Elements are diagrammed similarly as in (**a**) except the sub-terminal regions are not colored. Different acquired gene fragments in *Mu1–Mu8* are shown as *colored* boxes. For *MuDR*, introns are shown as lines connecting exons. In *Mu4*, the *red triangles* indicate that a portion of the acquired gene fragment was duplicated, inverted, and became part of the TIR

species (Hoen et al. 2006; van Leeuwen et al. 2007). Thus it appears that both autonomous and non-autonomous elements are capable of acquiring or *transduplicating* genomic sequences. Nevertheless, the number of reported examples of acquisition by non-autonomous elements far exceeds that of autonomous elements, and this correlates with the diversity of sequences that they carry. This is likely because, in general, the copy number of non-autonomous DNA elements is higher than that of their autonomous elements. It may also imply that the acquisition process is largely a consequence of sequence exchange, i.e., the acquisition of new sequences (genomic sequences) is often accompanied with the complete or partial loss of the original sequences (such as transposase). Finally, the few cases where autonomous elements are associated with additional genomic sequences may reflect the fact that these sequences are beneficial to the transposition process and thus are retained within autonomous elements (van Leeuwen et al. 2007).

3.3.2 Gene Acquisition, Duplication and Exon Shuffling

The first incidence of gene acquisition in plants was reported more than 20 years ago, when the *Mu1/Mu1.7* elements were shown to contain part of a gene of unknown function called *MRS-A* (Talbert and Chandler 1988). Nevertheless, the scale of gene acquisition by this type of element, or by DNA transposons in general, was not realized until the release of the complete genome sequence of rice, where thousands of gene-carrying MULEs (called Pack-MULEs or trans-duplicates) were found (Jiang et al. 2004; Juretic et al. 2005). Hence, the study of gene duplication by TEs in plants was greatly accelerated by the availability of genomic sequences. On the other hand, TE sequences such as TIRs are only recognizable for a few million years (MY), which consequently leads to the underestimation of the number of TE-mediated gene duplication events.

3.3.2.1 Elements Involved in Gene Acquisition

With the advent of sequence technology and subsequent analysis, it has become clear that most of the major types of TEs in plants are capable of duplicating genes or gene fragments, albeit the frequency of gene acquisition varies with the element family/type. Even for the same type of element, great variation exists in the abundance of the elements carrying genes among different species. In this section, the known gene acquisition events in plants will be discussed based on the super-family to which the elements belong. The genes that provide templates for duplication or acquisition by TEs will be called *parental genes* or *parental forms*.

Retrotransposons

Retrogenes are formed when transcribed and spliced mRNAs are processed by the reverse-transcription machinery generated from retrotransposons and inserted into new genomic positions. For example, during the retrotransposition process of a LTR retrotransposon, virus-like particles are produced in which the element mRNAs are included (Boeke and Corces 1989). During this process, gene mRNAs can be fortuitously packed into the particle and therefore retrotransposed. These retrogenes are distinguished from their parental forms by the presence of a poly (A) tract, the lack of introns, and the presence of a target site duplication (Brosius 1991). Some of the retrogenes are associated with the relevant elements, i.e., gene fragments may be found between the two LTRs of single LTR element. Other retrogenes may be present independently, making it difficult to deduce the element responsible for their formation.

The *Bs1* element in maize represents the first gene-carrying retrotransposon found in plants (Bureau et al. 1994; Jin and Bennetzen 1994). This LTR element has transduced gene fragments from three different parental genes that lack introns (Elrouby and Bureau 2001). Like *Bs1*, 425 LTR elements containing gene fragments are found in the sequenced genome of maize cultivar B73 (Schnable et al. 2009). Given that there are over one million copies of LTR elements in the genome (Baucom et al. 2009; Schnable et al. 2009), this fraction of LTR elements carrying genes does not seem to be very high. In rice, there are a total of 1,235 retrogenes among which 27 are present within LTR elements (Wang et al. 2006). Gene duplication by a LTR element was also reported for tomato (Xiao et al. 2008), but a systematic survey for the abundance of such events is lacking.

DNA Transposons: CACTA Elements

CACTA elements (*En/Spm*) containing parental genes have been described in multiple plant species. In snapdragon (*Antirrhinum majus*), the internal region of *Tam4* is unrelated to that of its autonomous element, *Tam1* (Luo et al. 1991). Moreover, the internal region of *Tam4* is homologous to genes that encode specific transcription factors. In Japanese morning glory (*Ipomoea nil*), a CACTA element called *Tpn1* captured a HMG domain sequence (Takahashi et al. 1999). Further study indicated that *Tpn1* belongs to a large family of TEs with 500–1,000 copies in the genome, and most of the *Tpn1*-like elements carry different types of parental gene fragments (Kawasaki and Nitasaka 2004). Likewise, a soybean (*Glycine max*) CACTA element (called *Tgm*) harbors five unrelated gene fragments within the element (Zabala and Vodkin 2005). The insertion of *Tgm* into a flavanone 3-hydroxylase gene leads to the modification of flower color and seed weight (Zabala and Vodkin 2005). In maize, it was estimated that 155 CACTA elements carry gene fragments in their internal regions (Schnable et al. 2009). Because the number of CACTA elements is only 1% of that of LTR retrotransposons, the fraction of gene-carrying elements in the CACTA family is ~30 times higher than that of LTR retrotransposons.

DNA Transposons: Pack-MULEs

To date, Pack-MULEs have been reported in maize, *Arabidopsis*, rice and *Lotus japonicus* (Holligan et al. 2006; Jiang et al. 2004; Juretic et al. 2005; Talbert and Chandler 1988; Yu et al. 2000). Notably, there are thousands of these elements present in both rice (a monocot) and *Lotus japonicus* (an eudicot), suggesting that the mechanism of gene duplication by MULEs is widespread in flowering plants (Holligan et al. 2006). Despite the fact that maize and rice are close relatives, there are many more Pack-MULEs in rice (2,853 Pack-MULEs with 1,537 parental genes) than in maize (276 Pack-MULEs with 235 parental genes) (Jiang et al. 2011, and unpublished data), suggesting that the amplification and retention of Pack-MULEs is lineage-specific. Unlike CACTA elements, which have short TIRs (less than 30 bp), most Pack-MULEs are associated with extended TIRs (50–600 bp) and it is currently not clear whether the presence of the long TIRs promotes sequence acquisition by these elements.

DNA Transposons: *Helitrons*

Helitrons use a rolling-circle mechanism for transposition, so they are distinguished from classical, *cut and paste* DNA transposons (Kapitonov and Jurka 2001). The discovery of *Helitrons* in eukaryotes is relatively recent, yet the gene capture phenomenon has been already reported in maize, *Arabidopsis*, and rice (Du et al. 2009; Hollister and Gaut 2007; Lai et al. 2005; Lal et al. 2003; Morgante et al. 2005; Sweredoski et al. 2008; Yang and Bennetzen 2009). Unlike Pack-MULEs, gene-carrying *Helitrons* are more abundant in maize than in rice. Particularly, there are ~840 gene fragments captured by 1,194 intact *Helitron* elements in maize, in addition to many more truncated elements

(Du et al. 2009; Feschotte and Pritham 2009; Yang and Bennetzen 2009). In contrast, there are only 11 genes captured by *Helitron*s in rice (Sweredoski et al. 2008). Given the complimentary distribution of Pack-MULEs and gene-carrying *Helitron*s in maize and rice, it is tempting to speculate that gene duplication by one type of element may substitute the function of that from another type of element. Alternatively, this may be the consequence of dynamic interaction among different types of elements.

Other DNA Transposons

In maize, it was reported that two *Ds* elements contain fragments similar to other maize genomic sequence (Rubin and Levy 1997). In the recently sequenced B73 genome, there are 23 hAT like elements carrying gene fragments. This is in addition to 20 *PIF*-like and 2 *Mariner*-like elements associated with gene fragments (Schnable et al. 2009). Thus, it appears that all types of DNA elements have the potential to capture gene or gene fragments. Nevertheless, the abundance of such events was only reported for CACTA elements, Pack-MULEs, and *Helitron*s.

It is unclear why different types of TEs vary dramatically in the frequency of gene acquisition. In general, gene acquisition events are more prevalent among DNA elements than retrotransposons, and this might be explained by the preferential insertion of DNA elements into genic regions, which may have provided spatial convenience for gene acquisition. Nevertheless, this does not explain the difference among DNA elements. As discussed below (3.3.2.3 Acquisition Mechanisms), the long TIRs of Pack-MULEs were hypothesized to be the cause for sequence acquisition, yet long TIRs are not (or not frequently) associated with CACTA or Helitron elements. Hence different elements may possess different structural or mechanistic features that promote sequence acquisitions.

3.3.2.2 Fate of Acquired Sequences

The evolutionary fates of individual duplicated genes are extremely diverse, but most fall into three categories (Lawton-Rauh 2003; Zhang et al. 2003). Many of the duplicated genes become pseudogenes (pseudogenization) due to the accumulation of deletions and mutations. Other duplicated genes experience subfunctionalization, where each of the duplicated genes carries out part of the parental gene function. This may occur, for example, by expression in different tissues. A third possibility is that one of the duplicated genes evolves a novel function (neofunctionalization), which is one of the most important consequences of gene duplication (Conant and Wolfe 2008; Doyle et al. 2008; Flagel and Wendel 2009). Eventually, pseudogenes become extinct, and only genes retaining original gene function or those associated with novel functions will be persist in the genome.

Coding Capacity of Acquired Sequences

Unlike segmental duplication or genome-wide duplication, most sequences duplicated by TEs represent gene fragments and not complete genes, which reduces the probability of retaining original function. On the other hand, TEs may acquire multiple gene fragments within a single element, facilitating the shuffling of coding regions or shuffling between coding regions and regulatory regions. For example, about 25% of the rice Pack-MULEs contain sequences from multiple genes that are fused to form chimeric open reading frames (Hanada et al. 2009; Jiang et al. 2004). Similarly, 42% of the 1,235 retrogenes in rice have recruited new exons from flanking sequences, leading to the formation of chimeric open reading frames (Wang et al. 2006). In maize, as many as ten different gene fragments have been found in a single *Helitron* element (Yang and Bennetzen 2009). Consistent with the notion that exon shuffling by TEs can confer evolutionary advantages, it is observed that *Helitron*s with multiple gene fragments are overrepresented in the maize genome (Yang and Bennetzen 2009), and Pack-MULEs with multiple gene fragments are more likely expressed than those with a single gene fragment (Hanada et al. 2009). These features may promote the evolution of novel functions.

For a nucleotide sequence, nonsynonymous substitutions refer to mutations that lead to the alteration of amino acids, and synonymous substitutions refer to mutations that do not change amino acid sequence. The ratio of nonsynonymous to synonymous substitution rates should be smaller than 1 under purifying selection (Li 1997). In several genome-wide studies, it has been shown that selection pressure is significant between the acquired regions in TEs and their parental genes, despite the fact that most acquired regions only represent gene fragments. This is true for retrogenes and sequences inside Pack-MULEs and *Helitron*s (Jiang et al. 2004; Wang et al. 2006; Yang and Bennetzen 2009). However, purifying selection may reflect patterns inherited from the parental genes rather than on the derived fragment (Juretic et al. 2005). These alternatives have been teased apart by the construction of putative ancestral sequences upon duplication, and their comparison to acquired sequences and parental genes (Hanada et al. 2009). Selection on each lineage (the acquired fragment and the parental gene) can be tested by the comparison with putative ancestral sequences. Not surprisingly, parental genes are in general under stronger selection pressure than acquired fragments (Hanada et al. 2009). However, there is a subset of acquired fragments under significant selective pressure, suggesting they may have retained, at least to some degree, the coding function of the parental genes (Hanada et al. 2009).

Consistent with the results from computational analysis, experimental evidence about the function of individual genes duplicated by TEs is emerging. From the SUN example used in Fig. 3.2a, the tomato *Rider* LTR retrotransposon duplicated a 24.7 kb fragment and placed the fragment in a new genomic location (Xiao et al. 2008). The novel phenotype derived from this rearrangement is due to a new combination between regulatory regions from the TE-flanking sequence and the coding region duplicated. In contrast, the *Bs1* element in maize forms a chimeric open reading frame using its own TE sequences and sequence transduplicated from the genome (Elrouby and Bureau 2001). A recent study indicated that this chimeric gene is transcribed and translated in early ear development and is likely to have a function in the reproductive pathway (Elrouby and Bureau 2010). In soybean, gene fragments carried by *Tgm* elements are incorporated into the transcript of the F3H gene, providing additional exons of the gene (Zabala and Vodkin 2007). In rice, over 20% of Pack-MULEs are transcribed; the expressed portion is much higher than that of the pseudogenes in the same genome. This is in addition to the fact that at least 28 Pack-MULEs are translated (Hanada et al. 2009).

Regulatory Roles

Given the redundancy generated after duplication and the fragmented nature of acquired sequences, it is very likely that only a small part of gene-carrying TEs will evolve into independent protein coding genes. A more common consequence of gene acquisition by TEs is likely the alteration of epigenetic status of the parental genes through the interaction of homologous sequences between TEs and their parental genes. It is worth mentioning that such interactions would occur in *trans*, since in most cases the TE and the parental gene are at different locations in the genome.

In rice, most of the 3,000 Pack-MULEs are associated with small RNAs, and the small RNAs derived from Pack-MULEs are much more abundant than that from the parental genes (Hanada et al. 2009). Moreover, most Pack-MULEs share small RNAs with their parental genes, correlating with a reduced expression level of these parental genes. This suggests that after acquisition, the acquired and amplified fragments may negatively regulate the expression of their parental genes, through the formation of small RNAs. Interestingly, the number of small RNAs associated with each Pack-MULE declines with the age of the element (Hanada et al. 2009). The implication of this observation is that the regulatory role of Pack-MULEs on their parental genes decreases as the acquired fragments become more diverged from their parental genes.

Genetic Modification of Existing Genes

As discussed above and in 3.2.2, the insertion or amplification of TEs may cause epigenetic modification of their adjacent genes or their parental genes. In addition, TEs may also cause genetic modification of their adjacent genes or genes they

inserted into. For example, the transposition of reversed *Ac* ends resulted in the rearrangement of two linked paralogous genes and the formation of new, chimeric genes in maize (Zhang et al. 2006). In rice, 300 protein coding regions are directly modified by MITEs, where they serve as parts of the coding regions, transcription start sites, transcription termination sites or novel splicing sites (Oki et al. 2008).

In maize, *Mutator* elements preferentially insert into the 5′ end of genes (Dietrich et al. 2002; Lisch 2002; Liu et al. 2009; Robbins et al. 2008), and the same trend is true for Pack-MULEs (Jiang et al. 2011). When a Pack-MULE is located at the 5′ end of a gene, transcription is often initiated in the internal region of the *Mu* element and extends to the adjacent gene (see Fig. 3.2). In this case, the internal regions of the Pack-MULE, which contains the acquired fragment, serves as the 5′ untranslated region and/or coding region in the chimeric transcript. Interestingly, Pack-MULEs preferentially acquire GC-rich gene fragments in rice and maize (Jiang et al. 2011). As a consequence, most of the modified transcripts are associated with a negative GC gradient, which means a higher GC content at the 5′ end than at the 3′ end. This suggests that the amplification of Pack-MULEs may have contributed to the elevated fraction of genes with negative GC gradient in grasses (Jiang et al. 2011).

In summary, due to the fact that the majority of TE duplicated sequences are gene fragments, it is expected that many of them would eventually evolve into pseudogenes. However, increasing evidence supports the notion that a subset of resulting elements is functional, in one way or another. Due to the abundance and continuing activity of TEs, gene duplication by TEs represents a unique evolutionary force in the genome through their extraordinary ability to shuffle regulatory and coding sequences.

3.3.2.3 Acquisition Mechanisms

To understand the acquisition/transduplication mechanisms of TEs, an essential question is whether this process is associated with transposition. For retrotransposons, this process is likely to occur during transposition since the duplicated sequences are processed by the reverse transcription complex. For non-LTR elements such as *LINE1* in humans, it is known that the elements duplicate sequences by read-through of flanking sequences during the transcription process, and the flanking sequences were thereafter transposed to new locations together with the element (Moran et al. 1999). For LTR elements, as mentioned above, the process is initiated with the accidental packaging of gene mRNAs in the virus-like particles. Subsequent reverse transcription and integration into the genome results in an independent retrogene. If template switching occurs between the LTR element and the gene mRNA, one outcome is the presence of a gene fragment inside the element (Negroni and Buc 2001).

Compared to retrotransposons, less is known about how DNA elements acquire their internal sequences, including the involvement of transposition in the process. Gene fragments inside DNA elements sometimes contain introns, so it is likely that the duplication occurs at the DNA level, not RNA or cDNA levels.

One interesting observation concerns the orientation of gene fragments inside the elements. When a *Helitron* element harbors gene fragments from multiple genes, all fragments are placed in the same orientation with respect to the transcription of the parental genes (Brunner et al. 2005). Moreover, the gene fragments are also in the same orientation as the *Helitron* Rep/helicase gene. Subsequently, it was shown that the biased distribution is likely a combined consequence of preferential acquisition of gene fragments in the same orientation and selection against "antisense" gene fragments (Yang and Bennetzen 2009). A bias for gene orientation was also reported for the *Tpn1* element (Kawasaki and Nitasaka 2004). In contrast, the orientation of gene fragments in Pack-MULEs seems to be random, although elements carrying gene fragments oriented differentially are less likely to be expressed (Hanada et al. 2009). If the acquisition of gene fragments occurs at the DNA level, it is mysterious how the *Helitron* elements identify the transcriptional orientation of the fragments. The difference in the orientation of acquired fragments demonstrated by Pack-MULEs and *Helitron*s suggest that they employ distinct acquisition mechanisms or that selection operates differently on these two types of elements.

Three models have been proposed for the acquisition of genomic sequences inside Pack-MULEs or other DNA elements. According to the first model, the individual MULE TIR is mobile and when a pair of TIRs encompasses a gene sequence, the two TIRs and the gene sequence between the TIRs will move as a single unit (Talbert and Chandler 1988). However, there is no evidence that a single TIR is capable of moving. The second model considers that the presence of long TIRs causes sequence acquisition (Bennetzen and Springer 1994). If the element is in a single strand DNA status, the presence of the TIR will result in the formation of a stem loop and a DNA nick will be created by an endonuclease. At this point, an ectopic sequence will be used as a template to repair the nick resulting in the introduction of novel sequences into the element (Bennetzen and Springer 1994). This process does not seem to require the presence of the transposase. A third model assumes that sequence acquisition occurs during the gap repair process after TE excision. The double strand break formed after excision is often repaired using a sister chromatid or a homologous sequence as template via a synthesis-dependent strand annealing (SDSA) mechanism (Engels et al. 1990; Nassif et al. 1994). If template-switching occurs during this process, novel sequences will be introduced to the element

(Yamashita et al. 1999). Sequence rearrangements caused by abortive gap repair after excision have been reported with *Mutator, Ac/Ds,* and *Spm* elements (Hsia and Schnable 1996; Masson et al. 1987; Rubin and Levy 1997; Yan et al. 1999). In one of the cases, the derivatives of *Ac*, formed at the donor site after the excision of *Ac*, contains short pieces (52 and 96 bp, respectively) of non-*Ac* sequences with unknown origin (Rubin and Levy 1997), suggesting that it is possible to acquire non-TE sequences through this mechanism. Both models 2 and 3 predict that an element may acquire new sequences without transposing to a new genomic locus.

If the acquisition is not associated with transposition, one possible consequence is that adjacent sequences will be more frequently used as template. If this is the case, one would expect a linkage between elements and their parental copies. However, all Pack-MULEs and their parental genes in maize, rice and *Arabidopsis* were recently examined and such linkage was not observed (Jiang et al. 2011). This indicates that adjacent sequences are not preferential targets for acquisition, or acquisition is associated with transposition. Further biochemical and computational analyses are required to resolve the mystery of sequence acquisition by DNA transposons.

3.3.3 Domestication of TEs

As discussed above, TEs may contribute to the generation of new genes by duplicating and shuffling non-TE sequences. Yet TEs may also contribute to the generation of new genes by donating the true TE sequences, mostly transposases, for cellular function. This process is termed *TE domestication, cooption,* or *exaptation,* as the protein now shifts to a secondary function in which it did not primarily evolve to perform. Since most transposase proteins harbor DNA binding domains, which are essential features of gene regulation, it is not surprising that many TEs evolve into transcription factors.

Numerous examples of TE domestication have been reported in animals and fungi (reviewed in Feschotte and Pritham 2007). Those include the RAG genes responsible for rearrangement of immunoglobulin genes, as well as the JERKY gene in mice, which is involved in proper brain function (Agrawal et al. 1998; Toth et al. 1995). Recently, it was shown that α*3*, a gene responsible for mating type switching in yeast, is related to a MULE transposase (Barsoum et al. 2010). In contrast, there are relatively few examples reported in plants. The first and best-characterized example in plants arose from the analysis of the *FAR1* and *FHY3* genes in *Arabidopsis*. Mutations in these two genes cause defects in the response to far-red light. Sequence comparison indicated that the *FAR1* and *FHY3* genes exhibit a high degree of similarity to the transposase encoded by MULEs such as *MuDR* and *Jittery* in maize (Hudson et al. 2003), and a subsequent study has demonstrated that each of these genes acts as a *bona fide* transcription factor (Lin et al. 2007). Since the mutations in *FAR1* and *FHY3* genes result in a visible phenotype, and no other TE features (such TIRs and target site duplications) are found with the two genes, it is unambiguous that these genes have a well-defined cellular function and no longer represent TEs. In fact, *FAR1* and *FHY3* belong to the FAR1/FRS gene family with over ten members in *Arabidopsis* (Lin et al. 2007). Yet only *FAR1* and *FHY3* have been functionally characterized, due to their role in light signaling. Members of the *FAR1* subclade are present in multiple plant species, including both monocots and dicots, suggesting that the domestication event occurred prior to the monocot–dicot split. Similarly, the MUSTANG family of genes was shown to share extensive homology with MULEs without being associated with mobility, which is supported by the fact that they are present at syntenic sites within different plant species (Cowan et al. 2005). Both MUSTANG and *FAR1* have orthologs in rice that are distinct from most other MULE sequences in that they are each transcribed in a broad range of tissues (Jiao and Deng 2007). Thus, domesticated transposase genes could be distinguished from regular TEs through their expression pattern and conservation among species. Interestingly, analysis of zinc finger proteins and other transcription factors encoded by MULE transposases has led to the hypothesis that MULEs have been domesticated independently many times over the course of evolution for the production of transcription factor genes (Babu et al. 2006).

In addition to MULE related genes, two hAT-related domesticated genes have been reported in plants. The *DAYSLEEPER* gene is essential for the growth of *Arabidopsis*, and contains a zinc-dependent DNA binding domain that is similar to that of the hAT transposase (Bundock and Hooykaas 2005). Like *FAR1* and *FHY3*, *DAYSLEEPER* seems to have lost other features related to TEs. The proteins encoded by *DAYSLEEPER* are capable of binding to a motif present in the upstream region of the DNA repair gene *Ku70* and many other genes in *Arabidopsis*, suggesting that *DAYSLEEPER* is a transcription factor derived from a h*AT* transposase (Bundock and Hooykaas 2005). Likewise, *Gary* is a hAT-related gene that is located in a syntenic region of barley and wheat, and is present in other grass genomes such as rice, oat, rye, maize, sorghum and sugarcane (Muehlbauer et al. 2006). *Gary* is expressed in wheat and barley spikes, yet its exact function remains to be elucidated.

The large number of examples reporting the interactions between TEs and genes suggests a blurry line between the two groups of sequences. The capability of TEs to duplicate, transpose, and amplify gene sequences confers mobility to cellular genes, and TEs may also lose their mobility to become *normal* genes. During the process of switching their roles, TEs provide great potential for innovation and introducing variation into a dynamic genome.

3.4 Contribution of TEs to Genome Size Variation

TEs not only play a role in the formation and regulation of single genes. On a genome-wide level, TEs are responsible for the expansion of chromosomes, and are responsible for the C-value paradox, in which the size of any particular genome does not correlate with the number of genes, their regulatory complexity, or the organismal/developmental complexity.

3.4.1 Factors Limiting Genome Size and the Role of TEs

Theoretically, genome size is determined by the interaction of two types of factors: those that lead to emergence of new sequences and those responsible for the elimination of existing sequences. Apparently, genome size will increase if the mechanisms of sequence elimination fail to counterbalance that of expansion. As discussed in the last chapter, TE content is positively correlated with genome size, so it is well accepted that TE amplification is one of the most important factors leading to genome size expansion. A significant example comes from grasses, which include many important cereal crops such as rice, maize, wheat, barley and sorghum. Among them, maize and sorghum are only separated by 15–20 MY, and both have ten chromosome pairs with comparable gene number and order (Bennetzen 1998). Yet the maize genome is at least three times that of sorghum and this is largely due to the presence of retrotransposons in the intergenic regions of maize (Avramova et al. 1996; SanMiguel et al. 1996). Furthermore, most of these elements were amplified within last 2–6 MY (SanMiguel et al. 1998). In the genus of *Oryza*, the diploid species *O. australiensis* has diverged from cultivated rice for less than 10 MY (Ammiraju et al. 2008; Ge et al. 1999). Yet the genome of *O. australiensis* is almost three times that of rice. Most of the variation is attributed to the amplification of three families of LTR retrotransposons in *O. australiensis* within the last 3 MY (Piegu et al. 2006). Similarly, diploid members of *Gossypium* have a threefold genome size variation, and a large part of the difference is due to the differential amplification of a single *gypsy*-like LTR element, *Gorge3* (Hawkins et al. 2006). This suggests that bursts of TE amplification may lead to rapid, lineage-specific expansion of genome size.

Although the amplification of TEs can induce rapid increase of genome size, there are a variety of mechanisms that mitigate the consequence of TE amplification. It is known that unequal recombination occurs between the two LTRs of a LTR retrotransposon, and this leads to the formation of solo LTRs and deletion of internal sequences between the two LTRs. TE sequences may be deleted by other mechanisms such as illegitimate recombination, which removes more DNA sequences than unequal homologous recombination in *Arabidopsis* (Devos et al. 2002). In rice, it is estimated that 190 Mb of retrotransposon sequences, which is equivalent to nearly half of the rice genome, was eliminated in last 8 MY by various deletion mechanisms (Ma et al. 2004). The presence of various types of deletion processes explains why TEs are only recognizable for a few MY and why they tend to go extinct upon the loss of transposition activity.

Beside the elimination of TE sequences, TEs may lead to the deletion of non-TE sequences in the genome. Such deletion results if unequal homologous recombination occurs between two copies of homologous TEs located on the same chromosome. In this case, one TE copy and the intervening sequences will be removed from the genome. Conceivably, selection will only favor the events where the intervening sequences are not essential for host functions.

If deletion and recombination processes results in the removal of both TE and non-TE sequences, is that sufficient to counterbalance the expansion of genome size caused by TE amplification? In regard to this question, recent studies in cotton provide novel insights into the directionality of genome size change in plants (Hawkins et al. 2008a, b, 2009). Using a modeling approach, it was shown that the rate of DNA loss is higher in the smaller genomes, and this is sufficient to reverse the genome expansion caused by TE amplification (Hawkins et al. 2009). Thus, reduction in genome size does occur, and the rate of DNA removal varies among different plants, albeit the frequency of such events remains a question.

3.4.2 Contribution of Retro and DNA TEs

As discussed in the last chapter, the composition of TEs varies among different plants, and dramatic variation is found among closely related species. Figure 3.4 diagrams the contribution of retro and DNA TEs to the genome size among species with entire genome sequences and where most of the repetitive sequences are classified. The genome size of these plants ranges from 125 Mb (*Arabidopsis*) to 2,300 Mb (maize). As shown in Fig. 3.4, in all plant genomes characterized so far retrotransposons account for more sequence mass than DNA TEs. This is in contrast to the situation in some animals where DNA elements are dominant in some of the genomes. In the model organism *C. elegans* and its relatives, most repetitive sequences are DNA transposons (*C. elegans* Sequencing Consortium 1998; Stein et al. 2003). Moreover, the genomic fraction of retrotransposons is largely correlated with genome size in plants, whereas such a correlation does not exist for DNA elements. Among the retrotransposons in plants, LTR retrotransposons are the most abundant. For example, in maize 75% of the genome is occupied by LTR elements, while non-LTR retrotransposons (LINEs and SINEs) only account for 1% of the genome (Schnable et al. 2009). This is in comparison to the genomes of human and other mammals, where non-LTR elements are the most abundant component of the genomes (Lander et al. 2001; Lindblad-Toh et al. 2005; Pontius et al. 2007; Waterston et al. 2002). As a result, it is more appropriate to state that LTR retrotransposons are the major factors responsible for genome expansion in plants (Kumar and Bennetzen 1999).

Does this imply DNA TEs are not as successful as retrotransposons in plants? Not necessarily. Compared to the genome fraction, the copy numbers of elements are more accurate in reflecting the combined and net effects of rate of transposition and deletion. In *Arabidopsis* and soybean, the copy number of DNA TEs is comparable to that of retrotransposons (Arabidopsis Genome Initiative 2000; Schmutz et al. 2010), suggesting that DNA TEs can amplify as rapidly as their RNA counterparts in both small and big genomes. In rice, the number of DNA TEs is more than twice that of retrotransposons, due to the exceptional abundance of the small non-autonomous DNA elements called MITEs (International Rice Genome Sequencing Project 2005). Since most MITEs are smaller than 500 bp, which is about 1/10 of the size of an intact LTR retrotransposon, this explains why in rice DNA TEs fail to compete with retrotransposons in terms of contribution to genome size.

The failure of DNA TEs to occupy as large a portion of the genome as retrotransposons is likely due to the targeting mechanism of these elements. It is known that most DNA TEs preferentially insert into low copy sequences in genic regions, not highly heterochromatinized repetitive regions, as is the case with LTR retrotransposons (Gao et al. 2008; SanMiguel et al. 1998). Under these circumstances, the only DNA TEs capable of achieving high copy numbers are elements such as MITEs, since their small size results in relatively subtle effects on nearby genes (Naito et al. 2009). It is for this same reason that MITEs and other small DNA TEs fail to have dramatic impacts on genome size. Since the gene content of different plant genomes is largely comparable, the amount of target sequences for DNA elements in each genome are similar and this is why DNA TE copy numbers are not proportional to genome size (Fig. 3.4). In contrast, LTR retrotransposons preferentially insert into heterochromatin, where they are nested within each other

Fig. 3.4 The genomic fraction and copy number of DNA and RNA TEs in selected plant species with distinct genome sizes (Sources of original data: *Arabidopsis* (Arabidopsis Genome Initiative 2000; Pereira 2004; Zhang and Wessler 2004), *Brachypodium* (International Brachypodium Initiative 2010), rice (International Rice Genome Sequencing Project 2005), sorghum (Paterson et al. 2009), soybean (Schmutz et al. 2010), and maize (Schnable et al. 2009))

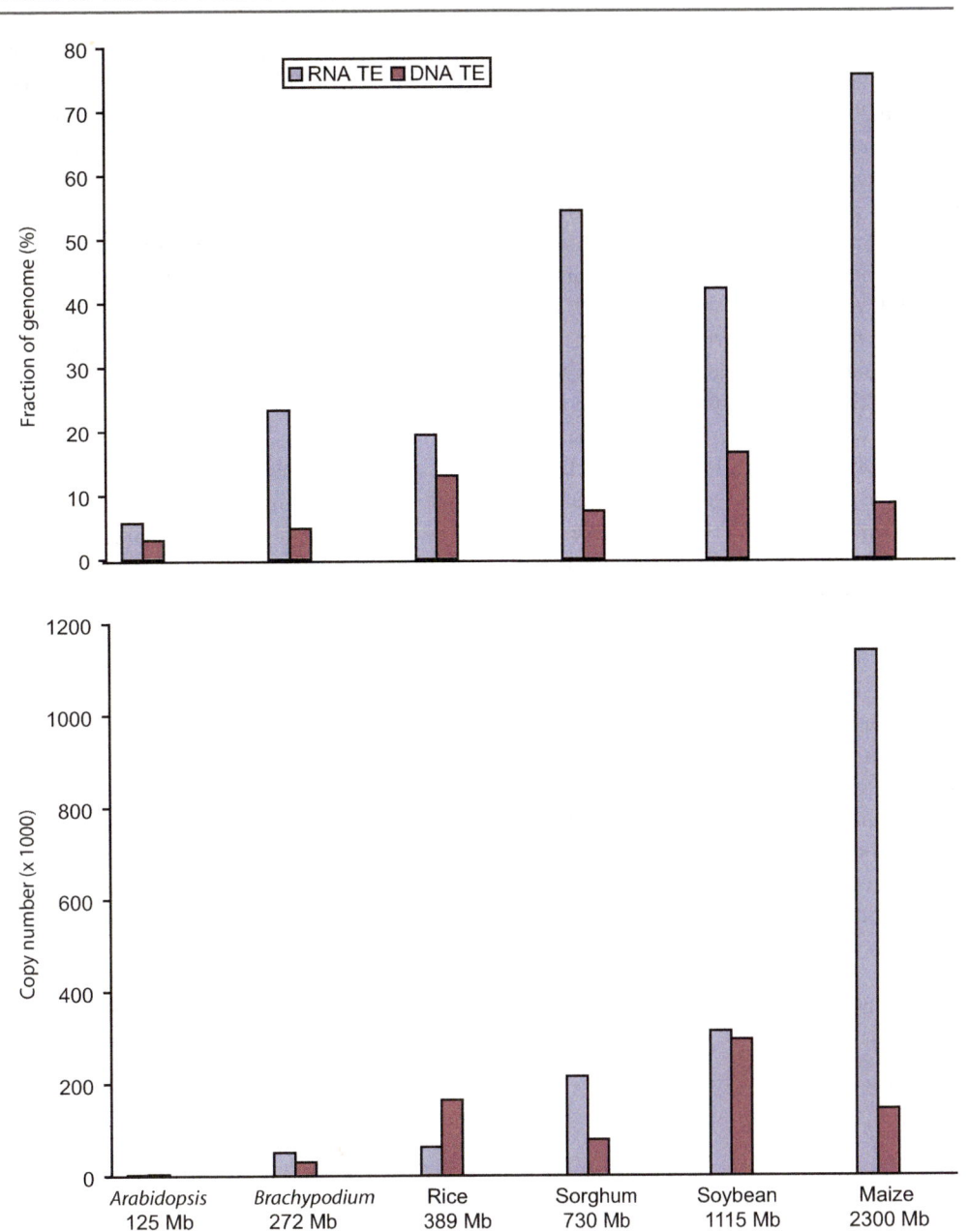

(Gao et al. 2008; SanMiguel et al. 1998). As a consequence, the amplification of LTR elements provides new targets for their future insertion, which forms a positive feedback pathway.

3.4.3 The Consequences of TE Amplification

Although amplification of LTR elements would seem to provide additional TE copies for even more replicative transposition, negative feedbacks may arise as the repetitive elements accumulate. Large genomes are phylogenetically widespread but relatively uncommon numerically, suggesting that there are negative fitness consequences to having

exceptionally large genomes (reviewed in Knight et al. 2005). According to this hypothesis, unchecked success of LTR retrotransposons may lead to the eventual failure of their host species.

In addition to the disadvantages caused by having a large genome size, TE amplification impacts chromatin structure, gene expression, the generation of new genes, and the rate of ectopic homologous recombination (Charlesworth et al. 1994; Kazazian 2004). If homologous copies are located on different chromosomes, recombination may results in translocations that create instability in the genome. The effect of TE amplification on the elements themselves is two fold. First, high copy number will promote the generation of transcripts having different orientations, thus providing precursors for the generation of small RNAs that trigger the silencing mechanism of the host (Feschotte and Pritham 2007; Lisch 2009). After a burst of TE amplification in a genome, defective forms often accumulate, which may titrate the transposition capacity of the functional copies (Hartl et al. 1997; Pinsker et al. 2001; Silva and Kidwell 2000; Slotkin et al. 2005). Both mechanisms lead to the decline of TE activity and the eventual extinction of the element. According to this model, all TEs experience a life cycle of *birth-burst-extinction* (Hartl et al. 1997), and the amplification of a single element is likely finite in a certain genome. Second, recombination and swapping of sequences between elements is dependent on the presence of multiple copies (Du et al. 2010; Jordan and McDonald 1998), and this favors the generation of new element families that may escape extinction. In maize, it was shown that expansion of the CRM1 retrotransposon family is due to the repeated formation of novel recombinant elements derived from two parental retrotransposon genotypes, which may have been brought together during the hybridization of two sympatric species that contribute to the current maize genome (Sharma et al. 2008). Based on this hypothesis, the activity of TEs of a super-family may fluctuate over evolutionary time, but may be retained indefinitely through the generation of new element families in the genome or the introduction of novel elements by horizontal transfer (Cheng et al. 2009; Diao et al. 2006; Feschotte and Pritham 2007; Gilbert et al. 2010; Robertson 2002; Roulin et al. 2008). These dynamics illustrate that the one-way ticket to genomic obesity (Bennetzen and Kellogg 1997) envisioned by unbridled TE proliferation is only transient, being counterbalanced by internal deletion and recombination forces at an evolutionary scale.

3.5 The Influence of TEs on the Evolution of Chromosomes

As described above, TEs can alter the regulation of a gene, create novel genes, and expand genome size. Thus, TEs can also play a role in the creation and regulation of not only suites of genes, but entire chromosomal regions. This role stems from the fact that TEs are not distributed equally in their respective genomes, but are instead clustered in specific chromosomal regions (Fig. 3.5a). This section explores the role of TE abundance on the function of the chromosome and centromere (see also chapter by Hirsch and Jiang 2012, this volume).

3.5.1 TE Composition of the Chromosome and Centromere

In plant genomes retrotransposons are the building blocks of heterochromatin, which is the dense staining and late-replicating portion of the chromosome that is transcriptionally repressed. Figure 3.5a displays the accumulation of

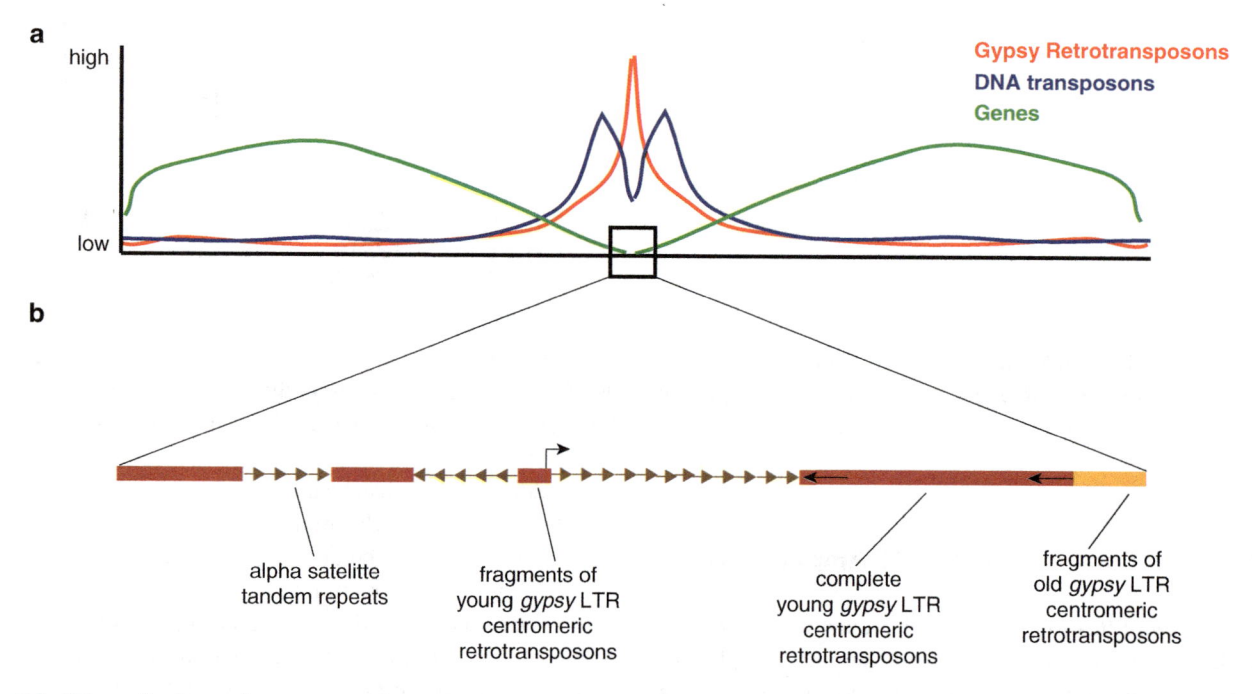

Fig. 3.5 TE contribution to chromosome and centromere structure. (**a**) The distribution of *gypsy* family LTR retrotransposons, DNA transposons and genes on a stereotypical plant chromosome. (**b**) The stereotypical layout of the internal region of a plant centromere. Alpha satellite repeats are the binding sites for the CENP-B protein. Fragments of *gypsy* family LTR retrotransposons drive the expression the satellite repeats (May et al. 2005). Full copies of centromeric retrotransposons, such as CRM and CRR are also located in this region. Older versions of *gypsy* retrotransposons, which no longer occupy the centromere core, are pushed to the pericentromere

gypsy family LTR retrotransposons and DNA transposons on a stereotypical plant chromosome. Of interest is the massive accumulation of both retrotransposons and DNA transposons at the centromere, although the exact location of their accumulation differs between the centromere core and the peripheral regions of the centromere (the pericentromere) (Fig. 3.5a) (Arabidopsis Genome Initiative 2000). This pattern of accumulation is opposite from the density of genes on the chromosome, segregating the chromosome into euchromatic regions (with high gene density and expression) and heterochromatic regions (with high TE density and low expression). However, there can be exceptions to this general trend, such as heterochromatic islands surrounded by gene-rich regions on the chromosomal arms. For example, the heterochromatic knob of *Arabidopsis* chromosome 4 is composed almost entirely of TEs and was formed by a large-scale inversion that translocated a portion of the centromeric heterochromatin out onto the chromosome arm (Fransz et al. 2000).

As mentioned above, plant genomes often use *gypsy* family LTR retrotransposons and alpha satellite repeats as the building blocks of the centromere (Fig. 3.5b) (see also chapter by Hirsch and Jiang 2012, this volume). However, which retrotransposon family is responsible for the expansion and composition of the centromere varies between genera. Two well-studied centromere-specific retrotransposons are the CRR element of rice and the CRM element of maize (Nagaki et al. 2005). These centromeric retrotransposons evolve and amplify quickly in relation to evolutionary time (Nagaki et al. 2003), inserting strictly into centromeric regions and usually into other copies of centromeric retrotransposons (Cheng et al. 2002). Even within the same genus or very closely related species, a different *gypsy* retrotransposon has amplified to take the structural role of the centromere (Gao et al. 2009; Wu et al. 2009). Facilitating this turnover are high rates of retrotransposition as well as satellite repeat expansion via segmental duplication (Ma and Jackson 2006). In addition, high rates of removal of other TEs from the centromere, mostly by illegitimate recombination (Devos et al. 2002), quicken the time a specific TE takes to occupy centromere. Older TE families that once composed the centromere can be found on the flanks in the pericentromere (Wolfgruber et al. 2009). These older TE families are no longer in command of the centromere and have given way to a centromere populated with a newer TE family (Fig. 3.5b). Thus, there is a dynamic and efficient turnover of the retrotransposon family that explodes in activity to take the role of composing the centromere core. Because of this rapid turnover at the centromere, and the essential role chromosome pairing plays in defining a species, the evolution of centromeric TE composition has been speculated to play a key role in the formation of new species (reviewed in Brown and O'Neill 2010).

3.5.2 Functions of TE-Derived Chromosomal Structures Such as Centromeres

The function of heterochromatic islands of TEs is controversial. Some heterochromatic structures, such as chromosomal knobs, have no known function. Other regions of chromosomal heterochromatin, such as centromeric or telomeric heterochromatin, play an essential role in chromosome biology. In yeast and animals, loss of centromere condensation by reduction of conserved condensin complexes results in improper segregation of chromosomes during meiosis (Chan et al. 2004a; Hartl et al. 2008; Yu and Koshland 2005). In fission yeast, the loss of small RNA-maintained heterochromatin condensation results in perturbed centromere function, lagging chromosomes, and potential sterility (Volpe et al. 2003). However, in plants there is less direct evidence for connection between centromere condensation and function. In one prime example, *Tripsacum* natural apomicts have loose centromere condensation (as measured by DAPI staining and histone H3 phosphorylation levels during meiosis) and display large numbers of unreduced gametes (Grimanelli et al. 2003). This finding suggests that the relationship between centromere condensation and function exists, but the effects may not be as apparent as plants are generally more resilient to chromosomal imbalances than animals.

In addition to the presence of CRR and CRM in the centromere (see above Sect. 3.5.1), an important question concerns the specific role that they perform. The CRM retrotransposon interacts with the centromere-specific histone variant CENH3 (Zhong et al. 2002), and this is a sufficient interaction to induce centromere identity and function. However, the mechanism and function of TEs at the centromere remains an enigma. Retrotransposon fragments and promoters drive centromere satellite DNA expression (May et al. 2005), and without this expression, condensation of the centromere is lost. CRM is transcribed (Topp et al. 2004), and it has been speculated that this expression is essential for CENH3 recruitment (Jiang et al. 2003a). However, at what level of regulation the CRM expression drives the CENH3 interaction remains a mystery. CRM transcripts could be turned over by an RdDM or RNAi-like pathway that leads to chromatin condensation, as in fission yeast (Volpe et al. 2002, 2003) and postulated in *Arabidopsis* (May et al. 2005). In addition, the CRM RNA could recruit CENH3 directly, similar to the Xist RNA recruitment of macroH2 in human cells (reviewed in Plath et al. 2002). The CRM *gag* protein could also facilitate the formation of heterochromatin and the kinetochore complex itself, as described in yeast and *Drosophila* (Rashkova et al. 2002; Xie et al. 2001).

The function and accumulation pattern of TEs in the centromere presents a chicken-and-egg dilemma: Is TE silencing required to initiate condensation of the centromere, or does the centromeric location of TEs result in the initiation of their silencing and condensation? Another way to potentially think about this question is whether the genome is being resourceful by using TE islands as *scrap metal* sequence, building buffers between the genes and the centromere, or if the TEs have a specific role in initiating and assembling essential non-protein coding structures such as centromeres? Recent data suggests that new centromeres (or neocentromeres) are devoid of TEs (Gong et al. 2009; Topp et al. 2009). This suggests that TE occupation of centromeres occurs after the initial centromeric activity, potentially to enhance the stability of the centromere by maintaining associations with centromere-specific histone variants (Topp et al. 2009) (see also chapter by Hirsch and Jiang 2012, this volume). If and how TE accumulation is initiating or maintaining centromeric identity is an essential question that needs to be resolved in the coming years.

One function of heterochromatic islands that is understood is their role in silencing invading active TEs. The older TEs that comprise heterochromatin are maintained in a silenced state by RdDM pathways that generate small RNAs. Young invading TEs that match these small RNAs will be silenced, providing the organism with a sort of cellular memory of past TE invasions, and the tools to prevent new invasions. These theories are also used to explain piRNA clusters of TEs in animal genomes (reviewed in Malone and Hannon 2009).

Conclusion

How TEs are viewed is influenced by the timescale. On the generation-to-generation timescale, TEs are efficient mutagens and cellular parasites, which the cell spends energy to silence. On an evolutionary timescale, however, TEs are innovators, producing novel structures and regulation. This chapter has attempted to highlight that plant TEs provide fodder from which novel genes and gene regulation evolve. The contribution of TEs to producing new genes, new gene regulation, and essential chromosomal structures are conserved among plant species and often beyond the plant kingdom. However, key questions remain, such as how centromeric TEs contribute to centromeric function? This question and others will undoubtedly be answered in the coming years. Facilitating this research, the recent explosion of genome sequence and expression data are beginning to make it possible to understand the overall contribution of TEs to genome control. It will be very interesting to determine how many total genes, and how much gene and chromosome regulation, are derived from the repetitive fraction of the genome. From this future research, we expect that TEs will be increasingly appreciated for their dynamic contributions to the evolutionary process, as constructive forces acting to shape plant genomes.

Acknowledgements We thank Cedric Feschotte, Jonathan F. Wendel, Veronica Vallejo and Ann A. Ferguson for critical reading of the manuscript, and Brandon Gaut for sharing data pre-publication. We also thank Steve Jacobsen, Hidetoshi Saze, Laurent Torregrosa, José M. Martinez-Zapater, Esther van der Knaap, Adnane Boualem, and Abdelhafid Bendahmane for sending images of their research. N.J. was supported by grant DBI 0607123 from the National Science Foundation. R.K.S's laboratory is supported by grant MCB-1020499 from the National Science Foundation.

References

Agrawal A, Eastman QM, Schatz DG (1998) Transposition mediated by RAG1 and RAG2 and its implications for the evolution of the immune system. Nature 394:744–751

Ammiraju JS, Lu F, Sanyal A, Yu Y, Song X, Jiang N, Pontaroli AC, Rambo T, Currie J, Collura K, Talag J, Fan C, Goicoechea JL, Zuccolo A, Chen J, Bennetzen JL, Chen M, Jackson S, Wing RA (2008) Dynamic evolution of oryza genomes is revealed by comparative genomic analysis of a genus-wide vertical data set. Plant Cell 20:3191–3209

Arabidopsis Genome Initiative (2000) Analysis of the genome sequence of the flowering plant *Arabidopsis thaliana*. Nature 408:796–815

Avramova Z, Tikhonov A, SanMiguel P, Jin YK, Liu C, Woo SS, Wing RA, Bennetzen JL (1996) Gene identification in a complex chromosomal continuum by local genomic cross-referencing. Plant J 10:1163–1168

Babu MM, Iyer LM, Balaji S, Aravind L (2006) The natural history of the WRKY-GCM1 zinc fingers and the relationship between transcription factors and transposons. Nucleic Acids Res 34:6505–6520

Barkan A, Martienssen RA (1991) Inactivation of maize transposon Mu suppresses a mutant phenotype by activating an outward-reading promoter near the end of Mu1. Proc Natl Acad Sci USA 88: 3502–3506

Barsoum E, Martinez P, Astrom SU (2010) Alpha3, a transposable element that promotes host sexual reproduction. Genes Dev 24:33–44

Baucom RS, Estill JC, Chaparro C, Upshaw N, Jogi A, Deragon JM, Westerman RP, Sanmiguel PJ, Bennetzen JL (2009) Exceptional diversity, non-random distribution, and rapid evolution of retroelements in the B73 maize genome. PLoS Genet 5:e1000732

Bennetzen JL (1998) The structure and evolution of angiosperm nuclear genomes. Curr Opin Plant Biol 1:103–108

Bennetzen JL, Kellogg EA (1997) Do plants have a one-way ticket to genomic obesity? Plant Cell 9:1509–1514

Bennetzen JL, Springer PS (1994) The generation of mutator transposable element subfamilies in maize. Theor Appl Genet 87:657–667

Boeke JD, Corces VG (1989) Transcription and reverse transcription of retrotransposons. Annu Rev Microbiol 43:403–434

Britten R, Davidson E (1969) Gene regulation for higher cells: a theory. Science 165:349–357

Brosius J (1991) Retroposons—seeds of evolution. Science 251:753

Brown JD, O'Neill RJ (2010) Chromosomes, conflict, and epigenetics: chromosomal speciation revisited. Annu Rev Genomics Hum Genet 11:291–316

Brunner S, Pea G, Rafalski A (2005) Origins, genetic organization and transcription of a family of non-autonomous helitron elements in maize. Plant J 43:799–810

Bundock P, Hooykaas P (2005) An *Arabidopsis* hAT-like transposase is essential for plant development. Nature 436:282–284

Bureau TE, White SE, Wessler SR (1994) Transduction of a cellular gene by a plant retroelement. Cell 77:479–480

C. elegans Sequencing Consortium (1998) Genome sequence of the nematode *C. elegans*: a platform for investigating biology. Science 282:2012–2018

Cao X, Aufsatz W, Zilberman D, Mette MF, Huang MS, Matzke M, Jacobsen SE (2003) Role of the DRM and CMT3 methyltransferases in RNA-directed DNA methylation. Curr Biol 13:2212–2217

Chan RC, Severson AF, Meyer BJ (2004a) Condensin restructures chromosomes in preparation for meiotic divisions. J Cell Biol 167:613–625

Chan SW, Zilberman D, Xie Z, Johansen LK, Carrington JC, Jacobsen SE (2004b) RNA silencing genes control de novo DNA methylation. Science 303:1336

Charlesworth B, Sniegowski P, Stephan W (1994) The evolutionary dynamics of repetitive DNA in eukaryotes. Nature 371:215–220

Cheng Z, Dong F, Langdon T, Ouyang S, Buell CR, Gu M, Blattner FR, Jiang J (2002) Functional rice centromeres are marked by a satellite repeat and a centromere-specific retrotransposon. Plant Cell 14:1691–1704

Cheng X, Zhang D, Cheng Z, Keller B, Ling HQ (2009) A new family of Ty1-copia-like retrotransposons originated in the tomato genome by a recent horizontal transfer event. Genetics 181:1183–1193

Conant GC, Wolfe KH (2008) Turning a hobby into a job: how duplicated genes find new functions. Nat Rev Genet 9:938–950

Cowan RK, Hoen DR, Schoen DJ, Bureau TE (2005) MUSTANG is a novel family of domesticated transposase genes found in diverse angiosperms. Mol Biol Evol 22:2084–2089

Devos KM, Brown JK, Bennetzen JL (2002) Genome size reduction through illegitimate recombination counteracts genome expansion in *Arabidopsis*. Genome Res 12:1075–1079

Diao X, Freeling M, Lisch D (2006) Horizontal transfer of a plant transposon. PLoS Biol 4:e5

Dietrich CR, Cui F, Packila ML, Li J, Ashlock DA, Nikolau BJ, Schnable PS (2002) Maize Mu transposons are targeted to the 5' untranslated region of the gl8 gene and sequences flanking Mu target-site duplications exhibit nonrandom nucleotide composition throughout the genome. Genetics 160:697–716

Doyle JJ, Flagel LE, Paterson AH, Rapp RA, Soltis DE, Soltis PS, Wendel JF (2008) Evolutionary genetics of genome merger and doubling in plants. Annu Rev Genet 42:443–461

Du C, Fefelova N, Caronna J, He L, Dooner HK (2009) The polychromatic Helitron landscape of the maize genome. Proc Natl Acad Sci USA 106:19916–19921

Du J, Tian Z, Bowen NJ, Schmutz J, Shoemaker RC, Ma J (2010) Bifurcation and enhancement of autonomous-nonautonomous retrotransposon partnership through LTR Swapping in soybean. Plant Cell 22:48–61

Elrouby N, Bureau TE (2001) A novel hybrid open reading frame formed by multiple cellular gene transductions by a plant long terminal repeat retroelement. J Biol Chem 276:41963–41968

Elrouby N, Bureau TE (2010) Bs1, a new chimeric gene formed by retrotransposon-mediated exon shuffling in maize. Plant Physiol 153:1413–1424

Engels WR, Johnson-Schlitz DM, Eggleston WB, Sved J (1990) High-frequency P element loss in *Drosophila* is homolog dependent. Cell 62:515–525

Fernandez L, Torregrosa L, Segura V, Bouquet A, Martinez-Zapater JM (2010) Transposon-induced gene activation as a mechanism generating cluster shape somatic variation in grapevine. Plant J 61:545–557

Feschotte C, Pritham EJ (2007) DNA transposons and the evolution of eukaryotic genomes. Annu Rev Genet 41:331–368

Feschotte C, Pritham EJ (2009) A cornucopia of Helitrons shapes the maize genome. Proc Natl Acad Sci USA 106:19747–19748

Flagel LE, Wendel JF (2009) Gene duplication and evolutionary novelty in plants. New Phytol 183:557–564

Fransz PF, Armstrong S, de Jong JH, Parnell LD, van Drunen C, Dean C, Zabel P, Bisseling T, Jones GH (2000) Integrated cytogenetic map of chromosome arm 4S of *A. thaliana*: structural organization of heterochromatic knob and centromere region. Cell 100:367–376

Fujimoto R, Kinoshita Y, Kawabe A, Kinoshita T, Takashima K, Nordborg M, Nasrallah ME, Shimizu KK, Kudoh H, Kakutani T (2008) Evolution and control of imprinted FWA genes in the genus *Arabidopsis*. PLoS Genet 4:e1000048

Gao X, Hou Y, Ebina H, Levin HL, Voytas DF (2008) Chromodomains direct integration of retrotransposons to heterochromatin. Genome Res 18:359–369

Gao D, Gill N, Kim HR, Walling JG, Zhang W, Fan C, Yu Y, Ma J, SanMiguel P, Jiang N, Cheng Z, Wing RA, Jiang J, Jackson SA (2009) A lineage-specific centromere retrotransposon in *Oryza brachyantha*. Plant J 60:820–831

Ge S, Sang T, Lu BR, Hong DY (1999) Phylogeny of rice genomes with emphasis on origins of allotetraploid species. Proc Natl Acad Sci USA 96:14400–14405

Gehring M, Bubb KL, Henikoff S (2009) Extensive demethylation of repetitive elements during seed development underlies gene imprinting. Science 324:1447–1451

Geyer PK, Corces VG (1992) DNA position-specific repression of transcription by a *Drosophila* zinc finger protein. Genes Dev 6:1865–1873

Gilbert C, Schaack S, Pace JK 2nd, Brindley PJ, Feschotte C (2010) A role for host–parasite interactions in the horizontal transfer of transposons across phyla. Nature 464:1347–1350

Girard L, Freeling M (2000) Mutator-suppressible alleles of rough sheath1 and liguleless3 in maize reveal multiple mechanisms for suppression. Genetics 154:437–446

Gong Z, Yu H, Huang J, Yi C, Gu M (2009) Unstable transmission of rice chromosomes without functional centromeric repeats in asexual propagation. Chromosome Res 17:863–872

Grimanelli D, Garcia M, Kaszas E, Perotti E, Leblanc O (2003) Heterochronic expression of sexual reproductive programs during apomictic development in *Tripsacum*. Genetics 165:1521–1531

Hanada K, Vallejo V, Nobuta K, Slotkin RK, Lisch D, Meyers BC, Shiu SH, Jiang N (2009) The functional role of pack-MULEs in rice inferred from purifying selection and expression profile. Plant Cell 21:25–38

Hartl DL, Lozovskaya ER, Nurminsky DI, Lohe AR (1997) What restricts the activity of mariner-like transposable elements? Trends Genet 13:197–201

Hartl TA, Sweeney SJ, Knepler PJ, Bosco G (2008) Condensin II resolves chromosomal associations to enable anaphase I segregation in *Drosophila* male meiosis. PLoS Genet 4:e1000228

Hawkins JS, Kim H, Nason JD, Wing RA, Wendel JF (2006) Differential lineage-specific amplification of transposable elements is responsible for genome size variation in *Gossypium*. Genome Res 16:1252–1261

Hawkins JS, Grover CE, Wendel JF (2008a) Repeated big bangs and the expanding universe: directionality in plant genome size evolution. Plant Sci 174:557–562

Hawkins JS, Hu G, Rapp RA, Grafenberg JL, Wendel JF (2008b) Phylogenetic determination of the pace of transposable element proliferation in plants: copia and LINE-like elements in *Gossypium*. Genome 51:11–18

Hawkins JS, Proulx SR, Rapp RA, Wendel JF (2009) Rapid DNA loss as a counterbalance to genome expansion through retrotransposon proliferation in plants. Proc Natl Acad Sci USA 106:17811–17816

Hayashi K, Yoshida H (2009) Refunctionalization of the ancient rice blast disease resistance gene Pit by the recruitment of a retrotransposon as a promoter. Plant J 57:413–425

Hirsch C, Jiang J (2012) Centromeres: sequences, structure, and biology. In: Wendel JF (ed) Plant genome diversity, vol 1, Plant genomes, their residents, and their evolutionary dynamics. Springer, Wien, New York

Hoen DR, Park KC, Elrouby N, Yu Z, Mohabir N, Cowan RK, Bureau TE (2006) Transposon-mediated expansion and diversification of a family of ULP-like genes. Mol Biol Evol 23:1254–1268

Holligan D, Zhang X, Jiang N, Pritham EJ, Wessler SR (2006) The transposable element landscape of the model legume *Lotus japonicus*. Genetics 174:2215–2228

Hollister JD, Gaut BS (2007) Population and evolutionary dynamics of Helitron transposable elements in *Arabidopsis thaliana*. Mol Biol Evol 24:2515–2524

Hollister JD, Gaut BS (2009) Epigenetic silencing of transposable elements: a trade-off between reduced transposition and deleterious effects on neighboring gene expression. Genome Res 19:1419–1428

Hollister JD, Smith LA, Guo Y-L, Ott F, Weigel D, Gaut BS (2011) Transposable elements and small RNAs contribute to gene expression divergence between *Arabidopsis thaliana* and *Arabidopsis lyrata*. Proc Natl Acad Sci USA 108:2322–2327

Hsia AP, Schnable PS (1996) DNA sequence analyses support the role of interrupted gap repair in the origin of internal deletions of the maize transposon, MuDR. Genetics 142:603–618

Hsieh TF, Ibarra CA, Silva P, Zemach A, Eshed-Williams L, Fischer RL, Zilberman D (2009) Genome-wide demethylation of *Arabidopsis* endosperm. Science 324:1451–1454

Hudson ME, Lisch DR, Quail PH (2003) The FHY3 and FAR1 genes encode transposase-related proteins involved in regulation of gene expression by the phytochrome A-signaling pathway. Plant J 34:453–471

International Brachypodium Initiative (2010) Genome sequencing and analysis of the model grass *Brachypodium distachyon*. Nature 463:763–768

International Rice Genome Sequencing Project (2005) The map-based sequence of the rice genome. Nature 436:793–800

Jacobsen SE, Sakai H, Finnegan EJ, Cao X, Meyerowitz EM (2000) Ectopic hypermethylation of flower-specific genes in *Arabidopsis*. Curr Biol 10:179–186

Jiang J, Birchler JA, Parrott WA, Dawe RK (2003a) A molecular view of plant centromeres. Trends Plant Sci 8:570–575

Jiang N, Bao Z, Zhang X, Hirochika H, Eddy SR, McCouch SR, Wessler SR (2003b) An active DNA transposon family in rice. Nature 421:163–167

Jiang N, Bao Z, Zhang X, Eddy SR, Wessler SR (2004) Pack-MULE transposable elements mediate gene evolution in plants. Nature 431:569–573

Jiang N, Gao D, Xiao H, van der Knaap E (2009) Genome organization of the tomato sun locus and characterization of the unusual retrotransposon Rider. Plant J 60:181–193

Jiang N, Ferguson AA, Slotkin RK, Lisch D (2011) Pack-Mutator-like transposable elements (Pack-MULEs) induce directional modification of genes through biased insertion and DNA acquisition. Proc Natl Acad Sci USA 108:1537–1542

Jiao Y, Deng XW (2007) A genome-wide transcriptional activity survey of rice transposable element-related genes. Genome Biol 8:R28

Jin YK, Bennetzen JL (1994) Integration and nonrandom mutation of a plasma membrane proton ATPase gene fragment within the Bs1 retroelement of maize. Plant Cell 6:1177–1186

Jones JD, Dangl JL (2006) The plant immune system. Nature 444:323–329

Jordan IK, McDonald JF (1998) Evidence for the role of recombination in the regulatory evolution of *Saccharomyces cerevisiae* Ty elements. J Mol Evol 47:14–20

Juretic N, Hoen DR, Huynh ML, Harrison PM, Bureau TE (2005) The evolutionary fate of MULE-mediated duplications of host gene fragments in rice. Genome Res 15:1292–1297

Kapitonov VV, Jurka J (2001) Rolling-circle transposons in eukaryotes. Proc Natl Acad Sci USA 98:8714–8719

Kawasaki S, Nitasaka E (2004) Characterization of Tpn1 family in the Japanese morning glory: En/Spm-related transposable elements capturing host genes. Plant Cell Physiol 45:933–944

Kazazian HH Jr (2004) Mobile elements: drivers of genome evolution. Science 303:1626–1632

Kejnovsky E, Hawkins J, Feschotte C (2012) Plant transposable elements: biology and evolution. In: Wendel JF (ed) Plant genome diversity, vol 1, Plant genomes, their residents, and their evolutionary dynamics. Springer, Wien, New York

Kikuchi K, Terauchi K, Wada M, Hirano HY (2003) The plant MITE mPing is mobilized in anther culture. Nature 421:167–170

Kinoshita T, Miura A, Choi Y, Kinoshita Y, Cao X, Jacobsen SE, Fischer RL, Kakutani T (2004) One-way control of FWA imprinting in *Arabidopsis* endosperm by DNA methylation. Science 303:521–523

Kloeckener-Gruissem B, Freeling M (1995) Transposon-induced promoter scrambling: a mechanism for the evolution of new alleles. Proc Natl Acad Sci USA 92:1836–1840

Knight CA, Molinari NA, Petrov DA (2005) The large genome constraint hypothesis: evolution, ecology and phenotype. Ann Bot 95:177–190

Kumar A, Bennetzen JL (1999) Plant retrotransposons. Annu Rev Genet 33:479–532

Lai J, Li Y, Messing J, Dooner HK (2005) Gene movement by Helitron transposons contributes to the haplotype variability of maize. Proc Natl Acad Sci USA 102:9068–9073

Lal SK, Giroux MJ, Brendel V, Vallejos CE, Hannah LC (2003) The maize genome contains a helitron insertion. Plant Cell 15:381–391

Lander ES, Linton LM, Birren B, Nusbaum C, Zody MC, Baldwin J, Devon K, Dewar K, Doyle M, FitzHugh W, Funke R, Gage D, Harris K, Heaford A, Howland J, Kann L, Lehoczky J, LeVine R, McEwan P, McKernan K, Meldrim J, Mesirov JP, Miranda C, Morris W, Naylor J, Raymond C, Rosetti M, Santos R, Sheridan A, Sougnez C, Stange-Thomann N, Stojanovic N, Subramanian A, Wyman D, Rogers J, Sulston J, Ainscough R, Beck S, Bentley D, Burton J, Clee C, Carter N, Coulson A, Deadman R, Deloukas P, Dunham A, Dunham I, Durbin R, French L, Grafham D, Gregory S, Hubbard T, Humphray S, Hunt A, Jones M, Lloyd C, McMurray A, Matthews L, Mercer S, Milne S, Mullikin JC, Mungall A, Plumb R, Ross M, Shownkeen R, Sims S, Waterston RH, Wilson RK, Hillier LW, McPherson JD, Marra MA, Mardis ER, Fulton LA, Chinwalla AT, Pepin KH, Gish WR, Chissoe SL, Wendl MC, Delehaunty KD, Miner TL, Delehaunty A, Kramer JB, Cook LL, Fulton RS, Johnson DL, Minx PJ, Clifton SW, Hawkins T, Branscomb E, Predki P, Richardson P, Wenning S, Slezak T, Doggett N, Cheng JF, Olsen A, Lucas S, Elkin C, Uberbacher E, Frazier M, Gibbs RA, Muzny DM, Scherer SE, Bouck JB, Sodergren EJ, Worley KC, Rives CM, Gorrell JH, Metzker ML, Naylor SL, Kucherlapati RS, Nelson DL, Weinstock GM, Sakaki Y, Fujiyama A, Hattori M, Yada T, Toyoda A, Itoh T, Kawagoe C, Watanabe H, Totoki Y, Taylor T, Weissenbach J, Heilig R, Saurin W, Artiguenave F, Brottier P, Bruls T, Pelletier E, Robert C, Wincker P, Smith DR, Doucette-Stamm L, Rubenfield M, Weinstock K, Lee HM, Dubois J, Rosenthal A, Platzer M, Nyakatura G, Taudien S, Rump A, Yang H, Yu J, Wang J, Huang G, Gu J, Hood L, Rowen L, Madan A, Qin S, Davis RW, Federspiel NA, Totoki Y, Taylor T, Weissenbach J, Heilig R, Saurin W, Artiguenave F, Brottier P, Bruls T, Pelletier E, Robert C, Wincker P, Smith DR, Doucette-Stamm L, Rubenfield M, Weinstock K, Lee HM, Dubois J, Rosenthal A, Platzer M, Nyakatura G, Taudien S,

Rump A, Yang H, Yu J, Wang J, Huang G, Gu J, Hood L, Rowen L, Madan A, Qin S, Davis RW, Federspiel NA, Totoki Y, Taylor T, Weissenbach J, Heilig R, Saurin W, Artiguenave F, Brottier P, Bruls T, Pelletier E, Robert C, Wincker P, Smith DR, Doucette-Stamm L, Rubenfield M, Weinstock K, Lee HM, Dubois J, Rosenthal A, Platzer M, Nyakatura G, Taudien S, Rump A, Yang H, Yu J, Wang J, Huang G, Gu J, Hood L, Rowen L, Madan A, Qin S, Davis RW, Federspiel NA, Abola AP, Proctor MJ, Myers RM, Schmutz J, Dickson M, Grimwood J, Cox DR, Olson MV, Kaul R, Raymond C, Shimizu N, Kawasaki K, Minoshima S, Evans GA, Athanasiou M, Schultz R, Roe BA, Chen F, Pan H, Ramser J, Lehrach H, Reinhardt R, McCombie WR, de la Bastide M, Dedhia N, Blocker H, Hornischer K, Nordsiek G, Agarwala R, Aravind L, Bailey JA, Bateman A, Batzoglou S, Birney E, Bork P, Brown DG, Burge CB, Cerutti L, Chen HC, Church D, Clamp M, Copley RR, Doerks T, Eddy SR, Eichler EE, Furey TS, Galagan J, Gilbert JG, Harmon C, Hayashizaki Y, Haussler D, Hermjakob H, Hokamp K, Jang W, Johnson LS, Jones TA, Kasif S, Kaspryzk A, Kennedy S, Kent WJ, Kitts P, Koonin EV, Korf I, Kulp D, Lancet D, Lowe TM, McLysaght A, Mikkelsen T, Moran JV, Mulder N, Pollara VJ, Ponting CP, Schuler G, Schultz J, Slater G, Smit AF, Stupka E, Szustakowski J, Thierry-Mieg D, Thierry-Mieg J, Wagner L, Wallis J, Wheeler R, Williams A, Wolf YI, Wolfe KH, Yang SP, Yeh RF, Collins F, Guyer MS, Peterson J, Felsenfeld A, Wetterstrand KA, Patrinos A, Morgan MJ, de Jong P, Catanese JJ, Osoegawa K, Shizuya H, Choi S, Chen YJ (2001) Initial sequencing and analysis of the human genome. Nature 409:860–921

Lawton-Rauh A (2003) Evolutionary dynamics of duplicated genes in plants. Mol Phylogenet Evol 29:396–409

Li W-H (1997) Molecular evolution. Sinauer, Sunderland

Lin R, Ding L, Casola C, Ripoll DR, Feschotte C, Wang H (2007) Transposase-derived transcription factors regulate light signaling in *Arabidopsis*. Science 318:1302–1305

Lindblad-Toh K, Wade CM, Mikkelsen TS, Karlsson EK, Jaffe DB, Kamal M, Clamp M, Chang JL, Kulbokas EJ 3rd, Zody MC, Mauceli E, Xie X, Breen M, Wayne RK, Ostrander EA, Ponting CP, Galibert F, Smith DR, DeJong PJ, Kirkness E, Alvarez P, Biagi T, Brockman W, Butler J, Chin CW, Cook A, Cuff J, Daly MJ, DeCaprio D, Gnerre S, Grabherr M, Kellis M, Kleber M, Bardeleben C, Goodstadt L, Heger A, Hitte C, Kim L, Koepfli KP, Parker HG, Pollinger JP, Searle SM, Sutter NB, Thomas R, Webber C, Baldwin J, Abebe A, Abouelleil A, Aftuck L, Ait-Zahra M, Aldredge T, Allen N, An P, Anderson S, Antoine C, Arachchi H, Aslam A, Ayotte L, Bachantsang P, Barry A, Bayul T, Benamara M, Berlin A, Bessette D, Blitshteyn B, Bloom T, Blye J, Boguslavskiy L, Bonnet C, Boukhgalter B, Brown A, Cahill P, Calixte N, Camarata J, Cheshatsang Y, Chu J, Citroen M, Collymore A, Cooke P, Dawoe T, Daza R, Decktor K, DeGray S, Dhargay N, Dooley K, Dooley K, Dorje P, Dorjee K, Dorris L, Duffey N, Dupes A, Egbiremolen O, Elong R, Falk J, Farina A, Faro S, Ferguson D, Ferreira P, Fisher S, FitzGerald M, Foley K, Foley C, Franke A, Friedrich D, Gage D, Garber M, Gearin G, Giannoukos G, Goode T, Goyette A, Graham J, Grandbois E, Gyaltsen K, Hafez N, Hagopian D, Hagos B, Hall J, Healy C, Hegarty R, Honan T, Horn A, Houde N, Hughes L, Hunnicutt L, Husby M, Jester B, Jones C, Kamat A, Kanga B, Kells C, Khazanovich D, Kieu AC, Kisner P, Kumar M, Lance K, Landers T, Lara M, Lee W, Leger JP, Lennon N, Leuper L, LeVine S, Liu J, Liu X, Lokyitsang Y, Lokyitsang T, Lui A, Macdonald J, Major J, Marabella R, Maru K, Matthews C, McDonough S, Mehta T, Meldrim J, Melnikov A, Meneus L, Mihalev A, Mihova T, Miller K, Mittelman R, Mlenga V, Mulrain L, Munson G, Navidi A, Naylor J, Nguyen T, Nguyen N, Nguyen C, Nguyen T, Nicol R, Norbu N, Norbu C, Novod N, Nyima T, Olandt P, O'Neill B, O'Neill K, Osman S, Oyono L, Patti C, Perrin D, Phunkhang P, Pierre F, Priest M, Rachupka A, Raghuraman S, Rameau R, Ray V, Raymond C, Rege F, Rise C, Rogers J, Rogov P, Sahalie J, Settipalli S, Sharpe T, Shea T, Sheehan M, Sherpa N, Shi J, Shih D, Sloan J, Smith C, Sparrow T, Stalker J, Stange-Thomann N, Stavropoulos S, Stone C, Stone S, Sykes S, Tchuinga P, Tenzing P, Tesfaye S, Thoulutsang D, Thoulutsang Y, Topham K, Topping I, Tsamla T, Vassiliev H, Venkataraman V, Vo A, Wangchuk T, Wangdi T, Weiand M, Wilkinson J, Wilson A, Yadav S, Yang S, Yang X, Young G, Yu Q, Zainoun J, Zembek L, Zimmer A, Lander ES (2005) Genome sequence, comparative analysis and haplotype structure of the domestic dog. Nature 438:803–819

Lippman Z, Gendrel AV, Black M, Vaughn MW, Dedhia N, McCombie WR, Lavine K, Mittal V, May B, Kasschau KD, Carrington JC, Doerge RW, Colot V, Martienssen R (2004) Role of transposable elements in heterochromatin and epigenetic control. Nature 430:471–476

Lisch D (2002) Mutator transposons. Trends Plant Sci 7:498–504

Lisch D (2009) Epigenetic regulation of transposable elements in plants. Annu Rev Plant Biol 60:43–66

Liu J, He Y, Amasino R, Chen X (2004) siRNAs targeting an intronic transposon in the regulation of natural flowering behavior in *Arabidopsis*. Genes Dev 18:2873–2878

Liu S, Yeh CT, Ji T, Ying K, Wu H, Tang HM, Fu Y, Nettleton D, Schnable PS (2009) Mu transposon insertion sites and meiotic recombination events co-localize with epigenetic marks for open chromatin across the maize genome. PLoS Genet 5:e1000733

Luck JE, Lawrence GJ, Finnegan EJ, Jones DA, Ellis JG (1998) A flax transposon identified in two spontaneous mutant alleles of the L6 rust resistance gene. Plant J 16:365–369

Luo D, Coen ES, Doyle S, Carpenter R (1991) Pigmentation mutants produced by transposon mutagenesis in *Antirrhinum majus*. Plant J 1:59–69

Ma J, Jackson SA (2006) Retrotransposon accumulation and satellite amplification mediated by segmental duplication facilitate centromere expansion in rice. Genome Res 16:251–259

Ma J, Devos KM, Bennetzen JL (2004) Analyses of LTR-retrotransposon structures reveal recent and rapid genomic DNA loss in rice. Genome Res 14:860–869

Malone CD, Hannon GJ (2009) Small RNAs as guardians of the genome. Cell 136:656–668

Martienssen RA, Barkan A, Freeling M, Taylor WC (1989) Molecular cloning of a maize gene involved in photosynthetic membrane organization that is regulated by Robertson's Mutator. EMBO J 8:1633–1639

Martienssen R, Barkan A, Taylor WC, Freeling M (1990) Somatically heritable switches in the DNA modification of Mu transposable elements monitored with a suppressible mutant in maize. Genes Dev 4:331–343

Martin A, Troadec C, Boualem A, Rajab M, Fernandez R, Morin H, Pitrat M, Dogimont C, Bendahmane A (2009) A transposon-induced epigenetic change leads to sex determination in melon. Nature 461:1135–1138

Masson P, Surosky R, Kingsbury JA, Fedoroff NV (1987) Genetic and molecular analysis of the Spm-dependent a-m2 alleles of the maize a locus. Genetics 117:117–137

May BP, Lippman ZB, Fang Y, Spector DL, Martienssen RA (2005) Differential regulation of strand-specific transcripts from *Arabidopsis* centromeric satellite repeats. PLoS Genet 1:e79

McClintock B (1956) Controlling elements and the gene. Cold Spring Harbor Symp Quant Biol 21:197–216

McClintock B (1971) The contribution of one component of a control system to versatility of gene expression. Carnegie Inst Wash Year Book 70:5–17

Moran JV, DeBerardinis RJ, Kazazian HH Jr (1999) Exon shuffling by L1 retrotransposition. Science 283:1530–1534

Morgante M, Brunner S, Pea G, Fengler K, Zuccolo A, Rafalski A (2005) Gene duplication and exon shuffling by helitron-like transposons generate intraspecies diversity in maize. Nat Genet 37:997–1002

Muehlbauer GJ, Bhau BS, Syed NH, Heinen S, Cho S, Marshall D, Pateyron S, Buisine N, Chalhoub B, Flavell AJ (2006) A hAT superfamily transposase recruited by the cereal grass genome. Mol Genet Genomics 275:553–563

Nagaki K, Song J, Stupar RM, Parokonny AS, Yuan Q, Ouyang S, Liu J, Hsiao J, Jones KM, Dawe RK, Buell CR, Jiang J (2003) Molecular and cytological analyses of large tracks of centromeric DNA reveal the structure and evolutionary dynamics of maize centromeres. Genetics 163:759–770

Nagaki K, Neumann P, Zhang D, Ouyang S, Buell C, Cheng Z, Jiang J (2005) Structure, divergence, and distribution of the CRR centromeric retrotransposon family in rice. Mol Biol Evol 22:845–855

Naito K, Zhang F, Tsukiyama T, Saito H, Hancock CN, Richardson AO, Okumoto Y, Tanisaka T, Wessler SR (2009) Unexpected consequences of a sudden and massive transposon amplification on rice gene expression. Nature 461:1130–1134

Nakazaki T, Okumoto Y, Horibata A, Yamahira S, Teraishi M, Nishida H, Inoue H, Tanisaka T (2003) Mobilization of a transposon in the rice genome. Nature 421:170–172

Nassif N, Penney J, Pal S, Engels WR, Gloor GB (1994) Efficient copying of nonhomologous sequences from ectopic sites via P-element-induced gap repair. Mol Cell Biol 14:1613–1625

Negroni M, Buc H (2001) Retroviral recombination: what drives the switch? Nat Rev Mol Cell Biol 2:151–155

Oki N, Yano K, Okumoto Y, Tsukiyama T, Teraishi M, Tanisaka T (2008) A genome-wide view of miniature inverted-repeat transposable elements (MITEs) in rice, Oryza sativa ssp. japonica. Genes Genet Syst 83:321–329

Paterson AH, Bowers JE, Bruggmann R, Dubchak I, Grimwood J, Gundlach H, Haberer G, Hellsten U, Mitros T, Poliakov A, Schmutz J, Spannagl M, Tang H, Wang X, Wicker T, Bharti AK, Chapman J, Feltus FA, Gowik U, Grigoriev IV, Lyons E, Maher CA, Martis M, Narechania A, Otillar RP, Penning BW, Salamov AA, Wang Y, Zhang L, Carpita NC, Freeling M, Gingle AR, Hash CT, Keller B, Klein P, Kresovich S, McCann MC, Ming R, Peterson DG, Mehboob-ur R, Ware D, Westhoff P, Mayer KF, Messing J, Rokhsar DS (2009) The Sorghum bicolor genome and the diversification of grasses. Nature 457:551–556

Pereira V (2004) Insertion bias and purifying selection of retrotransposons in the Arabidopsis thaliana genome. Genome Biol 5:R79

Piegu B, Guyot R, Picault N, Roulin A, Saniyal A, Kim H, Collura K, Brar DS, Jackson S, Wing RA, Panaud O (2006) Doubling genome size without polyploidization: dynamics of retrotransposition-driven genomic expansions in Oryza australiensis, a wild relative of rice. Genome Res 16:1262–1269

Pillot M, Baroux C, Vazquez MA, Autran D, Leblanc O, Vielle-Calzada JP, Grossniklaus U, Grimanelli D (2010) Embryo and endosperm inherit distinct chromatin and transcriptional states from the female gametes in Arabidopsis. Plant Cell 22:307–320

Pinsker W, Haring E, Hagemann S, Miller WJ (2001) The evolutionary life history of P transposons: from horizontal invaders to domesticated neogenes. Chromosoma 110:148–158

Plath K, Mlynarczyk-Evans S, Nusinow DA, Panning B (2002) Xist RNA and the mechanism of X chromosome inactivation. Annu Rev Genet 36:233–278

Pontius JU, Mullikin JC, Smith DR, Lindblad-Toh K, Gnerre S, Clamp M, Chang J, Stephens R, Neelam B, Volfovsky N, Schaffer AA, Agarwala R, Narfstrom K, Murphy WJ, Giger U, Roca AL, Antunes A, Menotti-Raymond M, Yuhki N, Pecon-Slattery J, Johnson WE, Bourque G, Tesler G, O'Brien SJ (2007) Initial sequence and comparative analysis of the cat genome. Genome Res 17: 1675–1689

Raffaele S, Farrer RA, Cano LM, Studholme DJ, MacLean D, Thines M, Jiang RH, Zody MC, Kunjeti SG, Donofrio NM, Meyers BC, Nusbaum C, Kamoun S (2010) Genome evolution following host jumps in the Irish potato famine pathogen lineage. Science 330:1540–1543

Rashkova S, Karam SE, Pardue ML (2002) Element-specific localization of Drosophila retrotransposon Gag proteins occurs in both nucleus and cytoplasm. Proc Natl Acad Sci USA 99:3621–3626

Robbins ML, Sekhon RS, Meeley R, Chopra S (2008) A Mutator transposon insertion is associated with ectopic expression of a tandemly repeated multicopy Myb gene pericarp color1 of maize. Genetics 178:1859–1874

Robertson HM (2002) Evolution of DNA transposons in eukaryote. In: Craig N, Craigie R, Gellert M, Lambowitz A (eds) Mobile DNA II. ASM Press, Washington, DC, pp 1093–1110

Roulin A, Piegu B, Wing RA, Panaud O (2008) Evidence of multiple horizontal transfers of the long terminal repeat retrotransposon RIRE1 within the genus Oryza. Plant J 53:950–959

Rubin E, Levy AA (1997) Abortive gap repair: underlying mechanism for Ds element formation. Mol Cell Biol 17:6294–6302

SanMiguel P, Tikhonov A, Jin YK, Motchoulskaia N, Zakharov D, Melake-Berhan A, Springer PS, Edwards KJ, Lee M, Avramova Z, Bennetzen JL (1996) Nested retrotransposons in the intergenic regions of the maize genome. Science 274:765–768

SanMiguel P, Gaut BS, Tikhonov A, Nakajima Y, Bennetzen JL (1998) The paleontology of intergene retrotransposons of maize. Nat Genet 20:43–45

Saze H, Kakutani T (2007) Heritable epigenetic mutation of a transposon-flanked Arabidopsis gene due to lack of the chromatin-remodeling factor DDM1. EMBO J 26:3641–3652

Schmutz J, Cannon SB, Schlueter J, Ma J, Mitros T, Nelson W, Hyten DL, Song Q, Thelen JJ, Cheng J, Xu D, Hellsten U, May GD, Yu Y, Sakurai T, Umezawa T, Bhattacharyya MK, Sandhu D, Valliyodan B, Lindquist E, Peto M, Grant D, Shu S, Goodstein D, Barry K, Futrell-Griggs M, Abernathy B, Du J, Tian Z, Zhu L, Gill N, Joshi T, Libault M, Sethuraman A, Zhang XC, Shinozaki K, Nguyen HT, Wing RA, Cregan P, Specht J, Grimwood J, Rokhsar D, Stacey G, Shoemaker RC, Jackson SA (2010) Genome sequence of the palaeopolyploid soybean. Nature 463:178–183

Schnable PS, Ware D, Fulton RS, Stein JC, Wei F, Pasternak S, Liang C, Zhang J, Fulton L, Graves TA, Minx P, Reily AD, Courtney L, Kruchowski SS, Tomlinson C, Strong C, Delehaunty K, Fronick C, Courtney B, Rock SM, Belter E, Du F, Kim K, Abbott RM, Cotton M, Levy A, Marchetto P, Ochoa K, Jackson SM, Gillam B, Chen W, Yan L, Higginbotham J, Cardenas M, Waligorski J, Applebaum E, Phelps L, Falcone J, Kanchi K, Thane T, Scimone A, Thane N, Henke J, Wang T, Ruppert J, Shah N, Rotter K, Hodges J, Ingenthron E, Cordes M, Kohlberg S, Sgro J, Delgado B, Mead K, Chinwalla A, Leonard S, Crouse K, Collura K, Kudrna D, Currie J, He R, Angelova A, Rajasekar S, Mueller T, Lomeli R, Scara G, Ko A, Delaney K, Wissotski M, Lopez G, Campos D, Braidotti M, Ashley E, Golser W, Kim H, Lee S, Lin J, Dujmic Z, Kim W, Talag J, Zuccolo A, Fan C, Sebastian A, Kramer M, Spiegel L, Nascimento L, Zutavern T, Miller B, Ambroise C, Muller S, Spooner W, Narechania A, Ren L, Wei S, Kumari S, Faga B, Levy MJ, McMahan L, Van Buren P, Vaughn MW, Ying K, Yeh CT, Emrich SJ, Jia Y, Kalyanaraman A, Hsia AP, Barbazuk WB, Baucom RS, Brutnell TP, Carpita NC, Chaparro C, Chia JM, Deragon JM, Estill JC, Fu Y, Jeddeloh JA, Han Y, Lee H, Li P, Lisch DR, Liu S, Liu Z, Nagel DH, McCann MC, SanMiguel P, Myers AM, Nettleton D, Nguyen J, Penning BW, Ponnala L, Schneider KL, Schwartz DC, Sharma A, Soderlund C,

Springer NM, Sun Q, Wang H, Waterman M, Westerman R, Wolfgruber TK, Yang L, Yu Y, Zhang L, Zhou S, Zhu Q, Bennetzen JL, Dawe RK, Jiang J, Jiang N, Presting GG, Wessler SR, Aluru S, Martienssen RA, Clifton SW, McCombie WR, Wing RA, Wilson RK (2009) The B73 maize genome: complexity, diversity, and dynamics. Science 326:1112–1115

Sharma A, Schneider KL, Presting GG (2008) Sustained retrotransposition is mediated by nucleotide deletions and interelement recombinations. Proc Natl Acad Sci USA 105:15470–15474

Silva JC, Kidwell MG (2000) Horizontal transfer and selection in the evolution of P elements. Mol Biol Evol 17:1542–1557

Skibbe DS, Fernandes JF, Medzihradszky KF, Burlingame AL, Walbot V (2009) Mutator transposon activity reprograms the transcriptomes and proteomes of developing maize anthers. Plant J 59:622–633

Slotkin RK, Martienssen R (2007) Transposable elements and the epigenetic regulation of the genome. Nat Rev Genet 8:272–285

Slotkin RK, Freeling M, Lisch D (2005) Heritable transposon silencing initiated by a naturally occurring transposon inverted duplication. Nat Genet 37:641–644

Song WY, Pi LY, Wang GL, Gardner J, Holsten T, Ronald PC (1997) Evolution of the rice Xa21 disease resistance gene family. Plant Cell 9:1279–1287

Soppe WJ, Jacobsen SE, Alonso-Blanco C, Jackson JP, Kakutani T, Koornneef M, Peeters AJ (2000) The late flowering phenotype of fwa mutants is caused by gain-of-function epigenetic alleles of a homeodomain gene. Mol Cell 6:791–802

Spanu PD, Abbott JC, Amselem J, Burgis TA, Soanes DM, Stuber K, Loren V, van Themaat E, Brown JK, Butcher SA, Gurr SJ, Lebrun MH, Ridout CJ, Schulze-Lefert P, Talbot NJ, Ahmadinejad N, Ametz C, Barton GR, Benjdia M, Bidzinski P, Bindschedler LV, Both M, Brewer MT, Cadle-Davidson L, Cadle-Davidson MM, Collemare J, Cramer R, Frenkel O, Godfrey D, Harriman J, Hoede C, King BC, Klages S, Kleemann J, Knoll D, Koti PS, Kreplak J, Lopez-Ruiz FJ, Lu X, Maekawa T, Mahanil S, Micali C, Milgroom MG, Montana G, Noir S, O'Connell RJ, Oberhaensli S, Parlange F, Pedersen C, Quesneville H, Reinhardt R, Rott M, Sacristan S, Schmidt SM, Schon M, Skamnioti P, Sommer H, Stephens A, Takahara H, Thordal-Christensen H, Vigouroux M, Wessling R, Wicker T, Panstruga R (2010) Genome expansion and gene loss in powdery mildew fungi reveal tradeoffs in extreme parasitism. Science 330:1543–1546

Stein LD, Bao Z, Blasiar D, Blumenthal T, Brent MR, Chen N, Chinwalla A, Clarke L, Clee C, Coghlan A, Coulson A, D'Eustachio P, Fitch DH, Fulton LA, Fulton RE, Griffiths-Jones S, Harris TW, Hillier LW, Kamath R, Kuwabara PE, Mardis ER, Marra MA, Miner TL, Minx P, Mullikin JC, Plumb RW, Rogers J, Schein JE, Sohrmann M, Spieth J, Stajich JE, Wei C, Willey D, Wilson RK, Durbin R, Waterston RH (2003) The genome sequence of Caenorhabditis briggsae: a platform for comparative genomics. PLoS Biol 1:E45

Stokes TL, Richards EJ (2002) Induced instability of two Arabidopsis constitutive pathogen-response alleles. Proc Natl Acad Sci USA 99:7792–7796

Sweredoski M, DeRose-Wilson L, Gaut BS (2008) A comparative computational analysis of nonautonomous helitron elements between maize and rice. BMC Genomics 9:467

Takahashi S, Inagaki Y, Satoh H, Hoshino A, Iida S (1999) Capture of a genomic HMG domain sequence by the En/Spm-related transposable element Tpn1 in the Japanese morning glory. Mol Gen Genet 261:447–451

Talbert LE, Chandler VL (1988) Characterization of a highly conserved sequence related to mutator transposable elements in maize. Mol Biol Evol 5:519–529

Topp CN, Zhong CX, Dawe RK (2004) Centromere-encoded RNAs are integral components of the maize kinetochore. Proc Natl Acad Sci USA 101:15986–15991

Topp CN, Okagaki RJ, Melo JR, Kynast RG, Phillips RL, Dawe RK (2009) Identification of a maize neocentromere in an oat-maize addition line. Cytogenet Genome Res 124:228–238

Toth M, Grimsby J, Buzsaki G, Donovan GP (1995) Epileptic seizures caused by inactivation of a novel gene, jerky, related to centromere binding protein-B in transgenic mice. Nat Genet 11:71–75

van Leeuwen H, Monfort A, Puigdomenech P (2007) Mutator-like elements identified in melon, Arabidopsis and rice contain ULP1 protease domains. Mol Genet Genomics 277:357–364

Vaughn MW, Tanurdzic M, Lippman Z, Jiang H, Carrasquillo R, Rabinowicz PD, Dedhia N, McCombie WR, Agier N, Bulski A, Colot V, Doerge RW, Martienssen RA (2007) Epigenetic natural variation in Arabidopsis thaliana. PLoS Biol 5:e174

Volpe TA, Kidner C, Hall IM, Teng G, Grewal SI, Martienssen RA (2002) Regulation of heterochromatic silencing and histone H3 lysine-9 methylation by RNAi. Science 297:1833–1837

Volpe T, Schramke V, Hamilton GL, White SA, Teng G, Martienssen RA, Allshire RC (2003) RNA interference is required for normal centromere function in fission yeast. Chromosome Res 11:137–146

Wang GL, Ruan DL, Song WY, Sideris S, Chen L, Pi LY, Zhang S, Zhang Z, Fauquet C, Gaut BS, Whalen MC, Ronald PC (1998) Xa21D encodes a receptor-like molecule with a leucine-rich repeat domain that determines race-specific recognition and is subject to adaptive evolution. Plant Cell 10:765–779

Wang W, Zheng H, Fan C, Li J, Shi J, Cai Z, Zhang G, Liu D, Zhang J, Vang S, Lu Z, Wong GK, Long M, Wang J (2006) High rate of chimeric gene origination by retroposition in plant genomes. Plant Cell 18:1791–1802

Waterston RH, Lindblad-Toh K, Birney E, Rogers J, Abril JF, Agarwal P, Agarwala R, Ainscough R, Alexandersson M, An P, Antonarakis SE, Attwood J, Baertsch R, Bailey J, Barlow K, Beck S, Berry E, Birren B, Bloom T, Bork P, Botcherby M, Bray N, Brent MR, Brown DG, Brown SD, Bult C, Burton J, Butler J, Campbell RD, Carninci P, Cawley S, Chiaromonte F, Chinwalla AT, Church DM, Clamp M, Clee C, Collins FS, Cook LL, Copley RR, Coulson A, Couronne O, Cuff J, Curwen V, Cutts T, Daly M, David R, Davies J, Delehaunty KD, Deri J, Dermitzakis ET, Dewey C, Dickens NJ, Diekhans M, Dodge S, Dubchak I, Dunn DM, Eddy SR, Elnitski L, Emes RD, Eswara P, Eyras E, Felsenfeld A, Fewell GA, Flicek P, Foley K, Frankel WN, Fulton LA, Fulton RS, Furey TS, Gage D, Gibbs RA, Glusman G, Gnerre S, Goldman N, Goodstadt L, Grafham D, Graves TA, Green ED, Gregory S, Guigo R, Guyer M, Hardison RC, Haussler D, Hayashizaki Y, Hillier LW, Hinrichs A, Hlavina W, Holzer T, Hsu F, Hua A, Hubbard T, Hunt A, Jackson I, Jaffe DB, Johnson LS, Jones M, Jones TA, Joy A, Kamal M, Karlsson EK, Karolchik D, Kasprzyk A, Kawai J, Keibler E, Kells C, Kent WJ, Kirby A, Kolbe DL, Korf I, Kucherlapati RS, Kulbokas EJ, Kulp D, Landers T, Leger JP, Leonard S, Letunic I, Levine R, Li J, Li M, Lloyd C, Lucas S, Ma B, Maglott DR, Mardis ER, Matthews L, Mauceli E, Mayer JH, McCarthy M, McCombie WR, McLaren S, McLay K, McPherson JD, Meldrim J, Meredith B, Mesirov JP, Miller W, Miner TL, Mongin E, Montgomery KT, Morgan M, Mott R, Mullikin JC, Muzny DM, Nash WE, Nelson JO, Nhan MN, Nicol R, Ning Z, Nusbaum C, O'Connor MJ, Okazaki Y, Oliver K, Overton-Larty E, Pachter L, Parra G, Pepin KH, Peterson J, Pevzner P, Plumb R, Pohl CS, Poliakov A, Ponce TC, Ponting CP, Potter S, Quail M, Reymond A, Roe BA, Roskin KM, Rubin EM, Rust AG, Santos R, Sapojnikov V, Schultz B, Schultz J, Schwartz MS, Schwartz S, Scott C, Seaman S, Searle S, Sharpe T, Sheridan A, Shownkeen R, Sims S, Singer JB, Slater G, Smit A, Smith DR, Spencer B, Stabenau A, Stange-Thomann N, Sugnet C, Suyama M, Tesler G, Thompson J,

Torrents D, Trevaskis E, Tromp J, Ucla C, Ureta-Vidal A, Vinson JP, Von Niederhausern AC, Wade CM, Wall M, Weber RJ, Weiss RB, Wendl MC, West AP, Wetterstrand K, Wheeler R, Whelan S, Wierzbowski J, Willey D, Williams S, Wilson RK, Winter E, Worley KC, Wyman D, Yang S, Yang SP, Zdobnov EM, Zody MC, Lander ES (2002) Initial sequencing and comparative analysis of the mouse genome. Nature 420:520–562

Wei F, Wing RA, Wise RP (2002) Genome dynamics and evolution of the Mla (powdery mildew) resistance locus in barley. Plant Cell 14:1903–1917

Wolfgruber TK, Sharma A, Schneider KL, Albert PS, Koo DH, Shi J, Gao Z, Han F, Lee H, Xu R, Allison J, Birchler JA, Jiang J, Dawe RK, Presting GG (2009) Maize centromere structure and evolution: sequence analysis of centromeres 2 and 5 reveals dynamic Loci shaped primarily by retrotransposons. PLoS Genet 5:e1000743

Wu J, Fujisawa M, Tian Z, Yamagata H, Kamiya K, Shibata M, Hosokawa S, Ito Y, Hamada M, Katagiri S, Kurita K, Yamamoto M, Kikuta A, Machita K, Karasawa W, Kanamori H, Namiki N, Mizuno H, Ma J, Sasaki T, Matsumoto T (2009) Comparative analysis of complete orthologous centromeres from two subspecies of rice reveals rapid variation of centromere organization and structure. Plant J 60:805–819

Xiao H, Jiang N, Schaffner E, Stockinger EJ, van der Knaap E (2008) A retrotransposon-mediated gene duplication underlies morphological variation of tomato fruit. Science 319:1527–1530

Xie W, Gai X, Zhu Y, Zappulla DC, Sternglanz R, Voytas DF (2001) Targeting of the yeast Ty5 retrotransposon to silent chromatin is mediated by interactions between integrase and Sir4p. Mol Cell Biol 21:6606–6614

Yamashita S, Takano-Shimizu T, Kitamura K, Mikami T, Kishima Y (1999) Resistance to gap repair of the transposon Tam3 in *Antirrhinum majus*: a role of the end regions. Genetics 153:1899–1908

Yan X, Martinez-Ferez IM, Kavchok S, Dooner HK (1999) Origination of Ds elements from Ac elements in maize: evidence for rare repair synthesis at the site of Ac excision. Genetics 152:1733–1740

Yang L, Bennetzen JL (2009) Distribution, diversity, evolution, and survival of Helitrons in the maize genome. Proc Natl Acad Sci USA 106:19922–19927

Yu Z, Wright SI, Bureau TE (2000) Mutator-like elements in *Arabidopsis thaliana*. Structure, diversity and evolution. Genetics 156:2019–2031

Yu HG, Koshland D (2005) Chromosome morphogenesis: condensin-dependent cohesin removal during meiosis. Cell 123:397–407

Zabala G, Vodkin LO (2005) The wp mutation of *Glycine max* carries a gene-fragment-rich transposon of the CACTA superfamily. Plant Cell 17:2619–2632

Zabala G, Vodkin LO (2007) Novel exon combinations generated by alternative splicing of gene fragments mobilized by a CACTA transposon in *Glycine max*. BMC Plant Biol 7:38

Zhang J, Peterson T (2005) A segmental deletion series generated by sister-chromatid transposition of Ac transposable elements in maize. Genetics 171:333–344

Zhang X, Wessler SR (2004) Genome-wide comparative analysis of the transposable elements in the related species *Arabidopsis thaliana* and *Brassica oleracea*. Proc Natl Acad Sci USA 101:5589–5594

Zhang P, Gu Z, Li WH (2003) Different evolutionary patterns between young duplicate genes in the human genome. Genome Biol 4:R56

Zhang J, Zhang F, Peterson T (2006) Transposition of reversed Ac element ends generates novel chimeric genes in maize. PLoS Genet 2:e164

Zhong CX, Marshall JB, Topp C, Mroczek R, Kato A, Nagaki K, Birchler JA, Jiang J, Dawe RK (2002) Centromeric retroelements and satellites interact with maize kinetochore protein CENH3. Plant Cell 14:2825–2836

Centromeres: Sequences, Structure, and Biology

4

Cory D. Hirsch and Jiming Jiang

Contents

4.1 Introduction

Although technological advances have continued to change the speed, cost, and number of plant genomes sequenced (see Flagel and Blackman 2012, this volume), parts of genomes remain to be sequenced and explored. Even the best-sequenced plant genomes, including *Arabidopsis thaliana* and rice, are missing 7–8% of their total genomic information (Kaul et al. 2000; Goff et al. 2002; Yu et al. 2002). One chromosomal region not often sequenced in genome projects is the centromere. Centromeres of almost all higher eukaryotes contain large stretches (up to several megabases) of tandemly repeated arrays of satellite DNA and retrotransposons. Such long arrays of highly homogenized repetitive DNA sequences cannot readily be cloned, sequenced, and assembled using the currently available cloning and sequencing technologies.

The centromere is a chromosomal site for the assembly of the kinetochore, to which spindle fibers attach during cell division. Thus, centromeres play a key role in chromosome segregation and transmission. The simplest centromeres are the 'point centromeres' found in budding yeasts, which are encompassed within a single nucleosome and consist of approximately 125 bp of DNA (Meluh et al. 1998). Most eukaryotic organisms, including all multicellular species, contain centromeres that are made up of large amounts of DNA, called 'regional centromeres', ranging in length from a few kilobases to several 100 kb. Cytologically, centromeres appear as the primary constrictions on somatic metaphase chromosomes. At the molecular level centromeres can be described as the chromosomal domain where canonical histone H3 is replaced with a centromere-specific histone 3 variant, CENH3. CENH3 was first identified as CENP-A in humans (Earnshaw and Rothfield 1985), and has since been found in all eukaryotes (Torras-Llort et al. 2009). CENP-A nucleosomes directly recruit numerous other proteins to the centromere (Foltz et al. 2006; Okada et al. 2006).

C.D. Hirsch (✉)
Department of Horticulture, University of Wisconsin-Madison, 1575 Linden Drive, Madison 53706, WI, USA
e-mail: cdhirsch@wisc.edu

J.F. Wendel et al. (eds.), *Plant Genome Diversity Volume 1*,
DOI 10.1007/978-3-7091-1130-7_4, © Springer-Verlag Wien 2012

The function of centromeres is highly conserved among eukaryotes. Often when function is conserved the DNA sequence involved in the function is highly conserved as well, but the DNA sequence at centromeres (centromeric DNA) differs greatly between species. How can such a foundational process have such a variant underlying DNA sequence? This question has been one of the major driving forces for centromere research.

In this chapter we will focus on the composition of plant centromeres, including both centromeric DNA sequences and a few key centromeric proteins, and the structure and evolution of plant centromeres.

4.2 DNA Composition of Plant Centromeres

4.2.1 Satellite Repeats

The centromeres of most plant species contain arrays of tandemly repeated sequences, often called satellite repeats (Jiang et al. 2003). In *A. thaliana*, a 178 bp tandem repeat, pAL1, was isolated (Martinez-Zapater et al. 1986) and found to be localized to centromeres (Maluszynska and Heslop-Harrison 1991; Murata et al. 1994). Each of the five *A. thaliana* centromeres contain megabase-sized arrays of this 178 bp repeat (Heslop-Harrison et al. 1999). However, cytological mapping suggested that only part of the pAL1 repeat arrays associates with CENH3 (Shibata and Murata 2004; Zhang et al. 2008). Thus, the 178 bp repeats in *Arabidopsis* occupy both centromeric and pericentromeric domains.

Arabidopsis arenosa, a close relative of *A. thaliana*, contains a centromeric tandem repeat ranging in size from 166 to 179 bp called pAa, which shares 50–80% sequence similarity with pAL1 (Kamm et al. 1995). Likewise, other members of the *Arabidopsis* family, *A. pumila* and *A. griffithiana*, contain the species-specific tandem repeats, pApKB2 and pAgKB1, respectively (Heslop-Harrison et al. 2003). Most diploid *Arabidopsis* species contain only a single centromeric tandem repeat. One exception to this was found in *A. halleri* (2n = 2x = 16). Not only does *A. halleri* contain pAa of *A. arenosa*, it also contains two unique subfamilies, pAge1 and pAge2. The repeats pAge1, pAge2, pAa, and pAL1 share an approximately 50 bp conserved region; thus, these repeats appear to be derived from the same ancestral element (Kawabe and Nasuda 2005).

The centromeres of several grass species have been extensively studied, especially in *Oryza*. Cultivated rice, *O. sativa*, contains a 155 bp centromeric tandem repeat called CentO (Dong et al. 1998; Nonomura and Kurata 1999; Cheng et al. 2002). CentO is highly specific to rice centromeres. However, the amount of CentO varies significantly in different centromeres, ranging from approximately 65 to 2,000 kb (Cheng et al. 2002), while the CENH3-binding domains in rice centromeres span 400–800 kb (Yan et al. 2008). Thus, the entire CentO arrays in several rice centromeres are included in the functional centromeres; while in other rice centromeres the CentO repeat occupies both centromeric and pericentromeric domains.

Most plant species studied containing centromeric tandem repeats have monomer sizes of approximately 150–200 bp, which are intriguingly similar in size to a single nucleosome. However, there are exceptions to this typical centromeric satellite repeat. Barley centromeres contain a specific 6 bp GC-rich satellite (Hudakova et al. 2001; Houben et al. 2007), *Solanum bulbocastanum* centromeres have long arrays of telomere similar sequences (Tek and Jiang 2004), and *Torenia fournieri* and *bailonii* contain a 52 bp centromeric tandem repeat (Kikuchi et al. 2005) to name a few.

4.2.2 Centromeric Retrotransposons

In addition to tandem repeats, retroelements are also abundant in plant centromeres. Retrotransposons make up a large proportion of the DNA in many higher eukaryotes, which has been accomplished through their copy-and-paste mechanism of proliferation (see chapters by Kejnovsky et al. 2012; Slotkin et al. 2012, this volume). Throughout a genome some retrotransposons are dispersed in a near-random fashion, while other families are confined to specific chromosomal regions (Kordis 2005). One type of family, called centromere retrotransposons (CRs), are mostly restricted to centromeric regions of plant chromosomes (Miller et al. 1998; Presting et al. 1998). CRs belong to the *Ty3-gypsy* class of retrotransposons and have been reported in a number of plant species, including rice (CRR) (Dong et al. 1998; Cheng et al. 2002), maize (CRM) (Ananiev et al. 1998; Zhong et al. 2002), wheat (CRW) (Liu et al. 2008), barley (*cereba*) (Presting et al. 1998; Hudakova et al. 2001), and sugarcane (CRS) (Nagaki and Murata 2005). The CR family was initially thought to be specific to grass species; however, it recently has been shown that CR-related elements are present in soybean (Du et al. 2010a). Two CR families in soybean were associated with soybean satellite repeats by their propensity to be located in the same whole genome shotgun sequences. Based on this physical association it appears that the enrichment of CR elements into centromeric regions occurred much earlier than originally thought. Therefore, the function of CR elements emerged before monocots and eudicots diverged, approximately 140–150 million years ago (Du et al. 2010a).

CR elements are intermingled with centromeric satellite repeats (Cheng et al. 2002) and nested CR elements can organize into long arrays of CR sequences (Cheng et al.

2002; Jin et al. 2004; Liu et al. 2008). The association between CR elements and CENH3 has been demonstrated in several cereal species (Zhong et al. 2002; Nagaki et al. 2004; Liu et al. 2008). These results showed that CR elements preferentially target the CENH3-associated chromatin domain of centromeres. CR elements can be divided into full-sized autonomous elements and non-autonomous, which have a deletion causing a loss of enzymatic functions (Langdon et al. 2000). Non-autonomous CRR elements in rice are less restricted to centromere regions and were likely mobilized through the retrotransposition machinery from autonomous CRR elements (Nagaki et al. 2005).

4.2.3 Transcription of Centromeric Repeats

Repetitive DNA elements, including retrotransposons, typically are transcriptionally inactive. Thus, the centromere has been viewed as a silent chromosomal domain because it consists mainly of repetitive DNA sequences. Recently, evidence from numerous species is refuting this thought. The centromere satellites of humans and mice have both been shown to have transcribed sequences (Fukagawa et al. 2004; Bouzinba-Segard et al. 2006). In rice and maize, transcription of both centromere satellites and CR elements has been reported (Topp et al. 2004; Lee et al. 2006; Neumann et al. 2007), although not all centromere repetitive elements are able to escape the host's silencing machinery (Neumann et al. 2007). Single-stranded RNA transcripts from centromeric repeat elements have been shown to bind to centromeric proteins (Topp et al. 2004; Wong et al. 2007; Du et al. 2010b). Telomerase RNA has been shown to be important not only for protein targeting, but also as a scaffold for other proteins (Zappulla and Cech 2004). It has been hypthesized that RNA from centromeric DNA could function in a similar manner (Topp et al. 2004).

Although a range of sizes of RNAs from centromeric repeats have been found, presumably from different processing mechanisms, both plants and animals contain small, centromere-derived RNAs of approximately 40 bp (Topp et al. 2004; Lee et al. 2006; Carone et al. 2009). Carone et al. (2009) found that the transcription of the RNA precursors to these small RNAs in animals, which could possibly be conserved in plants, is facilitated by the promoter capability of retroviral LTRs. Interestingly, stressed cells don't process centromeric RNAs into small RNAs, therefore, these cells accumulate a large amount of full-length centromeric RNA which impairs the proper formation of centromeric proteins and centromere function (Bouzinba-Segard et al. 2006). These data on centromeric DNA transcription point to the integral nature of small RNAs in the epigenetic formation of centromeres, mediated by small RNA binding and stabilization of centromere proteins.

4.3 Centromeric Proteins

4.3.1 CENH3

The centromeres in all eukaryotes studied contain the centromeric histone 3 variant CENH3 (Torras-Llort et al. 2009). CENH3 contains a N-terminal tail and a histone fold domain (HFD). In humans, insertion of the loop 1 and α2 helix of the HFD of CENH3/CENP-A into canonical H3 allows H3 to gain CENH3/CENP-A function, leading this region to be called the CENP-A targeting domain (CATD) (Black et al. 2004). However, CENH3 deposition in plants differs from humans, which was demonstrated in *Arabidopsis* in which inserting the CATD domain from CENH3/HTR12 into H3 does not confer CENH3/HTR12 function (Ravi et al. 2010).

CENH3 is encoded by a single copy gene in all diploid plant species studied (Cooper and Henikoff 2004), including maize, which is an ancient polyploid. However, two recent studies demonstrated that multiple *CENH3* genes can coexist in tetraploid plant species (Hirsch et al. 2009; Nagaki et al. 2009). The HFD sequences of plant CENH3 proteins show more divergence compared to the HFD sequences of canonical H3 proteins. The major distinguishing feature between H3 and CENH3 proteins is their N-terminal tail domains. The N-terminal tail domain of H3 proteins is highly conserved between species, whereas the N-terminal domain of CENH3 proteins is not conserved in sequence or length (Table 4.1). The N-terminal tail domain of *A. thaliana* and maize CENH3 proteins cannot even be aligned. The divergence of CENH3 sequence between species is puzzling. Often proteins with conserved function between species also have good sequence conservation, which is seen for canonical H3 as it is among the most conserved proteins across species.

Why is the sequence of a functionally conserved protein so different among species? The function of CENH3 is to bind the underlying DNA at centromeres, and its presence is required to recruit other essential proteins to form the functional kinetochore where microtubules attach. The problem CENH3 faces, to perpetuate proper function, is that the DNA sequence of centromeres, with respect to both tandem repeats and retrotransposons, is highly variable among species. For a protein to maintain proper binding to such dynamic DNA the protein itself most likely needs to co-evolve with the DNA sequence. Malik and Henikoff (2001) showed that CID (CENH3 in *Drosophila melanogaster*) is adaptively evolving in its N-terminal tail domain. Similarly, investigations of CENH3 evolution in the Brassicaceae determined not only significant adaptive evolution in the N-terminal tail domain, but also in the HFD (Cooper and Henikoff 2004). These findings led to an arms race hypothesis between the dynamic centromeric DNA and the

Table 4.1 Completely sequenced *CENH3* genes from plants and other representative eukaryotes

Species	gDNA (bp)	CDS (bp)	AA	Tail (AA)	HFD (AA)	Reference
Nicotiana tabacum	–	468	156	60	96	Nagaki et al. (2009)
Nicotiana tabacum	–	471	157	61	96	Nagaki et al. (2009)
Zea mays	–	471	157	62	95	Zhong et al. (2002)
Glycine max	–	474	158	62	96	Tek et al. (2010)
Oryza sativa	2,163	492	164	69	95	Nagaki et al. (2004)
Oryza australiensis	1,655	492	164	68	96	Hirsch et al. (2009)
Oryza alta	1,661	492	164	68	96	Hirsch et al. (2009)
Oryza alta	2,691	498	166	71	95	Hirsch et al. (2009)
Oryza punctata	2,216	498	166	71	95	Hirsch et al. (2009)
Oryza minuta	2,236	498	166	71	95	Hirsch et al. (2009)
Oryza minuta	2,694	498	166	71	95	Hirsch et al. (2009)
Oryza rhizomatis	2,691	498	166	71	95	Hirsch et al. (2009)
Luzula nivea	–	501	167	76	91	Nagaki et al. (2005)
Olimarabidopsis pumila	1,713	522	174	77	97	Cooper and Henikoff (2004)
Arabidopsis arenosa	1,684	528	176	79	97	Talbert et al. (2002)
Arabidopsis lyrata	1,721	528	176	79	97	Cooper and Henikoff (2004)
Crucihimalaya himalaica	1,682	531	177	80	97	Cooper and Henikoff (2004)
Arabidopsis thaliana	–	534	178	81	97	Talbert et al. (2002)
Homo sapiens	–	420	140	39	101	Sullivan et al. (1994)
Drosophila melanogaster	–	675	225	125	100	Henikoff et al. (2000)
Saccharomyces cerevisiae	–	687	229	130	99	Stoler et al. (1995)
Caenorhabditis elegans	–	864	288	192	96	Buchwitz et al. (1999)

adaptively evolving CENH3 (Henikoff et al. 2001; Malik and Henikoff 2001). In this hypothesis, the expansion of centromeric repeats could result in a meiotic drive in which a chromsome with an expanded centromere could be preferentially transmitted to the egg, because of the asymmetric transmission of homologous chromosomes during female meiosis. A meiotic drive of this type would be associated with deleterious effects. Thus, suppressors of such a drive would be expected to evolve, including *CENH3* alleles, with altered DNA-binding preferences to restore meiotic parity (Malik and Henikoff 2002). In polyploid *Oryza* species, CENH3 has evolved adaptively in a lineage-specific manner that has been maintained in the polyploid species, showing that the evolutionary track of CENH3 has not been disrupted by polyploidization events (Hirsch et al. 2009).

One of the most puzzling questions in centromere biology is how CENH3 is deposited and maintained in centromeres. In animals, CENH3/CENP-A loading is initiated late in mitosis, but primarily takes place during the G1 phase of the cell cycle (Jansen et al. 2007; Hemmerich et al. 2008). However, in *A. thaliana*, CENH3/HTR12 loading takes place during G2 (Lermontova et al. 2006), which is different than H3 loading (S phase) and CENH3 loading in animals (Lermontova et al. 2007). The mechanism of CENH3 loading is not well understood. Aspects of the CENH3 loading mechanism are starting to come to light, most noticeably in humans and yeast. Notably, the kinetochore assembly proteins Mis16 (RbAp46/48 in humans) and Mis18 in yeast and humans are required for targeting CENH3/Cnp1/CENP-

A nucleosomes to centromeres (Hayashi et al. 2004; Fujita et al. 2007). In yeast, Scm3 coimmunoprecipitates with Mis16, Mis18, and CENH3/Cnp1 and has been proven essential for localizing newly synthesized CENH3/Cnp1 (Pidoux et al. 2009; Williams et al. 2009). In addition, in humans the Holliday junction-recognition protein (HJURP) has been identified as a CENH3/CENP-A binding partner (Shuaib et al. 2010) and HJURP depletion results in a loss of CENH3/CENP-A from centromeres (Dunleavy et al. 2009; Foltz et al. 2009). Although some important proteins required for CENH3 loading have been identified, their precise roles and actions remain not completely understood.

4.3.2 CENP-C

CENP-C was first identified as a centromere protein in humans in the same assay that yielded CENH3/CENP-A (Earnshaw and Rothfield 1985), and homologs have since been found in plant species (Talbert et al. 2004). CENP-Cs are divergent proteins, but can be identified based on a conserved 24 amino acid motif that is shared between animal and plant species (Talbert et al. 2004). CENP-C proteins bind centromeric DNA (Yang et al. 1996), but the process of CENP-C binding to centromeric DNA is not well understood. In humans, CENP-C associates with RNA from alpha satellites (human centromeric satellite repeats). Studies using chromatin immunoprecipitation (ChIP), RNase treatments, and RNA replenishing assays proved centromeric satellite RNA in humans is critical for the assembly of CENP-C to centromeres (Wong et al. 2007). In maize CENP-C, the domain responsible for its DNA binding properties has been deduced down to a continuous 122 amino acid region. In addition, single-stranded RNA helps to stabilize CENP-C binding to centromeric DNA and centromeric RNA helps in the recruitment process (Du et al. 2010a). With CENH3, these are the only two proteins shown to bind centromeric DNA in both plants and animals.

CENH3 is not adaptively evolving in all species. Since CENP-C also binds to centromeric DNA it was hypothesized and shown to be adaptively evolving in those species to replace the role of CENH3 in evolution (Talbert et al. 2004). In grasses, CENP-C proteins have undergone numerous rounds of duplications of exons 9–12. The duplicated region in maize is responsible for the binding property of CENP-C in maize to centromeric DNA (Du et al. 2010a). Duplicated genes can undergo a myriad of selection pressures and likewise so can duplicate regions within a gene. Therefore, it is not surprising that in some grass species the duplicate regions are under positive selection (wheat and barley), while some regions are under negative selection (maize, sugarcane, and *Sorghum*) (Talbert et al. 2004).

4.4 Structure and Organization of Centromeric DNA

As already stated, the centromeres of most plant species contain two main types of sequences, tandem repeat arrays and transposable elements. In *Arabidopsis*, the centromere of chromosome 5 (Cen5) has a 2.9 Mb 'central domain'

(Kumekawa et al. 2000). This central domain consists mainly of arrays of the 178 bp tandem repeat, pAL1, that are interrupted by the *Athila* retrotransposon and possibly other repetitive elements. However, ChIP analysis suggested that only the 178 bp repeat is associated with CENH3/HTR12 (Nagaki et al. 2003). It was shown that CENH3/HTR12 binds to only a portion of the pAL1 repeat array present at *A. thaliana* centromeres (Shibata and Murata 2004). Similarly, the central domain of *A. thaliana* Cen4 is 2.7 Mb, consisting mainly of the 178 bp repeat arrays. The flanking regions of Cen5 and Cen4 are composed of a variety of transposons including, LINES, Mu elements, other *gypsy* and *copia*-type elements, and Ac/Ds elements to name a few (Kumekawa et al. 2001).

Maize centromeres contain long arrays of intermingled CentC satellite repeats and CRM retrotransposons (Fig. 4.1). The length of intermingled CentC/CRM repeats have been estimated at 300–700 kb for maize chromosomes 2, 3, 4, 6, and 9, while chromosomes 1 and 7 intermingled arrays are much larger, estimated at 1,700 and 2,800 kb respectively (Jin et al. 2004). Both CentC and CRM sequences are significantly enriched in CENH3-associated chromatin (Zhong et al. 2002). Thus, these two DNA elements are the main DNA components of maize centromeres. The centromere of the maize B chromosome has also been well characterized. The maize B centromere contains a B-specific satellite repeat, ZmB (Alfenito and Birchler 1993). The functional B centromere, which is associated with CENH3, spans approximately 700 kb of DNA consisting of five separate arrays of the ZmB repeat (Jin et al. 2005). Each of the five ZmB arrays is flanked by intermingled CentC satellite repeats and CRM retrotransposons. Both sides of the 700 kb core domain,

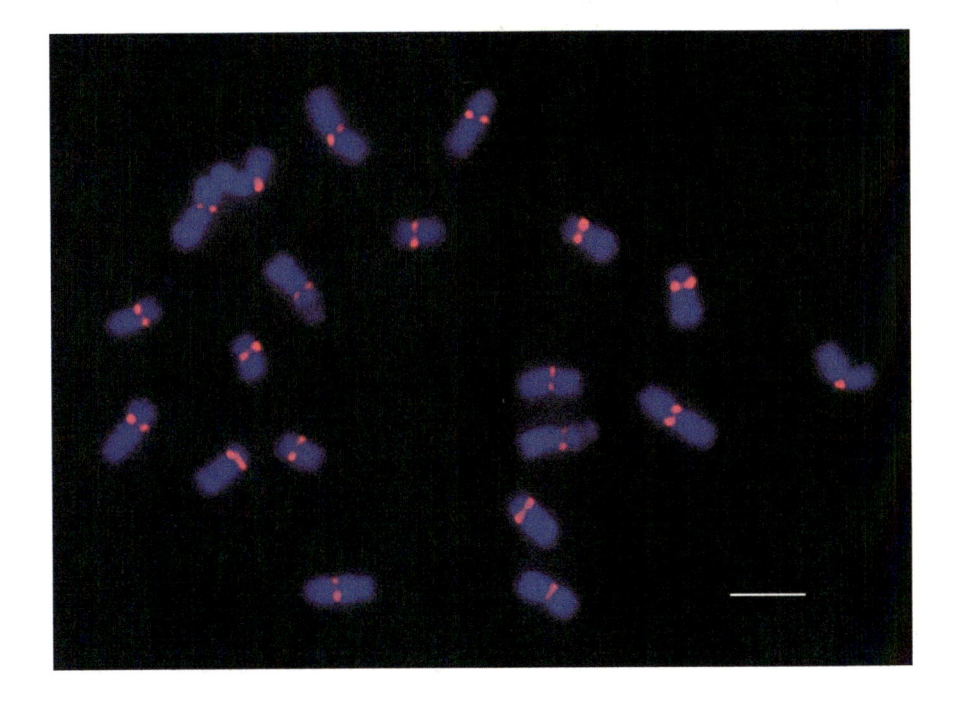

Fig. 4.1 Fluorescence in situ hybridization (FISH) mapping centromeric repeats in maize. CentC satellite repeats and CRM retrotransposons were mixed and labelled as a single FISH probe. The FISH signals are highly restricted in the centromeres of all maize chromosomes. Scale bar = 5 μm. The FISH image is courtesy of Dal-Hoe Koo

which appears to be associated with CENH3, are flanked by megabase-sized arrays of the ZmB repeats.

The structure and organization of centromeres in most plant species may resemble those of *A. thaliana* and maize. These centromeres may contain up to several megabases of satellite repeats or an intermingling of satellite repeats and retrotransposons. However, CENH3 may be associated with only portions of such long arrays of repetitive DNA. Several rice centromeres though, represent an exception to this centromeric DNA composition pattern. The main DNA components of rice centromeres are long arrays of intermingled CentO satellites and CRR retroelements. Nevertheless, such CentO/CRR arrays can be relatively short, spanning less than 200 kb, in several rice centromeres (Cheng et al. 2002). The CENH3 domains of several rice centromeres extend beyond the CentO/CRR arrays and included various types of DNA sequences, including active genes (see below).

4.5 Centromere Evolution

4.5.1 Dynamics of Centromeric Repeats

Genetic crossing over is suppressed in centromeres. However, centromeres are evolutionarily dynamic regions that undergo rapid changes of their underlying sequences. Comparative studies of centromeric DNA sequences in several *Oryza* species have provided the best example of the dynamics of plant centromere DNA.

Rice centromeres contain CentO satellite repeats and CRR retroelements. *Oryza brachyantha*, which diverged from rice less than 10 million years ago (Guo and Ge 2005), is devoid of CentO, which has been replaced with a completely unrelated tandem repeat, CentO-F (Lee et al. 2005). In addition, *O. brachyantha,* surprisingly, does not contain CR elements in its centromeres, representing a rare exception in which the CR elements were eliminated from the centromeres of a grass species. Instead, a new *Ty3-gypsy* type element named FRetro3 dominates in the centromeres of *O. brachyantha* and appears to be associated with CENH3 (Lee et al. 2005; Gao et al. 2009). FRetro3 has been amplified in the genome of *O. brachyantha* in the last few millions years, and also has a high level of removal based on the ratio of solo-LTRs and full-length elements. The rapid mobility of this element has been proposed to have played a role in the removal of CRR elements in *O. brachyantha* that are found throughout other *Oryza* species (Gao et al. 2009).

Oryza rhizomatis (CC genome) has diverged from cultivated rice for approximately 5 million years (Zhang and Ge 2007). The centromeres of *O. rhizomatis* contain tandem repeats specific to the CC genome, CentO-C1 and C2, but not all chromosomes have completely lost CentO

(Lee et al. 2005; Bao et al. 2006). Interestingly, CentO-C2 is not confined to centromeric regions as it is found more abundantly in subtelomeric regions. It is likely that CentO-C2 was brought into the centromeres of *O. rhizomatis* from its originating subtelomeric region (Lee et al. 2005). Thus, within 5 million years since its divergence from *O. sativa* two new repeats have emerged and the amount of CentO has been dramatically reduced in *O. rhizomatis*. These cross species comparisons highlight the activity of centromeric DNA that has occurred over an evolutionarily short period of time.

The molecular mechanisms of centromeric DNA dynamics are unclear, but highly intriguing. Genetic crossing over is suppressed in centromeres. Thus, it is unlikely that unequal crossing over plays a role in the expansion/contraction and homogenization of centromeric satellite repeat arrays. In some species, such as maize, CR elements are highly mingled with the centromeric satellites (Jin et al. 2004). Therefore, retrotransposons may mediate the dynamics of the centromeric satellite. However, this cannot explain the fact that in some species centromeres are composed almost exclusively of satellite arrays that are rarely disrupted by retrotransposons. Future research on the mechanisms of centromeric DNA dynamics will be essential to understand the evolution of centromeres.

4.5.2 From Neocentromeres to Mature Centromeres

Neocentromeres are new centromeres that have emerged from non-centromeric regions. There are two types of neocentromeres: the first type is associated with chromosome rearrangements, which results in the loss of the original centromere and the formation of a neocentromere on the remaining chromosome; the second type is due to the inactivation of the original centromere coincident with the formation of a new centromere on the same chromosome (Marshall et al. 2008). For the more uncommon, second type of neocentromere, the DNA sequences of the original centromere remain on the chromosome, but CENH3 is only associated with the neocentromere (Marshall et al. 2008). This type of neocentromere may survive and result in a 'centromere repositioning' (Brown and O'Neill 2010). The extensive research on neocentromere formation and centromere repositioning suggests that all centromeres may have evolved from neocentromeres.

Although more than 100 cases of neocentromeres have been discovered in humans, only a few neocentromeres have been reported in plants, all in cereal species (Nasuda et al. 2005; Gong et al. 2009; Topp et al. 2009). These neocentromeres are found on abnormal chromosomes devoid of their normal centromere region; therefore, they are also missing the repetitive DNA found at the original centromeres. Nasuda et al. (2005) were the first to demonstrate a neocentromere in plants on barley chromosome 7 H. This barley chromosome was induced using a gametocidal system which causes chromosomal mutations in barley chromosomes added to common wheat. Using this system neocentromeres were found on chromosomes lacking both barley specific centromere repeats and retroelements. Similarly, Topp et al. (2009) and Gong et al. (2009) reported neocentromeres on maize chromosome 3 and rice chromosome 8 respectively, which contained no centromere satellites or retroelements. All three of these plant neocentromeres were associated with CENH3, but their transmittance varied. Interestingly, compared to barley neocentromeres a reduced stability of transmittance and a wide variation in the size of the maize neocentromere was seen. This suggests the stability of neocentromeres is enhanced in subsequent generations through the reinforcement of CENH3 from previous cell cycles and through the recruitment of repetitive elements forming heterochromatic regions. Although repetitive elements are not necessary for neocentromere function it could possibly aid in stopping CENH3 removal from these new centromere sites. The increased size of the heterochromatic region could make for a highly favorable atmosphere for CENH3 incorporation and lead to the mature state of centromeres present at large repetitive regions (Topp et al. 2009).

4.5.3 Rice Centromere 8, a Case Study for Plant Centromere Evolution

The centromere of rice chromosome 8 (Cen8) contains only approximately 65 kb of the CentO repeat (Cheng et al. 2002; Nagaki et al. 2004). The lack of a large CentO array allowed a minimum tiling path of BAC clones, including and extending past the Cen8 CentO array, to be sequenced and assembled into the first fully sequenced centromere in any multicellular eukaryote. The CENH3 binding domain of Cen8 was localized to a 750 kb region (Nagaki et al. 2004). The most important and unexpected discovery in the sequencing of Cen8 was the identification of actively transcribing genes within the centromeric region (Nagaki et al. 2004). CENH3 binding throughout Cen8 is not continuous, but intermingled with H3 containing nucleosomes. A more detailed mapping of CENH3 binding across Cen8 was obtained using next-generation sequencing of enriched centromeric DNA from ChIP (ChIP-Seq) with CENH3 antibodies (Yan et al. 2008). By exploiting the relatively low amount of repetitive elements in Cen8, CENH3 binding was defined to 6 subdomains within the previously defined broader CENH3 binding domain (Fig. 4.2). Interestingly, the genes in Cen8 were all located within subdomains that are devoid of CENH3 (Fig. 4.2).

It was proposed that Cen8 in rice represents an intermediate stage in centromere evolution. This intermediate stage of centromere evolution is characterized by the lack of extensive amounts of satellite repeats and the retention of active genes. The active genes in such evolving centromeres are associated with H3 nucleosomes, thus, their regulation and epigenetic modifications are likely similar to those located outside of the centromeres. Such centromeres may gradually accumulate a large amount of satellite repeats and retrotransposons and eventually push the genes to outside of the centromeres (Yan et al. 2006). Therefore, evolutionarily intermediate centromeres like Cen8 in rice represent a link from young neocentromeres formed in euchromatic regions to mature centromeres that have amassed large amounts of transposons and tandem repeats.

4.6 Epigenetic Landscape of Centromeres

Neocentromere formation is the best evidence that the establishment and maintenance of eukaryotic centromeres are not defined by the underlying DNA sequences, but rather are determined by epigenetic mechanisms. However, there has been limited information as to what epigenetic modifications are important or specific to centromeric chromatin. DNA methylation and histone modifications potentially provide such epigenetic marks for the differentation between centromeric and pericentromeric chromatin.

Cytological mapping of 5-methylcytosine on highly stretched meiotic pachytene chromosomes revealed that the satellite repeats in CENH3-associated chromatin (CEN chromatin) in both *Arabidopsis thaliana* and maize are hypomethylated (Zhang et al. 2008; Koo and Jiang 2009). In contrast, the same satellite repeats located in pericentromeric regions were hypermethylated. Thus, the same satellite repeats can be either hypomethylated or hypermethylated, depending on their association with CEN chromatin or with pericentromeric heterochromatin. Zhang et al. (2008) cloned the 178 bp centromeric repeats associated with CENH3. Sequence analysis of the cloned repeats associated with CENH3 revealed a distinct distribution pattern of CG and CHG sites compared to the 178 bp repeats not assoceated with CEN chromatin. Thus, although all 178 bp repeats in *A. thaliana* share high sequence similarity, a unique CG/CHG composition may be sufficient to provide the foundation for the differential methylation of the CENH3-associated repeats from those located in the pericentromeric regions (Zhang et al. 2008).

Several recent studies in animal species have shown that histone modifications may play a role in centromere identity. The CEN chromatin in humans and *D. melanogaster* displayed histone modification patterns that are distinct from those of both euchromatin and heterochromatin

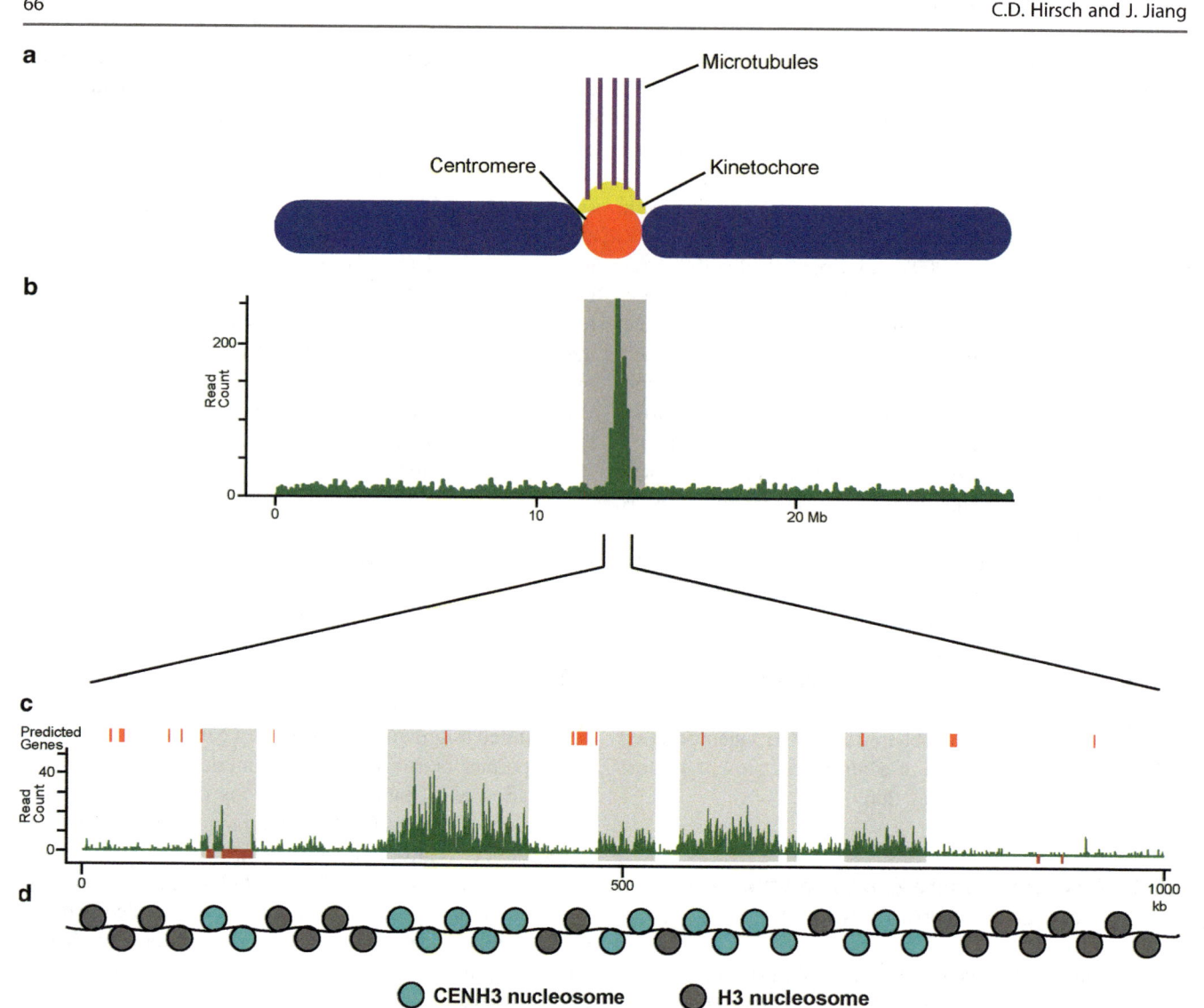

Fig. 4.2 Characterization of the centromere of rice chromosome 8. (**a**) General structure of a chromosome displaying chromosome arms (*blue*), centromere (*red*), kinetochore (*yellow*), and microtubules (*purple*). Only one sister chromatid is shown for simplicity. (**b**) Mapping of 454 sequence reads derived from ChIP-Seq against CENH3 across the entire chromosome 8 of rice. *Green bars* are the number of reads mapped to each location. The *gray box* indicates the crossing over-suppressed region of the chromosome. Enrichment of CENH3 binding can be seen within the crossing over-suppressed region by an increase in reads mapped (*large green peaks*). (**c**) Close-up of the CENH3-binding domain. Each of the six CENH3 binding subdomains are shown by *gray boxes*, locations of predicted genes are shown, as are the locations of the CentO arrays (*red boxes* beneath the mapped reads). Notice a high frequency of predicted genes is located outside of the mapped CENH3 binding subdomains. (**b**) and (**c**) adapted from Yan et al. (2008). (**d**) Representation of nucleosome pattern across Cen8 in rice. *Gray circles* represent nucleosomes with canonical histone H3 and *cyan circles* represent CENH3 containing nucleosomes (not drawn to scale)

(Sullivan and Karpen 2004). Specifically, CEN chromatin is enriched with H3 dimethylation at K4 (H3K4me2), an euchromatin epigenetic mark associated with permissive transcription, but it lacks other euchromatin marks, including H3K4me3 and acetylated H3 and H4, which are linked with active transcription. The CENH3-associated chromatin also lacks the heterochromatic marks H3K9me2 and H3K9me3. In contrast, the pericentromeric chromatin that flanks the CEN chromatin is enriched with H3K9me2 (Sullivan and Karpen 2004). This distinct histone modification pattern was proposed to contribute to the three-

dimensional structure of the centromere and to the epigenetic marking of centromere identity (Sullivan and Karpen 2004). In *A. thaliana*, CEN chromatin is associated with a significantly reduced level of H3K9me2 compared with the level of H3K9me2 in the pericentromeric heterochromatin. However, this low level of H3K9me2 is clearly visible in the CENH3-associated domains on pachytene chromosomes (Zhang et al. 2008). The presence of H3K9me2 in centromeres was also observed in both normal chromosomes and the supernumerary B chromosomes in maize (Shi and Dawe 2006; Jin et al. 2008). In addition, H3K4me2 was not observed in the maize centromeres (Shi and Dawe 2006; Jin et al. 2008). These results demonstrated that the abundance of H3K4 methylation and the absence of H3K9 methylation, which is characteristic of animal centromeres, is not typical in plant centromeres. Thus, the epigenetic modification patterns of centromeric chromatin are not universal among different eukaryotes, and need to be investigated across more diversified species.

4.7 Closing

Centromeres are complex regions of chromosomes which are an intricate arrangement and interaction of DNA, RNA, proteins, and epigenetic marks. Many questions remain unanswered in centromere research, but recent research has begun to unravel the potential evolutionary pathway of centromeres, from neocentromeres lacking repetitve elements to mature centromeres rich in repetitive elements. The ongoing sequencing of an ever more diverse set of plant species in the genomics era is allowing for new centromere components to be identified, more precise mapping of binding proteins and epigenetic modifications, and investigations into centromere evolution. Future centromere research in these areas will aid in unraveling the mysteries surrounding such an integral and necessary component for species survival.

Acknowledgments Research on rice and maize centromeres in the authors' laboratory have been supported by Grants DBI-0603927, DBI-0923640, and DBI-0922703 from the National Science Foundation.

References

Alfenito MR, Birchler JA (1993) Molecular characterization of a maize B chromosome centric sequence. Genetics 135:589–597

Ananiev EV, Phillips RL, Rines HW (1998) Chromosome-specific molecular organization of maize (*Zea mays* L.) centromeric regions. Proc Natl Acad Sci USA 95:13073–13078

Bao W, Zhang W, Yang Q, Zhang Y, Han B, Gu M, Xue Y, Cheng Z (2006) Diversity of centromeric repeats in two closely related wild rice species, *Oryza officinalis* and *Oryza rhizomatis*. Mol Genet Genomics 275:421–430

Black BE, Foltz DR, Chakravarthy S, Luger K, Woods VL, Cleveland DW (2004) Structural determinants for generating centromeric chromatin. Nature 430:578–582

Bouzinba-Segard H, Guais A, Francastel C (2006) Accumulation of small murine minor satellite transcripts leads to impaired centromeric architecture and function. Proc Natl Acad Sci USA 103:8709–8714

Brown JD, O'Neill RJ (2010) Chromosomes, conflict, and epigenetics: chromosomal speciation revisited. Annu Rev Genomics Hum Genet 11:291–316

Buchwitz BJ, Ahmad K, Moore LL, Roth MB, Henikoff S (1999) Cell division: a histone-H3-like protein in *C. elegans*. Nature 401:547–548

Carone D, Longo M, Ferreri G, Hall L, Harris M, Shook N, Bulazel K, Carone B, Obergfell C, O'Neill M, O'Neill R (2009) A new class of retroviral and satellite encoded small RNAs emanates from mammalian centromeres. Chromosoma 118:113–125

Cheng Z, Dong F, Langdon T, Ouyang S, Buell CR, Gu M, Blattner FR, Jiang J (2002) Functional rice centromeres are marked by a satellite repeat and a centromere-specific retrotransposon. Plant Cell 14:1691–1704

Cooper JL, Henikoff S (2004) Adaptive evolution of the histone fold domain in centromeric histones. Mol Biol Evol 21:1712–1718

Dong F, Miller JT, Jackson SA, Wang G-L, Ronald PC, Jiang J (1998) Rice (*Oryza sativa*) centromeric regions consist of complex DNA. Proc Natl Acad Sci USA 95:8135–8140

Du J, Tian Z, Hans CS, Laten HM, Cannon SB, Jackson SA, Shoemaker RC, Ma J (2010a) Evolutionary conservation, diversity and specificity of LTR-retrotransposons in flowering plants: insights from genome-wide analysis and multi-specific comparison. Plant J 63:584–598

Du Y, Topp CN, Dawe RK (2010b) DNA binding of centromere protein C (CENPC) is stabilized by single-stranded RNA. PLoS Genet 6:e1000835

Dunleavy EM, Roche D, Tagami H, Lacoste N, Ray-Gallet D, Nakamura Y, Daigo Y, Nakatani Y, Almouzni-Pettinotti G (2009) HJURP Is a cell-cycle-dependent maintenance and deposition factor of CENP-A at centromeres. Cell 137:485–497

Earnshaw WC, Rothfield N (1985) Identification of a family of human centromere proteins using autoimmune sera from patients with scleroderma. Chromosoma 91:313–321

Flagel L, Blackman B (2012) The first ten years of plant genome sequencing and prospects for the next decade. In: Wendel JF (ed) Plant genome diversity, vol 1, Plant genomes, their residents, and their evolutionary dynamics. Springer, Wien, New York

Foltz DR, Jansen LET, Black BE, Bailey AO, Yates JR, Cleveland DW (2006) The human CENP-A centromeric nucleosome-associated complex. Nat Cell Biol 8:458–469

Foltz DR, Jansen LET, Bailey AO, Yates Iii JR, Bassett EA, Wood S, Black BE, Cleveland DW (2009) Centromere-specific assembly of CENP-A nucleosomes is mediated by HJURP. Cell 137:472–484

Fujita Y, Hayashi T, Kiyomitsu T, Toyoda Y, Kokubu A, Obuse C, Yanagida M (2007) Priming of centromere for CENP-A recruitment by human hMis18α, hMis18β, and M18BP1. Dev Cell 12:17–30

Fukagawa T, Nogami M, Yoshikawa M, Ikeno M, Okazaki T, Takami Y, Nakayama T, Oshimura M (2004) Dicer is essential for formation of the heterochromatin structure in vertebrate cells. Nat Cell Biol 6:784–791

Gao D, Gill N, Kim HR, Walling JG, Zhang W, Fan C, Yu Y, Ma J, SanMiguel P, Jiang N, Cheng Z, Wing RA, Jiang J, Jackson SA (2009) A lineage-specific centromere retrotransposon in *Oryza brachyantha*. Plant J 60:820–831

Goff SA, Ricke D, Lan TH, Presting G, Wang RL, Dunn M, Glazebrook J, Sessions A, Oeller P, Varma H, Hadley D, Hutchinson D, Martin C, Katagiri F, Lange BM, Moughamer T, Xia Y, Budworth P, Zhong JP, Miguel T, Paszkowski U, Zhang SP, Colbert M, Sun WL, Chen LL, Cooper B, Park S, Wood TC, Mao L, Quail P, Wing R, Dean R, Yu YS, Zharkikh A, Shen R, Sahasrabudhe S, Thomas A, Cannings R, Gutin A, Pruss D, Reid J, Tavtigian S, Mitchell J, Eldredge G, Scholl T, Miller RM, Bhatnagar S, Adey N, Rubano T, Tusneem N, Robinson R, Feldhaus J, Macalma T, Oliphant A, Briggs S (2002) A draft sequence of the rice genome (*Oryza sativa* L. ssp *japonica*). Science 296:92–100

Gong Z, Yu H, Huang J, Yi C, Gu M (2009) Unstable transmission of rice chromosomes without functional centromeric repeats in asexual propagation. Chromosome Res 17:863–872

Guo Y-L, Ge S (2005) Molecular phylogeny of Oryzeae (Poaceae) based on DNA sequences from chloroplast, mitochondrial, and nuclear genomes. Am J Bot 92:1548–1558

Hayashi T, Fujita Y, Iwasaki O, Adachi Y, Takahashi K, Yanagida M (2004) Mis16 and Mis18 are required for CENP-A loading and histone deacetylation at centromeres. Cell 118:715–729

Hemmerich P, Weidtkamp-Peters S, Hoischen C, Schmiedeberg L, Erliandri I, Diekmann S (2008) Dynamics of inner kinetochore assembly and maintenance in living cells. J Cell Biol 180:1101–1114

Henikoff S, Ahmad K, Platero JS, van Steensel B (2000) Heterochromatic deposition of centromeric histone H3-like proteins. Proc Natl Acad Sci USA 97:716–721

Henikoff S, Ahmad K, Malik HS (2001) The centromere paradox: stable inheritance with rapidly evolving DNA. Science 293:1098–1102

Heslop-Harrison JS, Murata M, Ogura Y, Schwarzacher T, Motoyoshi F (1999) Polymorphisms and genomic organization of repetitive DNA from centromeric regions of *Arabidopsis* chromosomes. Plant Cell 11:31–42

Heslop-Harrison JS, Brandes A, Schwarzacher T (2003) Tandemly repeated DNA sequences and centromeric chromosomal regions of *Arabidopsis* species. Chromosome Res 11:241–253

Hirsch CD, Wu YF, Yan HH, Jiang JM (2009) Lineage-specific adaptive evolution of the centromeric protein CENH3 in diploid and allotetraploid *Oryza* species. Mol Biol Evol 26:2877–2885

Houben A, Schroeder-Reiter E, Nagaki K, Nasuda S, Wanner G, Murata M, Endo TR (2007) CENH3 interacts with the centromeric retrotransposon cereba and GC-rich satellites and locates to centromeric substructures in barley. Chromosoma 116:275–283

Hudakova S, Michalek W, Presting GG, Rt H, Kd S, Jasencakova Z, Schubert I (2001) Sequence organization of barley centromeres. Nucleic Acids Res 29:5029–5035

Jansen LET, Black BE, Foltz DR, Cleveland DW (2007) Propagation of centromeric chromatin requires exit from mitosis. J Cell Biol 176:795–805

Jiang JM, Birchler JA, Parrott WA, Dawe RK (2003) A molecular view of plant centromeres. Trends Plant Sci 8:570–575

Jin WW, Melo JR, Nagaki K, Talbert PB, Henikoff S, Dawe RK, Jiang JM (2004) Maize centromeres: organization and functional adaptation in the genetic background of oat. Plant Cell 16:571–581

Jin W, Lamb JC, Vega JM, Dawe RK, Birchler JA, Jiang J (2005) Molecular and functional dissection of the maize B chromosome centromere. Plant Cell 17:1412–1423

Jin W, Lamb J, Zhang W, Kolano B, Birchler J, Jiang J (2008) Histone modifications associated with both A and B chromosomes of maize. Chromosome Res 16:1203–1214

Kamm A, Galasso I, Schmidt T, Heslop-Harrison JS (1995) Analysis of a repetitive DNA family from *Arabidopsis arenosa* and relationships between *Arabidopsis* species. Plant Mol Biol 27:853–862

Kaul S, Koo HL, Jenkins J, Rizzo M, Rooney T, Tallon LJ, Feldblyum T, Nierman W, Benito MI, Lin XY, Town CD, Venter JC, Fraser CM, Tabata S, Nakamura Y, Kaneko T, Sato S, Asamizu E, Kato T, Kotani H, Sasamoto S, Ecker JR, Theologis A, Federspiel NA, Palm CJ, Osborne BI, Shinn P, Conway AB, Vysotskaia VS, Dewar K, Conn L, Lenz CA, Kim CJ, Hansen NF, Liu SX, Buehler E, Altafi H, Sakano H, Dunn P, Lam B, Pham PK, Chao Q, Nguyen M, Yu GX, Chen HM, Southwick A, Lee JM, Miranda M, Toriumi MJ, Davis RW, Wambutt R, Murphy G, Dusterhoft A, Stiekema W, Pohl T, Entian KD, Terryn N, Volckaert G, Salanoubat M, Choisne N, Rieger M, Ansorge W, Unseld M, Fartmann B, Valle G, Artiguenave F, Weissenbach J, Quetier F, Wilson RK, de la Bastide M, Sekhon M, Huang E, Spiegel L, Gnoj L, Pepin K, Murray J, Johnson D, Habermann K, Dedhia N, Parnell L, Preston R, Hillier L, Chen E, Marra M, Martienssen R, McCombie WR, Mayer K, White O, Bevan M, Lemcke K, Creasy TH, Bielke C, Haas B, Haase D, Maiti R, Rudd S, Peterson J, Schoof H, Frishman D, Morgenstern B, Zaccaria P, Ermolaeva M, Pertea M, Quackenbush J, Volfovsky N, Wu DY, Lowe TM, Salzberg SL, Mewes HW, Rounsley S, Bush D, Subramaniam S, Levin I, Norris S, Schmidt R, Acarkan A, Bancroft I, Brennicke A, Eisen JA, Bureau T, Legault BA, Le QH, Agrawal N, Yu Z, Copenhaver GP, Luo S, Pikaard CS, Preuss D, Paulsen IT, Sussman M, Britt AB, Selinger DA, Pandey R, Mount DW, Chandler VL, Jorgensen RA, Pikaard C, Juergens G, Meyerowitz EM, Dangl J, Jones JDG, Chen M, Chory J, Somerville MC, Ar Gen I (2000) Analysis of the genome sequence of the flowering plant *Arabidopsis thaliana*. Nature 408:796–815

Kawabe A, Nasuda S (2005) Structure and genomic organization of centromeric repeats in *Arabidopsis* species. Mol Genet Genomics 272:593–602

Kejnovsky E, Hawkins J, Feschotte C (2012) Plant transposable elements: biology and evolution. In: Wendel JF (ed) Plant genome diversity, vol 1, Plant genomes, their residents, and their evolutionary dynamics. Springer, Wien, New York

Kikuchi S, Kishii M, Shimizu M, Tsujimoto H (2005) Centromere-specific repetitive sequences from *Torenia*, a model plant for interspecific fertilization, and whole-mount FISH of its interspecific hybrid embryos. Cytogenet Genome Res 109:228–235

Koo D-H, Jiang J (2009) Super-stretched pachytene chromosomes for fluorescence in situ hybridization mapping and immunodetection of DNA methylation. Plant J 59:509–516

Kordis D (2005) A genomic perspective on the chromodomain-containing retrotransposons: chromoviruses. Gene 347:161–173

Kumekawa N, Hosouchi T, Tsuruoka H, Kotani H (2000) The size and sequence organization of the centromeric region of *Arabidopsis thaliana* chromosome 5. DNA Res 7:315–321

Kumekawa N, Hosouchi T, Tsuruoka H, Kotani H (2001) The size and sequence organization of the centromeric region of *Arabidopsis thaliana* chromosome 4. DNA Res 8:285–290

Langdon T, Seago C, Mende M, Leggett M, Thomas H, Forster JW, Thomas H, Jones RN, Jenkins G (2000) Retrotransposon evolution in diverse plant genomes. Genetics 156:313–325

Lee HR, Zhang WL, Langdon T, Jin WW, Yan HH, Cheng ZK, Jiang JM (2005) Chromatin immunoprecipitation cloning reveals rapid evolutionary patterns of centromeric DNA in *Oryza* species. Proc Natl Acad Sci USA 102:11793–11798

Lee HR, Neumann P, Macas J, Jiang JM (2006) Transcription and evolutionary dynamics of the centromeric satellite repeat CentO in rice. Mol Biol Evol 23:2505–2520

Lermontova I, Schubert V, Fuchs J, Klatte S, Macas J, Schubert I (2006) Loading of *Arabidopsis* centromeric histone CENH3 occurs mainly during G2 and requires the presence of the histone fold domain. Plant Cell 18:2443–2451

Lermontova I, Fuchs Jr, Schubert V, Schubert I (2007) Loading time of the centromeric histone H3 variant differs between plants and animals. Chromosoma 116:507–510

.01w?>Liu Z, Yue W, Li D, Wang R, Kong X, Lu K, Wang G, Dong Y, Jin W, Zhang X (2008) Structure and dynamics of retrotransposons at wheat centromeres and pericentromeres. Chromosoma 117:445–456

Malik HS, Henikoff S (2001) Adaptive evolution of cid, a centromere-specific histone in drosophila. Genetics 157:1293–1298

Malik HS, Henikoff S (2002) Conflict begets complexity: the evolution of centromeres. Curr Opin Genet Dev 12:711–718

Maluszynska J, Heslop-Harrison J (1991) Localization of tandemly repeated DMA sequences in *Arabidopsis thaliana*. Plant J 1:159–166

Marshall OJ, Chueh AC, Wong LH, Choo KHA (2008) Neocentromeres: new insights into centromere structure, disease development, and karyotype evolution. Am J Hum Genet 82:261–282

Martinez-Zapater JM, Estelle MA, Somerville CR (1986) A highly repeated DNA sequence in *Arabidopsis thaliana*. Mol Gen Genet 204:417–423

Meluh PB, Yang PR, Glowczewski L, Koshland D, Smith MM (1998) Cse4p is a component of the core centromere of *Saccharomyces cerevisiae*. Cell 94:607–613

Miller JT, Dong F, Jackson SA, Song J, Jiang J (1998) Retrotransposon-related DNA sequences in the centromeres of grass chromosomes. Genetics 150:1615–1623

Murata M, Ogura Y, Motoyoshi F (1994) Centromeric repetitive sequences in *Arabidopsis thaliana*. Jap J Genet 69:361–370

Nagaki K, Murata M (2005) Characterization of CENH3 and centromere-associated DNA sequences in sugarcane. Chromosome Res 13:195–203

Nagaki K, Talbert PB, Zhong CX, Dawe RK, Henikoff S, Jiang JM (2003) Chromatin immunoprecipitation reveals that the 180-bp satellite repeat is the key functional DNA element of *Arabidopsis thaliana* centromeres. Genetics 163:1221–1225

Nagaki K, Cheng ZK, Ouyang S, Talbert PB, Kim M, Jones KM, Henikoff S, Buell CR, Jiang JM (2004) Sequencing of a rice centromere uncovers active genes. Nat Genet 36:138–145

Nagaki K, Neumann P, Zhang D, Ouyang S, Buell CR, Cheng Z, Jiang J (2005) Structure, divergence, and distribution of the CRR centromeric retrotransposon family in rice. Mol Biol Evol 22:845–855

Nagaki K, Kashihara K, Murata M (2009) A centromeric DNA sequence colocalized with a centromere-specific histone H3 in tobacco. Chromosoma 118:249–257

Nasuda S, Hudakova S, Schubert I, Houben A, Endo TR (2005) Stable barley chromosomes without centromeric repeats. Proc Natl Acad Sci USA 102:9842–9847

Neumann P, Yan HH, Jiang JM (2007) The centromeric retrotransposons of rice are transcribed and differentially processed by RNA interference. Genetics 176:749–761

Nonomura KI, Kurata N (1999) Organization of the 1.9-kb repeat unit RCE1 in the centromeric region of rice chromosomes. Mol Gen Genet MGG 261:1–10

Okada M, Cheeseman IM, Hori T, Okawa K, McLeod IX, Yates JR, Desai A, Fukagawa T (2006) The CENP-H-I complex is required for the efficient incorporation of newly synthesized CENP-A into centromeres. Nat Cell Biol 8:446–457

Pidoux AL, Choi ES, Abbott JKR, Liu X, Kagansky A, Castillo AG, Hamilton GL, Richardson W, Rappsilber J, He X, Allshire RC (2009) Fission yeast Scm3: a CENP-A receptor required for integrity of subkinetochore chromatin. Mol Cell 33:299–311

Presting GG, Malysheva L, Fuchs J, Schubert I (1998) ATY3/GYPSY retrotransposon-like sequence localizes to the centromeric regions of cereal chromosomes. Plant J 16:721–728

Ravi M, Kwong PN, Menorca RMG, Valencia JT, Ramahi JS, Stewart JL, Tran RK, Sundaresan V, Comai L, Chan SW-L (2010) The rapidly evolving centromere-specific histone has stringent functional requirements in *Arabidopsis thaliana*. Genetics: genetics.110.120337

Shi J, Dawe RK (2006) Partitioning of the maize epigenome by the number of methyl groups on histone H3 lysines 9 and 27. Genetics 173:1571–1583

Shibata F, Murata M (2004) Differential localization of the centromere-specific proteins in the major centromeric satellite of *Arabidopsis thaliana*. J Cell Sci 117:2963–2970

Shuaib M, Ouararhni K, Dimitrov S, Hamiche A (2010) HJURP binds CENP-A via a highly conserved N-terminal domain and mediates its deposition at centromeres. Proc Natl Acad Sci USA 107:1349–1354

Slotkin R, Nuthikattu S, Jiang N (2012) The impact of transposable elements on gene and genome evolution. In: Wendel JF (ed) Plant genome diversity, vol 1, Plant genomes, their residents, and their evolutionary dynamics. Springer, Wien, New York

Stoler S, Keith KC, Curnick KE, Fitzgerald-Hayes M (1995) A mutation in CSE4, an essential gene encoding a novel chromatin-associated protein in yeast, causes chromosome nondisjunction and cell cycle arrest at mitosis. Genes Dev 9:573–586

Sullivan BA, Karpen GH (2004) Centromeric chromatin exhibits a histone modification pattern that is distinct from both euchromatin and heterochromatin. Nat Struct Mol Biol 11:1076–1083

Sullivan KF, Hechenberger M, Masri K (1994) Human CENP-A contains a histone H3 related histone fold domain that is required for targeting to the centromere. J Cell Biol 127:581–592

Talbert PB, Masuelli R, Tyagi AP, Comai L, Henikoff S (2002) Centromeric localization and adaptive evolution of an *Arabidopsis* histone H3 variant. Plant Cell 14:1053–1066

Talbert P, Bryson T, Henikoff S (2004) Adaptive evolution of centromere proteins in plants and animals. J Biol 3:18

Tek AL, Jiang J (2004) The centromeric regions of potato chromosomes contain megabase-sized tandem arrays of telomere-similar sequence. Chromosoma 113:77–83

Tek A, Kashihara K, Murata M, Nagaki K (2010) Functional centromeres in soybean include two distinct tandem repeats and a retrotransposon. Chromosome Res 18:337–347

Topp CN, Zhong CX, Dawe RK (2004) Centromere-encoded RNAs are integral components of the maize kinetochore. Proc Natl Acad Sci USA 101:15986–15991

Topp CN, Okagaki RJ, Melo JR, Kynast RG, Phillips RL, Dawe RK (2009) Identification of a maize neocentromere in an oat-maize addition line. Cytogenet Genome Res 124:228–238

Torras-Llort M, Moreno-Moreno O, Azorin F (2009) Focus on the centre: the role of chromatin on the regulation of centromere identity and function. EMBO J 28:2337–2348

Williams JS, Hayashi T, Yanagida M, Russell P (2009) Fission yeast Scm3 mediates stable assembly of Cnp1/CENP-A into centromeric chromatin. Mol Cell 33:287–298

Wong LH, Brettingham-Moore KH, Chan L, Quach JM, Anderson MA, Northrop EL, Hannan R, Saffery R, Shaw ML, Williams E, Choo KHA (2007) Centromere RNA is a key component for the assembly of nucleoproteins at the nucleolus and centromere. Genome Res 17:1146–1160

Yan HH, Ito H, Nobuta K, Ouyang S, Jin WW, Tian SL, Lu C, Venu RC, Wang GL, Green PJ, Wing RA, Buell CR, Meyers BC, Jiang JM (2006) Genomic and genetic characterization of rice Cen3 reveals extensive transcription and evolutionary implications of a complex centromere. Plant Cell 18:2123–2133

Yan H, Talbert PB, Lee H-R, Jett J, Henikoff S, Chen F, Jiang J (2008) Intergenic locations of rice centromeric chromatin. PLoS Biol 6: e286

Yang C, Tomkiel J, Saitoh H, Johnson D, Earnshaw W (1996) Identification of overlapping DNA-binding and centromere-targeting domains in the human kinetochore protein CENP-C. Mol Cell Biol 16:3576–3586

Yu J, Hu SN, Wang J, Wong GKS, Li SG, Liu B, Deng YJ, Dai L, Zhou Y, Zhang XQ, Cao ML, Liu J, Sun JD, Tang JB, Chen YJ, Huang XB, Lin W, Ye C, Tong W, Cong LJ, Geng JN, Han YJ, Li L, Li W, Hu GQ, Huang XG, Li WJ, Li J, Liu ZW, Liu JP, Qi QH, Liu JS, Li T, Wang XG, Lu H, Wu TT, Zhu M, Ni PX, Han H, Dong W, Ren XY, Feng XL, Cui P, Li XR, Wang H, Xu X, Zhai WX, Xu Z, Zhang JS, He SJ, Zhang JG, Xu JC, Zhang KL, Zheng XW, Dong JH, Zeng WY, Tao L, Ye J, Tan J, Ren XD, Chen XW, He J, Liu DF, Tian W, Tian CG, Xia HG, Bao QY, Li G, Gao H, Cao T, Zhao WM, Li P, Chen W, Wang XD, Zhang Y, Hu JF, Liu S, Yang J, Zhang GY, Xiong YQ, Li ZJ, Mao L, Zhou CS, Zhu Z, Chen RS, Hao BL, Zheng WM, Chen SY, Guo W, Li GJ, Liu SQ, Tao M, Zhu LH, Yuan LP, Yang HM (2002) A draft sequence of the rice genome (*Oryza sativa* L. ssp. *indica*). Science 296:79–92

Zappulla DC, Cech TR (2004) Yeast telomerase RNA: a flexible scaffold for protein subunits. Proc Natl Acad Sci USA 101:10024–10029

Zhang L-B, Ge S (2007) Multilocus analysis of nucleotide variation and speciation in *Oryza officinalis* and its close relatives. Mol Biol Evol 24:769–783

Zhang W, Lee H-R, Koo D-H, Jiang J (2008) Epigenetic modification of centromeric chromatin: hypomethylation of DNA sequences in the CENH3-associated chromatin in *Arabidopsis thaliana* and maize. Plant Cell 20:25–34

Zhong CX, Marshall JB, Topp C, Mroczek R, Kato A, Nagaki K, Birchler JA, Jiang JM, Dawe RK (2002) Centromeric retroelements and satellites interact with maize kinetochore protein CENH3. Plant Cell 14:2825–2836

Telomeres and Their Biology

5

Maria F. Siomos and Karel Riha

Contents

5.1 A Historical Perspective

In contrast to prokaryotic chromosomes and plasmids, which are usually circular, most eukaryotic chromosomes are composed of linear DNA. The organisation of genomes into linear chromosomes poses two major challenges for chromosome metabolism. First, conventional DNA replication processes are unable to completely replicate the $3'$ ends of linear chromosomes (known as the 'end replication problem') and second, the natural ends of linear chromosomes must be distinguished from DNA double-strand breaks. This is necessary so that they are protected from being recognised as DNA damage and being inappropriately repaired by, for example, the formation of chromosome fusions. To overcome these problems, eukaryotes have evolved specialised nucleoprotein complexes at the ends of linear chromosomes called telomeres. Telomeres have their own replication mechanism and serve as a protective chromosome end capping structure.

The notion that chromosome ends have special properties that set them apart from other chromosomal regions stems from experiments undertaken by Hermann J. Müller and Barbara McClintock in the 1920s and 1930s. When observing chromosomes of the fruit fly *Drosophila melanogaster* that had undergone rearrangements as a result of exposure to X-rays, which can induce double-strand breaks in DNA, Müller noticed that the ensuing rearranged chromosomes always had natural chromosome ends: "...fragments, even though provided with a centromere, die if their ends are broken ends, i.e., if their ends do not consist of natural termini, derived from the same or another chromosome" (Muller 1938). In recognition of the special properties of natural chromosome ends, Müller and Darlington (and Haldane independently) named the 'free end' of chromosomes the 'telomere' (Muller 1938), from the Greek 'Τέλος' (telos—end) and 'μέρος' (meros—part).

McClintock's work demonstrated the importance of intact chromosome ends for genome stability. She was able

K. Riha (✉)
Austrian Academy of Sciences, Gregor Mendel Institute of Molecular Plant Biology, Dr. Bohr-Gasse 3, 1030 Vienna, Austria
e-mail: Karel.Riha@gmi.oeaw.ac.at

J.F. Wendel et al. (eds.), *Plant Genome Diversity Volume 1*,
DOI 10.1007/978-3-7091-1130-7_5, © Springer-Verlag Wien 2012

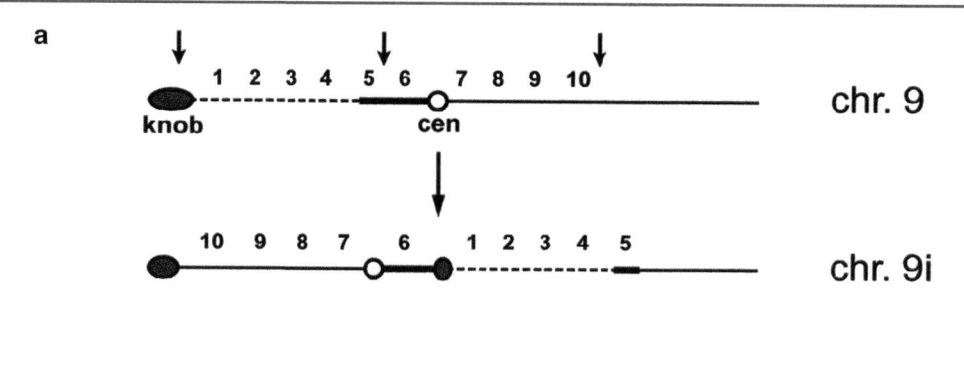

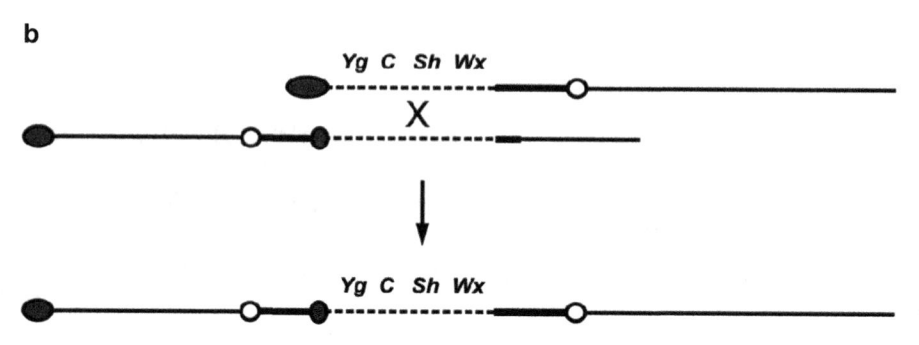

Fig. 5.1 Barbara McClintock's strategy for making dicentric chromosomes in maize. (**a**) McClintock used a rearranged chromosome (9i) derived from chromosome 9 using X-ray mutagenesis. The arrows point to positions of presumed breaks in chromosome 9, the rejoining of which generated a complex pericentric inversion. The heterochromatic knob and the centromere are indicated. (**b**) Meiotic recombination between chromosomes 9 and 9i in region 1–5 generates a dicentric chromosome in metaphase I. Four genes, *Yg*, *C*, *Sh* and *Wx*, are located in the translocated segment 1–5 that is positioned between the centromeres in the dicentric chromosome. Loss of the genes leads to easily detectable phenotypes in plants (*Yg*—normal green plant, *yg*—yellow green plant) or in endosperm (*C*—coloured aleuron, *c*—colourless aleuron, *Sh*—normal endosperm, *sh*—shrunken endosperm, *Wx*—normal starch in endosperm, *wx*—waxy endosperm) that were used to monitor the fate of the dicentric chromosomes in developing seeds and plants (Adapted from McClintock 1939)

to induce the formation of a dicentric chromosome in maize (*Zea mays*) during the first meiotic division, by recombination between chromosome 9 and its rearranged derivative that contained a complex pericentric inversion (Fig. 5.1) (McClintock 1939). The dicentric chromatid could be attached to both poles of the spindle and, as a result, when the chromosomes were separated at anaphase, the force of the mitotic spindle caused the dicentric chromosome to break, resulting in the formation of broken ends. Following DNA replication in S phase, such broken ends tended to fuse, forming a bridge, which, in the next round of mitosis, led to a new breakage at anaphase and the formation of new broken ends. This cycle of events, which is perpetuated when a broken chromosome end enters mitosis, is known as the breakage–fusion–bridge cycle. McClintock observed that while this cycle was continuous in maize endosperm, broken chromosome ends became stable and did not go through the breakage–fusion–bridge cycle during mitotic nuclear divisions in maize embryos (McClintock 1941). She, thus, postulated that it was possible for broken chromosome ends to 'heal'. It is now known that McClintock's 'healing' corresponds to the *de novo* formation of telomeres at the ends of chromosomes, as a result of the activity of the enzyme telomerase, which restores the protective capping

structure. Furthermore, the fact that McClintock observed healing in the embryo and not in the endosperm is due to the differential expression of telomerase in different tissues (Kilian et al. 1998).

The field of telomere biology has advanced significantly since Müller and McClintock's initial work in fruit flies and maize. Research in a whole host of additional model organisms, including ciliated protozoans, yeast, mouse, human cell lines, nematode worms and the thale cress *Arabidopsis thaliana*, has contributed to understanding the functional mechanisms and evolution of telomeres. The 2009 Nobel Prize in Physiology or Medicine was awarded to Elizabeth Blackburn, Carol Greider and Jack Szostak "for the discovery of how chromosomes are protected by telomeres and the enzyme telomerase".

5.2 Telomere Structure

5.2.1 Telomeric DNA

Telomeres are composed of specialised chromatin that assembles at the ends of chromosomes. The first telomeric DNA sequences to be identified were in the ciliated protozoan *Tetrahymena thermophila*. The somatic nucleus of *T. thermophila* contains a large number of linear minichromosomes. Due to the existence and abundance of these minichromosomes (<100 kb), it was possible to determine the sequence of their ends (Blackburn and Gall 1978). Each end of a minichromsome has about 50 tandemly-organised hexanucleotide TTGGGG repeats. The telomeric sequences of numerous organisms, including many plant species, have since been defined (Table 5.1). They are remarkably conserved and usually consist of regular tetra– to octanucleotide GT-rich tandem repeats. A notable exception is in the budding yeast *Saccharomyces cerevisiae*, in which telomeres have irregular TG_{1-3} repeats.

The first telomeric repeat to be sequenced in a higher eukaryote was that of the model plant *A. thaliana*, in which the canonical repeat TTTAGGG was discovered (Richards and Ausubel 1988). This sequence is found in the majority of plant species, from the moss *Physcomitrella patens* and gymnosperms to crop species such as maize, barley and peas. The only exception in higher plants is a clade of Asparagales, which includes over 6,000 species, that contains the vertebrate-like telomeric sequence, TTAGGG (Fajkus et al. 2005). It is likely that this change in sequence is due to a mutation that occurred around 80 Mya (Adams et al. 2001), which altered the RNA template of the RNA subunit of telomerase. Several plant species, such as onion and some members of the family Solanaceae, lack canonical telomeric repeats, and the precise structure of telomeres in these species has yet to be identified (Pich et al. 1996; Sykorova et al. 2003). The TTTAGGG sequence is also present in the unicellular green alga *Chlorella vulgaris*, whereas another green alga, *Chlamydomonas reinhardtii*, has an extra T in the telomeric repeat (TTTTAGGG) (Petracek et al. 1990).

While telomeric sequences are highly conserved among eukaryotes, there is notable variation in the length of telomeric repeat arrays. In general, unicellular species with a small genome tend to have short telomeres, while species with larger genomes have substantially longer telomeres (Table 5.1). This is particularly obvious in the plant kingdom, where unicellular algae have telomeres of 0.5 kb, while telomeres in some higher plants, such as tobacco and barley, exceed 100 kb (Petracek et al. 1990; Fajkus et al. 1995; Kilian et al. 1995). Significant variation in telomere length has also been uncovered within individual species. For example, average telomere length in maize inbred lines ranges from 2 to 20 kb, and a fivefold difference in telomere length is found among different *Arabidopsis* ecotypes (Burr et al. 1992; Shakirov and Shippen 2004). It currently is unclear whether this variation in telomere length has functional relevance.

5.2.2 Secondary Structure of Telomeric DNA

Telomeric sequences are characterised by clusters of guanine and complementary cytosine bases, which give rise to G-rich and C-rich telomeric strands, respectively. The G-rich strand is always oriented with its $3'$ end towards the chromosome terminus and is longer than the complementary C-rich strand, thus resulting in a single-stranded overhang. In *Arabidopsis* and another dicot plant, *Silene latifolia*, single-stranded overhangs are estimated to be around 20–30 nucleotides in length (Riha et al. 2000). Whereas G-overhangs of a roughly similar size were detected in budding yeast and ciliates, G-overhangs are substantially longer in humans, reaching up to 300 nucleotides in length (Wright et al. 1997). Single-stranded G-rich overhangs can form a telomere loop (t-loop) (Fig. 5.2) by invading a duplex

Table 5.1 Telomeric repeat sequences in various species

Species	Sequence repeat	Telomere length
Non-plants		
Tetrahymena thermophila	TTGGGG	20–70 repeats
Homo sapiens	TTAGGG	10–15 kb at birth
Mus musculus	TTAGGG	>40 kb (20–50 kb)
Saccharomyces cerevisiae	TG_{1-3}	250–350 bp
Schizosaccharomyces pombe	TTACAGG	300 bp
Caenorhabditis elegans	TTAGGC	2–9 kb
Plants		
Arabidopsis thaliana	TTTAGGG	2–9 kb
Zea mays	TTTAGGG	2–40 kb
Pisum sativum	TTTAGGG	10–80 kb
Hordeum vulgare (barley)	TTTAGGG	20–80 kb
Chlorella vulgaris	TTTAGGG	0.5 kb
Chlamydomonas reinhardtii	TTTAGGGG	0.5–1 kb
Asparagales	TTAGGG	Not determined
Cestrum (Solanaceae)	No canonical repeat	
Allium cepa (onion)	No canonical repeat	

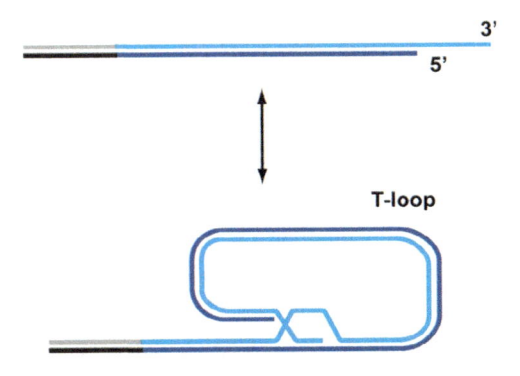

Fig. 5.2 Structure of telomeric DNA. The telomeric G-rich strand is oriented with its $3'$ end towards the chromosome terminus and is longer than the complementary C-rich strand, thus forming a $3'$ G-overhang. The G-overhang can invade duplex telomeric DNA, in a process resembling the first steps of intrachromatid recombination, creating a t-loop

region of telomeric repeats, thereby base-pairing to a complementary sequence on the C-rich strand and displacing the respective G-rich strand to form a displacement loop (D-loop). T-loops have been visualised by electron microscopy in a range of organisms, including mammals (Griffith et al. 1999), plants (Cesare et al. 2003), yeast (Cesare and Reddel 2008) and nematodes (Raices et al. 2008). It is proposed that the chromosome terminus, by sequestering the chromosome end in a t-loop, is protected from being recognised and processed by the DNA damage machinery. In some plants, including *Arabidopsis* and *Silene*, only half of the telomeres possess detectable 3' G-rich overhangs, which suggests that other end structures may be present in plants at a subset of chromosome ends (Riha et al. 2000). There is a similar situation in the nematode *Caenorhabditis elegans*, where a portion of chromosome ends contains protruding 5' C-overhangs instead of 3' G-overhangs (Raices et al. 2008).

5.2.3 Telomere Binding Proteins

5.2.3.1 The Shelterin Complex

Numerous proteins associate with telomeres and play an integral role in regulating telomere synthesis and chromosome end protection (Table 5.2). These proteins are evolutionarily highly conserved. Nevertheless, comparative analyses in multiple model organisms have demonstrated that their exact function in telomere metabolism may, in some cases, vary significantly. In mammals, the bulk of telomeric DNA is bound by the telomeric protein complex shelterin, which is composed of six telomere-specific proteins—TRF1 (telomeric repeat-binding factor 1), TRF2 (telomeric repeat-binding factor 2), POT1 (protection of telomeres 1), TIN2 (TRF2- and TRF1-interacting nuclear protein 2), TPP1 and RAP1 (repressor/activator protein 1) (Fig. 5.3). TRF1 and TRF2 directly bind to TTAGGG repeats in the double-stranded region of the telomere through their C-terminal Myb/SANT domain. Both of these proteins also bind to TIN2, which acts as a bridge between them and TPP1/POT1, which binds to single-stranded TTAGGG sequences via POT1. RAP1 binds to TRF2. The architecture of mammalian shelterin, thus, provides a link between double-stranded and single-stranded regions of telomeres. The shelterin complex appears to mediate several important functions at mammalian telomeres. The TRF2 protein is essential for the inhibition of chromosome end-to-end fusions and also promotes the formation and stabilisation of t-loops (van Steensel et al. 1998; Griffith et al. 1999; Wang et al. 2004; Poulet et al. 2009). POT1 represses DNA damage signalling from telomeres, presumably by binding to G-overhangs and preventing activation of the ATR checkpoint kinase (Denchi and de Lange 2007).

If the mammalian shelterin complex is compared to the telomeric protein complexes found in the fission and

Table 5.2 Telomeric components in human, budding yeast and *Arabidopsis*

Telomere component	Human	*S. cerevisiae*	*A. thaliana*
Shelterin	TRF1/TRF2	Tbf1 (lost telomeric function)	Group 1: TRB1, TRB2, TRB3
			Group 2: TRFL3, TRFL5, TRFL6, TRFL7, TRFL8, TRFL10
			Group 3: TBP1, TRP1, TRFL1, TRFL2, TRFL4, TRFL9
	RAP1	Rap1	Not identified
	POT1	–	POT1A, POT1B, POT1C
CST complex	CTC1	Cdc13	CTC1
	STN1	Stn1	STN1
	TEN1	Ten1	TEN1
DNA repair	KU70/KU86	Ku70/Ku80	KU70/KU80
	MRE11/RAD50/NBS1	Mre11/Rad50/Xrs2	MRE11/RAD50/NBS1
	ATM	Tel1	ATM
	ATR	Mec1	ATR
Telomerase	TERT	Est2	TERT
	TR (RNA subunit)	TLC1	TER1, TER2
	Dyskerin	–	Dyskerin
	SMG5, SMG6	Est1	SMG7

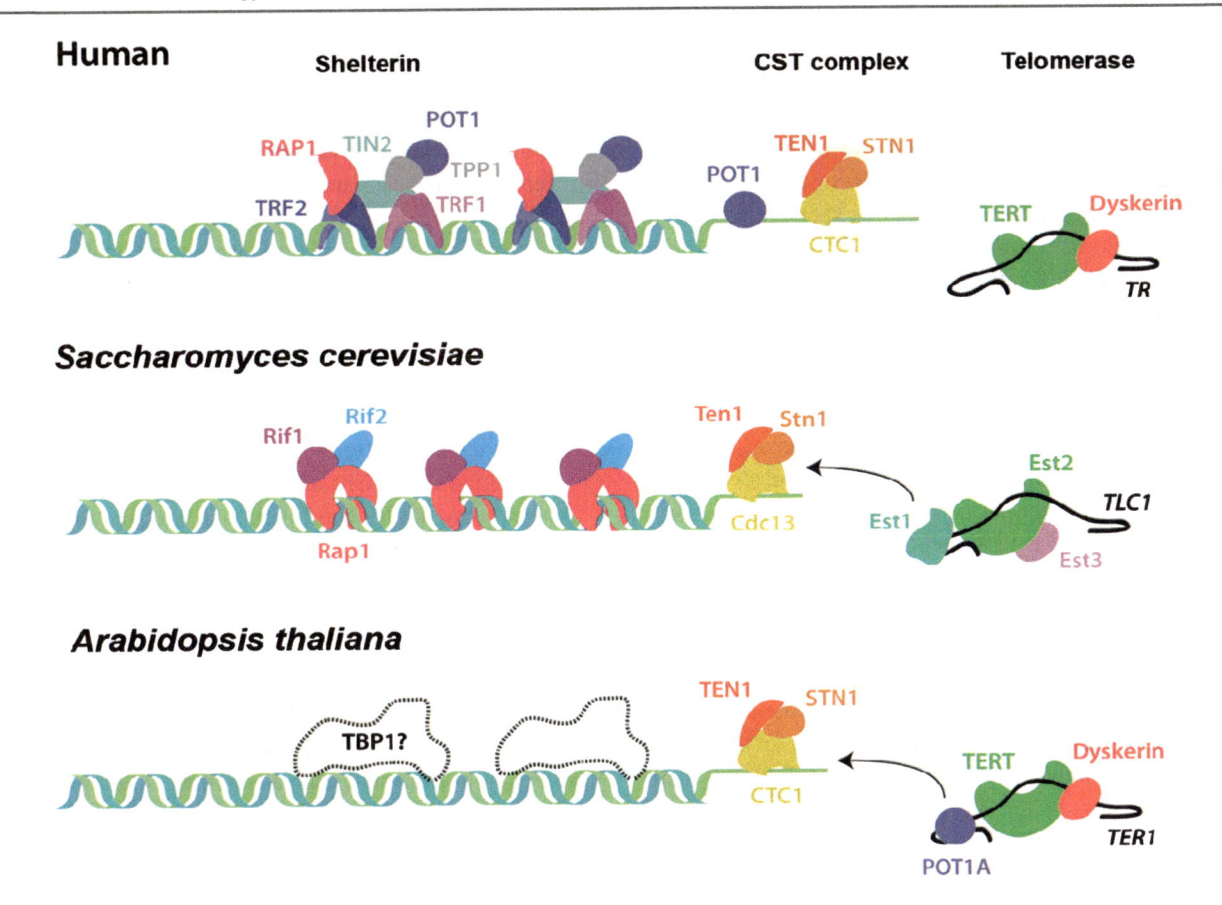

Fig. 5.3 Major telomere binding proteins and telomerase components in human, the budding yeast *Saccharomyces cerevisiae* and *Arabidopsis thaliana*. Proteins of the plant shelterin complex have not yet been identified. Arrows indicate a putative telomerase recruitment function of Est1 and POT1A in *S. cerevisiae* and *Arabidopsis*, respectively

budding yeasts, *Schizosaccharomyces pombe* and *S. cerevisiae*, respectively, there is a large degree of conservation with the *S. pombe* telomere-binding complex but the *S. cerevisiae* telomere-binding proteins are quite different. *S. cerevisiae* telomeric DNA is directly bound by Rap1 (Fig. 5.3), which is the only telomere-binding protein in budding yeast with homology to a shelterin component. It recruits several other proteins to telomeric chromatin, such as Rif1 and Rif2, which act to inhibit telomere extension by telomerase, and Sir2 histone deacetylase that promotes heterochromatin formation in telomere-adjacent regions (Bianchi and Shore 2008). Although there is a TRF-like protein in *S. cerevisiae* called Tbf1 that can bind the vertebrate telomeric motif TTAGGG in vitro, it does not bind to yeast telomeres. It rather plays an important role in transcriptional control linked to ribosome biogenesis (Preti et al. 2010).

As telomeric sequences are highly conserved between plants and vertebrates, it is likely that a shelterin-like complex also exists in plants. Bioinformatic and biochemical means have been employed in attempts to identify functional homologues of shelterin proteins in plants (reviewed in Watson and Riha 2010). However, although proteins with certain conserved features of shelterin components have been found in plant genomes (Table 5.2), functional components directly equivalent to shelterin proteins have yet to be confirmed. Plant genomes encode numerous TRF-like proteins that harbour the conserved Myb/SANT domain at their C-terminus. At least 12 such proteins are encoded in the *Arabidopsis* genome but not all of them are capable of binding telomeric DNA in vitro. Biochemical experiments have revealed that an additional 40 amino acid extension of the conserved Myb/SANT domain is required for efficient binding (Karamysheva et al. 2004). Structural analysis has shown that while the third helix of the Myb/SANT domain mediates the interaction with the bases located in the major groove of the DNA helix, the extension motif likely confers specificity to the plant telomere by forming a loop that extends direct contact to the minor groove (Ko et al. 2008). Despite this detailed knowledge about the binding mechanism, it is currently unclear which of the *Arabidopsis* TRF-like proteins represent functional orthologues of mammalian TRF1/TRF2 proteins and do indeed bind to telomeres in vivo, and which are transcription factors that bind to telomeric DNA motifs that are present at the promoters of many *Arabidopsis* genes. A promising candidate is the TBP1 protein (Fig. 5.3), the inactivation of which leads to altered telomere length homeostasis (Hwang and Cho 2007). Telomere phenotypes have also been observed upon mutation or downregulation of TBP1-related proteins in rice and tobacco (Yang et al. 2004; Hong et al. 2007). Nevertheless, none of the phenotypes is as strong as would be expected for the loss of the major telomere binding protein. This may be due to functional redundancy between similar proteins encoded by several related genes.

5.2.3.2 POT1

The only other component of the shelterin complex identified so far in plants is POT1. *POT1* was discovered in fission yeast as an essential gene required for chromosome end protection. The POT1 protein binds to single-stranded telomeric DNA (Baumann and Cech 2001). POT1 family proteins are characterised by two N-terminal oligosaccharide/oligonucleotide binding motifs (OB-folds) that mediate DNA interaction. As discussed above, in mammals and in fission yeasts, POT1 is tethered to telomeres as part of the shelterin complex. It is the binding of POT1 to single-stranded telomeric DNA that prevents the DNA damage response from being triggered (Denchi and de Lange 2007; Carneiro et al. 2010). While POT1 proteins can be readily identified by homology searches in plant genomes, their role in telomere metabolism appears to have diversified during plant evolution. The first plant POT1 proteins to be identified were in *Arabidopsis*, which possesses three *POT1* genes, namely *POT1A*, *POT1B* and *POT1C*. *POT1C* encodes a truncated protein consisting of only the conserved N-terminal domain. Interestingly, *Arabidopsis* POT1 proteins do not bind to telomeric DNA (Surovtseva et al. 2007; Shakirov et al. 2009a); instead, they appear to associate with the RNA subunit of telomerase (Cifuentes-Rojas et al. 2011). In accordance with these data, functional analysis in *Arabidopsis* has shown that POT1A is not required for chromosome end protection but is instead important for telomere replication by telomerase (Surovtseva et al. 2007). The original function in chromosome end protection may, nevertheless, still be retained in some plant lineages. Analysis of recombinant proteins from different plant groups has demonstrated the ability of POT1 to bind to single-stranded telomeric DNA in maize, asparagus, the green alga *Ostreococcus lucimarinus* and the moss *P. patens* (Shakirov et al. 2009b, 2010). Furthermore, knocking out *P. patens* POT1 results in immediate chromosome end-to-end fusions and an increased amount of single-stranded telomeric DNA, unveiling that POT1 is essential for chromosome end protection in moss (Shakirov et al. 2010). Thus, the evolution of POT1 in plants is an example of a remarkable functional switch of a highly conserved telomeric component.

5.2.3.3 The CST Complex

Telomere protection in budding yeast, which lacks shelterin and POT1 proteins, is dependent on the CST complex. The CST complex is a heterotrimer, consisting of Cdc13, Stn1 and Ten1, and is structurally related to the general single-strand DNA binding factor RPA (Gao et al. 2007). Cdc13 specifically binds to telomeric G-overhangs, where it serves as a loading platform for either telomerase or the Stn1/Ten1 complex that forms a telomere protective cap (Pennock et al. 2001; Puglisi et al. 2008). Furthermore, the CST complex physically associates with primase, indicating that it may also coordinate fill-in synthesis of the complementary C-strand after elongation of the G-strand by telomerase (Qi and Zakian 2000; Grossi et al. 2004).

Proteins structurally and functionally related to the yeast CST complex have also been identified in *Arabidopsis*. STN1 and TEN1 were found in *Arabidopsis* by homology searches using yeast orthologues as query sequences (Song et al. 2008; Price et al. 2010). In contrast, the *Arabidopsis* genome does not contain an obvious homologue of Cdc13. A putative functional Cdc13 orthologue was found by genetic mapping of a mutation that displays a set of telomere deprotection phenotypes, such as telomere shortening, chromosome end-to-end fusions and increased recombination at telomeres (Surovtseva et al. 2009). The locus encodes a protein that was named conserved telomere maintenance component 1 (CTC1). The CTC1 protein harbours two OB-folds related to RPA70, and one OB-fold homologous to POT1. CTC1 interacts with STN1 and this complex localises to telomeres, although it has not yet been shown whether *Arabidopsis* CTC1/STN1 can bind to G-rich overhangs. Plants carrying a disrupted *STN1* gene exhibit a similar set of phenotypes as *ctc1* mutants, and the telomere defects are not further exaggerated in *stn1 ctc1* double mutants. These data suggest that CTC1 and STN1 form a functional complex analogous to the CST complex in yeast that is important for telomere protection or maintenance. However, unlike the yeast CST complex, its *Arabidopsis* counterpart is not essential for plant survival, from which it can be inferred that other components of telomeric chromatin can still provide sufficient protection in the absence of the CST complex.

5.2.4 DNA Repair Factors

Another major group of proteins that are present at telomeres throughout the different branches of the tree of life are DNA repair proteins. A major function of telomeric DNA and telomere-associated proteins is to protect chromosome ends from being recognised as DNA double-strand breaks. Double-strand breaks are usually repaired either by a homologous recombination mechanism or by non-homologous end joining. Interestingly, DNA repair proteins involved in both of these types of repair are found at telomeres: (1) the double-strand break repair and signalling proteins Mre11, Rad50 (components of the MRN/X complex) and ATM and ATR kinases that are involved in homologous recombination and DNA damage recognition; (2) the Ku complex (consisting of the Ku70 and Ku80 subunits) that is involved in non-homologous end joining. DNA repair proteins may be present at telomeres as it is thought that chromosome ends are transiently recognised as double-strand breaks following DNA replication. This may be important both for telomere extension by telomerase and the formation of a protective capping structure (Gilson and Geli 2007; Verdun and Karlseder 2007). *Arabidopsis* has an MRN complex as well as ATM and ATR checkpoint kinases that appear to play a vital role in telomere function. The lack of MRE11 or RAD50 in *Arabidopsis* causes a range of chromosomal aberrations that include chromosome end-to-end fusions (Puizina et al. 2004; Vannier et al. 2006). Although inactivation of MRE11 or RAD50 does not lead to overtly altered telomere length homeostasis, the absence of telomerase further exacerbates the occurrence of chromosome end-to-end fusions that appear to be caused by the sporadic loss of the entire telomere on a small subset of chromosome arms (Vannier et al. 2006). One interpretation of this finding is that the MRN complex is important for the proper repair of replication forks, which may pause or stall more frequently while traversing the G/C-rich telomeric region.

The Ku70/Ku80 heterodimer is another DNA repair factor, the function of which is closely linked to telomeres. Inactivation of Ku leads to a range of telomere-related phenotypes in the majority of eukaryotic organisms in which it has been examined (Fisher and Zakian 2005; Riha et al. 2006). Studies in rice and *Arabidopsis* have shown that mutations in the *KU70* or *KU80* gene lead to telomere elongation, nucleolytic chromosome end resection and an increased level of telomeric recombination (Riha and Shippen 2003; Zellinger et al. 2007; Hong et al. 2010). These phenotypes are consistent with the idea that Ku controls the access of telomerase, nucleases and recombinases to chromosome termini. Since Ku also promotes non-homologous end joining DNA repair at intrachromosomal DNA double-strand breaks, an important question that remains to be answered is how Ku activity is modulated at telomeres to promote end protection but to inhibit non-homologous end joining.

5.2.5 Telomeric Chromatin

Despite the high density of telomere binding proteins, the bulk of telomeric DNA is wrapped around nucleosomes. In the majority of eukaryotic organisms, chromosome termini are devoid of functional genes, and telomere-adjacent sequences often consist of satellite repeats and transposable elements. Initial chromatin studies on plant and mammalian telomeres revealed that telomeric nucleosomes are unusually closely spaced (160 bp) in comparison to the rest of the genome (180 bp) (Tommerup et al. 1994; Fajkus et al. 1995; Sykorova et al. 2001). Nucleosomes bound to these regions usually exhibit features of constitutive heterochromatin that may share similarities with pericentromeric chromatin (Pryde et al. 1997; Perrod and Gasser 2003; Blasco 2007). Transgenes inserted in the vicinity of telomeres are transcriptionally silenced, a phenomenon known as the telomere positioning effect. Data obtained primarily in mouse indicate that telomeric heterochromatin has a functional significance, as disrupting it results in increased recombination and telomere elongation (Garcia-Cao et al. 2004; Gonzalo et al. 2005; Blasco 2007).

Studies on chromatin structures at plant telomeres have, so far, focused on *Arabidopsis*. However, telomere organisation in this model plant appears to be rather unusual, as it lacks subtelomeric repetitive sequences and active genes are immediately adjacent to telomeres (Vrbsky et al. 2010). Nevertheless, *Arabidopsis* telomeres, which are on average 3 kb long, still exhibit heterochromatic features, such as the presence of repressive histone H3 marks and cytosine methylation, and these marks extend several hundred base pairs into subtelomeric regions (Grafi et al. 2007; Vrbsky et al. 2010; Cokus et al. 2008). The histone methyltransferases SUVH4, 5 and 6 have been implicated in the maintenance of histone methylation (Grafi et al. 2007; Vaquero-Sedas et al. 2011). DNA methylation at plant telomeres likely represents a plant-specific phenomenon because the cytosines present in the context of the CCCTAAA sequence are exclusively in non-symmetric CNN positions. Cytosine methylation at telomeres is mediated by the RNA-dependent DNA methylation pathway, which converts non-coding telomeric RNA transcripts into siRNAs that are, in turn, utilised to target *de novo* DNA methyltransferases to telomeres (Vrbsky et al. 2010).

5.3 Telomere Maintenance

5.3.1 The End Replication Problem

During DNA replication, which takes place before every cell division, each of the two DNA strands acts as a template for generating two new strands of DNA. As each daughter cell inherits one template strand and one newly synthesised strand, DNA replication is said to be semi-conservative. Due to the antiparallel nature of the base-pairing of the two DNA strands in the double helix, there is a difference in how DNA is replicated from the two old strands. This is because DNA polymerases, which catalyse DNA synthesis, only do so in a 5′–3′ direction, as they require a 3′-OH end of a polynucleotide chain onto which to add a new nucleotide. As a result, one of the new strands—the leading strand—can be synthesised continuously, while the other—the lagging strand—is synthesised in a discontinuous manner, in short fragments called Okazaki fragments. In linear genomes, the DNA end synthesised by the leading strand mechanism is replicated to the very end. For the DNA end replicated by the lagging strand mechanism, the RNA primer of the last Okazaki fragment cannot be replaced by DNA synthesis from the downstream Okazaki fragment, which results in a gap. This failure to completely replicate the 3′ end of chromosomes is known as the end replication problem and leads to a loss, on average, of 4–100 nucleotides from a linear chromosome end at every cell division. This cumulative loss of DNA from telomeres over multiple rounds of replication could ultimately lead to the loss of essential genes.

The gradual attrition of the ends of linear chromosomes due to the end replication problem can be counteracted by the action of the specialised DNA polymerase enzyme, telomerase, a reverse transcriptase which adds telomeric DNA repeat units to the 3′ ends of chromosomes using an RNA template. Thus, the sustained proliferation of eukaryotic cells depends on a mechanism that counteracts the end replication problem and assures full genome synthesis.

5.3.2 Telomerase

Telomerase (Fig. 5.3) is a ribonucleoprotein with a core complex consisting of a catalytic telomerase reverse transcriptase (TERT) and a telomerase RNA (TER) subunit that includes the template domain. The TERT subunit is evolutionarily highly conserved from yeast to human and carries similarity to reverse transcriptases from non-LTR retrotransposons (Nakamura et al. 1997). TERT proteins have been identified in numerous plants including *Arabidopsis*, and plant TERTs appear to be more similar to vertebrate TERTs than to those in yeast and ciliates (Fitzgerald et al. 1999; Oguchi et al. 1999; Sykorova et al. 2006).

Although the RNA subunit of telomerase was first discovered over 20 years ago in *T. thermophila* (Greider and Blackburn 1989), TERs have only been cloned in a few evolutionarily distant groups of organisms (ciliates, yeasts, vertebrates). There has been rapid divergence in the sequence and length of the TERs, which has precluded their identification across species by sequence homology. Structural and mutational analyses have revealed several functionally conserved elements in TERs, including a 3′ the template domain that typically corresponds to one and a half of the telomeric repeat unit (Theimer and Feigon

2006). Vertebrate TERs carry the box H/ACA motif that binds dyskerin, an evolutionarily conserved component of H/ACA snoRNPs (Chen et al. 2000). Dyskerin is also a component of vertebrate telomerase and is required for RNP maturation and localisation to Cajal bodies (Collins 2006). Recently, two TER genes, *TER1* and *TER2*, were identified in *Arabidopsis* through biochemical purification of the telomerase RNP (Cifuentes-Rojas et al. 2011). *TER1* and *TER2* encode ~750 nucleotide RNAs that share a 220 nucleotide segment of homology, which also contains the template region (CTAAACCCTA). While both TER1 and TER2 can reconstitute telomerase activity when combined with the TERT protein in vitro, only TER1 appears to be essential for telomere synthesis in vivo.

Besides TERT and TER, two other components of *Arabidopsis* telomerase have been identified (Fig. 5.3). As in vertebrates, *Arabidopsis* telomerase associates with dyskerin (Kannan et al. 2008). This interaction is RNA-dependent and likely involves the box H/ACA motif that is present in both TER1 and TER2. A mutation in human dyskerin causes dyskeratosis congenita, a progeria syndrome that is associated with a reduced level of telomerase and aberrant telomere maintenance. If the exact same mutation that causes dyskeratosis congenita is introduced into *Arabidopsis* dyskerin, a similar phenotype is manifested, namely a decreased level of telomerase activity and shorter telomeres (Kannan et al. 2008). This indicates that biogenesis of the telomerase RNP particle is similar in plants and vertebrates. Another and rather unexpected *Arabidopsis* telomerase subunit is POT1A. As discussed above, POT1 proteins bind single-stranded telomeric DNA and are primarily involved in chromosome end protection. However, *Arabidopsis* POT1 proteins have lost this DNA binding activity; instead, POT1A associates with TER1 (Surovtseva et al. 2007; Cifuentes-Rojas et al. 2011). Although POT1A is not required for the assembly of biochemically active telomerase, its disruption leads to gradual telomere shortening, a phenotype that is identical to inactivation of TERT (Surovtseva et al. 2007). This indicates that POT1A acquired a novel function that may include activation or recruitment of telomerase to its substrate, the 3′ end of the telomere.

5.3.3 Telomere Replication

The telomerase enzyme sequentially adds deoxynucleotide triphosphates onto the 3′ single-stranded G-rich end of telomeric DNA sequences. It does this by attaching itself to the 3′ end of the chromosome, partly through base-pairing between the portion of template domain of TER and the single-stranded telomeric G-overhang. The 3′ end of the chromosome is used as a primer for reverse transcription of the TER template sequence by TERT. Repetition of this process leads to extension of the G-rich strand of the telomere. The *de novo* synthesis of the 3′ chromosome end is coupled with the replication of the complementary telomeric strand and this process likely requires a new priming event catalysed by primase. Telomere replication is best studied in budding yeast, where Cdc13 appears to coordinate recruitment of both telomerase and primase (Qi and Zakian 2000; Grossi et al. 2004; Bianchi and Shore 2008). The situation in other organisms is poorly understood in comparison and the key players involved in telomerase recruitment and regulation as well as in fill-in synthesis remain to be identified. The suggestion that *Arabidopsis* POT1A may have adopted a telomerase recruitment function relatively recently indicates that pathways regulating telomerase activity may be evolutionarily flexible (Shakirov et al. 2009b, 2010). On the other hand, the function of the CST complex in coordinating fill-in synthesis may be conserved as this complex associates with primase in both *Arabidopsis* and humans (Casteel et al. 2009; Price et al. 2010).

5.3.4 Telomerase Activity

Telomerase activity varies according to organism, cell type and stage of development. There are high levels of telomerase in unicellular eukaryotes, such as ciliates and yeast, in order that telomeres are maintained at the length necessary for chromosomal stability, thus compensating for the loss of telomeric DNA at each cycle of DNA replication. In multicellular organisms, in which different cell types proliferate at different rates at different stages of development, telomerase activity regulation is more complex (Gomes et al. 2010). In humans, telomerase is active in germline cells and early in embryogenesis, and in such cells telomeres are maintained at a length of about 15 kb. Telomerase is not expressed in most human somatic tissues, and, as a consequence, somatic tissues have substantially shorter telomeres. Telomerase is, however, reactivated in 85–90% of human tumour cells, thus enabling these cells to become immortal (Shay and Wright 2006). In plants, telomerase expression is correlated with cell proliferation. Tissues that contain a high proportion of dividing cells, such as floral buds and root tips, express high levels of telomerase, while low levels of telomerase are expressed in leaves and seeds (McKnight et al. 2002; Watson and Riha 2010). Studies in synchronised tobacco cells have shown that telomerase activity peaks in S-phase and is enhanced by auxin, a phytohormone that promotes cell proliferation (Tamura et al. 1999). Telomerase activity in plants appears to be primarily regulated by TERT transcription (Fitzgerald et al. 1999; Oguchi et al. 1999). A part of the regulatory cascade that controls TERT transcription has been identified in *Arabidopsis*. It involves the calmodulin binding protein, BT2, that directly activates TERT expression, and the zinc-finger transcriptional factor, TAC1, that regulates expression of BT2 and potentiates auxin responses (Ren et al. 2004, 2007).

5.4 Telomere Dysfunction

Appropriate telomere function and regulation in cells is vital for ensuring genome stability in organisms by maintaining the content of their genomes over multiple generations. Due to the end replication problem, essential genetic information would be lost if telomere replication were not to counterbalance attrition of telomeres. From an evolutionary perspective, telomere dysfunction can lead to end-to-end fusions of chromosomes, which may have contributed to the global rearrangements of chromosomes in genomes seen throughout evolution. Relicts of such events, in the form of intrachromosomally localised arrays of telomeric DNA, can be traced in the genomes of many organisms (Ijdo et al. 1991; Hartmann and Scherthan 2004; Mandakova et al. 2010).

Telomere length homeostasis is intimately linked to human health, and telomere regulation has been associated with ageing and cancer, as well as with inherited genetic diseases exhibiting features of premature ageing. Experiments performed by Leonard Hayflick and Paul Moorhead in the 1960s showed that human cells have a finite proliferative lifespan in culture (Hayflick and Moorhead 1961). Cultured human fibroblasts are able to grow for 50–60 population doublings before entering a state called replicative cell senescence, in which cells although viable cannot re-enter the cell cycle. As fibroblasts do not express telomerase, their proliferation is accompanied by gradual telomere shortening that eventually leads to loss of the chromosome end protective cap and induction of cell senescence (Harley et al. 1990; Bodnar et al. 1998). The downregulation of telomerase in the majority of human somatic cells is believed to act as a mechanism that limits cell proliferation capacity, thus acting to prevent tumour formation. By the same token, telomere shortening may impede tissue regeneration in advancing age and contribute to organismal ageing (Campisi 2005)

In *Arabidopsis*, disruption of telomerase leads to a gradual attrition of telomeres after propagating *tert* mutants for six to eight sexual generations (Riha et al. 2001). Shortening of telomeres from 2 to 4 kb to below 1 kb triggers chromosome end-to-end fusions, and ongoing chromosome rearrangements through the breakage–fusion–bridge cycle (Fig. 5.4) (Siroky et al. 2003; Heacock et al. 2004). This is accompanied by defects in growth and development, and

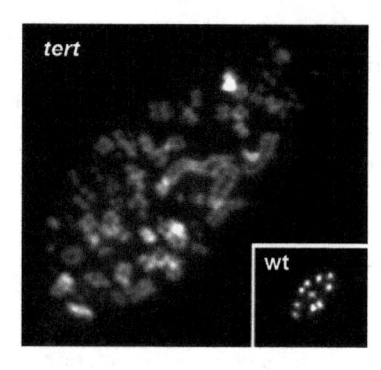

Fig. 5.4 A metaphase spread of *Arabidopsis* cells derived from telomerase-deficient plants that have lost all telomeric DNA after long term in vitro culture (Akimcheva et al. 2008). The karyotypes of these cells are highly polyploid, and contain circular as well as giant linear chromosomes generated by fusions of multiple chromosomes. A wild-type karyotype is shown in the inset

late generation mutant plants exhibit a dwarf stature, organ malformation and reduced fertility (Riha et al. 2001). A similar set of phenotypes is also observed in plants with a CST complex deficiency, whereby their onset is much faster, usually appearing already in first generation mutants (Song et al. 2008; Surovtseva et al. 2009). Despite the shortening telomeres and ongoing genome instabilities, *tert* plants can still be propagated for up to three more generations after the first manifestations of telomere dysfunction. This indicates that, in contrast to mammals, plants are more tolerant of telomere dysfunction, which may be due to their more relaxed DNA damage checkpoint mechanisms.

The absence of telomerase activity in most plant tissues invokes an interesting question as to whether telomeres in plants are more than guardians of chromosome stability and whether telomere biology plays a role in ageing and senescence. While there is an indication that telomere instability may contribute to reduced seed longevity (Osborne and Boubriak 2002), it does not seem that the function of telomeres in the ageing process in plants is as important as it is in humans (Watson and Riha 2011).

References

Adams SP, Hartman TP, Lim KY, Chase MW, Bennett MD, Leitch IJ, Leitch AR (2001) Loss and recovery of *Arabidopsis*-type telomere repeat sequences 5′-(TTTAGGG)(n)-3′ in the evolution of a major radiation of flowering plants. Proc Roy Soc Lond B Bio 268:1541–1546

Akimcheva S, Zellinger B, Riha K (2008) Genome stability in *Arabidopsis* cells exhibiting alternative lengthening of telomeres. Cytogenet Genome Res 122:388–395

Baumann P, Cech TR (2001) Pot1, the putative telomere end-binding protein in fission yeast and humans. Science 292:1171–1175

Bianchi A, Shore D (2008) How telomerase reaches its end: mechanism of telomerase regulation by the telomeric complex. Mol Cell 31:153–165

Blackburn EH, Gall JG (1978) A tandemly repeated sequence at the termini of the extrachromosomal ribosomal RNA genes in *Tetrahymena*. J Mol Biol 120:33–53

Blasco MA (2007) The epigenetic regulation of mammalian telomeres. Nat Rev Genet 8:299–309

Bodnar AG, Ouellette M, Frolkis M, Holt SE, Chiu CP, Morin GB, Harley CB, Shay JW, Lichtsteiner S, Wright WE (1998) Extension of life-span by introduction of telomerase into normal human cells. Science 279:349–352

Burr B, Burr FA, Matz EC, Romero-Severson J (1992) Pinning down loose ends: mapping telomeres and factors affecting their length. Plant Cell 4:953–960

Campisi J (2005) Senescent cells, tumor suppression, and organismal aging: good citizens, bad neighbors. Cell 120:513–522

Carneiro T, Khair L, Reis CC, Borges V, Moser BA, Nakamura TM, Ferreira MG (2010) Telomeres avoid end detection by severing the checkpoint signal transduction pathway. Nature 467:228–232

Casteel DE, Zhuang S, Zeng Y, Perrino FW, Boss GR, Goulian M, Pilz RB (2009) A DNA polymerase-{alpha}primase cofactor with homology to replication protein A-32 regulates DNA replication in mammalian cells. J Biol Chem 284:5807–5818

Cesare AJ, Reddel RR (2008) Telomere uncapping and alternative lengthening of telomeres. Mech Ageing Dev 129:99–108

Cesare AJ, Quinney N, Willcox S, Subramanian D, Griffith JD (2003) Telomere looping in *P. sativum* (common garden pea). Plant J 36:271–279

Chen JL, Blasco MA, Greider CW (2000) Secondary structure of vertebrate telomerase RNA. Cell 100:503–514

Cifuentes-Rojas C, Kannan K, Tseng L, Shippen DE (2011) Two RNA subunits and POT1a are components of *Arabidopsis* telomerase. Proc Natl Acad Sci USA 108:73–78

Cokus SJ, Feng S, Zhang X, Chen Z, Merriman B, Haudenschild CD, Pradhan S, Nelson SF, Pellegrini M, Jacobsen SE (2008) Shotgun bisulphite sequencing of the *Arabidopsis* genome reveals DNA methylation patterning. Nature 452:215–219

Collins K (2006) The biogenesis and regulation of telomerase holoenzymes. Nat Rev Mol Cell Biol 7:484–494

Denchi EL, de Lange T (2007) Protection of telomeres through independent control of ATM and ATR by TRF2 and POT1. Nature 448:1068–1071

Fajkus J, Kovarik A, Kralovics R, Bezdek M (1995) Organization of telomeric and subtelomeric chromatin in the higher plant *Nicotiana tabacum*. Mol Gen Genet 247:633–638

Fajkus J, Sykorova E, Leitch AR (2005) Telomeres in evolution and evolution of telomeres. Chromosome Res 13:469–479

Fisher TS, Zakian VA (2005) Ku: a multifunctional protein involved in telomere maintenance. DNA Repair 4:1215–1226

Fitzgerald MS, Riha K, Gao F, Ren S, McKnight TD, Shippen DE (1999) Disruption of the telomerase catalytic subunit gene from *Arabidopsis* inactivates telomerase and leads to a slow loss of telomeric DNA. Proc Natl Acad Sci USA 96:14813–14818

Gao H, Cervantes RB, Mandell EK, Otero JH, Lundblad V (2007) RPA-like proteins mediate yeast telomere function. Nat Struct Mol Biol 14:208–214

Garcia-Cao M, O'Sullivan R, Peters AH, Jenuwein T, Blasco MA (2004) Epigenetic regulation of telomere length in mammalian cells by the Suv39h1 and Suv39h2 histone methyltransferases. Nat Genet 36:94–99

Gilson E, Geli V (2007) How telomeres are replicated. Nat Rev Mol Cell Biol 8:825–838

Gomes NM, Shay JW, Wright WE (2010) Telomere biology in Metazoa. FEBS Lett 584:3741–3751

Gonzalo S, Garcia-Cao M, Fraga MF, Schotta G, Peters AH, Cotter SE, Eguia R, Dean DC, Esteller M, Jenuwein T, Blasco MA (2005) Role of the RB1 family in stabilizing histone methylation at constitutive heterochromatin. Nat Cell Biol 7:420–428

Grafi G, Ben-Meir H, Avivi Y, Moshe M, Dahan Y, Zemach A (2007) Histone methylation controls telomerase-independent telomere lengthening in cells undergoing dedifferentiation. Dev Biol 306:838–846

Greider CW, Blackburn EH (1989) A telomeric sequence in the RNA of *Tetrahymena* telomerase required for telomere repeat synthesis. Nature 337:331–337

Griffith JD, Comeau L, Rosenfield S, Stansel RM, Bianchi A, Moss H, de Lange T (1999) Mammalian telomeres end in a large duplex loop. Cell 97:503–514

Grossi S, Puglisi A, Dmitriev PV, Lopes M, Shore D (2004) Pol12, the B subunit of DNA polymerase alpha, functions in both telomere capping and length regulation. Genes Dev 18:992–1006

Harley CB, Futcher AB, Greider CW (1990) Telomeres shorten during ageing of human fibroblasts. Nature 345:458–460

Hartmann N, Scherthan H (2004) Characterization of ancestral chromosome fusion points in the *Indian muntjac* deer. Chromosoma 112:213–220

Hayflick L, Moorhead PS (1961) The serial cultivation of human diploid cell strains. Exp Cell Res 25:585–621

Heacock M, Spangler E, Riha K, Puizina J, Shippen DE (2004) Molecular analysis of telomere fusions in *Arabidopsis*: multiple pathways for chromosome end-joining. EMBO J 23:2304–2313

Hong JP, Byun MY, Koo DH, An K, Bang JW, Chung IK, An G, Kim WT (2007) Suppression of RICE TELOMERE BINDING PROTEIN 1 results in severe and gradual developmental defects accompanied by genome instability in rice. Plant Cell 19:1770–1781

Hong JP, Byun MY, An K, Yang SJ, An G, Kim WT (2010) OsKu70 is associated with developmental growth and genome stability in rice. Plant Physiol 152:374–387

Hwang MG, Cho MH (2007) *Arabidopsis thaliana* telomeric DNA-binding protein 1 is required for telomere length homeostasis and its Myb-extension domain stabilizes plant telomeric DNA binding. Nucleic Acids Res 35:1333–1342

Ijdo JW, Baldini A, Ward DC, Reeders ST, Wells RA (1991) Origin of human chromosome 2: an ancestral telomere-telomere fusion. Proc Natl Acad Sci USA 88:9051–9055

Kannan K, Nelson AD, Shippen DE (2008) Dyskerin is a component of the *Arabidopsis* telomerase RNP required for telomere maintenance. Mol Cell Biol 28:2332–2341

Karamysheva ZN, Surovtseva YV, Vespa L, Shakirov EV, Shippen DE (2004) A C-terminal Myb extension domain defines a novel family of double-strand telomeric DNA-binding proteins in *Arabidopsis*. J Biol Chem 279:47799–47807

Kilian A, Stiff C, Kleinhofs A (1995) Barley telomeres shorten during differentiation but grow in callus culture. Proc Natl Acad Sci USA 92:9555–9559

Kilian A, Heller K, Kleinhofs A (1998) Development patterns of telomerase activity in barley and maize. Plant Mol Biol 37:621–628

Ko S, Jun SH, Bae H, Byun JS, Han W, Park H, Yang SW, Park SY, Jeon YH, Cheong C, Kim WT, Lee W, Cho HS (2008) Structure of the DNA-binding domain of NgTRF1 reveals unique features of plant telomere-binding proteins. Nucleic Acids Res 36:2739–2755

Mandakova T, Joly S, Krzywinski M, Mummenhoff K, Lysak MA (2010) Fast diploidization in close mesopolyploid relatives of *Arabidopsis*. Plant Cell 22:2277–2290

McClintock B (1939) The behavior in successive nuclear divisions of a chromosome broken at meiosis. Proc Natl Acad Sci USA 25:405–416

McClintock B (1941) The stability of broken ends of chromosomes in *Zea mays*. Genetics 26:234–282

McKnight TD, Riha K, Shippen DE (2002) Telomeres, telomerase, and stability of the plant genome. Plant Mol Biol 48:331–337

Muller HJ (1938) The remaking of chromosomes. Collect Net 13:181–198

Nakamura TM, Morin GB, Chapman KB, Weinrich SL, Andrews WH, Lingner J, Harley CB, Cech TR (1997) Telomerase catalytic subunit homologs from fission yeast and human. Science 277:955–959

Oguchi K, Liu H, Tamura K, Takahashi H (1999) Molecular cloning and characterization of AtTERT, a telomerase reverse transcriptase homolog in *Arabidopsis thaliana*. FEBS Lett 457:465–469

Osborne DJ, Boubriak I (2002) Telomeres and their relevance to the life and death of seeds. Crit Rev Plant Sci 21:127–141

Pennock E, Buckley K, Lundblad V (2001) Cdc13 delivers separate complexes to the telomere for end protection and replication. Cell 104:387–396

Perrod S, Gasser SM (2003) Long-range silencing and position effects at telomeres and centromeres: parallels and differences. Cell Mol Life Sci 60:2303–2318

Petracek ME, Lefebvre PA, Silflow CD, Berman J (1990) *Chlamydomonas* telomere sequences are A + T-rich but contain three consecutive G-C base pairs. Proc Natl Acad Sci USA 87:8222–8226

Pich U, Fuchs J, Schubert I (1996) How do Alliaceae stabilize their chromosome ends in the absence of TTTAGGG sequences? Chromosome Res 4:207–213

Poulet A, Buisson R, Faivre-Moskalenko C, Koelblen M, Amiard S, Montel F, Cuesta-Lopez S, Bornet O, Guerlesquin F, Godet T, Moukhtar J, Argoul F, Declais AC, Lilley DM, Ip SC, West SC, Gilson E, Giraud-Panis MJ (2009) TRF2 promotes, remodels and protects telomeric Holliday junctions. EMBO J 28:641–651

Preti M, Ribeyre C, Pascali C, Bosio MC, Cortelazzi B, Rougemont J, Guarnera E, Naef F, Shore D, Dieci G (2010) The telomere-binding protein Tbf1 demarcates snoRNA gene promoters in *Saccharomyces cerevisiae*. Mol Cell 38:614–620

Price CM, Boltz KA, Chaiken MF, Stewart JA, Beilstein MA, Shippen DE (2010) Evolution of CST function in telomere maintenance. Cell Cycle 9:3157–3165

Pryde FE, Gorham HC, Louis EJ (1997) Chromosome ends: all the same under their caps. Curr Opin Genet Dev 7:822–828

Puglisi A, Bianchi A, Lemmens L, Damay P, Shore D (2008) Distinct roles for yeast Stn1 in telomere capping and telomerase inhibition. EMBO J 27:2328–2339

Puizina J, Siroky J, Mokros P, Schweizer D, Riha K (2004) Mre11 deficiency in *Arabidopsis* is associated with chromosomal instability in somatic cells and Spo11-dependent genome fragmentation during meiosis. Plant Cell 16:1968–1978

Qi H, Zakian VA (2000) The *Saccharomyces* telomere-binding protein Cdc13p interacts with both the catalytic subunit of DNA polymerase alpha and the telomerase-associated est1 protein. Genes Dev 14:1777–1788

Raices M, Verdun RE, Compton SA, Haggblom CI, Griffith JD, Dillin A, Karlseder J (2008) *C. elegans* telomeres contain G-strand and C-strand overhangs that are bound by distinct proteins. Cell 132:745–757

Ren S, Johnston JS, Shippen DE, McKnight TD (2004) TELOMERASE ACTIVATOR1 induces telomerase activity and potentiates responses to auxin in *Arabidopsis*. Plant Cell 16:2910–2922

Ren S, Mandadi KK, Boedeker AL, Rathore KS, McKnight TD (2007) Regulation of telomerase in *Arabidopsis* by BT2, an apparent target of TELOMERASE ACTIVATOR1. Plant Cell 19:23–31

Richards EJ, Ausubel FM (1988) Isolation of a higher eukaryotic telomere from *Arabidopsis thaliana*. Cell 53:127–136

Riha K, Shippen DE (2003) Ku is required for telomeric C-rich strand maintenance but not for end-to-end chromosome fusions in *Arabidopsis*. Proc Natl Acad Sci USA 100:611–615

Riha K, McKnight TD, Fajkus J, Vyskot B, Shippen DE (2000) Analysis of the G-overhang structures on plant telomeres: evidence for two distinct telomere architectures. Plant J 23:633–641

Riha K, McKnight TD, Griffing LR, Shippen DE (2001) Living with genome instability: plant responses to telomere dysfunction. Science 291:1797–1800

Riha K, Heacock ML, Shippen DE (2006) The role of the nonhomologous end-joining DNA double-strand break repair pathway in telomere biology. Annu Rev Genet 40:237–277

Shakirov EV, Shippen DE (2004) Length regulation and dynamics of individual telomere tracts in wild-type *Arabidopsis*. Plant Cell 16:1959–1967

Shakirov EV, McKnight TD, Shippen DE (2009a) POT1-independent single-strand telomeric DNA binding activities in Brassicaceae. Plant J 58:1004–1015

Shakirov EV, Song X, Joseph JA, Shippen DE (2009b) POT1 proteins in green algae and land plants: DNA-binding properties and evidence of co-evolution with telomeric DNA. Nucleic Acids Res 37:7455–7467

Shakirov EV, Perroud PF, Nelson AD, Cannell ME, Quatrano RS, Shippen DE (2010) Protection of Telomeres 1 is required for telomere integrity in the moss *Physcomitrella patens*. Plant Cell 22:1838–1848

Shay JW, Wright WE (2006) Telomerase and human cancer. Cold Spring Harbor Laboratory Press, Cold Spring Harbor

Siroky J, Zluvova J, Riha K, Shippen DE, Vyskot B (2003) Rearrangements of ribosomal DNA clusters in late generation telomerase-deficient *Arabidopsis*. Chromosoma 112:116–123

Song X, Leehy K, Warrington RT, Lamb JC, Surovtseva YV, Shippen DE (2008) STN1 protects chromosome ends in *Arabidopsis thaliana*. Proc Natl Acad Sci USA 105:19815–19820

Surovtseva YV, Shakirov EV, Vespa L, Osbun N, Song X, Shippen DE (2007) *Arabidopsis* POT1 associates with the telomerase RNP and is required for telomere maintenance. EMBO J 26:3653–3661

Surovtseva YV, Churikov D, Boltz KA, Song X, Lamb JC, Warrington R, Leehy K, Heacock M, Price CM, Shippen DE (2009) Conserved telomere maintenance component 1 interacts with STN1 and maintains chromosome ends in higher eukaryotes. Mol Cell 36:207–218

Sykorova E, Fajkus J, Ito M, Fukui K (2001) Transition between two forms of heterochromatin at plant subtelomeres. Chromosome Res 9:309–323

Sykorova E, Lim KY, Chase MW, Knapp S, Leitch IJ, Leitch AR, Fajkus J (2003) The absence of *Arabidopsis*-type telomeres in *Cestrum* and closely related genera *Vestia* and *Sessea* (Solanaceae): first evidence from eudicots. Plant J 34:283–291

Sykorova E, Leitch AR, Fajkus J (2006) Asparagales telomerases which synthesize the human type of telomeres. Plant Mol Biol 60:633–646

Tamura K, Liu H, Takahashi H (1999) Auxin induction of cell cycle regulated activity of tobacco telomerase. J Biol Chem 274:20997–21002

Theimer CA, Feigon J (2006) Structure and function of telomerase RNA. Curr Opin Struct Biol 16:307–318

Tommerup H, Dousmanis A, de Lange T (1994) Unusual chromatin in human telomeres. Mol Cell Biol 14:5777–5785

van Steensel B, Smogorzewska A, de Lange T (1998) TRF2 protects human telomeres from end-to-end fusions. Cell 92:401–413

Vannier JB, Depeiges A, White C, Gallego ME (2006) Two roles for Rad50 in telomere maintenance. EMBO J 25:4577–4585

Vaquero-Sedas MI, Gamez-Arjona FM, Vega-Palas MA (2011) *Arabidopsis thaliana* telomeres exhibit euchromatic features. Nucleic Acids Res 39(6):2007–2017

Verdun RE, Karlseder J (2007) Replication and protection of telomeres. Nature 447:924–931

Vrbsky J, Akimcheva S, Watson JM, Turne TL, Daxinger L, Vyskot B, Aufsatz W, Riha K (2010) siRNA-mediated methylation of *Arabidopsis* telomeres. PLoS Genet 6:e1000986

Wang RC, Smogorzewska A, de Lange T (2004) Homologous recombination generates T-loop-sized deletions at human telomeres. Cell 119:355–368

Watson JM, Riha K (2010) Comparative biology of telomeres: where plants stand. FEBS Lett 584:3752–3759

Watson JM, Riha K (2011) Telomeres, aging, and plants: from weeds to Methuselah—a mini-review. Gerontology 57:129–136

Wright WE, Tesmer VM, Huffman KE, Levene SD, Shay JW (1997) Normal human chromosomes have long G-rich telomeric overhangs at one end. Genes Dev 11:2801–2809

Yang SW, Kim SK, Kim WT (2004) Perturbation of NgTRF1 expression induces apoptosis-like cell death in tobacco BY-2 cells and implicates NgTRF1 in the control of telomere length and stability. Plant Cell 16:3370–3385

Zellinger B, Akimcheva S, Puizina J, Schirato M, Riha K (2007) Ku suppresses formation of telomeric circles and alternative telomere lengthening in *Arabidopsis*. Mol Cell 27:163–169

The Biology and Dynamics of Plant Small RNAs

6

Tzuu-fen Lee, Pingchuan Li, and Blake C. Meyers

Contents

6.1 Introduction

For decades, RNA molecules were only thought to be the intermediate messengers in the central dogma of molecular biology, carrying the biological information encoded in DNA molecules and subsequently providing the code for translation into proteins. The discovery of short, non-protein-coding RNA molecules in the past decade has fundamentally changed our understanding of the regulatory roles of RNAs and their influence on cell processes. Small RNAs are usually 18–25 nucleotides (nt) long and they went largely undetected for many years due to their small sizes compared to mRNA transcripts. To date, thousands of small RNAs have been identified from plants and animals, involved in the regulation of numerous cellular processes, including the maintenance of genome integrity, developmental transitions and patterning, responses to abiotic and biotic stresses, and even diseases in humans. Hailed as the breakthrough of the year for the journal *Science* in 2002 (Couzin 2002), the discovery of small RNAs opened an exciting new field of research in post-transcriptional control and epigenetic regulation.

Many interesting but puzzling plant biological phenomena that were described decades ago are actually the result of small RNA-directed gene regulation and gene silencing. In 1928, Wingard observed that when tobacco plants were infected with tobacco ringspot virus, the leaves initially infected at the bottom of the plant showed necrotic disease symptoms, while upper, more distant leaves showed no disease symptoms and became resistant to secondary infection (Wingard 1928). Although the mechanism of this "recovery" phenotype of virus-infected tobacco plants could not be explained at that time, this is an example of an RNA silencing mechanism in plants which specifically targets viral RNA and results in suppression of viral replication. In another well-documented case, Jorgensen and coworkers introduced additional copies of a gene encoding chalcone synthase, a key enzyme for pigment synthesis, into petunia plants with the intent to darken already-purple

B.C. Meyers (✉)
Department of Plant and Soil Sciences & Delaware Biotechnology Institute, University of Delaware, Newark, DE 19711, USA
e-mail: meyers@dbi.udel.edu

J.F. Wendel et al. (eds.), *Plant Genome Diversity Volume 1*,
DOI 10.1007/978-3-7091-1130-7_6, © Springer-Verlag Wien 2012

flowers (Napoli et al. 1990). To their surprise, the transgenic flowers became completely white, showing the first described case of what is now known to be a typical transgene-induced silencing phenomenon. The endogenous copies of the same gene triggered the RNA-directed gene silencing, also called RNA interference (RNAi), which led to simultaneous repression of both ectopic and endogenous copies. Whether induced by virus or transgene, "post-transcriptional gene silencing" (PTGS) is a common defense mechanism employed by various organisms to counteract the invasion of virus or foreign DNA in the genome.

The exploration of the RNAi phenomena has led to the breakthrough studies which revealed the underlying mechanisms of RNA-directed gene silencing and identified numerous biologically-active classes of small RNAs. In *Caenorhabditis elegans*, Fire, Mello, and their colleagues injected double-stranded RNA with sequence similarity to an endogenous gene, and this led to more effective inhibition of the endogenous genes compared to injection of either sense or antisense RNA (Fire et al. 1998). Sequence-specific dsRNA-induced gene silencing was shown to be systemic and heritable across generations, with the dsRNA functioning in silencing by reducing the target transcripts. To identify the RNA molecules behind the RNAi phenomenon, Hamilton and Baulcombe described a class of ~25 nt short-interference RNAs (siRNA) that could be resolved on polyacrylamide gels (Hamilton and Baulcombe 1999). They further demonstrated a link between siRNA and PTGS triggered by viruses or transgenes in plants. These studies revealed the underlying mechanism of RNAi. Specifically, virus RNA or transgenes are first converted into dsRNA, then processed into siRNA, which in turn directs the degradation of the target transcripts and represses gene expression. Another breakthrough came from a study in *C. elegans* by Ambrose and colleagues (Lee et al. 1993; Wightman et al. 1993). The *lin-4* locus is essential for proper timing of larval development by negatively regulating the expression of another regulator, the *lin-14* gene. Surprisingly, *lin-4* appeared to encode a non-protein sequence. *lin-4* produces 22 and 61 nt RNAs that are complementary to multiple sites within the 3′-untranslated region of the *lin-14* gene. Subsequent studies demonstrated that another conserved 21 nt RNA *(let-7)*, now known as a microRNA (miRNA), works along with *lin-4* to regulate developmental timing in *C. elegans* (Reinhart et al. 2000). Thus, these miRNAs and siRNAs were found to regulate gene expression through sequence-specific RNA-RNA interactions.

In this chapter, we introduce the types of small RNAs, their biogenesis pathways, and their roles in regulating various biological processes in plants. Next, we summarize recent advances in understanding the mechanisms of small RNA-directed gene silencing, including post-transcriptional silencing by RNA cleavage, translational repression, and transcriptional repression through RNA-directed DNA methylation and chromatin modification. Then, we discuss the diversification and specification of small RNA regulatory mechanisms which provide flexibility and precision for regulation of gene expression during plant development. Lastly, we discuss the origins and evolution of small RNAs, including a current model of small RNA biogenesis and evolution.

6.2 Small RNA Biogenesis

6.2.1 Small RNA Classification

In general, small RNAs are generated from double-stranded RNA (dsRNA) precursors by the action of RNase III ribonuclease DICER-LIKE (DCL) proteins that cleave and generate small RNA duplexes. A selected strand of the duplex is then bound by an Argonaute (AGO) protein to form a functional silencing complex, which represses target gene expression by various mechanisms. Based on their distinct biogenesis pathways and originating loci, small RNAs can be divided into several major categories: miRNAs, siRNAs, trans-acting siRNAs (ta-siRNAs), and piwi-interacting RNAs (piRNAs) (summarized in Table 6.1). miRNAs are processed from the hairpin structures of a long, single-stranded RNA (ssRNA) which generates a predominant miRNA species. siRNAs are derived from double-stranded RNA (dsRNA) precursors, which generate multiple siRNA species from both strands. In flies and mammals, piRNAs are derived from ssRNA precursors in a Dicer-independent manner. In this section, we will focus on the discussion of plant small RNA biogenesis and the known effectors involved in these pathways. Table 6.2 summarizes the genes mentioned in this chapter that are involved in small RNA biogenesis and their silencing activity.

6.2.2 Small RNA Biogenesis Pathways

6.2.2.1 miRNA

The biogenesis of miRNAs is a complex process which includes three major steps: (1) transcription of precursor genes; (2) maturation by processing, modification, and intracellular translocation; (3) incorporation into the RNA-induced silencing complex (RISC) (reviewed in Jones-Rhoades et al. 2006; Carthew and Sontheimer 2009). Eukaryotic genomes contain hundreds of miRNA coding regions called *MIRNA* loci, which are transcribed by a DNA-dependent RNA polymerase II (Pol II) in the same way as protein-coding genes. The *MIRNA* transcription produces a primary miRNA transcript (pri-miRNA) which is also 5′ capped and 3′ polyadenylated like protein-coding mRNAs, but the pri-miRNA contains a typically imperfect

Table 6.1 Classification of small RNAs

Class	Full name	Originating loci	Biogenesis	Function
miRNA	microRNA	*MIRNA* genes	The foldback structures of long ssRNA transcripts are cleaved by dicers	Repress target gene expression through mRNA cleavage and translational repression
siRNA	short-interfering RNA	Repeats, transposons, and retroelements (endogenous). Transgenes and viral RNAs (exogenous)	RdRP-generated dsRNAs are cleaved by dicers	Silence repeats and transposons through RNA-dependent DNA methylation and chromatin modification
ta-siRNA	trans-acting siRNA	*TAS* loci	*TAS* transcripts are cleaved by miRNAs, transcribed by RdRP into dsRNA, and then processed by dicers	Repress target gene expression through mRNA cleavage
nat-siRNA	natural antisense transcript-derived siRNA	Loci producing pairs of sense-antisense transcripts	The dsRNA derived from overlapping transcripts is cleaved by dicers	Stressed-induced nat-siRNA to repress target gene expression through mRNA cleavage
piRNA	piwi-interacting RNA	Repeats, transposons, and retroelements	ssRNA derived from transposons is cleaved by PIWI protein	Germ line-specific piRNA to suppress repeats and transposons in flies and mammals

inverted repeat sequence, resulting in a fold back to form a stem-loop structure (Lee et al. 2004). In plants, the stem-loop structure is first excised from pri-miRNA by Dicer-like 1 (DCL1), an endonuclease of RNase III family, to form the pre-miRNA. The pre-miRNA is further processed by DCL1 to form a mature miRNA duplex with a miRNA and a miRNA* strand (Park et al. 2002; Reinhart et al. 2002; Xie et al. 2005). The miRNA strand (the guide strand) will target mRNAs by complementary base-pairing, while the miRNA* strand ("microRNA-star," the passenger strand) is discarded. Once produced, the miRNA:miRNA* duplex is further modified by other effectors and transported from nucleus into cytoplasm, where the selected miRNA strand is quickly incorporated into RISC (Fig. 6.1a).

Many effectors are involved in miRNA biogenesis, processing and maturation. DAWDLE (DDL), a FHA domain containing nuclear RNA-binding protein, plays a key role in stabilizing the pri-miRNA through pri-miRNA binding (Yu et al. 2008). DDL interacts with DCL1 along with other RNA-binding proteins, including a double-stranded RNA binding protein called HYPONASTIC LEAVES1 (HYL1), and a zinc-finger binding protein SERRATE (SE). The interaction between HYL1, SE, and DCL1 is important for proper processing of pri-miRNA into miRNA (Fig. 6.1a). In *hyl1* and *se* mutants, the pri-miRNA is enriched and miRNA level is reduced (Kurihara et al. 2006). Furthermore, DCL1, HYL1 and SE are co-localized in a nuclear dicing-body called a D-body, working together to process pri-miRNA into mature miRNA (Fang and Spector 2007). Finally, a heterodimeric cap-binding complex (CBC) binds the 5′ methyl-guanosine (m7G) cap of the pri-miRNA transcript. Interestingly, CBC and SE appear to have dual functions in proper pri-miRNA processing and also mRNA splicing, which distinguishes them from the other specialized miRNA biogenesis factors DCL1 and HYL1 (Laubinger et al. 2008).

In plants, mature miRNA:miRNA* duplexes need to be further modified before they are exported to the cytoplasm and loaded into RISC. HUA ENHANCER 1 (HEN1) can transfer a methyl group to the 3′ terminus of a small RNA, thus protecting them from uridylation and subsequent degradation (Li et al. 2005; Yang et al. 2006). miRNAs are then exported from the nucleus to cytoplasm by an exporting factor, HASTY (Park et al. 2005). Unlike its animal homolog exportin-5, HASTY functions specifically for miRNA export but not for other RNA species like tRNA and siRNA. It is not clear whether the cargo of HASTY is the miRNA:miRNA* duplex or just the miRNA, nor is it known how the exporting complex works. Animal studies have indicated that the miRNAs or siRNAs are kept in a duplex structure until they are incorporated into RISC, suggesting that HASTY may export an RNA duplex.

6.2.2.2 ta-siRNA

Trans-acting siRNAs (ta-siRNAs) are a class of endogenous siRNAs which, as with miRNAs, regulate target genes in *trans*. ta-siRNAs were first discovered to require RNA-dependent RNA polymerase 6 (RDR6) and SUPPRESSOR OF GENE SILENCING 3 (SGS3), which are both involved in transgene-induced gene silencing. Studies of mutants defective in miRNA and siRNA biogenesis have shown that ta-siRNAs are a special class of endogenous siRNA. The biogenesis of a tasiRNA is quite unique, in that it requires both a miRNA and proteins such as RDR6 and SGS3.

The generation of ta-siRNA is a cooperative effort of different small RNA biogenesis pathways. *Arabidopsis* has four families of ta-siRNA generating loci, *TAS* loci (*TAS1-4*), which are also targets of miRNAs (Fig. 6.1b). Both *TAS1* and *TAS2* are targeted by miR173, *TAS3* by miR390 and *TAS4* by miR828 (reviewed by Chapman and Carrington 2007; Chen 2009). The non-coding transcripts from *TAS* are

Table 6.2 List of small RNA biogenesis effectors discussed in this chapter

Gene	Full name	Activity	Function	Domains
AGO1	ARGONAUTE 1	RNA slicer	Cleavage the RNA target of miRNA	PIWI, PAZ
AGO4	ARGONAUTE 4	Nucleic acid binding protein	24 nt siRNA-binding protein, RdDM	PIWI, PAZ
AGO6	ARGONAUTE 6	Nucleic acid binding protein	siRNA accumulation, RdDM	PIWI, PAZ
AGO7	ARGONAUTE 7	Nucleic acid binding protein	Biogenesis of ta-siRNA from TAS3	PIWI, PAZ
CLSY1	CLASSY1	Chromatin remodeling factor	Act with Pol IV to transcribe the methylated DNA in heterochromatin	SNF2
CMT	CHROMOMETHYLASE3	DNA methylation transferase	DNA cytosine methylation (on CNG sequence)	Methyltransferase (MTase)
DCL1	Dicer-like 1	RNase III	Process the dsRNA into 21 nt microRNA	PAZ, RNaseIII
DCL2	Dicer-like 2	RNase III	Process the dsRNA into 22 nt siRNA	PAZ, RNaseIII
DCL3	Dicer-like 3	RNase III	Process the dsRNA into 24 nt siRNA	PAZ, RNaseIII
DCL4	Dicer-like 4	RNase III	Process the dsRNA into 21 nt siRNA or tasiRNA	PAZ, RNaseIII
DDL	DAWDLE	RNA binding protein	Stabilize the pri-miRNA	Forkhead associated domain (FHA)
DMS3	DEFECTIVE IN MERISTEM SILENCING 3	Structural-maintenance-of-chromosome protein	RdDM involving secondary siRNAs	Hinge
DRD1	Defective in RNA-directed DNA methylation 1	Chromatin remodeling protein	Act with Pol V in RdDM	SNF2
DRM1	DOMAINS REARRANGED METHYLTRANSFERASE	DNA methylation transferase	Homolog to DRM2 with weaker activity	Rearranged catalytic domains
DRM2	DOMAINS REARRANGED METHYLTRANSFERASE	DNA methylation transferase	RdDM, de novo methylation of CG, CNG, CNN sequences	Rearranged catalytic domains
HEN1	Hua Enhancer 1	RNA methylase	Methylation of 21–24 nt sRNA	MTase, double-stranded RNA binding (dsRBD) motif 2
HST	HASTY	Nucleocytoplasmic transporter	Export the miRNA duplex from nucleus to cytoplasm	Exportin-1 like domain
HYL	HYPONASTIC LEAVES 1	dsRNA-binding protein	Act on the biogenesis pathway of miRNA	dsRBD
RDR2	RNA-dependent RNA polymerase 2	RNA-dependent RNA Polymerase	Synthesis the dsRNA in the siRNA biogenesis pathway	RNA-dependent RNA polymerase (RdRP) domain
RDR6	RNA-dependent RNA polymerase 6	RNA-dependent RNA Polymerase	Synthesis the dsRNA in the siRNA biogenesis pathway, especially for the ta-siRNA	RdRP
SDN	Small RNA Degrading Nuclease	Exo-ribonuclease	Degradation of single-stranded small RNA	$3'$–$5'$ exonuclease domain
SE	SERRATE	Zinc finger protein	pre-miRNA processing	C_2H_2-zinc finger
SGS3	SUPPRESSOR OF GENE SILENCING 3	RNA binding protein	Stabilize the non-coding of TAS gene	zinc finger, XH, XS, zf-XS
SUVH2	SUPRESSOR OF VARIEGATION 3-9 HOMOLOGE 2	Methylated cytosine-binding protein	RdDM effector	SET, SRA
SUVH9	SUPRESSOR OF VARIEGATION 3-9 HOMOLOGE 9	Methylated cytosine-binding protein	RdDM effector	SET, SRA
Protein complex				
CBP80 and CBP20	Cap-binding complex	RNA-binding protein complex	Binding the cap of miRNA precursor, pri-miRNA processing	Complex
NRPD1	Pol IV	Largest subunit of RNA polymerase IV	Generation of 24 nt heterochromatic siRNAs	Complex
NRPD2/ NRPE2	Pol IV/Pol V	Second largest subunit shared by Pol IV and Pol V	RdDM, generation of 24 nt heterochromatic siRNAs	Complex
NRPE1	Pol V	Largest subunit of RNA polymerase V	RdDM, generation of 24 nt heterochromatic siRNAs	Complex

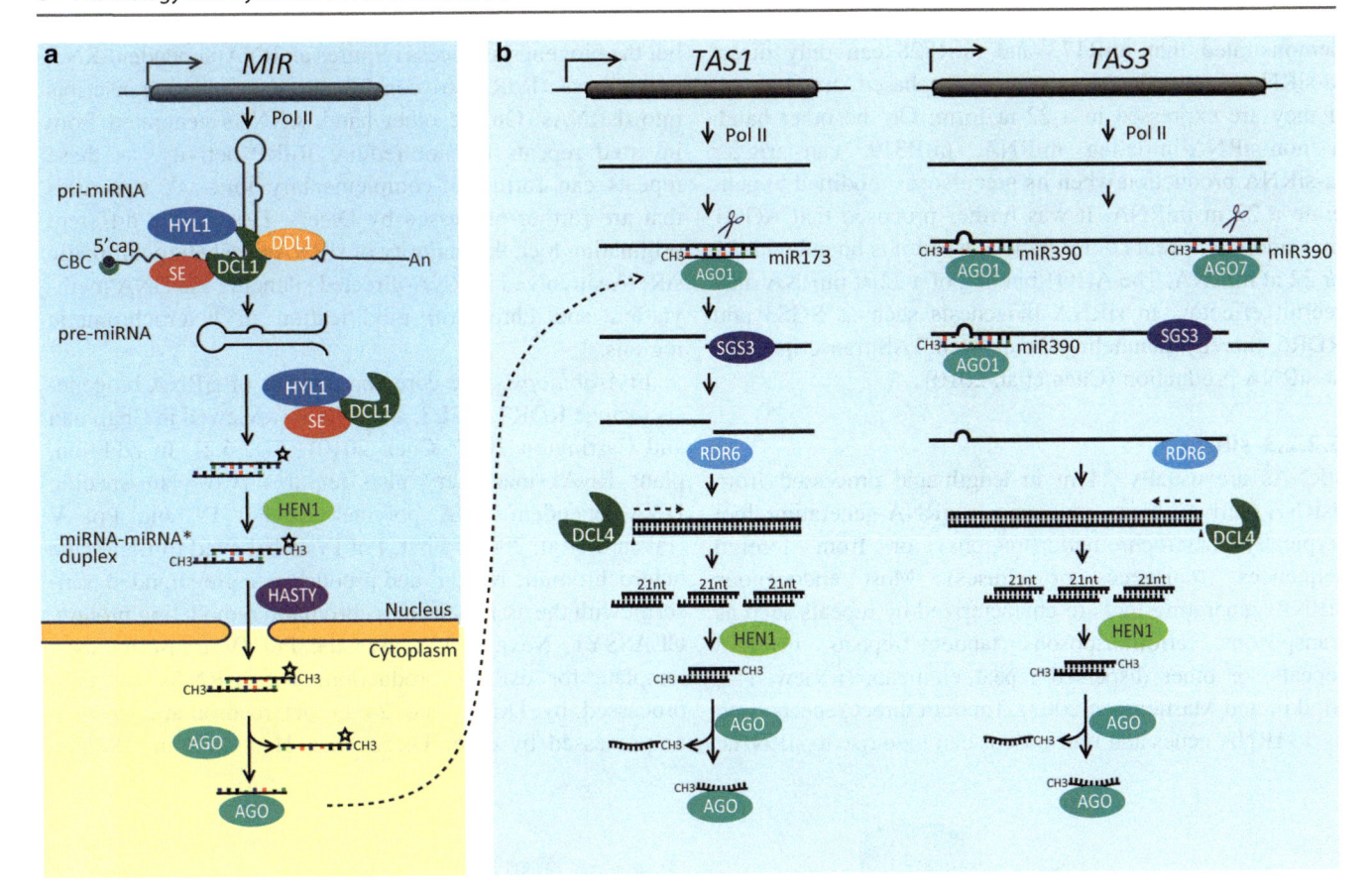

Fig. 6.1 Biogenesis of miRNAs and ta-siRNAs. (**a**) The *MIRNA* gene is transcribed by DNA-dependent RNA polymerase II (Pol II) into a pri-miRNA. The foldback structure of pri-miRNA is stabilized by DAWDLE (DDL), and processed into the stem-looped pre-miRNA and eventually a miRNA:miRNA* duplex by DICER-LIKE1 (DCL1). The processing of pri-miRNA into miRNA also requires the cooperative action of HYPONASTIC LEAVES1 (HYL1), SERRATE (SE), and nuclear cap-binding complex (CBC). The miRNA duplex is methylated by HUA ENHANCER1 (HEN1), and then exported by HASTY from the nucleus to cytoplasm. Only the miRNA, from the guide strand, is incorporated into Argonautes in RISC. (**b**) ta-siRNA generating loci, TAS1 and TAS3, are transcribed by Pol II. Transcripts of TAS1 and TAS3 loci are cleaved by different miRNA/AGO complexes. The cleaved transcripts are used by RDR6 as templates to form double-stranded products, which are processed by DCL4 into ta-siRNA in a phased pattern. SUPPRESSOR OF GENE SILENCING 3 (SGS3) stabilizes the miRNA-cleaved TAS transcripts, and HEN1 methylates the ta-siRNA duplex

transcribed by Pol II and then cleaved by specific miRNA in RISC. The cleaved RNA fragments are bound and stabilized by SGS3, which also protects cleaved RNAs from degradation. RDR6 uses the cleaved fragment as template to transcribe the complimentary strand and form dsRNA product. Next, the dsRNA precursor is digested by DCL4 progressively in a 21-nt register from the miRNA cleavage site, hence creating a "phased" pattern. Finally, ta-siRNAs are loaded into an effector complex, which likely contains either AGO1 or AGO7 to regulate their target gene expression post-transcriptionally through mRNA cleavage (Peragine et al. 2004; Vazquez et al. 2004).

The characterization of the target genes of *Arabidopsis* ta-siRNAs has suggested that they have roles in transcription regulatory networks. *TAS1* and *TAS2* generate ta-siRNAs that target genes encoding pentatricopeptide repeat (PPR) proteins (Peragine et al. 2004; Vazquez et al. 2004; Allen et al. 2005), RNA-binding proteins involved in post-transcriptional processes such as RNA editing and splicing in mitochondria and chloroplasts (Schmitz-Linneweber and Small 2008). The *TAS3* ta-siRNAs target the mRNA of *Arabidopsis AUXIN RESPONSE FACTORs* (*ARF3* and *ARF4*), which work together as the tasiR-ARF regulatory module to regulate phase transition and lateral root growth (Fahlgren et al. 2006; Marin et al. 2010). The *TAS4* tasiRNAs target several mRNAs encoding MYB transcription factors involved in phenolpropanoid biosynthesis and leaf senescence (Rajagopalan et al. 2006).

Although diverse mRNA transcripts are targeted and cleaved by *Arabidopsis* miRNAs, only cleaved *TAS* transcripts are routed to the ta-siRNA biogenesis pathway. Two recent studies have shed light on how ta-siRNA production may be triggered (Chen et al. 2010; Cuperus et al. 2010). These results suggested that AGO1-associated 22 nt miRNAs but not the typical 21 nt miRNAs can trigger RDR6-dependent ta-siRNA biogenesis. Experiments demonstrated that miR173 and miR828 can only direct ta-siRNA production in the typical phased arrangement if they are expressed in a 22 nt form. On the other hand, a non-siRNA-initiating miRNA, miR319, can trigger ta-siRNA production when its precursor is modified to generate a 22 nt miRNA. It was further proposed that AGO1 may adopt different conformations when it is bound with 21 or 22 nt miRNA. The AGO1 binding of a 22 nt miRNA may recruit effectors in siRNA biogenesis such as SGS3 and RDR6, thereby channeling the cleaved TAS transcripts into ta-siRNA production (Chen et al. 2010).

6.2.2.3 siRNA

siRNAs are usually 24 nt in length and processed from dsRNA derived from endogenous siRNA-generating loci (typically heterochromatic regions) or from foreign sequences (transgenes or viruses). Most endogenous siRNA-generating loci are characterized by repeats such as transposons, retrotransposons, tandem repeats, inverted repeats, or other dispersed repeat elements (reviewed in Slotkin and Martienssen 2007). Tandem direct repeats such as 5S rRNA genes and transposons can give rise to siRNAs,

but the biogenesis process requires an RNA-dependent RNA polymerase (RdRP) to convert single-stranded transcripts into dsRNAs. On the other hand, siRNAs generated from inverted repeats do not require RdRP activity, as these repeats can form self-complementary foldback structures that are further processed by Dicers. Despite the different originating loci, the majority of siRNAs are heterochromatic siRNAs involved in RNA-directed silencing via DNA methylation and chromatin modification in heterochromatic regions.

In *Arabidopsis*, the core components of siRNA biogenesis include RDR2, DCL3, and AGO4 (reviewed in Chapman and Carrington 2007; Chen 2010) (Fig. 6.2). In addition, plant RNAi machinery also requires two plant-specific DNA-dependent RNA polymerases Pol IV and Pol V (Pikaard et al. 2008). First, Pol IV is believed to transcribe heterochromatic regions and produce a single-stranded transcript with the assistance of a chromatin remodeling protein, CLASSY1. Next, RDR2 uses the Pol IV transcript as a template for dsRNA production. The dsRNAs are then processed by DCL3 into 24 nt heterochromatic siRNA. or processed by other Dicers into 21 and 22 nt siRNAs.

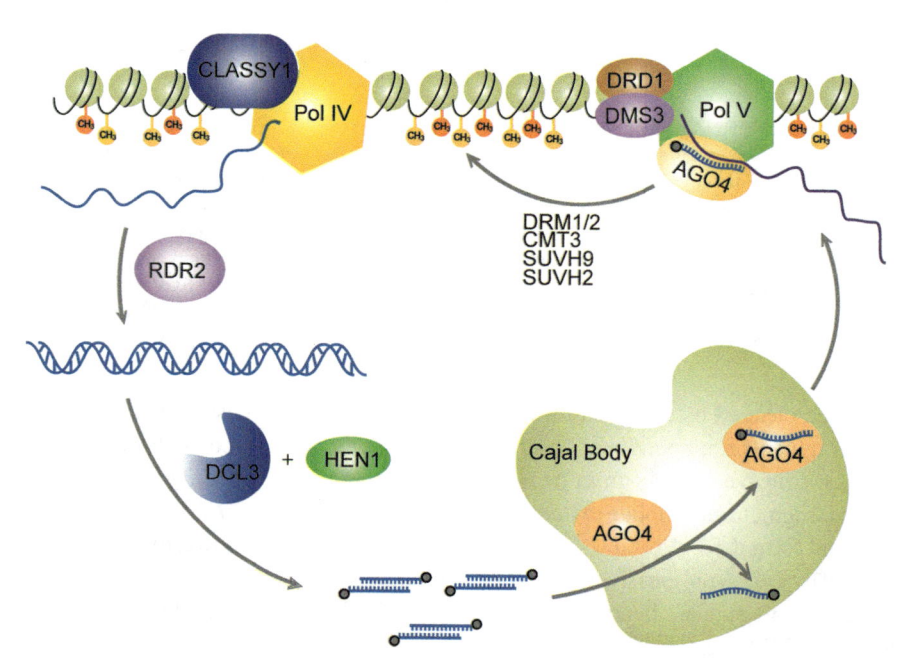

Fig. 6.2 Biogenesis of heterochromatin siRNAs. At a heterochromatic region, Pol IV transcribes a non-coding RNA which is converted into dsRNA by RNA-dependent RNA polymerase 2 (RDR2) with the assistance of CLASSY1. The dsRNA is then cleaved by DCL3 into 24 nt siRNA duplexes which are then methylated by HEN1 at the 3′ end. The mature siRNA duplex is bound to AGO4 in a Cajal body, where the guide strand is retained and the other strand is discarded. During RNA-directed DNA methylation (RdDM), Pol V likely transcribes a siRNA-targeting loci or intergenic region with the assistance of RdDM effectors, including DEFECTIVE IN RNA-DIRECTED DNA METHYLATION 1 (DRD1), DEFECTIVE IN MERISTEM SILENCING 3 (DMS3), and RNA-DIRECTED DNA METHYLATION 1 (RDM1).

The Pol V transcript may serve as a scaffold to recruit AGO4/siRNA and other effectors to form an effector complex at the siRNA-targeting loci. Through an unknown mechanism, the effector complex directs DNA and histone modification at the target loci by the action of DNA and histone methyltransferases, including DOMAINS REARRANGED METHYLASE1 and 2 (DRM1 and DRM2), CHROMOMETHYLASE3 (CMT3), SUPRESSOR OF VARIEGATION 3-9 HOMOLOG 9 (SUVH9) and SUVH2, which leads to heterochromatin formation and silencing of target loci. Lollipops with different colors mark the methylation of cytosines of DNA or lysine residues of the histone core

DCL-generated siRNA pairs have a two-nucleotide 3′ over-hang, which are subsequently methylated by the methyltransferase HEN1 to protect siRNAs from degradation (Li et al. 2005; Yang et al. 2006). Finally, siRNAs are bound with AGO4 in RISC to target and silence particular genomic loci (see below).

Pol IV and Pol V are closely related to Pol II, with many shared subunits in their multi-subunit complexes (Ream et al. 2009), but they work exclusively in siRNA biogenesis and RNAi pathways with non-redundant functions. The major difference from Pol II is that Pol IV and Pol V have different largest subunits (NRPD1 and NRPE1, respectively), and the second largest subunit (NRPD2/NRPE2) that is shared in *Arabidopsis* is also different from the second largest subunit of Pol II. Although Pol IV and Pol V are considered nonessential polymerases due to the absence of obvious visible phenotypes in *nrpd1* and *nrpe1* mutants, emerging evidence suggests that they are functional polymerases which play important roles in epigenetic regulation in plants. While Pol IV is required for primary siRNA biogenesis, Pol V is involved in directing de novo DNA methylation and chromatin modification at the siRNA-targeted loci (Herr et al. 2005; Kanno et al. 2005; Onodera et al. 2005; Pontier et al. 2005).

Other than the repeat sequences, small RNAs derive from loci which produce overlapping bidirectional transcripts from opposing promoters. In plants, the natural antisense transcript-derived siRNA (nat-siRNA) is a new class of siRNA which is induced by abiotic and biotic stimuli. Studies in *Arabidopsis* show that biotic stresses can induce the transcription of partially overlapping dsRNAs, which are processed by Dicers into 24 nt nat-siRNA (Borsani et al. 2005; Katiyar-Agarwal et al. 2006). Other proteins in small RNA biogenesis pathways like Pol IV, RDR6, SGS3, and DCLs are also involved in nat-siRNA biogenesis. nat-siRNAs then act to repress the complementary transcript in the gene pair. One study showed that nat-siRNA down-regulated a negative regulator of defense through transcriptional repression (Katiyar-Agarwal et al. 2006). Nevertheless, a bioinformatics analysis for natural antisense transcripts and small RNA abundance in *Arabidopsis* suggested that the nat-siRNA contribution to small RNA-mediated regulation is limited in the absence of stress (Henz et al. 2007). Future studies will help to understand the function and mode of action of nat-siRNAs.

6.3 Mechanisms of Small RNA Function

In both plant and animal systems, mature, double-stranded small RNAs are incorporated into RISC during biogenesis. The presence of a mature small RNA duplex is quite transient. The duplex is unwound rapidly once it is bound by AGO, and then only one strand is retained while the other strand is lost by a mechanism that is not fully understood. The retained small RNA results in a functional RISC. Besides the distinct biogenesis pathways described above, different subclasses of small RNAs have varying requirements for strand selection, cofactors, and RISC assembly. Once the functional RISC is assembled, it searches for target transcripts based on sequence complementarity to the bound small RNA and executes silencing activity by one of three mechanisms: mRNA cleavage, translational repression, or transcriptional repression through RNA-directed DNA methylation and histone modification. miRNAs and tasiRNAs function to silence the target loci mainly through miRNA-directed RNA cleavage and translational repression (post-transcriptional events), while siRNAs function mainly through transcriptional repression (Fig. 6.3). Here, we discuss the current understanding of the molecular actions of these three mechanisms.

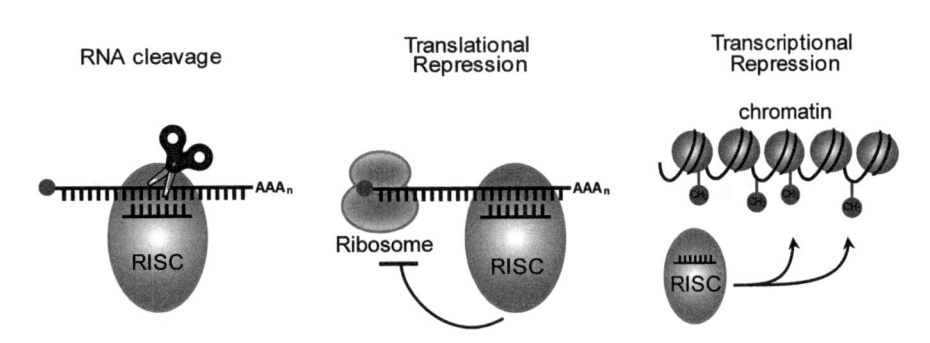

Fig. 6.3 Mechanisms of small RNA-directed gene silencing. Small RNAs are bound by an Argonaute protein and assembled into a functional RNA-induced silencing complex (RISC) with other effectors. For silencing activity, RISC can (1) direct endonucleolytic cleavage of the target mRNA, (2) repress translation of the target gene, or (3) repress target gene transcription through chromatin modification including RNA-directed DNA methylation and histone modification

6.3.1 Small RNA-Directed RNA Cleavage

The mechanism of direct RNA cleavage is the best-understood mode of action of small RNAs, having been extensively studied primarily in animal systems (reviewed in Jones-Rhoades et al. 2006; Valencia-Sanchez 2006; Carthew and Sontheimer 2009). RNA cleavage guided by miRNAs and siRNAs seems to share the same principle, although the key components of RISC may differ. In miRNA-directed RNA cleavage, plant small RNAs first guide the RISC to the complementary mRNA target, then a specific Argonaute (AGO1 in this case) cleaves a single phosphodiester bond opposite to the 10th and 11th nucleotides of the miRNA in the miRNA:mRNA duplex. This "slicer" activity is contributed by the endonuclease activity of the PIWI domain of AGO1, which generates fragments with 5'-monophosphate and 3'-hydroxyl termini. Next, the cleaved RNA fragments are released from RISC, and RISC is free to bind to another transcript for cleavage. For the mRNA target, once the initial cleavage is made, the RISC-cleaved fragments are directed through the general cellular mRNA degradation machinery for degradation (Orban and Izaurralde 2005). Many *Arabidopsis* mutants defective in miRNA functions such as *ago1* and *hen1* have increased accumulation of mRNAs targets, demonstrating that miRNAs regulate the abundance of their target transcripts (Jones-Rhoades et al. 2006).

The degradation of RISC-cleaved mRNA products appears to follow the general turnover pathways of other cellular mRNAs (reviewed in Valencia-Sanchez 2006). After the deadenylation step to remove the 3' poly(A) tail, cellular mRNAs may be degraded by 3' to 5' exonuclease activity in exosomes, or degraded by 5' to 3' exonucleases such as *Arabidopsis* XRN4 after being decapped (Souret et al. 2004). On the other hand, little is known about how a 5' fragment is degraded. It has been suggested that the 5' fragment can still be a substrate for exosome degradation, albeit a poor one (Shen and Goodman 2004). The cleaved end of the 5' fragment could be further modified by oligouridylation, which adds predominately U's to the fragments and promotes exosome degradation of poor substrates. Alternatively, the uridylated 5' fragment may become a target for a 5' decapping mechanism, which leads to 5' to 3' degradation.

6.3.2 Small RNA-Directed Translational Repression

The notion that small RNAs may translationally repress target genes comes from the finding of disproportional mRNA to protein level of the target genes. Studies have shown that miRNAs represses the amount of target protein without causing a substantially decreased level of target mRNA, indicating that miRNAs mediate translational repression of the target mRNA using mechanisms other than directing mRNA cleavage (reviewed in Valencia-Sanchez 2006; Chen 2009). The mechanism is more prevalent in animals, but a recent study in *Arabidopsis* suggests that the translational repression is also a widespread mode of action for plant miRNAs (Brodersen et al. 2008). These authors isolated mutants defective in miRNA-mediated translational repression, and showed that several known miRNA target genes have wild-type levels of mRNA but elevated protein levels. Other effectors are also required for translational repression, including AGO1, AGO10, a microtubule-severing enzyme, and components of "processing bodies" (P-bodies), the cytoplasmic sites for mRNA storage and degradation in yeast and mammals. Although the mechanism of miRNA-directed translational repression remains unclear, the finding of roles for Argonaute and target mRNA together in mammalian P-bodies has provided some hints (Liu et al. 2005; Sen and Blau 2005). It is speculated that miRNA along with AGO may direct target mRNA to the P-bodies, which sequester the mRNA from ribosome machinery and prevent their translation.

It has been proposed that the degree of miRNA:mRNA complementarity may be the key determinant of whether function is mediated by either RNA cleavage or translational repression (reviewed in Carthew and Sontheimer 2009). Perfect complementarity between miRNA:mRNA may allow endonucleolytic cleavage by AGO, while substantial bulges from mismatches in miRNA:mRNA pairing may block the cleavage site for AGO activity and direct translational repression. Of course, these two silencing mechanisms may not be mutually exclusive. A recent model has integrated these two mechanisms in addition to the attributes of target mRNA, based on animal studies (Valencia-Sanchez 2006). In this model, mRNA with a fast turnover rate is mainly targeted for translational repression, whereas long-lived mRNA is mainly targeted by mRNA cleavage. This model further illustrates the dynamic nature of miRNA-directed silencing mechanisms, since the expression of the same mRNA may be regulated differently in a spatiotemporal manner at any given developmental stage. Whether plant miRNAs work in the same way as animal miRNAs in this respect requires further investigation.

6.3.3 Small RNA-Directed Transcriptional Silencing

Other than the two mechanisms mentioned above, endogenous siRNAs are mainly responsible for the transcriptional silencing of target loci by DNA methylation and chromatin modification (reviewed in Henderson and Jacobsen 2007; Matzke et al. 2009; Law and Jacobsen 2010). One key role of siRNA is to establish and maintain the state of heterochromatin, the densely-packed chromosomal regions which are epigenetically marked by DNA or chromatin modification as transcriptionally silenced regions. RNA-directed DNA methylation (RdDM) occurs via the covalent addition of a methyl group to the cytosine residue in a CG dinucleotide or CHG and CHH trinucleotides (reviewed in Gehring and Henikoff 2007; Law and Jacobsen 2010). Histone modifications may vary, but typically include modifications to the side chain of the histone 3 nucleosomal subunit; examples include methylation of lysine residues of histone H3, such as the single methylation of histone 3 on lysine 9 residue (H3K9me1) (reviewed in Cazzonelli et al. 2009). Both DNA methylation and histone modification are considered heritable epigenetic marks, and they play important roles in regulating gene expression without changing the primary DNA sequences.

Recent advances have greatly expanded our understanding of siRNA-directed epigenetic regulation, such as the RdDM pathways in *Arabidopsis* (Fig. 6.2) (reviewed in Henderson and Jacobsen 2007; Matzke et al. 2009; Law and Jacobsen 2010). Downstream of siRNA biogenesis, Pol V is thought to assist heterochromatin formation by recruiting the silencing complex to the siRNA-targeted loci. Studies have shown that Pol V may generate non-coding transcripts from siRNA-targeting loci or intergenic regions outside of the loci with the assistance of other known RdDM components, including DEFECTIVE IN RNA-DIRECTED DNA METHYLATION 1 (DRD1), DEFECTIVE IN MERISTEM SILENCING 3 (DMS3), and RNA-DIRECTED DNA METHYLATION 1 (RDM1) (Wierzbicki et al. 2008; Gao et al. 2010). For certain heterochromatic loci, Pol II may substitute for Pol V and produce the non-coding transcripts (Zheng et al. 2009). Next, the Pol V-dependent nascent RNA may serve as a scaffold to recruit AGO4/siRNA and other effectors, which collectively form an effector complex and direct AGO4/siRNA to the siRNA-targeting loci (Wierzbicki et al. 2009; Law et al. 2010). Finally, through a mechanism which is still not fully understood, the effector complex directs DNA and histone modification at the target loci through the action of DNA and histone methyltransferases, including DOMAINS REARRANGED METHYLASE1 and 2 (DRM1 and DRM2), CHROMOMETHYLASE3 (CMT3), SUPRESSOR OF VARIEGATION 3-9 HOMOLOG 9 (SUVH9) and SUVH2, which leads to heterochromatin formation and silencing of target loci (Law and Jacobsen 2010).

6.4 Regulatory Roles of Small RNAs in Plants

Small RNA-directed gene silencing is a widely-used regulatory mechanism that plays an important role in almost every aspect of plant growth and development. Although their target genes and sequences are genetically and functionally diverse and their targeting mechanisms vary, the silencing activity of small RNAs greatly influences the transcriptome, proteome, and epigenome of many eukaryotic organisms. The most conserved miRNA families in insects, mammals, and plants target key regulators (primarily transcription factors) in growth, development and stress responses (reviewed in Bartel 2004; Jones-Rhoades et al. 2006). On the other hand, siRNAs have been implicated in maintaining genome integrity against invading sequences that may be either endogenous (transposon) or exogenous (virus) sources (Ruiz-Ferrer and Voinnet 2009). Here, we review recent progress in dissecting the regulatory roles of miRNAs, tasiRNAs, and siRNAs in plants.

The role of small RNAs in plant development first became evident when mutants defective in small RNA biogenesis or functions were identified. Mutations of key factors in miRNA biogenesis and function, including *DCL1*, *AGO1*, *HEN1*, and *HYL1* genes result in pleiotropic developmental defects (reviewed in Chen 2010). Among those mutants, *dcl1* mutations were isolated from multiple mutant screens, with mutant phenotypes ranging from severe early embryonic arrest in null mutants to abnormal flower organs and leaf morphology in partial loss-of-function mutants. In contrast to the pleiotropic phenotypes of mutants of genes involved in the miRNA pathway, mutants of siRNA biogenesis factors and silencing effectors in *Arabidopsis*, such as AGO4, RDR6, and DCL4, exhibit at most a weak developmental abnormality. This difference suggests that disruption of the miRNA-mediated pathway results in more severe biological consequences than in the siRNA pathway, in accordance with the functionality of the targets of miRNA or siRNA.

6.4.1 Timing and Phase Transition

In flowering plants, the transition between three major developmental phases (juvenile vegetative, adult vegetative and reproductive phases) is usually accompanied by distinct phenotypic characteristics such as changes in leaf morphology and cell identity (Poethig 1990). Mutants with either accelerated or delayed phase transitions have in many cases turned out to contain mutations in miRNA-mediated regulation (Willmann and Poethig 2005). miR156 negatively regulates the *SPL* gene family, a family of transcription factors whose expression increases from young to mature shoots (reviewed in Chen 2009; Chuck and O'Connor 2010). Mutation of the miR156 binding site in the *SPL15* gene causes an early transition from juvenile to adult phase. The role of miR156 in the regulation of phase change appears to be conserved among monocots and dicots, since a similar miR156 regulatory module is also found in maize.

Many miRNAs involved in the control of development timing also exhibit temporal expression in developmental phases. For example, *MIR164* gene has high expression level in the juvenile phase, but the level decreases toward the senescence stage. miR164 works together with two aging-related regulators, ORESARA1 (ORE1) and ETHYLENE INSENSITIVE2 (EIN2), to control aging-induced cell death in *Arabidopsis* leaves. During the juvenile phase, miR164 represses the transcriptional factor ORE1 which promotes aging-induced cell death in leaves. While leaves age, *MIR164* expression is repressed by EIN2, which leads to de-suppression of *ORE1* and subsequently leaf senescence (Kim et al. 2009).

6.4.2 Pattern Formation and Morphogenesis

A key differentiation event in leaf development is the determination of polarity, where cell fate is determined for adaxial (upper) or abaxial (lower) identities, leading to the polarized development of a flat lamina. The establishment and maintenance of abaxial and adaxial identities are tightly controlled by a complex regulatory network conserved across land plants, which includes transcription factors, miRNAs and ta-siRNAs (review by Barkoulas et al. 2007; Chen 2009; Chuck and O'Connor 2010; Kidner and Timmermans 2010). Specifically, expression of miR165/166 at the abaxial side of leaf primodia is critical for restricting the expression of HD-ZIPIII genes, encoding transcription factors involved in meristerm maintenance and polarity establishment, to the adaxial side. Mutations of miRNA target sites in the *Arabidopsis* and maize HD-ZIPIII genes abolish the miR165-directed cleavage of these transcripts, which subsequently result in increased meristem size and leaf adaxialization. Adaxialized leaves are also found in an *Arabidopsis serrate* mutant, which disrupts the biosynthesis of miRNA165/166 (Grigg et al. 2005).

Recent studies demonstrate that the target gene of miR390, *TAS3*, is also important for leaf polarity in *Arabidopsis*. In contrast to miR165/166, *TAS3* gene and its product (special tasi-RNAs named tasiR-ARFs after their target genes) are expressed on the adaxial surface. tasiR-ARFs promote adaxial identity by repressing the expression of *Arabidopsis ARF3* and *ARF4* and maize *ARF3a* and restricting their activity to the abaxial side. A similar phenomenon is found in maize, where a mutation in *leafbladeless1*, a homolog of the *Arabidopsis SGS3* gene, blocks tasi-RNA biogenesis, shows expanded expression of miR165/166 and abaxialization of leaves (Nogueira et al. 2007). Therefore, the opposing ta-siRNA and miRNA gradients seem to coordinate the adaxial-abaxial development, and mutual inhibition between their target genes expressed on each side refines the boundary. The miR165/166 and tasiR-ARFs networks appear to be a conserved regulatory mechanism in eudicot and monocot plants.

6.4.3 Hormone Signaling

Plant hormones, including abscisic acid (ABA), auxin, cytokinin, gibberellins (GA), ethylene, jasmonic acid and brassinosteroids, are key regulators of growth, development, and stress responses throughout the life cycle of plants (reviewed by Gray 2004). The biogenesis, reception, and signaling of phytohormones is a highly diversified and intricate network. To add to the complexity, recent work has demonstrated small RNA-regulation of hormone responses (reviewed by Liu and Chen 2009). One indication of this small RNA-hormone intersection is the fact that many miRNAs are up- or down-regulated by phytohormones. Here, we will focus on the miRNA involvement in auxin signaling, which has been the most extensively studied in recent years.

In the auxin response pathway, a family of transcription factors called auxin response factor (ARF) is responsible for activating auxin-responsive genes. ARFs are normally suppressed by Aux/IAA transcription factors. But in the presence of auxin, Aux/IAAs are degraded through proteolysis and ARF activity is restored to induce auxin-responsive genes. Many *ARFs* are shown to be regulated by miRNAs. For example, miR160 regulates *ARF10*, *ARF16*, and *ARF17* during shoot and root development post-transcriptionally by mRNA cleavage. Mutation in the miR160 binding sites of these *ARF* genes results in pleiotropic developmental defects, such as aberrant aerial organs and root cap formation (Mallory et al. 2005; Wang et al. 2005; Liu et al. 2007). Also, miR167 targets *ARF6* and *ARF8* which regulates ovule and anther development during flower development (Wu et al. 2006). Mutations in the miR167 binding sites affect both male and female fertility, leading to ovule abortion and anthers that fail to dehisce. In addition, miR164 is shown to regulate NAC1, a transcription activator responsible for auxin-regulated lateral root initiation (Guo et al. 2005). While *miR164* mutants show increased *NAC1* mRNA accumulation and more lateral roots, overexpression of *MIR164* reduced the emergence of lateral roots. Lastly, miR393 targets the F-box auxin receptor transport inhibitor response 1 (TIR1) and three other closely related F-box proteins (AFB1, AFB2 and AFB3), whose actions have been implicated in biotic stress responses in plants. Therefore, auxin responses involved a complex cascade of regulators including signaling molecules, transcriptional factors, and miRNAs.

6.4.4 VIGS and Biotic Stress

Virus-directed gene silencing is a small RNA-directed defense mechanism against viruses. Besides VIGS, recent evidence also indicates a crucial role of endogenous siRNAs in defense against bacterial pathogens (reviewed in Ruiz-Ferrer and Voinnet 2009). Pathogen-derived nucleic acids such as viral RNA can induce RNA-mediated silencing in host plants, which in turn suppresses pathogen gene expression and replication.

The presence of viral double-stranded RNA in virus-infected plants is thought to be the trigger of the antiviral silencing mechanism (reviewed in Ruiz-Ferrer and Voinnet 2009). The dsRNA also could be from the foldback structure of single-stranded viral transcripts from DNA viruses, RNA viruses, viroids, and satellite DNAs. Once produced, viral dsRNA is processed by DCL4 in *Arabidopsis* into 21 nt viral small RNAs (vsRNAs), or by DCL2 into 22 nt vsRNA if DCL4 activity is suppressed by a virus (Deleris et al. 2006; Garcia-Ruiz et al. 2010). vsRNAs are then incorporated into the AGO1 complex, which mediates cleavage or translational repression of viral RNA. Alternatively, the genome of DNA viruses could be repressed by DNA and histone methylation through DCL3-dependent vsRNAs. Alternatively, these 21 or 22 nt dsRNAs could act as primary signals that are further amplified by RDR6 and RDR1-dependent secondary vsRNA production (Wang et al. 2010). As a consequence, secondary vsRNAs suppress viral gene replication and virus accumulation.

To counteract host RNA silencing, pathogens are known to affect the plant cellular small RNA profiles by targeting key components in small RNA biogenesis and function (reviewed in Ruiz-Ferrer and Voinnet 2009). For example, the expression of *MIR393* is induced by bacteria flagellin-derived peptide flg22, a bacterial pathogen-associated molecular pattern (PAMP), and by a non-virulent strain of *Pseudomonas syringae* pv. tomato (*P.s.t.*) DC3000 *hrcC*. miR393 is found to negatively regulate the F-box auxin receptor TIR1 and F-box proteins (AFB2 and AFB3). Therefore, recognition of bacterial flg22 can lead to down-regulation of auxin signaling in *Arabidopsis* by miR393 through targeting auxin receptors (Navarro et al. 2006). In other words, suppressing plant auxin signaling may increase plant resistance to bacterial pathogens. In contrast to miR393, *P. s.t. hrc*C strain inoculation also results in down-regulation of several miRNA genes including *MIR825* (Fahlgren et al. 2007; Katiyar-Agarwal and Jin 2010). Surprisingly, these down-regulated miRNAs are found to target genes encoding positive defense regulators such as resistance (R) protein and *MIRNA* genes that suppress organ development (Lu et al. 2007). Therefore, the role of miRNAs in PAMP-triggered immunity is to maintain a balancing act between miRNA-mediated repression of positive and negative regulators.

6.4.5 Maintenance of Genome Integrity and Paramutation

siRNAs repress transcription by maintaining heterochromatin through DNA methylation and histone modifications.

A major function of this siRNA-mediated silencing is thought to be genome defense, where silencing of invading sequences from transposons and viruses helps to maintain genome integrity and prevent deleterious gene disruption by foreign sequences (reviewed in Gehring and Henikoff 2007; Zilberman 2008). As detailed in the chapters (this volume) by Feschotte and Kejnovsky (2.1.2) and Jiang and Slotkin (2.1.3), transposons are ubiquitously methylated (Lippman et al. 2004; Tran et al. 2005), with pericentromeric regions being particularly rich in highly methylated transposons and repeats (Zhang et al. 2006; Zilberman et al. 2006). The importance of repressing transposons is clear in *Arabidopsis ddm1* mutants: when DNA methylation is reduced, TE mobility increases, which results in abnormal phenotypes due to transposon transposition (Miura et al. 2001). Another example is the *Mutator* (*Mu*) transposon in maize, a highly mobile transposon that is silenced progressively as the plant ages. The *mop1* mutation causes reactivation of silenced *Mu* elements and loss of small RNA involved in RdDM (Woodhouse et al. 2006). Characterization of *mop1* revealed that *MOP1* encodes RDR2, which is in accordance with the role of small RNA-directed DNA methylation in maintaining genome integrity. One interesting observation about the maize *mop1* mutant is that the establishment and maintenance of paramutation of the *Booster 1* (*b1*) genes is also reduced (Alleman et al. 2006; Woodhouse et al. 2006). In maize, paramutation is a phenomenon where one allele heritably changes the expression of another homologous allele. An excellent example of paramutation is the *b1* locus, which encodes a transcription factor in the anthocyanin biosynthetic pathway, which can be monitored by the degree of purple pigmentation in mature plant tissues (Arteaga-Vazquez and Chandler 2010). The *mop1* study demonstrates the involvement of RDR2 in paramutation, indicating that paramutation involves RNA-mediated chromatin modification and RNAi pathways.

6.5 Evolution of Plant Small RNA

In the past few years, advances in high-throughput sequencing and its application to genome-scale small RNA profiling have greatly increased the speed of small RNA discovery and characterization. In addition to the discovery of numerous miRNAs and siRNAs in model organisms such as *C. elegans* and *Arabidopsis*, small RNAs have also been characterized from lower eukaryotes like ciliates and green algae, as well as from higher eukaryotes. The widespread presence of small RNAs suggests a deep evolutionary origin for small RNA-based regulation in eukaryotes. Many miRNAs, such as miR390 and the miR156 families are conserved among mosses, ferns, monocots, and eudicots. Not surprisingly, key factors involved in small RNA

Table 6.3 Small RNAs and components of RNAi machinery in five major groups of eukaryotes

Eukaryote supergroups[a]	Species	Dicer-like protein	AGO–PIWI like protein	RdRP	siRNA	RNAi	miRNA	piRNA
Excavata	Giardia (*Giardia intestinalis*)	✓[b]	✓	−	−	−	−	−
	Trypanosoma (*Trypanosoma brucei, T. cruzi, Leishmania major, L. braziliensis*)	✓	✓	−	✓	✓	−	−
Chromalveolata	Ciliates (*Paramecium tetraurelia, Tetrahymena thermophila, Plasmodium falciparum*)	✓	✓	✓	✓	✓	?	✓
	Oomycete (*Phytophthora infestans*)	?	✓	?	✓	✓	−	−
	Marine centric diatom (*Thalassiosira pseudonana*)	−	−	−	−	−	−	−
Archaeplastida	Red algae (*Cyanidioschyzon merolae*)	−	−	−	−	−	−	−
	Green algae (*Chlamydomonas reinhardtii*)	✓	✓	−	✓	✓	✓	
	Arabidopsis (*Arabidopsis thaliana*)	✓	✓	✓	✓	✓	✓	
	Rice (*Oryza sativa* ssp. *japonica*)	✓	✓	✓	✓	✓	✓	
Amoebozoa	Slime mold (*Dictyostelium discoideum*)	✓	✓	✓	✓	✓	✓	−
Opisthokonta	Brewer's yeast (*Saccharomyces cerevisiae*)	−	−	−	−	−	−	−
	Fission yeast (*Schizosaccharomyces pombe*)	✓	✓	✓	✓	✓	−	
	Fungi (*Neurospora crassa, Aspergillus nidulans*)	✓	✓	✓	✓	✓	✓	−
	Roundworm (*Caenorhabditis elegans*)	✓	✓	✓	✓	✓	✓	✓
	Fruit fly (*Drosophila melanogaster*)	✓	✓	✓	✓	✓	✓	✓
	Sea urchin (*Strongylocentrotus purpuratus*)	✓	✓	−	−	?	✓	−
	Zebrafish (*Danio rerio*)	✓	✓	−	?	✓	✓	−
	Human (*Homo sapiens*)	✓	✓	−	✓	✓	✓	✓

[a]The classification of eukaryotes is based on previous publication (Adl et al. 2005). Table 6.3 is adapted and updated from previous reviews (Axtell and Bowman 2008; Shabalina and Koonin 2008)

[b]The presence of miRNA, siRNA, piRNA, and small RNA biogenesis factors is based on current literature, and from miRBase release 16 (Griffiths-Jones et al. 2006; Axtell and Bowman 2008; Shabalina and Koonin 2008). The absence of evidence for small RNAs and factors is marked by minus sign (−)

biogenesis such as Argonautes, Dicers, and the dsRNA-triggered gene silencing machinery are also found across eukaryotic organisms (summarized in Table 6.3) (Adl et al. 2005; Cerutti and Casas-Mollano 2006). The prevalence of small RNAs, effectors, and RNAi implies that these genomic features evolved early in the evolution of eukaryotes.

Notwithstanding this broad pattern of evolutionary conservation, RNA-based genome regulation appears to be lost independently in several eukaryotic lineages. Some unicellular eukaryotes, including the brewer's yeast *Saccharomyces cerevisiae* and red algae, have completely lost the RNAi machinery, with a genomic absence of small RNAs, AGO, or Dicer homologs (Aravind et al. 2000; Cerutti and Casas-Mollano 2006). One possible explanation is that the RNA-based genome defense and expression regulation may be nonessential for unicellular eukaryotes which have simpler biology since they don't need to coordinate development across multiple diverse cell types. While some lower eukaryotes maintain RNA-based regulation for selective advantage, some find it dispensable and lose the function over evolutionary time. For multicellular organisms, RNA-based regulation and the RNAi mechanism has become essential for regulating various biological processes, especially in development or differentiation processes that are much more complex and specialized compared to that in unicellular ancestors.

One interesting note is that although prokaryotes appear to have miRNA or siRNA-like systems, the proteins involved in these processes are not homologous to the AGO or PIWI families in eukaryotes despite having similar RNA binding domains (Makarova et al. 2009). While these factors may function to protect prokaryotes against the exogenous DNA from phages and plasmids, the prokaryotic RNA-based silencing systems likely have evolved independently from the eukaryotic system.

6.5.1 Conservation and Diversification of Animal and Plant Small RNAs

Small RNAs are diverse in their size, sequence, biogenesis, and mode of action. For example, animal and plant miRNAs are quite distinct from each other, an explanation of which is described in the following paragraphs (reviewed by Chapman and Carrington 2007; Axtell and Bowman 2008; Shabalina and Koonin 2008).

1. *miRNA-generating loci*: Both animal and plant miRNA genes are usually located in intergenic regions, while some animal miRNA genes are found in the introns of protein-coding genes (Reinhart and Bartel 2002; Rodriguez et al. 2004). Animal miRNA genes often form clusters in the genome and are transcribed into one

primary transcript (poly-cistronic), whereas the majority of plant miRNA genes are transcribed individually (mono-cistronic) (Bartel 2004; Kim and Nam 2006). Plant miRNA precursors vary in length and secondary structure features such as stem loop position. In contrast, animal miRNA precursors are less diverse and have a defined loop size and stem-loop positions for their secondary structure.

2. *miRNA feature and biogenesis*: The biogenesis of animal miRNAs involves a two-step process. The miRNA precursor is first processed by the RNase III activity of Drosha and DGCR8 proteins into a stem-loop structure in the nucleus (Han et al. 2004). After the hairpin pre-miRNA is exported into the cytosol, Dicer completes the cleavage process and produces a miRNA:miRNA* duplex, which is incorporated into RISC. However, plants don't have a homolog of Drosha, and the enzymatic cleavage of miRNA precursor appears to be replaced by specialized Dicer proteins. Plant miRNA precursors are cleaved into a miRNA:miRNA* duplex by DCL1 in the nucleus (Kurihara and Watanabe 2004). The duplex is then methylated at the 3′ end by HEN1 and exported to the cytosol to bind with RISC.

3. *miRNA target and regulation mechanism*: Animal miRNAs have imperfect complementary to their target mRNA compared to their plant counterparts. Two to eight nucleotides of complete base-pairing in the 5′ region of miRNA to their target RNA are usually sufficient for miRNA action. Multiple target sites are mostly located in the 3′-untranslated regions of genes, and miRNA:mRNA pairing mainly leads to repression of translation. In contrast, plant miRNAs have near-perfect complementarity to their target mRNAs. The single target site is mostly located within the open reading frame of genes. miRNA:mRNA pairing mainly leads to posttranscriptional silencing with AGO-dependent cleavage of target mRNA in RISC (as discussed above).

Diversification of small RNAs is evident in the case of miRNA families, for which *MIRNA* genes are categorized based on sequence homology and their secondary structure. Animal genomes have a much larger total number of miRNA families compared to plant genomes. On the other hand, the average size of a miRNA family in animals is less than two gene members, whereas a plant family has about 2.5 gene members (Li and Mao 2007). These differences suggest distinct evolutionary paths for animal and plant miRNA families. The larger miRNA families in plants may result from duplication of paralogs, possibly due to genome duplication events, while new *MIRNA* genes apparently arise more frequently in animal genomes. Furthermore, the expansion of plant miRNA families also changes in a lineage- and species-specific manner. Currently, there are 133 miRNA families and 214 *MIRNA* genes predicted in *Arabidopsis*

(miRBase release 16) (Griffiths-Jones et al. 2006), while there are 28 miRNA families and 170 *MIRNA* genes predicted in maize genome. The high gene per family number (6.07 genes/family) suggests the extensive expansion of maize miRNA families (the lower total number of families is likely due to less extensive analysis in the recently-completed maize genome). For instance, the miR166 family has seven members in *Arabidopsis* but 14 members in maize. Interestingly, only 29 of more than 100 miRNA families from *Arabidopsis* are found in rice, indicating that many miRNA families evolved after the divergence of dicot and monocot lineages (reviewed in Axtell and Bowman 2008).

While animal and plant miRNAs appear to have gone through substantial diversification during evolution, many miRNA families remain conserved in particular lineages. Conserved miRNA genes are often represented by multiple loci in each genome; these could arise from the result of ancient or recent genome duplication events (Maher et al. 2006). For example, 39 miRNA families are present in two or more phylogenetically distant plant species (summarized in Axtell and Bowman 2008). While some miRNA families have a wide distribution in diverse angiosperms (the seed plants), a subset of them, including miR156 and miR390, are also present in early land plants, including gymnosperms (e.g. pine), lycopods (e.g. spikemoss), and bryophytes (e.g. moss), suggesting that they evolved much earlier. The number of conserved miRNA families expands from lower to higher eukaryotes, which may reflect an increasing demand for the regulatory capacity provided by small RNA-mediated regulation (Chapman and Carrington 2007; Shabalina and Koonin 2008).

Conserved miRNA genes are often represented by multiple loci in each genome; these could arise from the result of ancient or recent genome duplication events (Maher et al. 2006). More importantly, conserved miRNAs usually target genes that are also conserved and involved in important biological functions. One good example is that 13 out of 21 conserved miRNAs target genes encoding transcription factors for developmental control (Axtell and Bowman 2008). In addition, the two oldest miRNA families, miR165/166 in plants and let-7 in animals, also are involved in symmetry development and differentiation in plants and animals. On the other hand, non-conserved miRNAs have limited phylogenetic distribution in fewer species, but their numbers exceed the conserved miRNAs. These non-conserved miRNAs usually have just a single copy in the genome, and exhibit significant similarity to the target genes not only at the recognition sites but throughout the length of miRNA precursors (Fahlgren et al. 2007). The picture that emerges is one where a small set of miRNAs have an ancient origin while new miRNAs are constantly being generated and evolving to target various genes in different species. Once the new miRNAs become essential and are incorporated into regulatory networks, they become stabilized over evolutionary time. More details about the birth and evolution of small RNA will be discussed in the last section of this chapter.

6.5.2 Conservation and Diversification of Small RNA Biogenesis Factors

In eukaryotes, small RNAs and the core components of biogenesis and silencing pathways have been through extensive specialization and diversification. Variation on modules of the core RNAi machinery (e.g. biogenesis factors) creates a reprogrammable system and provides the potential to fine tune regulatory networks. Here, we will focus on the conservation and diversification of three key proteins involved in small RNAs biogenesis and RNAi, Argonaute-PIWI (AGO-PIWI), Dicer, and RNA-dependent RNA polymerase (RdRP) (reviewed in Chapman and Carrington 2007; Shabalina and Koonin 2008).

1. *Dicer*: Mammal and nematode genomes each contain only one Dicer protein, which processes and generates multiple classes of small RNAs, while plants delegate small RNA biogenesis among multiple Dicer proteins (reviewed in Carthew and Sontheimer 2009). Members of the Dicer families usually have specified function in processing either miRNA or siRNA. But in some cases, different Dicers work in hierarchy or they could have redundant functions. *Arabidopsis* has four Dicer-like proteins (DCLs): DCL1 is devoted to miRNA production and DCL2, DCL3, and DCL4 are involved in siRNA biogenesis. In addition, DCL4 is also required for producing 21 nt tasi-RNA, and it acts hierarchical to DCL2 in secondary siRNA production from viruses and transgenes (Deleris et al. 2006; Moissiard et al. 2007). One possible explanation for the proliferation of DCL proteins in plants is the evolution of antiviral mechanisms via RNA silencing. It has been suggested that multiple members of the DCL family could each contribute to plant defense against various viral pathogens, encouraging duplication and specification of DCL proteins through evolution (Havecker et al. 2010).

2. *Argonaute*: Argonautes are at the core of the RNA silencing machinery since they are the central components of RISC along with the small RNA guide. Argonautes can be divided into three groups: the AGO group that binds miRNAs and siRNAs; the PIWI group that binds piRNA; and a third group found only in nematodes that binds secondary siRNAs. Studies have shown that the Argonaute families have gone through extensive proliferation and diversification, resulting in various numbers and diverse functions of AGO members among organisms (reviewed in Vaucheret et al. 2004; Chapman and

Carrington 2007). Although some species like fission yeast (*Schizosaccharomyces pombe*) only have a single AGO protein, most organisms contain multiple Argonaute proteins. For example, *Drosophila* has two AGOs and three PIWI proteins whereas humans have four AGOs and four PIWIs. *Arabidopsis* and rice have 10 and 18 AGOs, respectively. One extreme is the roundworm (*C. elegans*), which has in total 27 Argonautes from all three groups. The need for multiple Argonautes with their distinct activities demonstrates the functional specialization of RNAi machinery. Furthermore, the spatiotemporal expression of Argonaute members has introduced an additional layer of specialization. For example, in a recent study in *Arabidopsis* of closely related AGO4, AGO6, and AGO9 proteins, the selective binding of sRNA with different AGOs is determined by the coincident expression and interaction of the AGO and their target loci in a tissue-specific manner (Havecker et al. 2010). The proliferation of eukaryotic Argonautes provides a perfect example of how the RNA-based silencing mechanism has expanded and functionally diversified.

3. *RNA-dependent RNA polymerase*: *Arabidopsis* has six RdRPs that play important roles in siRNA formation and silencing pathways, especially in antiviral defense mechanisms (reviewed in Ruiz-Ferrer and Voinnet 2009). While RDR2 participates in the production of 24 nt siRNAs from transposons and repeats via DCL3 activity, RDR6 is involved in secondary siRNA accumulation in transgene-induced gene silencing in a DCL4-dependent manner and is responsible for ta-siRNA biogenesis (Katiyar-Agarwal and Jin 2010). On the other hand, RdRP homologs are absent in certain animal genomes such as those of sea urchin and human, indicating that RdRP-dependent silencing amplification may not be required in these organisms (Shabalina and Koonin 2008).

Although prokaryotes don't have RNAi, many proteins contain homologous domains to the eukaryotic Argonaute, Dicer, and RdRP. For example, the RNase III and Superfamily II RNA helicase domains in Dicer proteins are common domains in prokaryotes, but the combination of these two domains into one Dicer protein can be only found in eukaryotes. The RNase III domain is frequently found in proteins involved in rRNA processing and mRNA degradation, and the helicase domain is related to the Hef protein, the Superfamily II helicase in archaea which is involved in DNA repair (Aravind et al. 1999; MacRae and Doudna 2007). Therefore, the eukaryotic RNAi machinery may evolve from the fusion of two enzymatic domains from different prokaryotic proteins. Similar principles are also evident for AGO-PIWI and RdRP (reviewed in Shabalina and Koonin 2008). Therefore, the eukaryotic RNAi and its components appear to have evolved from the fusion and shuffling of various catalytic building blocks that originated from prokaryotic and archaeal sources. Subsequently, the newly-evolved effectors and RNAi systems have proliferated and specialized by duplication or deletion over evolutionary time, resulting in the divergence of RNAi mechanisms in eukaryotes.

6.5.3 Evolution of miRNA and siRNA

6.5.3.1 Small RNA Biogenesis

How miRNAs and siRNAs originate has been an intriguing question since the discovery of small RNAs. One clue may be provided by the observation that many non-conserved miRNA genes, such as *MIR161* and *MIR163*, have extensive sequence similarity to their targets genes (protein-coding genes) beyond the recognition sites (Allen et al. 2004). This phenomenon led to the development of the "inverted duplication hypothesis" for small RNA genesis, where genes encoding small RNAs in plants originate from inverted duplication of the target genes (Fig. 6.4) (reviewed in Willmann and Poethig 2007; Axtell and Bowman 2008).

According to this model, transcripts from inverted duplication of target genes form a long, self-complementary hairpin that is further processed by dicers into siRNA with regulatory potential. These siRNAs may silence the progenitor genes or other closely-related sequences. If the regulatory role of the siRNA is selectively advantageous, the novel siRNA-progenitor gene pair will be maintained in the genome. Over time, the accumulated mutations and sequence changes would cause the small RNA gene to become more divergent from the original gene, with more mismatches in the foldback region except for the miRNA: miRNA* segment. As the gene evolves further, the hairpin structure with imperfect complementation would become more miRNA-like and eventually form miRNA, and the structural changes of the hairpin structure could render the dsRNA processing to different dicers.

The proposed model suggests several scenarios about how miRNA and siRNA arise and evolve. First, the model suggests that newly evolved miRNA genes will contain extensive sequence similarity to their target genes. Second, the relatively new, non-conserved miRNA genes will contain longer near-perfect complementary hairpin structure

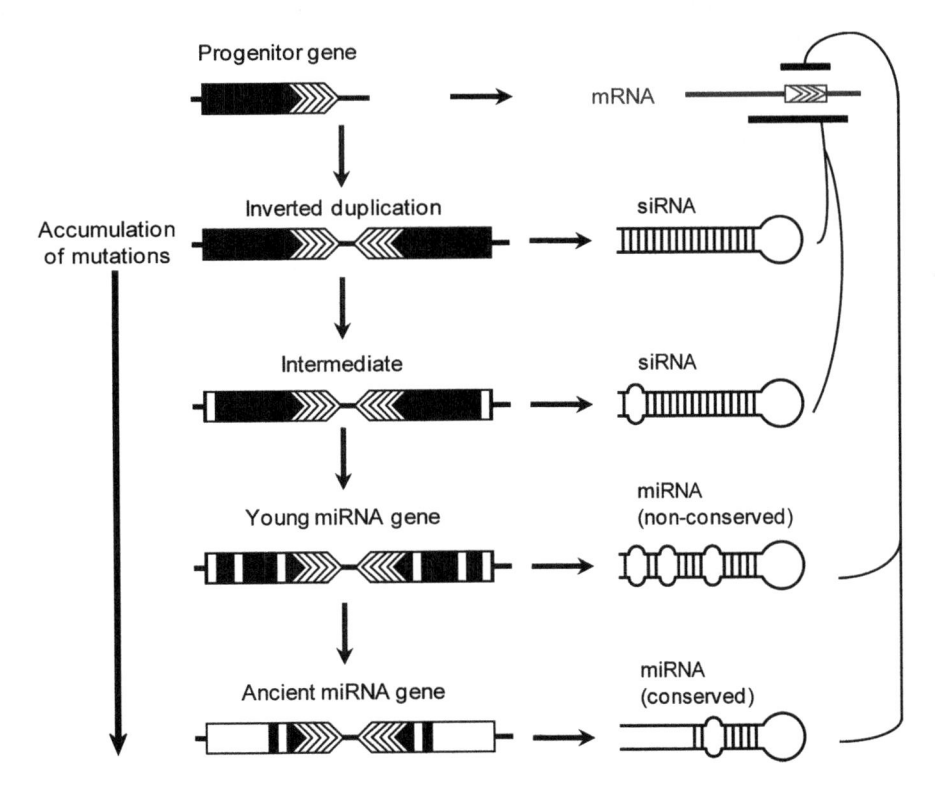

Fig. 6.4 The model of inverted duplication for miRNA genesis and evolution in plants. miRNAs can evolve from inverted duplication of their target genes (*black pentagon*). The transcript of the inverted duplication segment produces hairpin dsRNA which is further processed into siRNA-like small RNAs. siRNAs are able to silence target genes by targeting an extensive length of the target transcript. Over time, the accumulation of random mutations (*white*) in the repeat region leads to a more divergent small RNA, with more mismatches

and less complementarity to their target gene except for the core sequence which gives rise to the miRNA and miRNA* (*pointed stripes*). While the siRNA-like small RNA takes on a structure more similar to miRNAs, the small RNA processing is also shifted from DCL4 to DCL1. Young *MIRNA* genes still contain considerable similarity to the target sequences, whereas ancient *MIRNA* genes contain very little similarity to the target gene beyond the miRNA target site after further gene drift resulting in accumulated mutations while more ancient, conserved miRNA genes will contain imperfect hairpin structures with more mismatches due to the accumulated mutations (Fig. 6.4). Third, the precursors of near-perfect hairpins may be processed by other dicers involved in siRNA biogenesis rather than DCL1, which is specialized for processing imperfect hairpins in miRNAs biogenesis. Indeed, several large-scale small RNA sequencing projects have identified many small RNA-generating loci which generate self-complementary foldbacks with extensive similarity to their target genes (reviewed in Chapman and Carrington 2007). One recent study shows that nearly 80% of miRNA gene families are predicted to form by inverted duplication of the miRNA-related loci (Fahlgren et al. 2010). Some of these miRNA-generating loci, such as *ath-MIR822* and *ath-MIR839*, produce heterogeneous miRNAs that require DCL4 but not DCL1 activity. These miRNA are considered evolutionary intermediates, which may subsequently diverge into DCL1-dependent miRNA-biogenesis pathways. The evidence of miRNA evolution is not limited to plant lineages; these evolutionary intermediate miRNA genes can be found in the unicellular alga (*Chlamydomonas reinhardtii*) and *Drosophila*, where they are processed heterogeneously between factors in the miRNA and siRNA pathways (Chapman and Carrington 2007).

6.5.3.2 Evolution of Newly-Evolved Small RNAs

Newly-evolved, small RNA-generating loci could follow three different fates (reviewed by Chapman and Carrington 2007; Axtell and Bowman 2008). The most common fate for a small RNA-generating locus is the loss or elimination of its functionality through accumulation of mutations in the foldback sequences or miRNA binding site in the target sequence. This is evident when some non-conserved miRNAs have drifted to the point when they no longer target the progenitor gene transcripts anymore, resulting in the decay or death of the small RNA loci (Fahlgren et al. 2007; Fahlgren et al. 2010). The second possible fate is that the small RNA-generating locus is selected for its specificity toward an unrelated gene or gene family, possibly through genetic drift and change of the sequences. This could provide an explanation for the role of *Arabidopsis* miR395 in sulfate uptake and assimilation, in which miR395 targets transcripts of two unrelated genes encoding an ATP sulfurylase and a sulfate transporter, respectively (Jones-Rhoades and Bartel 2004; Allen et al. 2005). The third possible fate is that the small RNA locus is selected to regulate the progenitor gene or closely-related sequences, and that the small RNA-target gene pairs may co-evolve as a unit. The small RNA-target gene unit can gain new function during evolution by further duplication and specialization. For example, a miRNA family can be expanded by duplication of miRNA genes and gene clusters, such as the seven-

member *Arabidopsis* miR166 family (Maher et al. 2006). The duplication of miRNA families may translate into some degree of redundancy. But *MIRNA* gene duplication can also contribute to further specialization through changes in sequences so these *MIRNA* loci may obtain different expression patterns for tissue-specific gene targeting and regulation. A good example is the *Arabidopsis MIR164* family: while three *MIR164* genes work redundantly in developing shoots to regulate phyllotaxis in organ numbers and boundary formation, *MIR164C* is the predominant source of miR164 in regulation of flower development such as petal number (Sieber et al. 2007).

Conclusions

From the first discovery of eukaryotic small RNA pathways over a decade ago, much progress has been made in understanding the biogenesis of these molecules, their functions in gene expression regulation, and the evolution of both small RNAs and their biogenesis factors. Small RNA-directed gene silencing has now been shown to have essential regulatory roles in many aspects of plant biology. The role of small RNAs in transcriptional gene silencing has led to increased interest in epigenetic modifications. Notwithstanding this rapid growth in our understanding, many challenges remain. For example, the factors determining whether miRNAs function to direct mRNA cleavage or translational repression are poorly understood. In addition, many questions remain regarding the establishment of heterochromatin, as well as its maintenance and removal, particularly regarding the roles and relative contributions of siRNAs, Pol IV and Pol V, methyltransferases and demethylases, and other unknown factors. Future investigations and characterization of small RNA action will provide more insights into the functional roles of small RNA-directed genomic regulation and provide tools for future applications in basic and applied sciences.

Acknowledgments Work on plant small RNAs and epigenetics in the Meyers laboratory is supported by the NSF Plant Genome Research Program.

References

Adl SM, Simpson AG, Farmer MA, Andersen RA, Anderson OR, Barta JR, Bowser SS, Brugerolle G, Fensome RA, Fredericq S, James TY, Karpov S, Kugrens P, Krug J, Lane CE, Lewis LA, Lodge J, Lynn DH, Mann DG, McCourt RM, Mendoza L, Moestrup O, Mozley-Standridge SE, Nerad TA, Shearer CA, Smirnov AV, Spiegel FW, Taylor MF (2005) The new higher level classification of eukaryotes with emphasis on the taxonomy of protists. J Eukaryot Microbiol 52:399–451

Alleman M, Sidorenko L, McGinnis K, Seshadri V, Dorweiler JE, White J, Sikkink K, Chandler VL (2006) An RNA-dependent

RNA polymerase is required for paramutation in maize. Nature 442:295–298

Allen E, Xie Z, Gustafson AM, Sung GH, Spatafora JW, Carrington JC (2004) Evolution of microRNA genes by inverted duplication of target gene sequences in *Arabidopsis thaliana*. Nat Genet 36:1282–1290

Allen E, Xie Z, Gustafson AM, Carrington JC (2005) microRNA-directed phasing during trans-acting siRNA biogenesis in plants. Cell 121:207–221

Aravind L, Walker DR, Koonin EV (1999) Conserved domains in DNA repair proteins and evolution of repair systems. Nucleic Acids Res 27:1223–1242

Aravind L, Watanabe H, Lipman DJ, Koonin EV (2000) Lineage-specific loss and divergence of functionally linked genes in eukaryotes. Proc Natl Acad Sci U S A 97:11319–11324

Arteaga-Vazquez MA, Chandler VL (2010) Paramutation in maize: RNA mediated trans-generational gene silencing. Curr Opin Genet Dev 20:156–163

Axtell MJ, Bowman JL (2008) Evolution of plant microRNAs and their targets. Trends Plant Sci 13:343–349

Barkoulas M, Galinha C, Grigg SP, Tsiantis M (2007) From genes to shape: regulatory interactions in leaf development. Curr Opin Plant Biol 10:660–666

Bartel DP (2004) microRNAs: genomics, biogenesis, mechanism, and function. Cell 116:281–297

Borsani O, Zhu J, Verslues PE, Sunkar R, Zhu JK (2005) Endogenous siRNAs derived from a pair of natural *cis*-antisense transcripts regulate salt tolerance in *Arabidopsis*. Cell 123:1279–1291

Brodersen P, Sakvarelidze-Achard L, Bruun-Rasmussen M, Dunoyer P, Yamamoto YY, Sieburth L, Voinnet O (2008) Widespread translational inhibition by plant miRNAs and siRNAs. Science 320:1185–1190

Carthew RW, Sontheimer EJ (2009) Origins and mechanisms of miRNAs and siRNAs. Cell 136:642–655

Cazzonelli CI, Millar T, Finnegan EJ, Pogson BJ (2009) Promoting gene expression in plants by permissive histone lysine methylation. Plant Signal Behav 4:484–488

Cerutti H, Casas-Mollano JA (2006) On the origin and functions of RNA-mediated silencing: from protists to man. Curr Genet 50:81–99

Chapman EJ, Carrington JC (2007) Specialization and evolution of endogenous small RNA pathways. Nat Rev Genet 8:884–896

Chen X (2009) Small RNAs and their roles in plant development. Annu Rev Cell Dev Biol 35:21–44

Chen X (2010) Small RNAs—secrets and surprises of the genome. Plant J 61:941–958

Chen HM, Chen LT, Patel K, Li YH, Baulcombe DC, Wu SH (2010) 22-Nucleotide RNAs trigger secondary siRNA biogenesis in plants. Proc Natl Acad Sci U S A 107:15269–15274

Chuck G, O'Connor D (2010) Small RNAs going the distance during plant development. Curr Opin Plant Biol 13:40–45

Couzin J (2002) Breakthrough of the year. Small RNAs make big splash. Science 298:2296–2297

Cuperus JT, Carbonell A, Fahlgren N, Garcia-Ruiz H, Burke RT, Takeda A, Sullivan CM, Gilbert SD, Montgomery TA, Carrington JC (2010) Unique functionality of 22-nt miRNAs in triggering RDR6-dependent siRNA biogenesis from target transcripts in *Arabidopsis*. Nat Struct Mol Biol 17:997–1003

Deleris A, Gallego-Bartolome J, Bao J, Kasschau KD, Carrington JC, Voinnet O (2006) Hierarchical action and inhibition of plant Dicer-like proteins in antiviral defense. Science 313:68–71

Fahlgren N, Montgomery TA, Howell MD, Allen E, Dvorak SK, Alexander AL, Carrington JC (2006) Regulation of AUXIN RESPONSE FACTOR 3 by *TAS3* ta-siRNA affects developmental timing and patterning in *Arabidopsis*. Curr Biol 16:939–944

Fahlgren N, Howell MD, Kasschau KD, Chapman EJ, Sullivan CM, Cumbie JS, Givan SA, Law TF, Grant SR, Dangl JL, Carrington JC (2007) High-throughput sequencing of *Arabidopsis* microRNAs: evidence for frequent birth and death of *MIRNA* genes. PLoS One 2:e219

Fahlgren N, Jogdeo S, Kasschau KD, Sullivan CM, Chapman EJ, Laubinger S, Smith LM, Dasenko M, Givan SA, Weigel D, Carrington JC (2010) microRNA gene evolution in *Arabidopsis lyrata* and *Arabidopsis thaliana*. Plant Cell 22:1074–1089

Fang Y, Spector DL (2007) Identification of nuclear dicing bodies containing proteins for microRNA biogenesis in living *Arabidopsis* plants. Curr Biol 17:818–823

Fire A, Xu S, Montgomery MK, Kostas SA, Driver SE, Mello CC (1998) Potent and specific genetic interference by double-stranded RNA in *Caenorhabditis elegans*. Nature 391:806–811

Gao Z, Liu HL, Daxinger L, Pontes O, He X, Qian W, Lin H, Xie M, Lorkovic ZJ, Zhang S, Miki D, Zhan X, Pontier D, Lagrange T, Jin H, Matzke AJ, Matzke M, Pikaard CS, Zhu JK (2010) An RNA polymerase II- and AGO4-associated protein acts in RNA-directed DNA methylation. Nature 465:106–109

Garcia-Ruiz H, Takeda A, Chapman EJ, Sullivan CM, Fahlgren N, Brempelis KJ, Carrington JC (2010) *Arabidopsis* RNA-dependent RNA polymerases and dicer-like proteins in antiviral defense and small interfering RNA biogenesis during Turnip Mosaic Virus infection. Plant Cell 22:481–496

Gehring M, Henikoff S (2007) DNA methylation dynamics in plant genomes. Biochim Biophys Acta 1769:276–286

Gray WM (2004) Hormonal regulation of plant growth and development. PLoS Biol 2:E311

Griffiths-Jones S, Grocock RJ, van Dongen S, Bateman A, Enright AJ (2006) miRBase: microRNA sequences, targets and gene nomenclature. Nucleic Acids Res 34(Database issue):D140–D144

Grigg SP, Canales C, Hay A, Tsiantis M (2005) *SERRATE* coordinates shoot meristem function and leaf axial patterning in *Arabidopsis*. Nature 437:1022–1026

Guo HS, Xie Q, Fei JF, Chua NH (2005) microRNA directs mRNA cleavage of the transcription factor *NAC1* to downregulate auxin signals for *Arabidopsis* lateral root development. Plant Cell 17:1376–1386

Hamilton AJ, Baulcombe DC (1999) A species of small antisense RNA in posttranscriptional gene silencing in plants. Science 286:950–952

Han J, Lee Y, Yeom KH, Kim YK, Jin H, Kim VN (2004) The Drosha-DGCR8 complex in primary microRNA processing. Genes Dev 18:3016–3027

Havecker ER, Wallbridge LM, Hardcastle TJ, Bush MS, Kelly KA, Dunn RM, Schwach F, Doonan JH, Baulcombe DC (2010) The *Arabidopsis* RNA-directed DNA methylation argonautes functionally diverge based on their expression and interaction with target loci. Plant Cell 22:321–334

Henderson IR, Jacobsen SE (2007) Epigenetic inheritance in plants. Nature 447:418–424

Henz SR, Cumbie JS, Kasschau KD, Lohmann JU, Carrington JC, Weigel D, Schmid M (2007) Distinct expression patterns of natural antisense transcripts in *Arabidopsis*. Plant Physiol 144:1247–1255

Herr AJ, Jensen MB, Dalmay T, Baulcombe DC (2005) RNA polymerase IV directs silencing of endogenous DNA. Science 308:118–120

Jones-Rhoades MW, Bartel DP (2004) Computational identification of plant microRNAs and their targets, including a stress-induced miRNA. Mol Cell 14:787–799

Jones-Rhoades MW, Bartel DP, Bartel B (2006) microRNAs and their regulatory roles in plants. Annu Rev Plant Biol 57:19–53

Kanno T, Huettel B, Mette MF, Aufsatz W, Jaligot E, Daxinger L, Kreil DP, Matzke M, Matzke AJ (2005) Atypical RNA polymerase subunits required for RNA-directed DNA methylation. Nat Genet 37:761–765

Katiyar-Agarwal S, Jin H (2010) Role of small RNAs in host-microbe interactions. Annu Rev Phytopathol 48:225–246

Katiyar-Agarwal S, Morgan R, Dahlbeck D, Borsani O, Villegas A Jr, Zhu JK, Staskawicz BJ, Jin H (2006) A pathogen-inducible endogenous siRNA in plant immunity. Proc Natl Acad Sci U S A 103 (47):18002–18007

Kidner CA, Timmermans MC (2010) Signaling sides adaxial-abaxial patterning in leaves. Curr Top Dev Biol 91:141–168

Kim VN, Nam JW (2006) Genomics of microRNA. Trends Genet 22:165–173

Kim JH, Woo HR, Kim J, Lim PO, Lee IC, Choi SH, Hwang D, Nam HG (2009) Trifurcate feed-forward regulation of age-dependent cell death involving miR164 in Arabidopsis. Science 323:1053–1057

Kurihara Y, Watanabe Y (2004) Arabidopsis micro-RNA biogenesis through Dicer-like 1 protein functions. Proc Natl Acad Sci U S A 101:12753–12758

Kurihara Y, Takashi Y, Watanabe Y (2006) The interaction between DCL1 and HYL1 is important for efficient and precise processing of pri-miRNA in plant microRNA biogenesis. RNA 12:206–212

Laubinger S, Sachsenberg T, Zeller G, Busch W, Lohmann JU, Ratsch G, Weigel D (2008) Dual roles of the nuclear cap-binding complex and SERRATE in pre-mRNA splicing and microRNA processing in Arabidopsis thaliana. Proc Natl Acad Sci U S A 105:8795–8800

Law JA, Jacobsen SE (2010) Establishing, maintaining and modifying DNA methylation patterns in plants and animals. Nat Rev Genet 11:204–220

Law JA, Ausin I, Johnson LM, Vashisht AA, Zhu JK, Wohlschlegel JA, Jacobsen SE (2010) A protein complex required for polymerase V transcripts and RNA- directed DNA methylation in Arabidopsis. Curr Biol 20:951–956

Lee RC, Feinbaum RL, Ambros V (1993) The C. elegans heterochronic gene lin-4 encodes small RNAs with antisense complementarity to lin-14. Cell 75:843–854

Lee Y, Kim M, Han J, Yeom KH, Lee S, Baek SH, Kim VN (2004) microRNA genes are transcribed by RNA polymerase II. EMBO J 23:4051–4060

Li A, Mao L (2007) Evolution of plant microRNA gene families. Cell Res 17:212–218

Li J, Yang Z, Yu B, Liu J, Chen X (2005) Methylation protects miRNAs and siRNAs from a 3'-end uridylation activity in Arabidopsis. Curr Biol 15:1501–1507

Lippman Z, Gendrel AV, Black M, Vaughn MW, Dedhia N, McCombie WR, Lavine K, Mittal V, May B, Kasschau KD, Carrington JC, Doerge RW, Colot V, Martienssen R (2004) Role of transposable elements in heterochromatin and epigenetic control. Nature 430:471–476

Liu Q, Chen YQ (2009) Insights into the mechanism of plant development: interactions of miRNAs pathway with phytohormone response. Biochem Biophys Res Commun 384:1–5

Liu J, Valencia-Sanchez MA, Hannon GJ, Parker R (2005) microRNA-dependent localization of targeted mRNAs to mammalian P-bodies. Nat Cell Biol 7:719–723

Liu PP, Montgomery TA, Fahlgren N, Kasschau KD, Nonogaki H, Carrington JC (2007) Repression of AUXIN RESPONSE FACTOR 10 by microRNA160 is critical for seed germination and post-germination stages. Plant J 52:133–146

Lu S, Sun YH, Amerson H, Chiang VL (2007) microRNAs in loblolly pine (Pinus taeda L.) and their association with fusiform rust gall development. Plant J 51:1077–1098

MacRae IJ, Doudna JA (2007) Ribonuclease revisited: structural insights into ribonuclease III family enzymes. Curr Opin Struct Biol 17:138–145

Maher C, Stein L, Ware D (2006) Evolution of Arabidopsis microRNA families through duplication events. Genome Res 16:510–519

Makarova KS, Wolf YI, van der Oost J, Koonin EV (2009) Prokaryotic homologs of Argonaute proteins are predicted to function as key

components of a novel system of defense against mobile genetic elements. Biol Direct 4:29

Mallory AC, Bartel DP, Bartel B (2005) microRNA-directed regulation of Arabidopsis AUXIN RESPONSE FACTOR17 is essential for proper development and modulates expression of early auxin response genes. Plant Cell 17:1360–1375

Marin E, Jouannet V, Herz A, Lokerse AS, Weijers D, Vaucheret H, Nussaume L, Crespi MD, Maizel A (2010) miR390, Arabidopsis TAS3 tasiRNAs, and their AUXIN RESPONSE FACTOR targets define an autoregulatory network quantitatively regulating lateral root growth. Plant Cell 22:1104–1117

Matzke M, Kanno T, Daxinger L, Huettel B, Matzke AJM (2009) RNA-mediated chromatin-based silencing in plants. Curr Opin Cell Biol 21:367–376

Miura A, Yonebayashi S, Watanabe K, Toyama T, Shimada H, Kakutani T (2001) Mobilization of transposons by a mutation abolishing full DNA methylation in Arabidopsis. Nature 411:212–214

Moissiard G, Parizotto EA, Himber C, Voinnet O (2007) Transitivity in Arabioopsis can be primed, requires the redundant action of the antiviral Dicer-like 4 and Dicer-like 2, and is compromised by viral-encoded suppressor proteins. RNA 13:1268–1278

Napoli C, Lemieux C, Jorgensen R (1990) Introduction of a chimeric chalcone synthase gene into petunia results in reversible co-suppression of homologous genes in trans. Plant Cell 2:279–289

Navarro L, Dunoyer P, Jay F, Arnold B, Dharmasiri N, Estelle M, Voinnet O, Jones JDG (2006) A plant miRNA contributes to antibacterial resistance by repressing auxin signaling. Science 312:436–439

Nogueira FT, Madi S, Chitwood DH, Juarez MT, Timmermans MC (2007) Two small regulatory RNAs establish opposing fates of a developmental axis. Genes Dev 21:750–755

Onodera Y, Haag JR, Ream T, Nunes PC, Pontes O, Pikaard CS (2005) Plant nuclear RNA polymerase IV mediates siRNA and DNA methylation-dependent heterochromatin formation. Cell 120:613–622

Orban TI, Izaurralde E (2005) Decay of mRNAs targeted by RISC requires XRN1, the Ski complex, and the exosome. RNA 11:459–469

Park W, Li J, Song R, Messing J, Chen X (2002) CARPEL FACTORY, a Dicer homolog, and HEN1, a novel protein, act in microRNA metabolism in Arabidopsis thaliana. Curr Biol 12:1484–1495

Park MY, Wu G, Gonzalez-Sulser A, Vaucheret H, Poethig RS (2005) Nuclear processing and export of microRNAs in Arabidopsis. Proc Natl Acad Sci U S A 102:3691–3696

Peragine A, Yoshikawa M, Wu G, Albrecht HL, Poethig RS (2004) SGS3 and SGS2/SDE1/RDR6 are required for juvenile development and the production of trans-acting siRNAs in Arabidopsis. Genes Dev 18:2368–2379

Pikaard CS, Haag JR, Ream T, Wierzbicki AT (2008) Roles of RNA polymerase IV in gene silencing. Trends Plant Sci 13:390–397

Poethig RS (1990) Phase change and the regulation of shoot morphogenesis in plants. Science 250:923–930

Pontier D, Yahubyan G, Vega D, Bulski A, Saez-Vasquez J, Hakimi MA, Lerbs-Mache S, Colot V, Lagrange T (2005) Reinforcement of silencing at transposons and highly repeated sequences requires the concerted action of two distinct RNA polymerases IV in Arabidopsis. Genes Dev 19:2030–2040

Rajagopalan R, Vaucheret H, Trejo J, Bartel DP (2006) A diverse and evolutionarily fluid set of microRNAs in Arabidopsis thaliana. Genes Dev 20:3407–3425

Ream TS, Haag JR, Wierzbicki AT, Nicora CD, Norbeck AD, Zhu JK, Hagen G, Guilfoyle TJ, Pasa-Tolic L, Pikaard CS (2009) Subunit compositions of the RNA-silencing enzymes Pol IV and Pol V reveal their origins as specialized forms of RNA polymerase II. Mol Cell 33:192–203

Reinhart BJ, Bartel DP (2002) Small RNAs correspond to centromere heterochromatic repeats. Science 297:1831

Reinhart BJ, Slack FJ, Basson M, Pasquinelli AE, Bettinger JC, Rougvie AE, Horvitz HR, Ruvkun G (2000) The 21-nucleotide *let-7* RNA regulates developmental timing in *Caenorhabditis elegans*. Nature 403:901–906

Reinhart BJ, Weinstein EG, Rhoades MW, Bartel B, Bartel DP (2002) microRNAs in plants. Genes Dev 16:1616–1626

Rodriguez A, Griffiths-Jones S, Ashurst JL, Bradley A (2004) Identification of mammalian microRNA host genes and transcription units. Genome Res 14:1902–1910

Ruiz-Ferrer V, Voinnet O (2009) Roles of plant small RNAs in biotic stress responses. Annu Rev Plant Biol 60:485–510

Schmitz-Linneweber C, Small I (2008) Pentatricopeptide repeat proteins: a socket set for organelle gene expression. Trends Plant Sci 13:663–670

Sen GL, Blau HM (2005) Argonaute 2/RISC resides in sites of mammalian mRNA decay known as cytoplasmic bodies. Nat Cell Biol 7:633–636

Shabalina S, Koonin E (2008) Origins and evolution of eukaryotic RNA interference. Trends Ecol Evol 23:578–587

Shen B, Goodman HM (2004) Uridine addition after microRNA-directed cleavage. Science 306:997

Sieber P, Wellmer F, Gheyselinck J, Riechmann JL, Meyerowitz EM (2007) Redundancy and specialization among plant microRNAs: role of the *MIR164* family in developmental robustness. Development 134:1051–1060

Slotkin RK, Martienssen R (2007) Transposable elements and the epigenetic regulation of the genome. Nat Rev Genet 8:272–285

Souret FF, Kastenmayer JP, Green PJ (2004) AtXRN4 degrades mRNA in *Arabidopsis* and its substrates include selected miRNA targets. Mol Cell 15:173–183

Tran RK, Zilberman D, de Bustos C, Ditt RF, Henikoff JG, Lindroth AM, Delrow J, Boyle T, Kwong S, Bryson TD, Jacobsen SE, Henikoff S (2005) Chromatin and siRNA pathways cooperate to maintain DNA methylation of small transposable elements in *Arabidopsis*. Genome Biol 6:R90

Valencia-Sanchez MA (2006) Control of translation and mRNA degradation by miRNAs and siRNAs. Genes Dev 20:515–524

Vaucheret H, Vazquez F, Crete P, Bartel DP (2004) The action of *ARGONAUTE1* in the miRNA pathway and its regulation by the miRNA pathway are crucial for plant development. Genes Dev 18:1187–1197

Vazquez F, Vaucheret H, Rajagopalan R, Lepers C, Gasciolli V, Mallory AC, Hilbert JL, Bartel DP, Crete P (2004) Endogenous trans-acting siRNAs regulate the accumulation of *Arabidopsis* mRNAs. Mol Cell 16:69–79

Wang JW, Wang LJ, Mao YB, Cai WJ, Xue HW, Chen XY (2005) Control of root cap formation by microRNA-targeted auxin response factors in Arabidopsis. Plant Cell 17:2204–2216

Wang XB, Wu Q, Ito T, Cillo F, Li WX, Chen X, Yu JL, Ding SW (2010) RNAi-mediated viral immunity requires amplification of virus-derived siRNAs in *Arabidopsis thaliana*. Proc Natl Acad Sci U S A 107:484–489

Wierzbicki AT, Haag JR, Pikaard CS (2008) Noncoding transcription by RNA polymerase Pol IVb/Pol V mediates transcriptional silencing of overlapping and adjacent genes. Cell 135:635–648

Wierzbicki AT, Ream TS, Haag JR, Pikaard CS (2009) RNA polymerase V transcription guides ARGONAUTE4 to chromatin. Nat Genet 41:630–634

Wightman B, Ha I, Ruvkun G (1993) Posttranscriptional regulation of the heterochronic gene *lin-14* by *lin-4* mediates temporal pattern formation in *C. elegans*. Cell 75:855–862

Willmann MR, Poethig RS (2005) Time to grow up: the temporal role of small RNAs in plants. Curr Opin Plant Biol 8:548–552

Willmann MR, Poethig RS (2007) Conservation and evolution of miRNA regulatory programs in plant development. Curr Opin Plant Biol 10:503–511

Wingard SA (1928) Hosts and symptoms of ring spot, a virus disease of plants. J Agric Res 37:127–153

Woodhouse MR, Freeling M, Lisch D (2006) The *mop1* (*mediator of paramutation1*) mutant progressively reactivates one of the two genes encoded by the *MuDR* transposon in maize. Genetics 172:579–592

Wu MF, Tian Q, Reed JW (2006) Arabidopsis microRNA167 controls patterns of *ARF6* and *ARF8* expression, and regulates both female and male reproduction. Development 133:4211–4218

Xie Z, Allen E, Fahlgren N, Calamar A, Givan SA, Carrington JC (2005) Expression of Arabidopsis *MIRNA* genes. Plant Physiol 138:2145–2154

Yang Z, Ebright YW, Yu B, Chen X (2006) HEN1 recognizes 21–24 nt small RNA duplexes and deposits a methyl group onto the 2′ OH of the 3′ terminal nucleotide. Nucleic Acids Res 34:667–675

Yu B, Bi L, Zheng B, Ji L, Chevalier D, Agarwal M, Ramachandran V, Li W, Lagrange T, Walker JC, Chen X (2008) The FHA domain proteins DAWDLE in *Arabidopsis* and SNIP1 in humans act in small RNA biogenesis. Proc Natl Acad Sci U S A 105:10073–10078

Zhang X, Yazaki J, Sundaresan A, Cokus S, Chan SW, Chen H, Henderson IR, Shinn P, Pellegrini M, Jacobsen SE, Ecker JR (2006) Genome-wide high-resolution mapping and functional analysis of DNA methylation in *Arabidopsis*. Cell 126:1189–1201

Zheng B, Wang Z, Li S, Yu B, Liu JY, Chen X (2009) Intergenic transcription by RNA polymerase II coordinates Pol IV and Pol V in siRNA-directed transcriptional gene silencing in *Arabidopsis*. Genes Dev 23:2850–2860

Zilberman D (2008) The evolving functions of DNA methylation. Curr Opin Plant Biol 11:554–559

Zilberman D, Gehring M, Tran RK, Ballinger T, Henikoff S (2006) Genome-wide analysis of *Arabidopsis thaliana* DNA methylation uncovers an interdependence between methylation and transcription. Nat Genet 39:61–69

Variation in Rates of Molecular Evolution in Plants and Implications for Estimating Divergence Times

7

J. Gordon Burleigh

Contents

7.1 Introduction

For nearly as long as molecular sequence data have been available for plants, they have been used to construct phylogenetic hypotheses and date the origin and diversification of clades (e.g., Boulter et al. 1972; Ramshaw et al. 1972). These studies infer evolutionary history of plants from patterns of molecular variation, and consequently, they rely in part on assumptions about the processes that create this variation. The availability of more and more molecular data have revealed increasingly complex patterns of evolution, resulting largely from the pervasive and highly nuanced effects of selection at the molecular level. Methods of evolutionary inference now must confront this molecular complexity. In this chapter, I review some of the factors associated with molecular rate variation in plants and discuss how insights into molecular evolution both inform and confound our ability to infer divergence times from molecular data.

7.2 Molecular Rate Variation in Plants

7.2.1 First Studies

Although the rates and patterns of molecular evolution are undoubtedly intricate, they may be viewed as the result of two processes: mutation, which creates molecular variation, and substitution, which fixes the new variants in a population or species. Neutral theory provides a theoretical foundation and general-purpose null hypothesis for molecular evolution (Kimura 1983). It posits that most molecular variants are neutral, incurring no selective advantage or disadvantage, and that the rate of neutral molecular evolution depends only on the rate of mutation (Kimura 1983). The neutral theory does not necessarily imply a molecular clock, or constant rate of evolution among lineages; shifts in the mutation rate can result in rate variation. However, extensive rate variation

J.G. Burleigh (✉)
Department of Biology, University of Florida, P.O. Box 118526, Gainesville, FL 32611, USA
e-mail: gburleigh@ufl.edu

J.F. Wendel et al. (eds.), *Plant Genome Diversity Volume 1*,
DOI 10.1007/978-3-7091-1130-7_7, © Springer-Verlag Wien 2012

may be indicative of shifting selective constraints and non-neutral evolution.

The earliest studies of inter-specific molecular variation in plants revealed differences in rates of evolution both among genes and among the chloroplast, mitochondrial, and nuclear genomes (Wolfe et al. 1987, 1989a; Gaut et al. 1993). They also observed frequent deviations from a molecular clock among lineages (Ritland and Clegg 1987; Martin et al. 1989; Wolfe et al. 1989b; Bousquet et al. 1992; Gaut et al. 1992). Thus, even with limited data, these studies identified the presence of gene-specific and genome-specific effects on the rates of molecular evolution as well as variation in the process of molecular evolution across lineages. Perhaps most interesting, different genes often displayed similar patterns of rate variation across lineages (Gaut et al. 1993, 1996; Clegg et al. 1994; Muse and Gaut 1994; Muse 2000), especially with respect to synonymous substitutions (Eyre-Walker and Gaut 1997; Muse and Gaut 1997). This suggests the importance of lineage-wide factors, processes that affect all genes in an organism, on the observed patterns of molecular variation. Yet identifying and quantifying these lineage-specific effects and the mechanisms by which they act continues to challenge researchers. In an early example, the fast rates of molecular evolution observed in grasses (with the exception of woody bamboos) compared to some woody outgroups has been explained by shorter generation times, shifts in population size, a woody versus herbaceous habit, and by high speciation rates (e.g., Bousquet et al. 1992; Gaut et al. 1992, 1996, 1997; Clegg et al. 1994; Laroche et al. 1997). These factors often are correlated and not easy to parse.

7.2.2 Associations with Rate Variation in Plants

Many studies over the past decades have attempted to identify factors influencing, or at least associated with, molecular variation among plant species. The generation-time hypothesis is among the first and most often-cited lineage-specific effects of the rates of molecular evolution (e.g., Laird et al. 1969; Sarich and Wilson 1973; Ohta 1993). In the simplest case, we might expect an inverse relationship between generation time and rate of molecular evolution. In animals, this is usually explained by an increased number of germ line cell divisions per unit time (Laird et al. 1969; Ohta 1993). In plants, the germ-line is not as straightforward, as gametes ultimately arise from somatic cells and generation time also may be confounded by extreme plasticity and long-term seed banks (Whittle 2006). Still, molecular rate differences have been described between annuals and perennials (Bousquet et al. 1992; Gaut et al. 1992, 1996, 1997; Savard et al. 1994; Laroche et al. 1997; Aïnouche and Bayer 1999; Laroche and

Bousquet 1999; Andreasen and Baldwin 2001; Soria-Hernanz et al. 2008; Yue et al. 2010) and short-lived, herbaceous plants and long-lived, woody plants (Kay et al. 2006; Smith and Donoghue 2008; Korall et al. 2010).

Although few of these studies examined low-copy nuclear loci (but see Yue et al. 2010), the evidence linking generation time and rates of molecular evolution in plants is extensive. However, generation time does not predict relative rates of evolution in all analyses; several studies have reported instances in which support for the generation time hypothesis has been ambiguous (Gaut et al. 1997; Soria-Hernanz et al. 2008) or nonexistent (Whittle and Johnston 2003). At the very least, the lack of correlation among rates of different genes in generation time comparisons demonstrates that other factors also influence molecular rate variation (Gaut et al. 1997). Furthermore, the mechanism by which generation time affects rates of evolution in plants is not obvious. Since somatic and germ line mutations can be inherited in plants, there is not necessarily a direct link between generation time and germ line cell divisions (e.g., Gaut et al. 1996). This relationship is further obscured in organellar genomes, where most generation time effects have been observed, whose replication is not linked directly to cell divisions. Finally, it can be difficult to unravel the effects of generation time from other closely linked traits that can influence rates of molecular evolution, such as population size and speciation rate.

In several cases, elevated rates of molecular evolution in plants have been attributed to high speciation rates rather than increased generation time (Bousquet et al. 1992; Barraclough et al. 1996; Laroche et al. 1997; Laroche and Bousquet 1999; Jobson and Albert 2002; Webster et al. 2003; Lancaster 2010). Since it is difficult to parse speciation and extinction rates from net diversification simply using a phylogeny of extant species (see Rabonsky 2010), the link between speciation and molecular evolution is not easy to evaluate directly. Also, like the generation-time hypothesis, the mechanism by which speciation rate affects rates of molecular evolution is not clear. In fact, it is not even clear if molecular evolution drives speciation rates or speciation affects rates of molecular evolution.

Speciation may be associated with population bottlenecks, and population size may influence rates of molecular evolution through genetic drift. Although the substitution rate of neutral mutations should not be affected by population size (Kimura 1983), if a new mutation is mildly deleterious, it may have a greater chance of reaching fixation through genetic drift in a small population (Ohta 1992). In these cases, lineages with small effective population sizes, including lineages with frequent speciation bottlenecks, may have slightly elevated nonsynonymous substitution rates. Historical population sizes, like speciation rates, are difficult to infer and likely fluctuate through time, but there is evidence that species with large population sizes are subject to more purifying selection than species with smaller populations (e.g., Slotte et al. 2010). Also, species limited to small islands, and thus small populations, have slightly inflated nonsynonymous substitution rates compared to their mainland relatives (Woolfit and Bromham 2005). Similarly, mating system appears to have a small effect on rates of fixation of slightly deleterious mutations, as selfing decreases the effective population size (Charlesworth 1992; Charlesworth and Wright 2001). Perhaps more importantly, increased heterozygosity in selfing lineages may reduce the rates of indel-associated mutations, thereby decreasing rates of molecular evolution (Hollister et al. 2009).

While many of the studies of rates of plant molecular evolution have focused on associations with life history and species-level processes, numerous other context-specific factors have been associated with rates of evolution in plants. For example, a parasitic life-history may be associated with relaxed functional constraints, and consequently elevated rates of molecular evolution, in some genes (e.g., Wolfe et al. 1992; Nickrent and Starr 1994; dePamphilis et al. 1997; Duff and Nickrent 1997; Wolfe and dePamphilis 1998; Young and dePamphilis 2005). Some evidence suggests that the environmental context of a species may affect rates of molecular evolution, as species in high-energy or tropical environments appear to have elevated rates of molecular evolution (Davies et al. 2004; Wright et al. 2006). The genomic context of a species, such as GC composition, also can affect the observed rates and patterns of genome evolution (e.g., Zhang et al. 2002; Wang et al. 2004; DeRose-Wilson and Gaut 2007).

7.2.3 New and Future Perspectives on Rate Variation

Most studies of rates and patterns of molecular evolution in plants have focused on chloroplast or mitochondrial loci. They also have largely emphasized lineage or genome-specific effects on rates of molecular evolution. With the availability of large nuclear gene data sets across many, phylogenetically-diverse lineages, it is now possible to examine these putative lineage specific effects across a broad range of unlinked, nuclear loci. These studies also may be improved by new modeling approaches that jointly estimate molecular evolution and phenotypes, providing a potentially powerful statistical approach to quantify directly the effects of phenotypes on the rates and patterns of molecular evolution (Lartillot and Poujol 2011; Mayrose and Otto 2011).

The availability of nuclear gene data also provides the opportunity to explore in depth gene-specific processes of evolution and to examine the evolution in a functional genomic context. For example, are rates and patterns of evolution of a gene affected by its place in genetic pathway or network? In the anthocyanin pathway, upstream genes have lower rates of molecular evolution than downstream genes due to stronger purifying selection on the upstream genes (Rausher et al. 1999, 2008; Lu and Rausher 2003). In contrast, in the gibberellin pathway, upstream genes have higher rates of molecular evolution than downstream genes, again due to different levels of purifying selection (Yang et al. 2009). In other (non-plant) systems, evidence for links between rates of molecular evolution and a gene's place in a pathway has been ambiguous (e.g., Hahn et al. 2004), possibly suggesting that other functional traits, such as a gene's control of flux in a metabolic pathway, may be a better predictor of rates of evolution (Hahn et al. 2004; Yang et al. 2009). Rates and patterns of evolution also may be affected by the a gene's location in a genome and the associated rates of recombination (Yang and Gaut 2011). Furthermore, in plants, with generally high rates of gene and whole genome duplication, rates of sequence evolution may be affected by the existence of paralogs and whether these paralogs were generated by single gene or whole genome duplication (Zhang et al. 2002; Blanc and Wolfe 2004; Ganko et al. 2007; Yang and Gaut 2011). The role of many other factors, such as protein structure and interactions and epigenetic context, in molecular evolution are yet to be explored in depth.

While studies describing lineage-specific effects on rates of evolution can help reveal processes that generate and maintain broad-scale molecular variation, the phenotypic or functional consequences of this variation rarely is obvious. Several surveys of genomic data in plants have attempted to identify evidence of positive selection throughout plant genomes, the frequency of which also may be affected by lineage-specific factors (Slotte et al. 2010; Gossmann et al. 2010), but it is still difficult to link positive selection at the molecular level to selection on phenotypes. Analyses linking rates of evolution and gene ontology (GO) terms or gene expression data may have some potential to link rates of evolution with function and possibly reveal correlated functional evolution (e.g., Li and Rodrigo 2009; Yang and Gaut 2011). However, as large-scale genomic data bring the promise of identifying the molecular mechanisms of phenotypic and species diversity, it is likely that more studies will seek the molecular variation that affects phenotypes rather than the phenotypes that are associated with broad-scale molecular variation.

7.3 Dating Plant Phylogenetic Hypotheses

Dating a phylogenetic tree places the evolution of species and lineages in a historical context, relating the species history to geological or climatological events or the evolution of distantly related lineages. This historical context can be inferred in part from the fossil record, but the fossil record is incomplete and entirely missing for many plant clades. In contrast, dating based on molecular variation potentially can provide a historical record for all extant organisms. It is no wonder then that molecular dating analyses in plants have generated tremendous interest (see Sanderson et al. 2004; Renner 2005; Bell et al. 2010; Magallón 2010; Smith et al. 2010). Yet the complexities of rates and patterns of molecular evolution in plants greatly complicate and confound dating estimates, making molecular dating studies among the most contentious topics in evolutionary inference.

7.3.1 Early Molecular Clock Studies

The earliest molecular dating studies in plants dated divergence times by assuming a molecular clock, or constant rate of evolution. The molecular clock hypothesis is consistent with the neutral theory and implies selectively neutral variation and constant mutation rates (Zuckerkandl and Pauling 1965). There are two general approaches to date a tree based on the molecular clock assumption. First, with reliable estimates of substitution rates (substitutions per unit time) and the number of substitutions along a branch, it is straightforward to convert the molecular branch length into units of time. This approach often is invoked for animal mitochondrial genes (e.g., 2% divergence per million years; see Brown et al. 1979), but it also is used with plants (see Kay et al. 2006). In the absence of a reliable estimate of substitution rate, divergence times still can be obtained with one or more fossil-calibrated nodes. If the age of an ancestral node is known, then the substitution rate for the gene can be calculated from the number of substitutions between the dated node and the present. This rate can then be used to date nodes throughout the tree that lack fossil-calibrated dates. Both of these approaches assume that the substitution rate is constant through time and among lineages, and they can be affected by error or uncertainty in the topology, branch length estimates, or fossil calibrations.

The early molecular clock-based divergence time studies in plants are characterized by highly variable dating estimates that often greatly predated estimates from fossil records (e.g., Ramshaw et al. 1972; Martin et al. 1989; Wolfe et al. 1989b; Brandl et al. 1992; Savard et al. 1994; Laroche et al. 1995). Ramshaw et al. (1972) used molecular clock estimates based on amino acid divergence among cyctochrome c sequences to data the origin of seed plants at 400–520 mya and the most recent common ancestor of

angiosperms at 300–400 mya. Martin et al. (1989) and Brandl et al. (1992) estimated origin of angiosperms at more than 300 mya, while Wolfe et al. (1989b) and Laroche et al. (1995) placed the date at around 200 mya. Even the most recent dates precede the earliest fossil evidence of angiosperms by approximately 75 million years (e.g., Sun et al. 2002; Friis et al. 2006). This is not impossible; the ancient fossil record may be undiscovered or not exist at all. The probability of such large deviations from the fossil record depend on the probability of fossilization and the level of sampling (e.g., Huelsenbeck and Rannala 2000).

Many of the early molecular dating studies in plants were aware of violations of the molecular clock assumption and the high degree of uncertainty in dating estimates, and they used a variety of approaches to ameliorate these problems. For example, Wolfe et al. (1989a) emphasized the more clock-like synonymous rates of substitution rather than nonsynoymous rates, which likely are more susceptible to the vagaries of selection. Savard et al. (1994) and Laroche et al. (1995) used the results of relative rate tests to exclude lineages or gene regions with high molecular rate variation among lineages. Still, with increasing evidence of extensive molecular rate variation within and among lineages, genomes, and genes, dating plants based solely on molecular clock assumptions became less accepted.

7.3.2 Relaxed Clock Studies

Since these earliest studies, numerous methods have been developed to estimate divergence times while relaxing the molecular clock assumptions (Sanderson 1997, 2002; Thorne et al. 1998; Huelsenbeck et al. 2000; Kishino et al. 2001; Aris-Brosou and Yang 2002; Drummond et al. 2006; Lepage et al. 2006; Rannala and Yang 2007; Guindon 2010). Several recent reviews detail these methods, which often are conceptually and computationally complex (Renner 2005; Welch and Bromham 2005; Rutschmann 2006; Ho 2009). Relaxed clock methods can be divided into those that assume similar rates between adjacent branches, or autocorrelated rates (most of the methods above), and those that do not (e.g., Drummond et al. 2006; Drummond and Rambaut 2007). In both cases, the models allow lineage-specific rate variation, but this variation is not linked to a specific mechanism. The underlying assumption of the autocorrelated rates methods is that the factors that determine rates of evolution, like DNA repair mechanisms or demographic or life history traits, are similar in closely related lineages. This may be true in many cases, but it is not difficult to find examples of plant lineages with radically different rates of molecular evolution compared to their sister lineages (e.g., Cho et al. 2004). In plant studies, perhaps the most commonly used of these "relaxed clock" methods are the autocorrelated nonparametric rate smoothing and penalized likelihood, implemented in r8s (Sanderson 2003b), and, more recently, the uncorrelated Bayesian relaxed clock methods implemented in BEAST (Drummond and Rambaut 2007).

The computational burden of most relaxed clock methods is compounded by large-scale phylogenetic trees, and some relaxed clock dating methods may be intractable for extremely large data sets. Pathd8 represents a relaxed clock approach that is based on calculating the mean path length from an internal node to the descending leaves (Britton et al. 2007). The appeal of Pathd8 is its run-time, not its implicit model of molecular evolution. Still, it appears to perform well in many cases, if not always as well as other, parametric relaxed clock methods (Britton et al. 2007; Svennblad and Britton 2007; Svennblad 2008). Pathd8 has been used to date several plant trees with over 1,000 taxa (e.g., Smith and Donoghue 2008; Edwards and Smith 2010), and it may be the most feasible option for dating a comprehensive tree of plants without invoking a molecular clock.

While these sophisticated molecular dating methods may better reflect the patterns of molecular evolution than methods that assume a strict molecular clock, they have not produced consensus in the divergence time estimates of major plant clades. Rather, they often have highlighted high levels of uncertainty and lack of robustness associated with molecular dating analyses. Focusing on dating the origin of angiosperms, the first relaxed clock analysis estimated the origin of angiosperms at 160 mya, which is more consistent with the fossil record than the early molecular clock studies (Sanderson 1997). Although later relaxed clock studies often produced similar age estimates for angiosperms (e.g., Sanderson and Doyle 2001; Wikström et al. 2001; Bell et al. 2005; Magallón and Sanderson 2005), they also demonstrated that the dating estimates could vary tremendously based on the phylogeny, the genes or gene partitions used in the analysis, taxon sampling, the number and placement of fossil constraints, and the assumptions of dating method. Taken as a whole, these studies emphasized that dating analyses should be interpreted with caution. Most recently, three new studies with extensive taxonomic sampling, large molecular data sets, and uncorrelated relaxed clock dating methods estimated an age for angiosperms more similar to the first molecular clock studies. Whereas Bell et al. (2005) place the age of angiosperms at 167–199 mya, Magallón (2010) and Smith et al. (2010) date the angiosperms (with large confidence intervals) to the upper Triassic or even early Permian, well over 200 mya.

7.3.3 A New Molecular Clock?

With the availability of numerous relaxed-clock dating methods and some scathing critiques of molecular clock analyses (see Graur and Martin 2004), the strict molecular clock concept may seem outdated. Still, it is perhaps too useful to abandon completely. In fact, Renner (2005) found that only approximately 20% of the plant biogeographic studies from 1995 to 2005 that estimated divergence times used relaxed clock methods. For practical purposes, many plant groups, especially within the tropics, have limited or no fossil records to calibrate sequence divergence estimates. Thus, while it is possible to generate enormous molecular data sets for these clades and accurately estimate molecular divergence, there is no means to link this divergence directly to time. In these cases, a molecular clock approach may be the best option to date the tree.

One example of modern molecular clock analyses in plants is the ITS (internal transcribed spacer) clock. Kay et al. (2006) surveyed rates of evolution of nuclear ribosomal ITS in numerous plant clades and found relatively similar rates across among herbaceous lineages and also among woody lineages, with rates of molecular evolution in herbaceous lineages approximately twice as high as in woody lineages. This suggests that sequence divergence estimates from ITS can be used to calibrate plant trees, with local clocks (e.g., Yoder and Yang 2000) for woody and herbaceous lineages. With the potential for highly divergent paralogs and pseudogenes and complex processes of concerted evolution (see Álvarez and Wendel 2003), ITS must be used with caution. Also, ITS is a relatively short locus, and thus, dating estimates based on ITS likely will have high variance. Still, ITS is present in all plants, and easy to amplify, making it an extremely practical, if imperfect, option for dating divergence events in many plant clades. With the relative ease of producing molecular data sets with next-generation sequencing technologies, perhaps a more robust approach today would be to date trees based on the average molecular divergence across many loci.

7.3.4 The Future for Dating the Past

With frequent contradictory, and sometimes unbelievable, results, it would seem that our desire to estimate evolutionary divergence times often exceeds our ability to obtain accurate dating estimates. This does not mean that molecular dating is hopeless or that progress has not been made in the field. Rather, it speaks to the tremendous difficulty of interpreting the complex patterns and processes of molecular evolution.

In a broad sense, many of the methodological advances in molecular dating studies are based on generalizing the models of rate variation, freeing them further and further from the assumptions of clock-like, homogenous evolution. Yet to estimate dates from molecular data, there must exist some relationship between the amount of molecular variation and time. Simply relaxing the constraints of the molecular clock may ameliorate some of the biases associated with molecular clock dating, but it remains to be seen whether they can produce both accurate and precise estimates from empirical data. In plants, the credibility of the relaxed clock dating estimates is debatable, but even dating estimates from the largest data sets often have wide confidence intervals (e. g., Smith et al. 2010).

Given the differences in estimates among different methods and data sets and high levels of uncertainty, it may be useful to address directly the statistical identifiability of analyses, or, in other words, to address whether it is possible to estimate the desired divergence time given a specified model and data set. To this end, it would be useful to have diagnostics of identifiability that can be applied to analyses. Identifiability still does not guarantee that the estimates will be correct; an inadequate model, for example, may still mislead the inference. However, it will address whether an analysis can be informative.

Also, current models of rate variation among lineages generally are not based on any specific biological processes. This may be appropriate given the difficulty of explaining fully variation in rates of molecular evolution with any biological mechanism(s). Still, one way to increase the fit of the models without necessarily increasing complexity may be to use models that combine the evolution of either continuous or discrete characters with patterns and rates of molecular evolution (Lartillot and Poujol 2011; Mayrose and Otto 2011). With these models, we can directly link biological insights into the causes of molecular rate variation with the estimates of molecular rate variation. It remains to be seen whether any phenotype(s) can sufficiently predict rates of molecular evolution.

Another approach to the improving of dating methods may be to focus on the choice of data. Most molecular dating studies have taken an opportunistic approach to data assembly, often using the same loci used to infer phylogenies. Little work has been done to characterize the best data for dating, for example, the data that behaves in the most clock-like manner or a manner that best fits the models of rate variation. Indeed, dating methods need not be limited to molecular sequence data. Simple modeling approaches are available to estimate substitution rates for phenotypic or genomic characters (Lewis 2001), and these can be used in molecular clock or relaxed-clock dating methods (see Gray and Atkinson 2003). It is worth entertaining the idea of applying large-scale phenotypic data sets, perhaps gleaned from text or image data, or other, non-sequence, molecular characters for dating.

Most dating studies in plants use chloroplast, and to a lesser extent mitochondrial and nuclear rRNA sequences. Consequently, they largely have ignored gene-specific processes of evolution or considered the implications of estimating speciation dates from the divergence of genes. From a population genetic perspective, multiple copies of a single gene found in sister species originate from a single sequence in a historical population. Stepping back in time, the coalescence of the genes, representing the divergence of the gene sequences from the single gene copy ancestor, likely predates speciation, and coalescence dates will differ among unlinked loci (e.g., Nichols 2001). That is, each unlinked gene likely has a different set of divergence dates. Although coalescence is rarely considered in molecular dating studies of plants, there is much recent interest in accounting for coalescence in phylogenetics (e.g., Maddison and Knowles 2006; Liu and Pearl 2007; Liu et al. 2008; Kubatko et al. 2009). These analyses are based on the premise that, due to coalescence, the topology of the gene trees often differs from the topology of the species phylogeny in which they evolve. In other words, given an outgroup taxon and two sister species, the common ancestor of a gene from two sister species may predate the most recent common ancestor of the gene from the outgroup and one of the sister taxa. These situations should be most common when the time between speciation events is short and/or ancestral population sizes are large, and accurately dating species divergences in these situations may be difficult.

Dating speciation events with plant nuclear genes is further complicated by high rates of gene, and whole genome, duplication and gene loss. Thus, it may be difficult to determine if homologous nuclear loci from multiple species are orthologous or paralogous, or, in other words, if their common ancestor reflects a speciation or gene duplication event. Interestingly, studies that have used large nuclear gene data sets to date the origins of major plant clades have produced among the most ancient estimates of the age of land plants (>700 mya; Heckman et al. 2001; Zimmer et al. 2007) and angiosperms (>301 mya; Zimmer et al. 2007). One possible explanation for these ancient estimates is that the data contain many hidden paralogs, and the dates may represent, in part, ancient duplications that pre-date the origin of the clades (Sanderson 2003a). Thus, dating analyses that incorporate plant nuclear genes must carefully consider and account for processes of duplication and loss.

Conclusion

As increasing amounts of molecular data have revealed the complexity of plant molecular evolution, they also have highlighted the difficulty of estimating the dates that clades originated and diversified. In this sense, more data have not simplified evolutionary inference or increased the power and scale of possible analyses; instead, they have provided new perspectives on the challenges of evolutionary inference and new insights into how analyses can fail. Consequently, molecular dating studies have never been as frustrating, or as interesting, as they are today. Although molecular dating studies in plants have been done for over four decades, it is, more than ever, a new field with unaddressed challenges.

References

Aïnouche AK, Bayer RJ (1999) Phylogenetic relationships in *Lupinus* (Fabaceae: Papilionoideae) based on internal transcribed spacer sequences (ITS) of nuclear ribosomal DNA. Am J Bot 86:590–607

Álvarez I, Wendel JF (2003) Ribosomal ITS sequences and plant phylogenetic inference. Mol Phylogenet Evol 29:417–434

Andreasen K, Baldwin BG (2001) Unequal evolutionary rates between annual and perennial lineages of checker mallows (*Sidalcea*, Malvaceae): evidence from 18S-26S rDNA internal and external transcribed spacers. Mol Biol Evol 18:936–944

Aris-Brosou S, Yang Z (2002) Effects of models of rate evolution on estimation of divergence dates with special reference to the meta-zoan 18S ribosomal RNA phylogeny. Mol Biol Evol 20:1947–1954

Barraclough TG, Harvey PH, Nee S (1996) Rate of *rbcL* gene sequence evolution and species diversification in flowering plants. Proc Roy Soc Lond B Bio 263:589–591

Bell CD, Soltis DE, Soltis PS (2005) The age of angiosperms: a molecular timescale without a clock. Evolution 59:1245–1258

Bell CD, Soltis DE, Soltis PS (2010) The age and diversification of the angiosperms re-revisited. Am J Bot 97:1296–1303

Blanc G, Wolfe KH (2004) Functional divergence of duplicated genes formed by polyploidy during *Arabidopsis* evolution. Plant Cell 16:1679–1691

Boulter D, Ramshaw JAM, Thompson EW, Richardson M, Brown RH (1972) A phylogeny of higher plants based on the amino acid sequences of cytochrome c and its biological implications. Proc Roy Soc Lond B Bio 181:441–455

Bousquet J, Strauss SH, Doerksen AH, Price RA (1992) Extensive variation in evolutionary rate of *rbcL* gene sequence among seed plants. Proc Natl Acad Sci USA 89:7844–7848

Brandl R, Mann W, Sprinzl M (1992) Estimation of the monocot-dicot age through tRNA sequences from the chloroplast. Proc Roy Soc Lond B Bio 249:13–17

Britton T, Anderson CL, Jaquet D, Lundqvist S, Bremer K (2007) Estimating divergence times in large phylogenetic trees. Syst Biol 56:741–752

Brown WM, George M Jr, Wilson AC (1979) Rapid evolution of animal mitochondrial DNA. Proc Natl Acad Sci USA 76:1967–1971

Charlesworth B (1992) Evolutionary rates in partially self-fertilizing species. Am Nat 140:126–148

Charlesworth D, Wright SI (2001) Breeding systems and genome evolution. Curr Opin Genet Evol 11:685–690

Cho Y, Mower JP, Qiu Y-L, Palmer JD (2004) Mitochondrial substitution rates are extraordinarily elevated and variable in a genus of flowering plants. Proc Natl Acad Sci USA 51:17741–17746

Clegg MT, Gaut BS, Learn GH, Morton BR (1994) Rates and patterns of chloroplast DNA evolution. Proc Natl Acad Sci USA 91:6795–6801

Davies TJ, Savolainen V, Chase MW, Moat J, Barraclough TJ (2004) Environmental energy and evolutionary rates in flowering plants. Proc Roy Soc Lond B Bio 271:2195–2200

dePamphilis CW, Young ND, Wolfe AD (1997) Evolution of plastid gene *rps2* in a lineage of hemiparasitic and holoparasitic plants: many losses of photosynthesis and complex patterns of rate variation. Proc Natl Acad Sci USA 94:7367–7372

DeRose-Wilson LJ, Gaut BS (2007) Transcription-regulated mutations and GC content drive variation in nucleotide substitution rates across the genomes of *Arabidopsis thaliana* and *Arabidopsis lyrata*. BMC Evol Biol 7:66

Drummond AJ, Rambaut A (2007) BEAST: Bayesian evolutionary analysis by sampling trees. BMC Evol Biol 7:214

Drummond AJ, Ho SYM, Phillips MJ, Rambaut A (2006) Relaxed phylogenetics and dating with confidence. PLoS Biol 4:e88

Duff RJ, Nickrent DL (1997) Characterization of mitochondrial small subunit ribosomal RNAs from holoparasitic plants. J Mol Evol 45:631

Edwards EJ, Smith A (2010) Phylogenetic analyses reveal the shady history of C_4 grasses. Proc Natl Acad Sci USA 107:2532–2537

Eyre-Walker A, Gaut BS (1997) Correlated rates of synonymous site evolution across plant genomes. Mol Biol Evol 14:455–460

Friis EM, Pederson KR, Crane PR (2006) Cretaceous angiosperm flowers: innovation and evolution in plant reproduction. Palaeogeogr Palaeocl 232:251–293

Ganko EW, Meyers BC, Vision TJ (2007) Divergence in expression between duplicated genes in *Arabidopsis*. Mol Biol Evol 24:2298–2309

Gaut BS, Muse SV, Clark D, Clegg MT (1992) Relative rates of nucleotide substitution at the *rbcL* locus of monocotyledonous plants. J Mol Evol 35:292–303

Gaut BS, Muse SV, Clegg MT (1993) Relative rates of nucleotide substitution in the chloroplast genome. Mol Phylogenet Evol 2:89–96

Gaut BS, Morton BR, McCaig BC, Clegg MT (1996) Substitution rate comparisons between grasses and palms: synonymous differences at the nuclear gene *Adh* parallel rate differences at the plastid gene *rbcL*. Proc Natl Acad Sci USA 93:10274–10279

Gaut BS, Clark LG, Wendel JF, Muse SV (1997) Comparisons of the molecular evolutionary process at *rbcL* and *ndhF* in the grass family (Poaceae). Mol Biol Evol 14:769–777

Gossmann TI, Song B-H, Windsor AJ, Mitchell-Olds T, Dixon CJ, Kapralov MV, Filatov DA, Eyre-Walker A (2010) Genome wide analyses reveal little evidence for adaptive evolution in plant species. Mol Biol Evol 27:1822–1832

Graur D, Martin W (2004) Reading the entrails of chickens: molecular timescales of evolution and the illusion of precision. Trends Genet 20:80–86

Gray RD, Atkinson QD (2003) Language-tree divergence times support the Anatolian theory of Indo-European origin. Nature 426:435–439

Guindon S (2010) Bayesian estimation of divergence times from large sequence alignments. Mol Biol Evol 27:1768–1781

Hahn MW, Conant GC, Wagner A (2004) Molecular evolution in large genetic networks: does connectivity equal constraint? J Mol Evol 58:203–211

Heckman DS, Geiser DM, Eidell BR, Stauffer RL, Kardos NL, Hedges SB (2001) Molecular evidence for the early colonization of land by fungi and plants. Science 293:1129–1133

Ho SYW (2009) An examination of phylogenetic models of substitution rate variation among lineages. Biol Lett 5:421–424

Hollister JD, Ross-Ibarra J, Gaut BS (2009) Indel-associated mutation rate varies with mating system in flowering plants. Mol Biol Evol 27:409–416

Huelsenbeck JP, Rannala B (2000) Using stratigraphic information in phylogenetics. In: Wiens JJ (ed) Phylogenetic analysis of morphological data. Smithsonian Institution Press, Washington, DC, pp 165–191

Huelsenbeck JP, Larget B, Swofford D (2000) A compound poisson process for relaxing the molecular clock. Genetics 154:1879–1892

Jobson RW, Albert VA (2002) Molecular rates parallel diversification contrasts between carnivorous plant sister lineages. Cladistics 18:127–136

Kay KM, Whittall JB, Hodges SA (2006) A survey of nuclear ribosomal internal transcribed spacer substitution rates across angiosperms: an approximate molecular clock with life history effects. BMC Evol Biol 6:36

Kimura M (1983) The neutral theory of molecular evolution. Cambridge University Press, Cambridge

Kishino H, Thorne JL, Bruno WJ (2001) Performance of a divergence time estimation method under a probabilistic model of rate evolution. Mol Biol Evol 18:352–361

Korall P, Schuettpelz E, Pryer KM (2010) Abrupt deceleration of molecular evolution linked to the origin of arborescence in ferns. Evolution 64:2786–2792

Kubatko LS, Carstens BC, Knowles LL (2009) STEM: species tree estimation using maximum likelihood for gene trees under coalescence. Bioinformatics 25:971–973

Laird CD, McConaughy BL, McCarthy BJ (1969) Rate of fixation of nucleotide substitutions in evolution. Nature 224:149–154

Lancaster LT (2010) Molecular evolutionary rates predict both extinction and speciation in temperate angiosperm lineages. BMC Evol Biol 10:162

Laroche J, Bousquet J (1999) Evolution of mitochondrial *rps3* intron in perennial and annual angiosperms and homology to *nad5* intron 1. Mol Biol Evol 16:441–452

Laroche J, Li P, Bousquet J (1995) Mitochondrial DNA and the monocot-dicot divergence time. Mol Biol Evol 12:1151–1156

Laroche J, Li P, Maggia L, Bousquet J (1997) Molecular evolution of angiosperm mitochondrial introns and exons. Proc Natl Acad Sci USA 94:5722–5727

Lartillot N, Poujol R (2011) A phylogenetic model for investigating correlated evolution of substitution rates and continuous phenotypic characters. Mol Biol Evol 28:729–744

Lepage T, Tupper P, Bryant D, Lawi S (2006) Continuous and tractable models of the variation of evolutionary rates. Math Biosci 199:216–233

Lewis PO (2001) A likelihood approach to estimating phylogeny from discrete morphological data. Syst Biol 50:913–925

Li WL, Rodrigo AG (2009) Covariation of branch lengths in phylogenies of functionally related genes. PLoS One 4:e8487

Liu L, Pearl DK (2007) Species trees from gene trees: reconstructing Bayesian posterior distributions of a species phylogeny using estimated gene tree distributions. Syst Biol 56:504–514

Liu L, Pearl DK, Brumfield RT, Edwards SV (2008) Estimating species trees using multiple-allele DNA sequence data. Evolution 62:2080–2091

Lu Y, Rausher MD (2003) Evolutionary rate variation in anthocyanin pathway genes. Mol Biol Evol 20:1844–1853

Maddison WP, Knowles LL (2006) Inferring phylogeny despite incomplete lineage sorting. Syst Biol 55:21–30

Magallón S (2010) Using fossils to break long branches in molecular dating: a comparison of relaxed clocks applied to the origin of angiosperms. Syst Biol 59:384–399

Magallón S, Sanderson MJ (2005) Angiosperm divergence times: the effect of genes, codon positions, and time constraints. Evolution 59:1653–1670

Martin W, Gierl A, Saedler H (1989) Molecular evidence for pre-Cretaceous angiosperm origins. Nature 339:46–48

Mayrose I, Otto SP (2011) A likelihood model for detecting trait-dependent shifts in rate of molecular evolution. Mol Biol Evol 28:759–770

Muse SV (2000) Examining rates and patterns of nucleotide substitution in plants. Plant Mol Biol 42:25–43

Muse SV, Gaut BS (1994) A likelihood approach for comparing synonymous and nonsynonymous nucleotide substitution rates, with application to the chloroplast genome. Mol Biol Evol 11:715–724

Muse SV, Gaut BS (1997) Comparing patterns of nucleotide substitution rates among chloroplast loci using the relative ratio test. Genetics 146:392–399

Nichols R (2001) Gene trees and species trees are not the same. Trends Ecol Evol 16:358–364

Nickrent DL, Starr EM (1994) High rates of nucleotide substitution in nuclear small-subunit (18S) rDNA from holoparasitic flowering plants. J Mol Evol 39:62–70

Ohta T (1992) The nearly neutral theory of molecular evolution. Annu Rev Ecol Syst 23:263–286

Ohta T (1993) An examination of the generation-time effect on molecular evolution. Proc Natl Acad Sci USA 90:10676–10680

Rabonsky DL (2010) Extinction rates should not be estimated from molecular phylogenies. Evolution 64:1816–1824

Ramshaw JAM, Richardson DL, Meatyard BT, Brown RH, Richardson M, Thompson EW, Boulter D (1972) The time of origin of the flowering plants determined by using amino acid sequence data of cytochrome c. New Phytol 71:773–779

Rannala B, Yang Z (2007) Inferring speciation times under an episodic molecular clock. Syst Biol 56:453–466

Rausher MD, Miller RE, Tiffin P (1999) Patterns of evolutionary rate variation among genes of the anthocyanin biosynthetic pathway. Mol Biol Evol 16:266–274

Rausher MD, Lu Y, Meyer K (2008) Variation in constraint versus positive selection as an explanation for evolutionary rate variation among anthocyanin genes. J Mol Evol 67:137–144

Renner SS (2005) Relaxed molecular clocks for dating historical plant dispersal events. Trends Plant Sci 10:550–558

Ritland K, Clegg MT (1987) Evolutionary analysis of plant DNA sequences. Am Nat 130:S74–S100

Rutschmann F (2006) Molecular dating of phylogenetic trees: a brief review of current methods that estimate divergence times. Divers Distrib 12:35–48

Sanderson MJ (1997) A nonparametric approach to estimating divergence times in the absence of rate constancy. Mol Biol Evol 14:1218–1231

Sanderson MJ (2002) Estimating absolute rates of molecular evolution and divergence times: a penalized likelihood approach. Mol Biol Evol 19:101–109

Sanderson MJ (2003a) Molecular data from 27 proteins do not support a Precambrian origin of land plants. Am J Bot 90:954–956

Sanderson MJ (2003b) r8s: inferring absolute rates of molecular evolution and divergence times in the absence of a molecular clock. Bioinformatics 19:301–302

Sanderson MJ, Doyle JA (2001) Sources of error and confidence intervals in estimating the age of angiosperms from *rbcL* and 18S rDNA data. Am J Bot 88:1499–1516

Sanderson MJ, Thorne JL, Wikström N, Bremer K (2004) Molecular evidence on plant divergence times. Am J Bot 91:1656–1665

Sarich VM, Wilson AC (1973) Generation time and genomic evolution in primates. Science 179:1144–1147

Savard L, Li P, Strauss SH, Chase MW, Michaud M, Bousquet J (1994) Chloroplast and nuclear gene sequences indicate Late Pennsylvanian time for the last common ancestor of extant seed plants. Proc Natl Acad Sci USA 91:5163–5167

Slotte T, Foxe JP, Hazzouri KM, Wright SI (2010) Genome-wide evidence for efficient positive and purifying selection in *Capsella grandiflora*, a plant species with a large effective population size. Mol Biol Evol 27:1813–1821

Smith SA, Donoghue MJ (2008) Rates of molecular evolution are linked to life history in flowering plants. Science 322:86–89

Smith SA, Beaulieu JM, Donoghue MJ (2010) An uncorrelated relaxed-clock analysis suggests an earlier origin for flowering plants. Proc Natl Acad Sci USA 107:5897–5902

Soria-Hernanz DF, Fiz-Palacios O, Braverman JM, Hamilton MB (2008) Reconsidering the generation time hypothesis based on nuclear ribosomal ITS sequence comparisons in annual and perennial angiosperms. BMC Evol Biol 8:344

Sun G, Ji Q, Dilcher DL, Zheng S, Nixon KC, Wang X (2002) Archaefructaceae, a new basal angiosperm family. Science 296:899–904

Svennblad B (2008) Consistent estimation of divergence times in phylogenetic trees with local molecular clocks. Syst Biol 57:947–954

Svennblad B, Britton T (2007) Improving divergence time estimation in phylogenetics: more taxa vs. longer sequences. Stat Appl Genet Mol Biol 6:35

Thorne JL, Kishino H, Painter IS (1998) Estimating the rate of evolution of the rate of evolution. Mol Biol Evol 15:1647–1657

Wang H-c, Singer GAC, Hickey DA (2004) Mutational bias affects protein evolution in flowering plants. Mol Biol Evol 21:90–96

Webster AJ, Payne RJH, Pagel M (2003) Molecular phylogenies link rates of evolution and speciation. Science 301:478

Welch JJ, Bromham L (2005) Molecular dating when rates vary. Trends Ecol Evol 20:320–327

Whittle C-A (2006) The influence of environmental factors, the pollen: ovule ratio and seed bank persistence on molecular evolutionary rates in plants. J Evol Biol 19:302–308

Whittle C-A, Johnston MO (2003) Broad-scale analysis contradicts the theory that generation time affects molecular evolutionary rates in plants. J Mol Evol 56:223–233

Wikström N, Savolainen V, Chase MW (2001) Evolution of the angiosperms: calibrating the family tree. Proc Roy Soc Lond B Bio 268:2211–2220

Wolfe AD, dePamphilis CW (1998) The effect of relaxed functional constraints on the photosynthetic gene rbcL in photosynthetic and nonphotosynthetic parasitic plants. Mol Biol Evol 15:1243–1258

Wolfe KH, Li W-H, Sharp PM (1987) Rates of nucleotide substitution vary greatly among plant mitochondrial, chloroplast, and nuclear DNAs. Proc Natl Acad Sci USA 84:9054–9058

Wolfe KH, Sharp PM, Li W-H (1989a) Rates of synonymous substitution in plant nuclear genes. J Mol Evol 29:208–211

Wolfe KH, Gouy M, Yang Y-W, Sharp PM, Li W-H (1989b) Date of the monocot-dicot divergence estimated from chloroplast DNA sequence data. Proc Natl Acad Sci USA 86:6201–6205

Wolfe KH, Morden CW, Ems SC, Palmer JD (1992) Rapid evolution of the plastid translational apparatus in a nonphotosynthetic plant: loss or accelerated sequence evolution of tRNA and ribosomal protein genes. J Mol Evol 35:304–317

Woolfit M, Bromham L (2005) Population size and molecular evolution on islands. Proc Roy Soc Lond B Bio 272:2277–2282

Wright S, Keeling J, Gillman L (2006) The road from Santa Rosalia: a faster tempo of evolution in tropical climates. Proc Natl Acad Sci USA 103:7718–7722

Yang L, Gaut BS (2011) Factors that contribute to variation in evolutionary rate among Arabidopsis genes. Mol Biol Evol 28: 2359–2369

Yang Y-h, Zhang F-m, Ge S (2009) Evolutionary rate patterns of the gibberellin pathway genes. BMC Evol Biol 9:206

Yoder AD, Yang ZH (2000) Estimation of primate speciation dates using local molecular clocks. Mol Biol Evol 17:1081–1090

Young ND, dePamphilis CW (2005) Rate variation in parasitic plants: correlated and uncorrelated patterns among plastid genes of different function. BMC Evol Biol 5:16

Yue J-X, Li J, Wang D, Araki H, Yang S (2010) Genome-wide investigation reveals high evolutionary rates in annual model plants. BMC Plant Biol 10:242

Zhang L, Vision TJ, Gaut BS (2002) Patterns of nucleotide substitution among simultaneously duplicated gene pairs in Arabidopsis thaliana. Mol Biol Evol 19:1464–1473

Zimmer A, Lang D, Richardt S, Frank W, Reski R, Rensing SA (2007) Dating the early evolution of plants: detection and molecular clock analyses of orthologs. Mol Genet Genomics 278:393–402

Zuckerkandl E, Pauling L (1965) Evolutionary divergence and convergence in proteins. In: Bryson V, Vogel HJ (eds) Evolving genes and proteins. Academic, New York, pp 97–166

Conserved Noncoding Sequences in Plant Genomes 8

Sabarinath Subramaniam and Michael Freeling

Contents

8.1 Introduction and Definitions

Plant genomes carry a great diversity of all sorts of sequences and many have functions. Some of these sequences specify functions important to the plant by encoding RNA sequences, some of which (genes or coding domain sequences) encode proteins. Some sequences encode binding functions important to the plant, such as the DNA sites near genes that bind regulatory proteins (motifs), while others may function to block the binding or movement of chromosomal proteins (e.g. insulators). One of the goals of molecular biology studies is to discover the exact functions specified by the genome. However, this is not a simple task. A typical plant has about 30,000 genes, and this does not include genes that function largely selfishly such as most transposons. All of these about 30,000 genes encode one or more messenger RNAs (mRNAs) and many of these genes contain different parts: introns, exons, RNA binding sites, DNA binding sites and similar. Somewhere near the transcriptional units of a gene comprised of its coding regions are the chromosomal regulatory regions that enable the gene to be a part of one or more biological pathways or networks, transcription factor binding sites, enhancer sites, insulator sites and so forth. Added up, there are "millions" of specific DNA sequences that carry specific coding, binding or blocking functions important to gene function sequences with a chromosomal start, stop and strand. We know something functional, however vague, about several thousand of these sequences and almost nothing about the meaning of their combinations.

There is a way to infer that particular sequences functioned over evolutionary time, and presumably still function, even though the function itself remains unknown. This is because sequences that carry specific function will resist deletions, insertions, and base substitutions expected to alter function; these sequences are said to be "conserved." This conservation derives from the ongoing process of purifying selection, where most mutations in functional

S. Subramaniam (✉)
Department of Plant and Microbial Biology, University of California, Berkeley, CA 94720, USA
e-mail: shabari@berkeley.edu; sshabari@gmail.com

J.F. Wendel et al. (eds.), *Plant Genome Diversity Volume 1*,
DOI 10.1007/978-3-7091-1130-7_8, © Springer-Verlag Wien 2012

Table 8.1 Definitions involving CNSs and their identification in plants

Phylogenetic footprint	The most inclusive term for sequences that tend to stay the same over evolutionary time, while sequences within the genome known to be without specific function are becoming randomized. Thus, a phylogenetic footprint often marks a sequence with function
Plant CNSs (conserved noncoding sequences)	Plant CNSs, a subset of phylogenetic footprints, have a set definition that permits quantitative comparisons between "usefully diverged" homologous genes in all sorts of plants. A bl2seq (blastn-2-sequences with default settings; Tatusova and Madden 1999) hit between the nonprotein-coding sequences near usefully diverged, syntenic (either orthologous or homeologous) genes. A CNS pair must have an e-value equal to or more significant than a 15/15 exact nucleotide match (Kaplinsky et al. 2002; Inada et al. 2003) and must be syntenic (Freeling and Subramaniam 2009). "Usefully diverged" is of paramount importance, and will be independently defined in this glossary
Plant αCNS and orthologous CNSs	Orthologous CNSs, or just CNSs, may usefully be distinguished from paralogous CNSs, such as those generated by polyploidy. In *Arabidopsis*, these are r αCNSs (Thomas et al. 2007)
"usefully diverged"	In order to predict function, a pair of bonafide CNSs needs to have diverged for a long enough period of time to let sequences randomize or delete if not under selection. Otherwise, conservation is nothing but neutral carryover. For biological reasons, if divergence continues for too long, detection by sequence similarity becomes difficult. This "usefully diverged" window in plants is between approximately Ks 40% and 90% where Ks is the rate of synonymous base pair substitution in codons, used here as a measure of divergence time only

nucleotides confer some lowering of fitness and tend to not pass on to subsequent generations. Thus, there is a tendency for the DNA that encoded ancestral functions to be conserved over time. These blocks of conserved sequence are called "phylogenetic footprints" (Table 8.1). Exons, with their open reading frames, comprise a good portion of the longer phylogenetic footprints in plants, as expected.

As described already, some sequences with specific functions for the plant do not encode proteins, and these are sometimes resolved as phylogenetic footprints. Some encode those parts of the transcripts that are not translated, and some RNAs fold into functional gene products themselves. A specific category of phylogenetic footprints represent sequences that function by binding or blocking some molecule, so called conserved noncoding sequences, or CNSs; they have precise definitions that fit the biological realities of plants (Table 8.1). Since functional DNA that does not encode a product must function as a CNS, the annotation of protein-coding sequence and RNA product-coding sequence is a necessary part of CNS discovery. Hence every CNS requires validation following the release of each new version of a genome's annotations.

In order to visualize the optimal number of plant CNSs, much care has been given to the divergence times between the chromosomal regions being footprinted (compared). Among divisions of eukaryotes, flowering plants have the maximum frequency of successful polyploidy over the last 150 million years or so. Figure 8.1 shows a heavily pruned phylogenetic tree of plants with those genomes at the tips that have been sequenced and released. For comparison, Eutheria (a vertebrate clade about 125 million years old) have had no successful tetraploidies. Each ancient plant tetraploidy is a small starburst. Each tetraploidy has a doubled genome for only a short time, and then the chromosomes rearrange and one or the other of the two duplicate

(homeologous) genes tend to be lost; this post-polyploidy process is called "fractionation." Plant species too distantly related, like any dicot compared to any monocot (Fig. 8.1), present difficulties simply lining up the chromosomes into orthologous pairs. CNS discovery begins by comparing genes within chromosomes that are obviously matched, but diverged enough to ensure that functionless DNA has randomized. However, the divergence time separating the two chromosomal regions cannot be too recent either, because then sequences would be similar just by virtue of recent divergence (Fig. 8.1, Table 8.1). Sometimes there are no two sequenced species that are usefully diverged. However, sometimes the most recent tetraploidy happened a useful number of years ago, as with the most recent *Arabidopsis thaliana* tetraploidy, called "alpha" (notice the starburst in the window of Fig. 8.1). In such a case, CNSs can be discovered by comparing retained segments of syntenic, homologous (homeologous, Table 8.1 definitions) chromosomes within the same genome.

We use a local alignment algorithm, blastn, to find CNSs in usefully diverged, homologous DNA sequences. The settings and e-value cutoffs are specifically defined for plant CNS discovery (Table 8.1). Local alignment algorithms are preferred because noise is easily detected: noise includes alignments—called "blast hits"—that are not syntenous, and tend to be palindromic or overly simple. The CNS significance cutoff is set just above noise. The CNSs in Fig. 8.2, colored orange, exemplify CNS discovery between the grasses sorghum and rice. Note that the divergence between sorghum-like grasses and rice-like grasses happened within the window of useful divergence (Fig. 8.1). CNSs are sorted to their nearest gene pair. The overall result for two comparable genomes is a long list of CNSs. Each is a pair of homologous sequences, either orthologous or homeologous (Table 8.1), and each has been sorted to one

Fig. 8.1 Heavily pruned phylogenetic tree of plants with those genomes sequenced and released at the tips. The *colored bar* refers to the window of useful divergence for CNS discovery

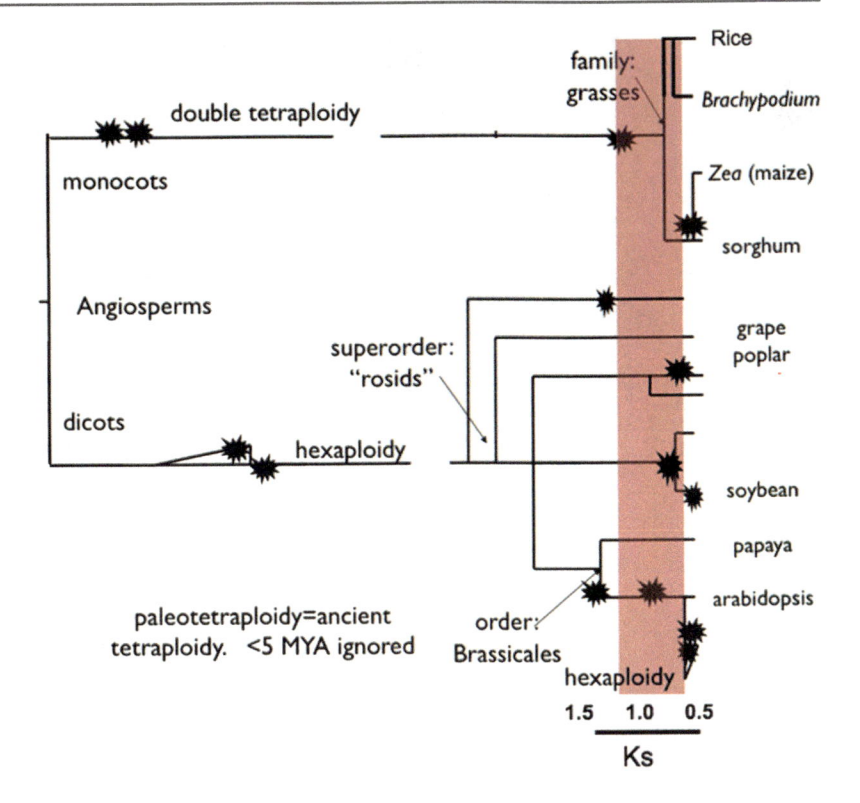

Fig. 8.2 GEvo panel comparison of orthologous regions of *Sorghum bicolor* (*top panel*) and *Oryza sativa* (*bottom panel*). *Red bars* indicate High scoring segment pairs (HSPs) between the two regions, some of them shown connected with *green lines*. HSPs called as CNSs are annotated in *purple bars* drawn below the HSP graphic in the *Sorghum* panel. This result can be regenerated at http://genomevolution.org/r/2xc0

gene in the genome. CNS discovery has been automated for plants (Fig. 8.2: purple rectangles on the rice model annotation line). Following genes by their CNS-richness, or by the DNA-binding motifs within their CNSs, is beginning to generate useful data on how CNSs—or clusters or families of CNSs—specify function.

Specific function is known for only a few plant CNSs. One is the cluster of CNSs in the longest intron of the grass homeolbox gene *knotted1*. When this gene is ectopically expressed in the leaves of grass plants, the leaf cells do not mature properly and do not stop proliferating. Figure 8.3

shows the region of a maize leaf between sheath and blade covered by grotesque finger-like projections, called knots; this is a dramatic version of the dominant *knotted1* phenotype. The laboratory of Sarah Hake, Plant Gene Expression Center, ARS, Albany California (cited in Fig. 8.3 legend), found that nine different transposon insertions happened in this large intron, and just in the CNSs, and that these could cause extopic expression and knots. We hypothesize that the transposon insertions prevent the binding of some product to the intron CNSs, so *knotted1* turns on in leaves. All monocot *knotted1* genes tested so far have these intron CNSs.

Fig. 8.3 One extreme example of the mutant phenotype conferred by dominant *knotted1* (*kn1*) mutants. This plant has grotesque finger-like projections of more-or-less normal leaf coming off its outer surface. These failures to stop dividing result from the ectopic expression of KN1 protein in leaves, an organ where this gene is normally not expressed. The Hake laboratory showed that phenotypes of this general type were made by multiple transposon insertions into specific regions of the *kn1* big intron (Greene et al. 1994) containing a number of clustered CNSs (Inada et al. 2003)

8.2 Detection of CNSs in Plant Genomes

The original paper that found CNSs between maize BACs and the rice genome defined and defended the blastn settings and cutoffs of Table 8.1 (Kaplinsky et al. 2002). Eric Lyons, the project lead and programmer of the CoGe suite of software tools (http://synteny.cnr.berkeley.edu/CoGe/index.pl), compared and contrasted various DNA sequence comparison algorithms for CNS resolution, purity and discovery speed. He concluded that blastn (or bl2seq) using our exact settings was an excellent compromise (Lyons and Freeling 2008). Of course, any algorithm that dips into nonsyntenic, repetitive noise will find some additional syntenic footprints. Fig. 8.4a, b show GEvo panels of blastn output from comparisons of homeologous (Table 8.1) regions of *Arabidopsis thaliana* settings defined in Table 8.1, (Fig. 8.4a) and similar

settings with a slightly lowered e-value cutoff (Fig. 8.4b and legend). The choice among alignment algorithms is less important than carefully setting the noise cutoff and far less important than picking alignments that are within the window of useful divergence (Lyons and Freeling 2008).

CNS discovery has been automated. The CNS Discovery Pipeline 1.0 (Woodhouse et al. 2010) entails repeat masking, tandem identification, orthologous pair finding, CNSs discovery, and CNS sorting to the nearest orthologous gene pair. This pipeline utilizes a new algorithm for finding bonafide orthologous pairs when comparing genomes with more than one possible syntenic partner, as is the case for all plant comparisons seeking syntenic runs of gene pairs (Tang et al. 2011). The rice-sorghum orthologous CNSs have been cataloged and form the starting point for hypothesis testing (Schnable et al. 2011). The rice genome model line in Fig. 8.2 is annotated with more than the exons, introns and transcriptional unit. The purple bars on this model line are the locations of the Version 1.0 (v1) CNS discovery pipeline automated sorghum-rice CNS calls. Also note the orange-sorghum-rice blast hits above the rice model line, and note how there is excellent correspondence between the v1 pipeline calls and the actual data, that data being the blast hits themselves (rectangles colored orange in Fig. 8.4). Thus, Fig. 8.2 exemplifies how the v1 CNS discovery pipeline was proofed for accuracy. This experiment can be visited at http://genomevolution.org/r/2xc0; the researcher may change settings, algorithms, dimensions and more and then rerun the experiment on-the-fly.

8.3 General Characteristics of Plant CNSs Suggest a Predominant Regulatory Function

The general conclusions from the original maize-rice orthologous CNS studies (Kaplinsky et al. 2002; Guo and Moose 2003; Inada et al. 2003) have been replicated—in general—in the two large scale homeologous CNS studies in *Arabidopsis* (Freeling et al. 2007; Thomas et al. 2007) and rice (Lin et al. 2008). These αCNS (defined in Table 8.1) data demonstrate that, in plants, αCNSs average from 20 to 30 bp in length. αCNSs are predominantly found in close proximity to one and only one gene pair.

In an effort to associate CNSs with function, genes, gene names or their Gene Ontology (GO) terms have been quantified for CNS-richness. To date, all studies have concluded that "regulatory genes" are CNS-rich, with genes encoding transcription factors generally being more CNS-rich than genes encoding protein kinases, and much richer than genes encoding ancient structural functions like ribosomal subunits, motors, or energy metabolic pathways. In one study of arabidopsis αCNSs (Freeling et al.

a

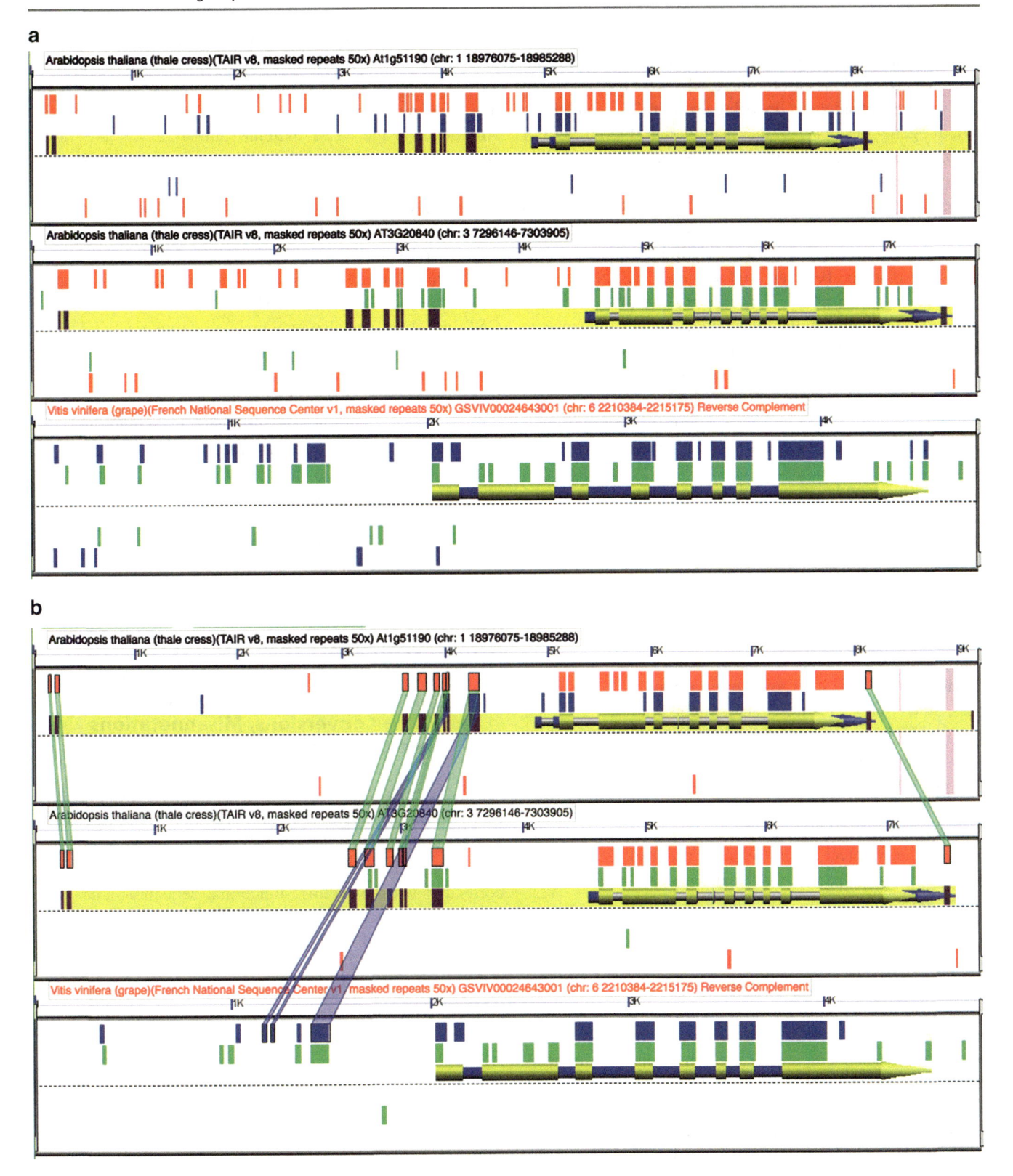

b

Fig. 8.4 GEvo panel comparison of homeologous regions of *Arabidopsis thaliana* (*top two panels*) and corresponding orthologous region in *Vitis vinifera* (*bottom panel*). (**a**) Using blastn at optimal CNS discovery settings (Table 8.1). (**b**) similar settings with a slightly lowered e-value cutoff. *Red* bars indicate High scoring segment pairs (HSPs) for each pairwise comparison, some of them shown connected with *purple lines*. HSPs called as CNSs are annotated in *purple bars* drawn below the HSP graphic in the *A. thaliana* panels

2007)—those homeologous CNSs retained from the most recent tetraploidy—246 gene pairs occupied an exceptionally long stretch of the chromosome, where the regions full of CNSs were conspicuous in being exon voids, sometimes called "gene deserts" in the mammalian literature. For *Arabidopsis*, >4 kb 5′ plus 3′ of CNS-defined genespace "exceptionally long," but this definition does not apply to all genomes. These genes were called "Bigfoot" and tended to be annotated with "response to..." GO biological function categories, these often being "transcription factor activity" as well. The distance between Bigfoot gene 5′ CNSs and the start of exon1 averaged 3.1 kb. This same study found a statistically significant over-abundance in CNS sequence of the most famous transcription factor binding site in plants, the G box palindrome (CACGTG). Several other known motifs, plus previously unknown 7mers, were significantly over-represented, but none to the extent of the G-box and Gbox-like motifs. CNSs in both *Arabidopsis* and rice occur 5′ and 3′ to genes and in introns, and are skewed toward the 5′ end (Lin et al. 2008). Some intron CNSs, like those found inside the grass *knotted1* gene, are noncoding (Inada et al. 2003), while some are certainly unannotated exons expressed in alternative transcripts. In general, intron CNSs are suspect, especially those that are touching an annotated exon; they are likely unannotted exons. The data for the 5′ and 3′ plant CNSs supports their role, in general, as binding sites for gene regulatory products.

Most CNSs contain known transcription factor binding sites, and CNSs generally are enriched in several such motifs. Most plant genes function in developmentally complicated ways in the absence of CNSs. It must be understood that known transcription factor binding motifs tend to be short, far too short to be detected by themselves as sequence, with the minimum CNS being a 15/15 exact match. CNSs may well only detect clusters of sites, riterated binding sites or other macrostructure associated with particular sorts of genes, about which we know little.

8.4   Plant Versus Vertebrate CNSs

Mammalian CNS research, now called CNE research, began years earlier than comparable research in plants (Hardison 2000). Among the early conclusions derived from early plant CNS research was that plant CNSs are considerably smaller and less numerous than those in mammals, and that plant CNSs do not generally "run together" on the chromosome (Kaplinsky et al. 2002; Guo and Moose 2003; Inada et al. 2003). This clustering of CNSs around gene pairs supported the feasibility of sorting a particular CNS to a particular gene using proximity alone, an activity not possible to do in

man-mouse comparisons. Such assignments are arbitrary, and some of these assignments are expected to be proved wrong, since some CNSs are expected to encode functions involving more than one gene in a chromosomal domain. Additionally, the most extreme animal CNSs are more conserved than are the most conserved plant CNSs. Several hundred ultraconserved elements (UCEs), deeply conserved vertebrate CNSs, have been found to be identical or near-identical for 200–500 bp in human, mouse, and rat, with a majority of them showing on the order of 96% identity with birds (chicken) that diverged from mammals approximately 310 MYA (Bejerano et al. 2004; Sakuraba et al. 2008). Plant CNSs are short, and also may mutate relatively quickly over evolutionary time, although the binding functions themselves may well be conserved. This problem of detecting divergent functional homologous binding sites by sequence only is called "binding site turnover" in animals (Moses et al. 2006). It is not yet clear whether or not binding site turnover mechanisms in animals apply to plants.

Some known, characterized cis-acting binding sites are CNSs, cases like that of the grass *knotted1* big intron CNS cluster already discussed (Table 8.2). These have been reviewed by Freeling and Subramaniam (2009).

8.5   Experiments That Address the Possibility That Some CNSs Are Artifacts, Mutation Coldspots, Gene Conversions, Misannotations and the Like

Because CNSs cluster around genes that are "response to" transcription factors—these are very often bigfoot genes—and because CNSs are enriched in known DNA binding motifs the indication is that CNSs as defined are positively correlated with cis-acting, functional sequences that bind regulatory molecules. However, that does not mean that each CNS is real or even that the average CNS is real. Artifacts are possible.

There is evidence against CNSs being mutational coldspots in animals (Drake et al. 2006; Kim and Pritchard 2007), but there is no such evidence for plants. However, it seems unlikely that replication or repair processes should be error-free. If one sequence copies over to another syntenic, homologous sequence—that is, suffers a gene conversion event over a short stretch of paired DNAs—then an artifactual CNS could be created essentially at any time in evolutionary history. Recent studies found many regions of gene conversion within the rice genome (Wang et al. 2007; Xu et al. 2008), including approximately 6 Mb of near-identical sequence at the ends of homeologous chromosomes 11 and

Table 8.2 CNSs characterized as cis-acting binding sites (Reprinted with permission From *Current Opinions in Plant Biology* (Freeling and Subramaniam 2009))

Type	Function associated with plant CNSs	Citation
G	Grass regulatory genes are rich in orthologous CNSs	Guo and Moose (2003), Inada et al. (2003)
G	*Arabidopsis* genes that are induced by stimuli and/or encode transcription factors are rich in homeologus CNSs, and are often "Bigfoot" genes. Japonica rice homeologous CNSs are similar	Thomas et al. (2007), Lin et al. (2008)
G	Arabidopsis homeologous CNSs are significantly enriched for several known transcription factor binding motifs, especially the G-box	Freeling et al. (2007)
S	Intron CNSs in a Class I homeobox gene (kn1) bind a negative cis regulator, a binding that is disrupted by Mu transposons, but only in Mu-active lines	Greene et al. (1994), Inada et al. (2003)
S	5′ CNSs contain conserved, known transcription factor binding motifs and motif-patterns: RAB16/17 in grasses, rbc a/b in dicots, and [a]proximal promoters in dicots	Buchanan et al. (2004), Vandepoele et al. (2006), Weeks et al. (2007)
S	Two 5′ CNSs of the shoot meristemless (STM) homeobox gene in dicots and monocots, and their binding motifs, cis- regulate repression/re-establishment of leaf expression	Uchida et al. (2007)
S	A QTL for flowering time in maize mapped to a cis-acting, regulatory grass CNS 70 bp upstream of an Ap2-like gene	Salvi et al. (2007)
S	Some 5′UTR grass CNSs are uORFs, one mechanism to down-regulate translation	Tran et al. (2008)

G general, *S* specific

[a]Vandepoele et al. (2006) used phylogenetic footprinting of homologous proximal promoters of dicot genes, along with TF binding motif over-representation and transcript co-expression, to infer functional regulatory modules composed of two or more transcription factor binding sites in close proximity. Being confined to noncoding space close to the start of transcription, this work addresses the transcription factor binding potential of a few CNSs, not CNSs in any general way, and specifically not Bigfoot gene CNSs

12. Since such gene conversions are not understood, they should not be discounted as possible sources of "CNSs" that have no function. Another source of artifactual CNSs are unannotated RNA genes (like *MIR, Ta-si-RNA* or targets for siRNAs; Lee et al. 2012, this volume); analyses of current data indicted that 1.5% of CNSs can be explained in this way (Thomas et al. 2007). Finally, CNSs could be undiscovered, small genes or exons that are used differentially. New genes are being discovered continuously, and any of these could conceivably be erroneously interpreted as a CNS. Proteogenomic studies in *Arabidopsis* resulted in the expansion of exons for 2446 TAIR7 models and called 838 new, usually short, genes (Castellana et al. 2008). We back-translated each of these unexpected peptides using tblastn and compared all sequences to each published Arabidopsis αCNS. None of the homeologous CNSs identified within *Arabidopsis thaliana* could be translated into any sequence on this new protein list. Even so, CNSs, and especially intron CNSs adjacent to an exon, not only could but should turn out to be protein coding. One of us (Freeling, unpublished data) has data from blastn analyses of intron CNSs to *distantly related* plants: some CNSs are conserved in such a way as to be best explained as alternatively spliced exons that are not yet annotated as such in *Arabidopsis*. Because of the purely computational nature of the average CNS's identity, there is always a chance that it will be artifactual. However, that the *bulk* of CNSs are functional binding sites is supported by the fact that CNSs are located near regulatory genes and especially "responds to" transcription factors, and that CNSs are enriched for known DNA-binding motifs. Those two results are not expected of any of the artifactual or trivial explanation.

8.6 CNSs Can Exist for Any Duplicate DNA Segment

It is possible to find CNS-like sequences among the noncoding regions of some tandem duplicates (unpublished). Unfortunately, there is no perfect way to know when the tandem array occurred since gene conversions are well-known in duplicate arrays (Gao and Innan 2004) and selection within tandem genes is expected to be reduced. For orthologous genes—between rice and sorghum for example—every gene diverged from its ortholog at the same time. Using similar logic, all alpha homeologous pairs in *Arabidopsis* happened contemporaneously. For an orthologous gene-pairs dataset, each gene's divergence frequency for any endpoint is comparable to the frequencies of other contemporaneous pairs. There is no contemporaneous control for tandem genes or segmental duplicates. Even so, it might well be possible to control for CNS discovery among these duplicates.

8.7 CNS-Rich Genes, Bigfoot Genes, Are Particularly Enigmatic

Genes high up in the regulatory cascade, especially those that respond to external stimuli, tend to be CNS-rich, while genes that encode housekeeping enzymes, for example, tend to be CNS-poor. CNS-richness seems a reasonable quantitative metric for at least one sort of gene regulation. Using this metric, *the most regulatory genes are also the most regulated.* When faced with enigma, metaphors are useful to some of us.

Here is a metaphor from the market. From the company's point of view, the Chief Executive Officer (CEO) is at the highest control level and the worker is near the lowest. However, ask any CEO about her job and hear about how changes in unions, markets, investor sentiment, government regulations, tax laws, unexpected lawsuits and a myriad of additional connections to "the system" limit and often override any control the CEO might hope to exert on behalf of company success. Perhaps the genes at the highest control level, like CEOs, are also those under the most control themselves. Bigfoot genes may be these enigmatic CEO genes.

8.8 Fresh Data on How CNS Richness Influences the Ability of a Gene to Be Retained Post-tetraploidy

As seen in Fig. 8.1, there is a complex phylogenetic relationship between the pre-grass tetraploidy, the radiation of grasses, and the tetraploidy that happened just after the maize and sorghum lineages split. The rice, sorghum and maize genomes are sequenced. For rice, it was possible to obtain a list of orthologous pairs with sorghum, each with a CNS collection obtained from the CNS Discovery Pipeline v1.0, and the rice genome can also be organized into two homeologous, highly fractionated genomes anchored by their homeologous pairs. Thus, it was possible for Schnable and coworkers (Schnable et al. 2011) to ask whether CNS richness of rice orthologous pairs correlated with having been retained after the pre-grass tetraploidy. Figure 8.5 (reprinted here) shows these data. As sorghum-rice

orthologous CNS-richness increases, the chances of having been retained as a pair from the last tetraploidy increases. When the ">15 CNSs/gene" bin of Fig. 8.5 is subdivided, retention levels go up. Six of the 15 rice–sorghum gene pairs with >28 CNSs possess a retained homeolog (40% retention) and 25 of the 56 gene pairs with 22–28 CNSs possess a retained homeolog (45% retention).

This result was not expected on the basis of prevailing theory on why genes are retained post-tetraploidy. Based on the robust gene balance hypothesis (Birchler and Veitia 2010) a generally applicable reason that genes are retained as pairs post-tetraploidy is that, were one of the duplicates removed, dosage imbalance would ensue and fitness would go down. Thus, for those gene networks where dosage balance is functionally preferred, purifying selection maintains the status quo by preventing fractionation of the pair. Most genes lose one or the other duplicate; that is, most genes fractionate. The consequences of gene dosage theory on gene and genome duplication research has been reviewed at length (Semon and Wolfe 2007; Edger and Pires 2009; Freeling 2009). Selection for gene dosage balance is usually envisioned at the level of gene product balance, as in the concentration of protomeric proteins assembling into larger complexes. The most advanced theoretical treatment of gene dosage sensitivity (haploinsufficiency) is at the protein assembly level (Veitia 2002), and this may explain why genes encoding components of ribosomes, motors, and proteosomal cores are over-retained post-tetraploidy. Since transcription factors may participate in complicated aggregates, enhancosomes (Levine 2010), the dose-sensitivity of genes encoding transcription factors might well be explained

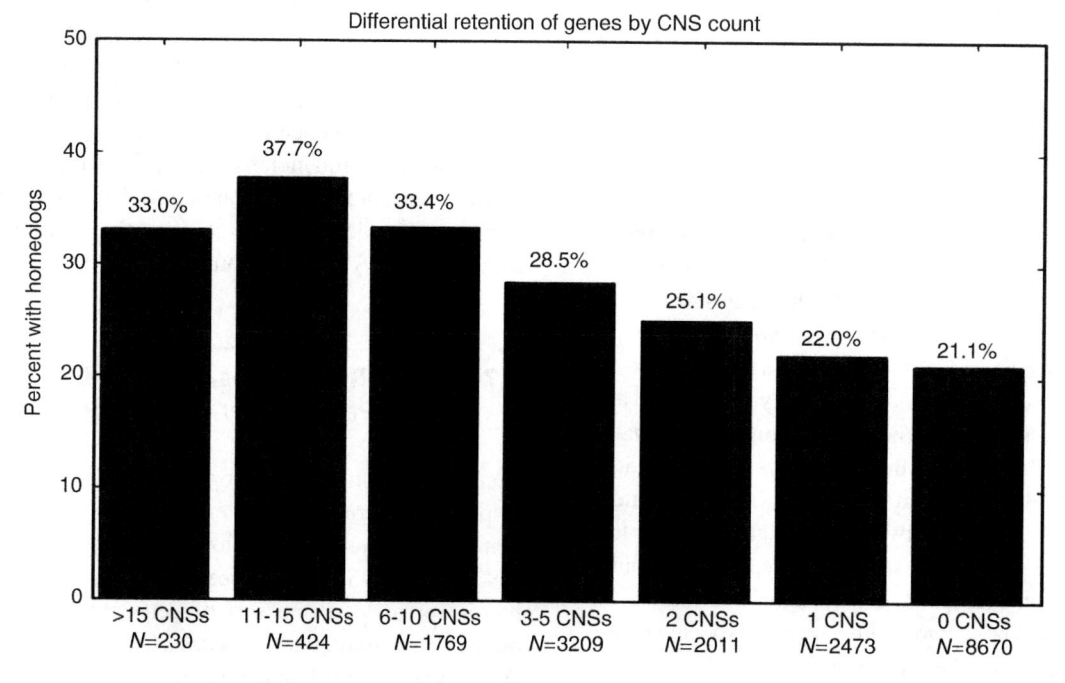

Fig. 8.5 Relationship between CNS richness and retention

at the protein-protein level as well. However, the data of Fig. 8.5 suggest an alternative explanation. Some genes have binding sites in their genespace, and for some of these genes the concentration of binding sites may be just as important as the concentration of protein that binds them. A bigfoot transcription factor gene, for example, might be dose-sensitive not because of any optimal protein-protein stoichiometries involving the gene's product, but because of DNA-protein stoichiometries being selectively important on its own flanks.

The newly sequenced maize genome (Schnable et al. 2009) permitted an experiment that begins to utilize the analytical power of phylogenetics. Rice homeologous pairs of bigfoot genes were analyzed for orthologous sorghum-rice CNS content. There were hundreds of cases where one of the homeologs had more CNSs than the other. There was a significant tendency for the homeolog with the greater CNS count to be retained in the maize lineage tetraploidy. Clearly, the differential CNS loss must have occured in the branch shared by all grasses. These data (Schnable et al. 2011) make sense if CNS-richness is directly related to gene dosage sensitivity, but this explanation is not proved; CNS-loss *could* switch the expression profile per organ, region or time leading to an avoidance of dose-sensitivity in the pregrass branch of the phylogenetic tree. In the absence of dose-sensitivity, loss is expected. More research is necessary on the molecular biology of gene dose sensitivity.

Conclusions

Plant CNSs often function as cis-acting regulatory binding sites, but published CNSs lists undoubtedly contain artifacts of a number of types. Genes that are particularly CNS-rich and take up a lot of chromosomal space—bigfoot genes—tend to be in a "responds to…" GO category and also tend to be transcription factors. These long genes are of particular interest. They are retained preferentially following ancient tetraploidies, and are retained less often if they lose CNSs over evolutionary time. This conclusion is from an experiment that marks the beginning of CNS-driven phylogenetic research. The relationship between CNS-richness and gene dosage sensitivity is complex, and begs for continued research.

The first paper on plant CNSs (Kaplinsky et al. 2002) suggested that CNSs may be particularly close arrays of transcription factor binding motifs that function as do enzymes, to mechanically facilitate the binding of proteins into enhancer (or insulator or transcription-factor) complexes. That idea, in the absense of relevant data, seems about as useful today as it was over a decade ago.

Deciphering the meaning of CNSs is an important part of understanding the language of gene regulation.

References

Bejerano G, Pheasant M, Makunin I, Stephen S, Kent WJ, Mattick JS, Haussler D (2004) Ultraconserved elements in the human genome. Science 304:1321–1325

Birchler JA, Veitia RA (2010) The gene balance hypothesis: implications for gene regulation, quantitative traits and evolution. New Phytol 186:54–62

Buchanan CD, Klein PE, Mullet JE (2004) Phylogenetic analysis of 5′-noncoding regions from the ABA-responsive rab16/17 gene family of soghum, maize and rice provides insight into the composition, organization and function of cis-regulatory modules. Genetics 168:1639–1654

Castellana N, Payne S, Shen Z, Stanke M, Bafna V, Briggs S (2008) Discovery and revision of *Arabidopsis* genes by proteogenomics. Proc Natl Acad Sci U S A 105:21034–21038

Drake JA, Bird C, Nemesh J, Thomas D, Newton-Cheh C, Raymond A, Excoffler L, Attar H, Antonarakis S, Dermitzakis E et al (2006) Conserved noncoding sequerces are selectively constrained and not mutation cold spots. Nat Genet 38:223–227

Edger PP, Pires JC (2009) Gene and genome duplications: the impact of dosage-sensitivity on the fate of nuclear genes. Chromosome Res 17:699–717

Freeling M (2009) Bias in plant gene content following different sorts of duplication: tandem, whole-genome, segmental, or by transposition. Annu Rev Plant Biol 60:433–453

Freeling M, Subramaniam S (2009) Conserved noncoding sequences (CNSs) in higher plants. Curr Opin Plant Biol 12:126–132

Freeling M, Rapaka L, Lyons E, Pedersen B, Thomas BC (2007) G-boxes, bigfoot genes and environmental response: characterization of intragenomic conserved noncoding sequences in *Arabidopsis*. Plant Cell 19:1441–1457

Gao L-z, Innan H (2004) Very low gene duplication rate in the yeast genome. Science 306:1367–1370

Greene B, Walko R, Hake S (1994) Mutator insertions in an intron of the maize knotted1 gene result in dominant suppressible mutations. Genetics 138:1275–1285

Guo H, Moose SP (2003) Conserved noncoding sequences among cultivated cereal genomes identify candidate regulatory sequence elements and patterns of promoter evolution. Plant Cell 15:1143–1158

Hardison RC (2000) Conserved noncoding sequences are reliable guides to regulatory elements. Trends Genet 16:369–372

Inada DC, Bashir A, Lee C, Thomas BC, Ko C, Goff SA, Freeling M (2003) Conserved noncoding sequences in the grasses. Genome Res 13:2030–2041

Kaplinsky NJ, Braun DM, Penterman J, Goff SA, Freeling M (2002) Utility and distribution of conserved noncoding sequences in the grasses. Proc Natl Acad Sci U S A 99:6147–6151

Kim S, Pritchard J (2007) Adaptive evolution of conserved noncoding elements in mammals. PLoS Genet 3:e147

Lee T-F, Li P, Meyers B (2012) The biology and dynamics of plant small RNAs. In: Wendel JF (ed) Plant genome diversity, vol 1, Plant genomes, their residents, and their evolutionary dynamics. Springer, Wien/New York

Levine M (2010) Transcriptional enhancers in animal development and evolution. Curr Biol 20:R754–R763

Lin H, Ouyang S, Egan A, Nobuta K, Haas BJ, Zhu W, Gu X, Silva JC, Meyers BC, Buell CR (2008) Characterization of paralogous protein families in rice. BMC Plant Biol 2008:18

Lyons E, Freeling M (2008) How to usefully compare homologous plant genes and chromosomes as DNA sequences. Plant J 53:661–673

Moses AM, Pollard DA, Nix DA, Iyer VN, Li XY, Biggin MD, Eisen MB (2006) Large-scale turnover of functional transcription factor binding sites in *Drosophila*. PLoS Comput Biol 2:e130

Sakuraba Y, Kimura T, Masuya H, Noguchi H, Sezutsu H, Takahasi KR, Toyoda A, Fukumura R, Murata T, Sakaki Y, Yamamura M, Wakana S, Noda T, Shiroishi T, Gondo Y (2008) Identification and characterization of new long conserved noncoding sequences in vertebrates. Mamm Genome 19:703–712

Salvi S, Sponza G, Morgante M, Tomes D, Niu X, Fengler KA, Meeley R, Ananiev EV, Svitashev S, Bruggemann E et al (2007) Conserved noncoding genomic sequences associated with a flowering-time quantitative trait locus in maize. Proc Natl Acad Sci U S A 104:11376–11381

Schnable PS, Ware D, Fulton RS, Stein JC, Wei F, Pasternak S, Liang C, Zhang J, Fulton L, Graves TA et al (2009) The B73 maize genome: complexity, diversity, and dynamics. Science 326: 1112–1115

Schnable JC, Pedersen BS, Subramaniam S, Freeling M (2011) Dose-sensitivity, conserved noncoding sequences and duplicate gene retention through multiple tetraploidies in the grasses. Frontiers in Plant Genetics and Genomics 2:1–7

Semon M, Wolfe KH (2007) Consequences of genome duplication. Curr Opin Genet Dev 17:505–512

Tang H, Lyons E, Pedersen, B, Schnable JC, Paterso AH, Freeling M (2011) Screening synteny blocks in pairwise genome comparisons through integer programming. BMC Bioinformatics 12:102

Tatusova TA, Madden TL (1999) BLAST 2 Sequences, a new tool for comparing protein and nucleotide sequences. FEMS Microbiol Lett 174:247–250

Thomas BC, Rapaka L, Lyons E, Pedersen B, Freeling M (2007) Intragenomic conserved noncoding sequences in *Arabidopsis*. Proc Natl Acad Sci U S A 104:3348–3353

Tran M, Schultz C, Baumann U (2008) Conserved upstream open reading frames in higher plants. BMC Genomics 9:361

Uchida N, Townsley B, Chung KH, Sinha N (2007) Regulation of SHOOT MERISTEMLESS genes via an upstream-conserved non-coding sequence coordinates leaf development. Proc Natl Acad Sci U S A 104:15953–15958

Vandepoele K, Casneuf T, Van de Peer Y (2006) Identification of novel regulatory modules in dicotyledonous plants using expression data and comparative genomics. Genome Biol 7:R103

Veitia RA (2002) Exploring the etiology of haploinsufficiency. Bioessays 24(2):175–184

Wang X, Tang H, Bowers JE, Feltus FA, Paterson AH (2007) Extensive concerted evolution of rice paralogs and the road to regaining independence. Genetics 177:1753–1763

Weeks KE, Chuzhanova NA, Donnison IS, Scott IM (2007) Evolutionary hierarchies of conserved blocks in 5′-noncoding sequences of dicot rbcS genes. BMC Evol Biol 7:51

Woodhouse MR, Schnable JC, Pedersen BS, Lyons E, Lisch D, Subramaniam S, Freeling M (2010) Following tetraploidy in maize, a short deletion mechanism removed genes preferentially from one of the two homeologs. PLoS Biol 8:e1000409

Xu S, Clark T, Zheng H, Vang S, Li R, Wong GK-S, Wang J, Zheng X (2008) Gene conversion in the rice genome. BMC Genomics 9:93

Plant Mitochondrial Genome Diversity: The Genomics Revolution

9

Jeffrey P. Mower, Daniel B. Sloan, and Andrew J. Alverson

Contents

J.P. Mower (✉)
Center for Plant Science Innovation and Department of Agronomy and Horticulture, University of Nebraska, Lincoln, NE, USA
e-mail: jmower2@unl.edu

9.1   Introduction

The mitochondrion originated some 1.5 billion years ago, most likely through the taming of an alpha-proteobacterial endosymbiont. Extensive evidence now unambiguously supports the endosymbiotic theory and a single origin of mitochondria for all eukaryotes (Lang et al. 1999). The most well-known function of mitochondria is the generation of chemical energy in the form of ATP through oxidative phosphorylation, although this is not the case for degenerate mitochondria such as mitosomes or hydrogenosomes (Tovar et al. 1999; Boxma et al. 2005; Stechmann et al. 2008). The mitochondria of plants (and other photosynthetic eukaryotes) share the cell with another energy-producing organelle, the plastid. In addition to the production of ATP, plant mitochondria also play a major role in the synthesis of metabolic precursors such as amino acids, vitamins, and lipids (Nunes-Nesi and Fernie 2007).

Most mitochondria contain a remnant of their bacterial heritage in the form of a highly reduced genome. The genome of the original endosymbiotic organism likely contained thousands of genes. As it became ever more dependent on the host cell, many essential genes became non-essential and were eliminated from the genome, and most remaining genes were transferred into the nuclear genome of its host. The mitochondrial genomes of extant eukaryotes typically contain some subset of the same ~100 genes, which suggests that the vast majority of genes were lost quite early in eukaryotic evolution, although parallel losses have undoubtedly occurred as well. Still, the number of genes retained in the mitochondrion varies widely among eukaryotes, from 97 protein and RNA genes in the jakobid *Reclinomonas americana* to only 3 protein genes and 2 fragmented sets of rRNA genes in *Plasmodium* and other apicomplexan parasites (Lang et al. 1999). In many species with degenerate mitochondria such as *Entamoeba histolytica*, the genome has been lost completely (e.g., Tovar et al. 1999).

J.F. Wendel et al. (eds.), *Plant Genome Diversity Volume 1*,
DOI 10.1007/978-3-7091-1130-7_9, © Springer-Verlag Wien 2012

The mitochondrial genomes of land plants, particularly flowering plants, are remarkable among eukaryotes. They contain the largest genomes of any organelle, up to several megabases in size in some species, with dramatic variations in size between closely related species (Ward et al. 1981; Alverson et al. 2010). Their structures are very fluid, having multiple genomic configurations in varying stoichiometries due to active recombination across repeats (Lonsdale et al. 1984; Palmer and Shields 1984; Maréchal and Brisson 2010). As a result, rearrangement rates are so high that gene order is rarely conserved among close relatives (Palmer and Herbon 1988). They have the propensity to integrate foreign DNA from various sources, including from the plastid and nuclear genomes via intracellular transfer and from other species via horizontal transfer (Stern and Lonsdale 1982; Schuster and Brennicke 1987; Vaughn et al. 1995). Their gene repertoires can vary considerably among species because the transfer of functional mitochondrial genes to the nucleus is still ongoing (Nugent and Palmer 1991; Covello and Gray 1992; Kobayashi et al. 1997; Adams et al. 2002a). They have nucleotide substitution rates that, for most species, are among the lowest known for any genome, yet some lineages have experienced dramatic rate accelerations that put them among the fastest evolving genomes from any organism (Wolfe et al. 1987; Mower et al. 2007; Sloan et al. 2009). They have numerous introns present within genes, some of which have been split and require *trans*-splicing to create contiguous transcripts (Chapdelaine and Bonen 1991; Wissinger et al. 1991). Their transcripts also undergo extensive modification from C-to-U and, in some lineages, U-to-C RNA editing (Giegé and Brennicke 1999; Picardi et al. 2010; Grewe et al. 2011). Together, these characteristics make the plant mitochondrial genome a fascinating topic of study for evolutionary research.

This review will highlight the early developments that led to the discovery of many of the plant mitochondrial features described above, discuss how the genomics revolution in plant mitochondria has advanced our understanding of the origin and evolution of these features, and stress how future genomic studies might focus on specific evolutionary questions to resolve several outstanding riddles.

9.2 The Plant Mitochondrial Genomics Revolution

The first mitochondrial genomes to be characterized, from yeasts and mammals, were hallmark achievements that revealed much information about the genomic structure and contents of mitochondria. The genomics revolution for plant mitochondria began shortly thereafter with the sequencing of fragments of the *Chlamydomonas reinhardtii* genome (Boer et al. 1985). However, even before these efforts it was clear that *Chlamydomonas* mtDNA was atypical relative to land plants, with a small size (<16 kb), linear structure, and gene-dense organization (Ryan et al. 1978). The first land plant mitochondrial genome to be completely sequenced was from the liverwort, *Marchantia polymorpha* (Oda et al. 1992). This genome sequence established some of the trends that were to be found in nearly all subsequently sequenced land plant mtDNAs: a large, circular-mapping structure with roughly 40 protein-coding genes (including many ribosomal protein genes), numerous introns, and an incomplete set of tRNAs. The first angiosperm to be sequenced, from the model plant *Arabidopsis thaliana* (Unseld et al. 1997), exhibited many additional characteristic features of land plant mtDNAs (specifically flowering plants) such as the presence of large, recombinationally active repeats, the incorporation of plastid DNA, *trans*-splicing of introns, and the existence of RNA editing, all of which were well known by this time. The field of plant mitochondrial genomics blossomed in the ensuing 20 years, with more than 60 genomes now available from evolutionarily diverse species of land plants and green algae. A summary of the diversity of structure and content among these sequenced genomes is given in Fig. 9.1.

Among land plants, mitochondrial sequencing efforts have focused largely on economically important crop species, including sugar beet (*Beta vulgaris*, Kubo et al. 2000), rice (*Oryza sativa*, Notsu et al. 2002), rapeseed (*Brassica napus*, Handa 2003), maize (*Zea mays*, Clifton et al. 2004), wheat (*Triticum aestivum*, Ogihara et al. 2005), tobacco (*Nicotiana tabacum*, Sugiyama et al. 2005), grape (*Vitis vinifera*, Goremykin et al. 2009), watermelon (*Citrullus lanatus*, Alverson et al. 2010), zucchini (*Cucurbita pepo*, Alverson et al. 2010), mung bean (*Vigna radiata*, Alverson et al. 2011), sorghum (*Sorghum bicolor*, unpublished), and papaya (*Carica papaya*, unpublished). Further genome sequencing efforts to understand genomic diversity within a crop species, with emphasis on understanding the molecular basis of cytoplasmic male sterility (CMS), have been pursued for sugar beet (Satoh et al. 2004), rice (Tian et al. 2006; Fujii et al. 2010), and maize (Allen et al. 2007). However, many recent sequencing projects are shifting focus to sample the phylogenetic diversity of land plants, including the gymnosperm *Cycas taitungensis* (Chaw et al. 2008), the lycophytes *Isoetes engelmannii* and *Selaginella moellendorffii* (Grewe et al. 2009; Hecht et al. 2011), the hornworts *Megaceros aenigmaticus* and *Phaeoceros laevis* (Li et al. 2009; Xue et al. 2010), the moss *Physcomitrella patens* (Terasawa et al. 2007), and the liverwort *Pleurozia purpurea* (Wang et al. 2009). The only major land plant group without a mitochondrial sequence is the fern lineage, although several genome projects are underway (JPM, unpublished results).

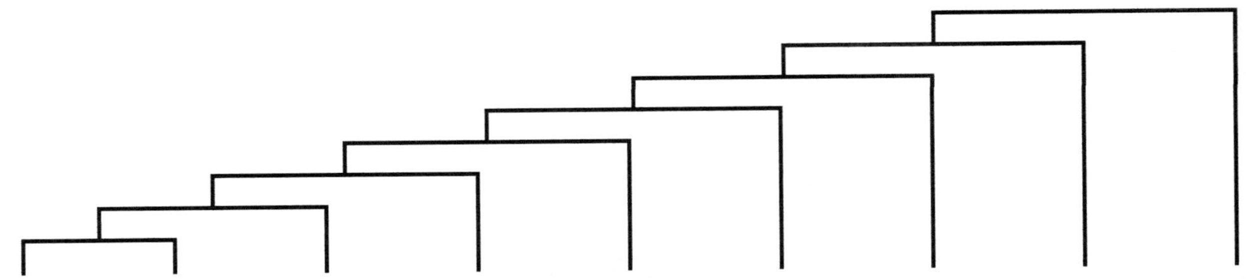

	Angiosperms	Gymnosperms	Lycophytes	Hornworts	Mosses	Liverworts	Charophytes	Chlorophytes	Rhodophytes
Genomes	30	1	1	2	1	2	4	21	3
Length (kb)	222-983	415	58+ [a]	185-209	105	169-187	42-202	13-96	26-37
GC Content (%)	43-45	47	49	45-46	41	42-45	32-41	22-57	27-33
Identified Genes	37-67	66	42	42-43 [b]	69	69-83 [b]	69-77	10-69	49-62
rRNAs	3	3	3	3	3	3	3	2-3	2-3
tRNAs	9-26	22	14	18-20	24	25-27	26-28	1-27	23-25
Proteins	24-38	40	24	20-21	40	41	36-39	7-36	23-34
Endonucleases	0	0	0	?	0	?-3	1-5	0-7	0
Maturases	1	1	1	?	2	?-9	0-6	0-1	0-3
Introns	19-25	26	30	30-34	27	31-32	7-27	0-18	0-2
Group I	0-1	0	3	0-1	3	7	4-14	0-18	0
Group II	19-25	26	27	30-33	24	24-25	2-14	0-2	0-2
Cis-spliced	13-20	21	29	30-34	27	31-32	5-27	0-18	0-2
Trans-spliced	5-6	5	1	0	0	0	0-2	0	0
RNA Edit Sites	189-835	1000+ [c]	1782	525-612 [c]	11	0-8+ [d]	0	0	0
Gene Order	rearranged	rearranged	rearranged	conserved	conserved	conserved	conserved/ rearranged	conserved/ rearranged	conserved

[a] Genome length for *Isoetes* does not include any repetitive content

[b] Gene counts for *Megaceros*, *Phaeoceros*, and *Pleurozia* do not include potential intron-encoded ORFs

[c] Edit site counts for *Cycas* and *Phaeoceros* are based on predicted editing positions

[d] Edit site count for *Pleurozia* is based on sites predicted to affect start and stop codons only

Fig. 9.1 Summary of mitochondrial genome content and structure across plants

Although not the focus of this review, the mitochondrial sequencing efforts in land plants are complemented by parallel efforts in related algal lineages. The streptophytic green algae (referred to throughout this review as "charophytes"), which constitute the closest relatives to land plants, have sequences available from four species: *Chaetosphaeridium globosum* (Turmel et al. 2002a), *Mesostigma viride* (Turmel et al. 2002b), *Chara vulgaris* (Turmel et al. 2003), and *Chlorokybus atmophyticus* (Turmel et al. 2007). The chlorophytic green algae have multiple representatives from all four subgroups: the Chlorophyceae, Trebouxiophyceae, Ulvophyceae, and Prasinophyceae (e.g., Turmel et al. 1999; Pombert et al. 2004, 2006; Robbens et al. 2007; Smith and Lee 2009; Smith et al. 2010a, b; Turmel et al. 2010). More distant relatives to the green plant lineage include the rhodophytes (red algae), with three complete mitochondrial sequences available (Leblanc et al. 1995; Ohta et al. 1998; Burger et al. 1999), and the glaucophytes, from which a single unpublished genome has recently been sequenced. The mitochondrial genomic diversity in these and more distant eukaryotic lineages has been the subject of several insightful reviews (Lang et al. 1999; Gray et al. 2004).

9.3 Plant Mitochondrial Genome Size

It has long been recognized that the mitochondrial genome is very large in plants. This was apparent from the first published restriction endonuclease digestions of purified plant mitochondrial DNA, though precisely how large was a source of considerable confusion at the time (reviewed in Leaver and Gray 1982). Other early studies established that genome size is also extremely variable among plants, from only 15.8 kb in the green alga *Chlamydomonas reinhardtii* (Ryan et al. 1978) to almost 3 Mb in muskmelon (Cucurbitaceae) (Ward et al. 1981). Evaluation of mitochondrial genome sizes of other cucurbits further demonstrated that sizes can vary dramatically, even among closely related species (Ward et al. 1981). A flurry of subsequent restriction-mapping studies measured genome size for a variety of plants (predominantly angiosperms), more fully documenting the wide range of sizes among both closely and distantly related plants.

With precise genome size information for >60 genomes today (Fig. 9.2), we have a much better understanding of the differences in size within and among major plant lineages. Within chlorophytes, *Polytomella capuana* (Chlamydomonadaceae) has a mitochondrial genome of only 13 kb in size, the smallest known from any green plant. This genome is only slightly smaller than the closely related *C. reinhardtii* genome, and it shares many of the same compact genomic features initially found in *C. reinhardtii* (Smith et al. 2010a).

At the opposite extreme, the mitochondrial genome of *Pseudendoclonium akinetum* is 96 kb, exhibiting a much greater proportion of intergenic and repetitive DNA (Pombert et al. 2004). Size variability of roughly twofold or more is also observed among more closely related species within each major chlorophyte subdivision.

Within streptophytes, genome size diversity is even greater (Fig. 9.2). The smallest genomes are generally found among charophytes, a paraphyletic group of green algae that includes *Mesostigma* (42 kb), *Chaetosphaeridium* (57 kb), and *Chara* (68 kb). *Chlorokybus* is also a member of this group, but it contains an unexpectedly large genome of 202 kb. Like *Pseudendoclonium*, the *Chlorokybus* mitochondrial genome exhibits much larger intergenic regions and more repetitive DNA than its closest algal relatives (Turmel et al. 2007).

Whereas this type of genome expansion appears to be the exception for green algae, it is the rule for land plants (Fig. 9.2). Complete sequences from the paraphyletic assemblage of non-vascular land plants, the bryophytes, range in size from 105–209 kb, with ratios of genic to intergenic DNA similar to that in *Chlorokybus* (Turmel et al. 2007). The vascular plants (tracheophytes) have taken this intergenic expansion even further. The genomes of nearly all species exceed 200 kb in size, less than 20% of which encodes for known protein or RNA products. The largest mitochondrial genome sequenced to date is from *Cucurbita pepo*, with known genes comprising only 7% of the 983 kb genome (Alverson et al. 2010). One glaring exception among tracheophytes is the sole lycophyte representative, *Isoetes engelmannii*, whose non-repetitive genome complexity is only 58 kb in length, the majority of which is composed of genes and introns (Grewe et al. 2009). Because this genome did not map circularly like other land plants, its genome size estimate does not account for repetitive DNA and thus remains a minimal estimate. Sequencing of additional lycophytes and ferns is necessary to clarify whether *Isoetes* is representative or atypical of species from these two groups.

9.4 Plant Mitochondrial Genetic Content

Much about the content of the plant mitochondrial genome was known well before the completion of the first genomic sequence. As in the yeast and mammalian genomes, plants were shown to contain large and small subunit rRNAs and numerous tRNAs (Bonen and Gray 1980). The first protein-coding gene identified, *cox2*, also revealed that plant mitochondria contain introns (Fox and Leaver 1981). But plants were also shown to contain mitochondrially encoded genes not found in yeast or humans, including a 5S rRNA

Fig. 9.2 Species with complete mitochondrial genome sequences. Genome sizes (in kb) are shown to the *right* of each taxon. The number of genomes available for a genus is given in *parentheses* (when more than one). Taxa in *bold* text have no sequenced species

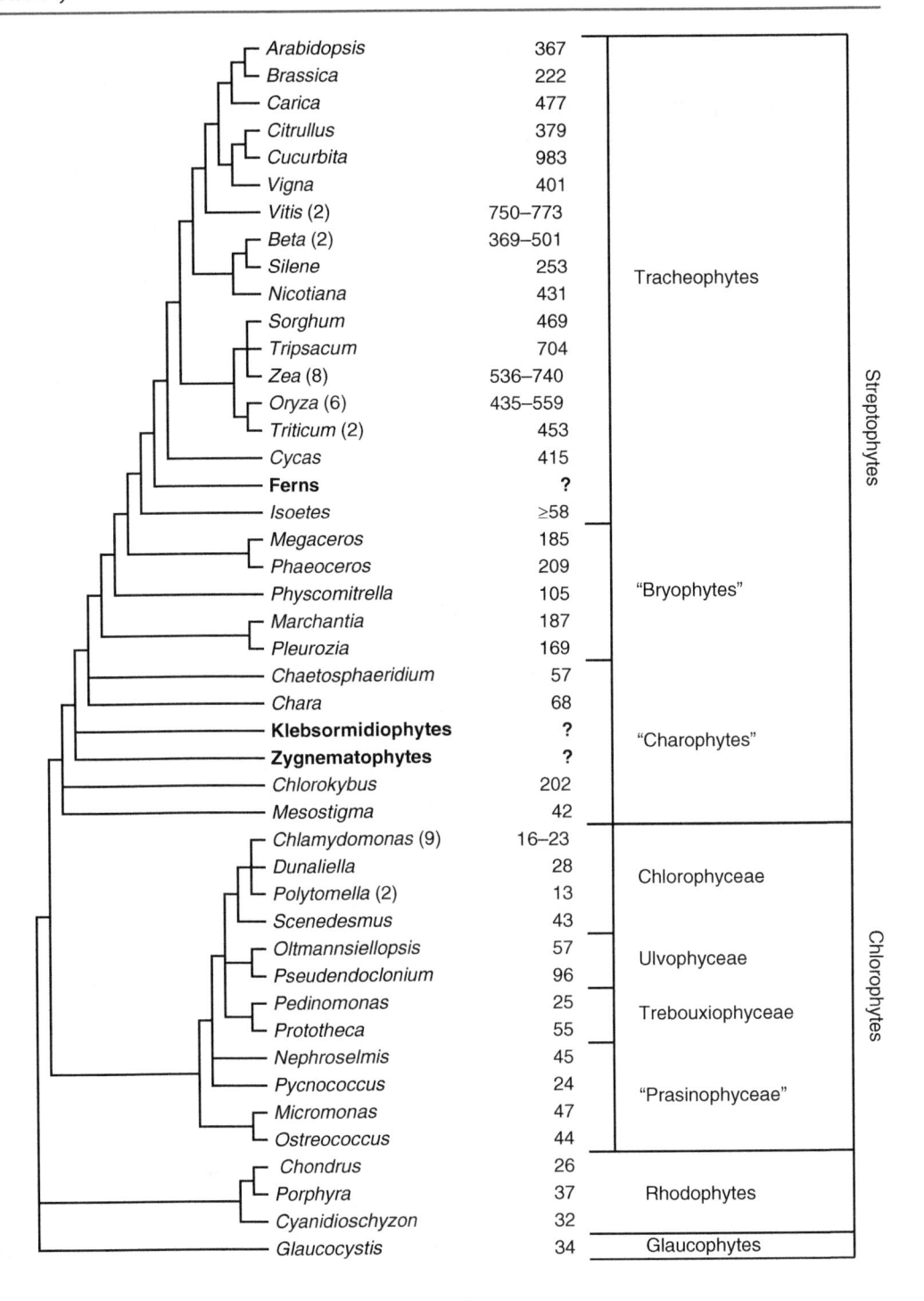

9.4.1 Protein Genes

9.4.1.1 NADH Dehydrogenase Subunits

and additional subunits of ATP synthase and the ribosome. Numerous subsequent reports proceeded to sequence and identify these genes, which also uncovered several more introns, some of which are fragmented onto separate transcripts and require *trans*-splicing for proper removal (Chapdelaine and Bonen 1991; Knoop et al. 1991; Wissinger et al. 1991). Today, the mitochondrial genome of plants is known to contain genes encoding products involved in electron transport, ATP synthesis, intron splicing, and the translation, maturation, and translocation of proteins.

NADH-ubiquinone oxidoreductase (complex I) is the NADH dehydrogenase complex of the electron transport system that oxidizes NADH generated by the TCA cycle, transfers electrons to ubiquinone, and actively pumps protons from the matrix across the inner membrane and into the intermembrane space. The complex was found to consist of at least 49 protein subunits in *Arabidopsis*

(Klodmann et al. 2010), of which nine are typically encoded by genes (*nad1*, *nad2*, *nad3*, *nad4*, *nad4L*, *nad5*, *nad6*, *nad7*, and *nad9*) in land plant mitochondrial genomes (Fig. 9.3). Of these genes, only *nad7* is variably present among land plants, with independent losses from the mitochondrial genomes of many liverworts and hornworts. These losses were likely accompanied by transfers of a functional copy to the nuclear genome, which has been confirmed for *Marchantia* (Kobayashi et al. 1997). Additional losses of these genes have been shown in several chlorophyte lineages, especially Chlorophyceae (Smith and Lee 2009; Smith et al. 2010a, b; Turmel et al. 2010). The *nad10* gene is found in the mitochondrion in some eukaryotes, including the charophyte

Chlorokybus and some prasinophytes (Turmel et al. 1999; Robbens et al. 2007; Turmel et al. 2007). However, it is absent from the mitochondria of all sequenced land plants and most green algae, suggesting several independent transfers to the nucleus during green plant evolution.

9.4.1.2 Succinate Dehydrogenase Subunits

Succinate dehydrogenase (complex II) is another electron transport complex that transfers electrons to ubiquinone; it also functions during the TCA cycle to oxidize succinate to fumarate. Eight protein subunits comprise the complex in *Arabidopsis* (Millar et al. 2004). Of these, *sdh3* and *sdh4* are the only genes found in green plant mitochondrial genomes

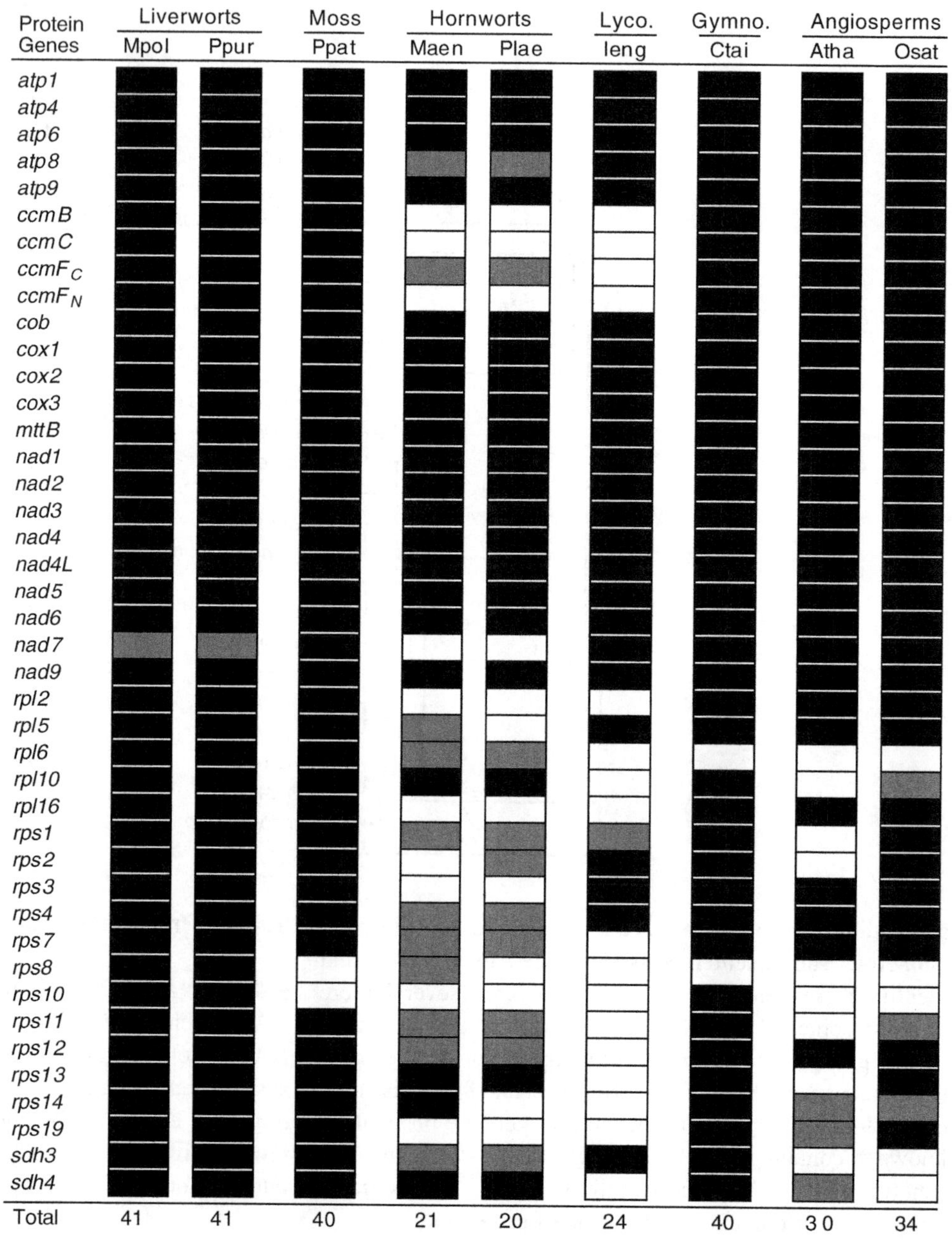

Fig. 9.3 Distribution of mitochondrial protein genes in sequenced land plants. Present (*black*), absent (*white*), pseudogene (*gray*)

(Fig. 9.3), and even these have been lost from the mitochondrial genome in a number of lineages. In many angiosperms, mitochondrial loss of *sdh3* and *sdh4* has been associated with functional transfer to the nucleus (Adams et al. 2002a), which may also be true for representatives of chlorophytes, hornworts, and lycophytes that have lost functional mitochondrial copies of one or both of these genes. Another complex II subunit, *sdh2*, is located in the mitochondria of all three sequenced red algae (Leblanc et al. 1995; Ohta et al. 1998; Burger et al. 1999), but its absence from all green plant sequences indicates an ancient transfer to the nucleus in the common ancestor of this group.

9.4.1.3 Cytochrome *bc*₁ Complex Subunits

The cytochrome bc_1 complex (complex III) also functions in the electron transport chain. Its role is to oxidize ubiquinone and then transfer the electrons to cytochrome *c*, and it is a second site of proton translocation across the inner membrane. It was shown to have ten protein subunits in potato (Braun et al. 1994). Apocytochrome *b* is the only component whose gene (*cob*) is present in the mitochondrial genome, and it is universally present in the mitochondrial genomes of all land plants (and green and red algae) sequenced to date (Fig. 9.3). Indeed, it is one of the few proteins encoded in the mitochondrial genomes of nearly all sequenced eukaryotes, with only rare exceptions in anaerobic organisms with degenerate mitochondria (Boxma et al. 2005; Stechmann et al. 2008).

9.4.1.4 Cytochrome *c* Oxidase Subunits

Cytochrome *c* oxidase (complex IV) accepts electrons from cytochrome *c* in order to reduce oxygen to water and also help shuttle protons into the intermembrane space. Of the 14 subunits of this complex identified in *Arabidopsis* (Millar et al. 2004), three are encoded by genes (*cox1*, *cox2*, and *cox3*) almost universally located in the mitochondrial genome of plants (Fig. 9.3). The one exception in streptophytes is *cox2*, which was transferred to the nuclear genome in some legumes (Nugent and Palmer 1991; Covello and Gray 1992; Alverson et al. 2011). Among chlorophytes, *cox2* and *cox3* have been lost from the highly reduced mitochondrial genomes of several species, including *Pedinomonas* and most Chlamydomonadaceae (Turmel et al. 1999; Smith and Lee 2009; Smith et al. 2010a, b).

9.4.1.5 ATP Synthase Subunits

ATP synthase (complex V) uses the proton gradient generated by the electron transport chain to catalyze the synthesis of ATP. In *Arabidopsis*, this complex is composed of 11 subunits (Heazlewood et al. 2003). Five subunits are encoded by mitochondrial genes (*atp1*, *atp4*, *atp6*, *atp8*, and *atp9*) for most land plants (Fig. 9.3). However, *atp8* was recently shown to be a pseudogene in the two sequenced

hornwort mitochondrial genomes, suggesting a functional copy resides instead in the nuclear genome. The *atp8* gene was also inferred to be lost from the mitochondrial genome of *Allium* (Adams et al. 2002a). In addition, the *atp4* gene has been lost from the lycophyte *Selaginella* (Hecht et al. 2011). Losses of one or more of these genes from the mitochondrial genome are more common in chlorophytes and rhodophytes; indeed, many Chlorophyceae have in fact lost all five genes (Smith and Lee 2009; Smith et al. 2010a, b).

9.4.1.6 Cytochrome *c* Maturation Proteins

Cytochrome *c* maturation proteins attach a heme prosthetic group to *c*-type cytochromes, which are essential electron transporters, but the complete pathway for cytochrome *c* maturation is not fully understood (reviewed in Giegé et al. 2008). In *Escherichia coli*, eight genes are involved (*ccmA-ccmH*). Homologs to six of these genes are found in *Arabidopsis*, of which the *ccmB*, *ccmC*, and *ccmF* homologs are located in the mitochondrion and the *ccmA*, *ccmE*, and *ccmH* homologs are nucleus-encoded. In land plants, the *ccmF* gene is split into two separate genes, *ccmF*ₙ and *ccmF*꜀, and in Brassicaceae (including *Arabidopsis*), *ccmF*ₙ is further subdivided into *ccmF*ₙ₁ and *ccmF*ₙ₂. Like in *Arabidopsis*, the homologs to *ccmB*, *ccmC*, and *ccmF* of other land plants are typically located in the mitochondrial genome, but in some lineages the entire suite of mitochondrial genes has been deleted (Fig. 9.3). The entire suite of genes is also absent from most algal mitochondrial genomes, except for the green alga *Chara vulgaris*, which retains homologs to *ccmB*, *ccmC*, and *ccmF* (Turmel et al. 2003), and the red alga *Cyanidioschyzon merolae*, which retains these three plus a *ccmA* homolog (Ohta et al. 1998). Whether these missing mitochondrial genes will be found in the nucleus is an open question. It is possible that many algae (and even some land plants) have adopted alternative strategies for heme attachment, such as those found in the plastid or in animal mitochondria (Giegé et al. 2008).

9.4.1.7 Ribosomal Proteins

By far the most diversity in gene content among plant mitochondrial genomes stems from the presence or absence of genes encoding protein components of the mitochondrial ribosomes (Fig. 9.3). Land plants typically contain some or all of five large ribosomal subunits (*rpl2*, *rpl5*, *rpl6*, *rpl10*, and *rpl16*) and 12 small ribosomal subunits (*rps1*, *rps2*, *rps3*, *rps4*, *rps7*, *rps8*, *rps10*, *rps11*, *rps12*, *rps13*, *rps14*, and *rps19*). Some groups, such as liverworts, retain the complete set of ribosomal protein genes, whereas other groups, like hornworts, retain only two or three functional genes. The most extensive loss documented to date at the genome level in angiosperms is from *Silene latifolia*, which retains only a single intact ribosomal protein gene (Sloan et al. 2010a), and the lycophyte *Selaginella* appears to have none (Hecht et al. 2011). Similar extremes of presence/absence are found across green and red algae. Several other subunits are present in algal lineages, such as *rpl14* in green algae and *rpl20* in red algae, but they have not been identified in any land plants.

Like the succinate dehydrogenase genes, ribosomal protein genes have been functionally transferred to the nuclear genome many times over the course of angiosperm evolution (Adams et al. 2002a). This is not always the case, however. Mitochondrial ribosomal proteins have occasionally been functionally replaced by homologous proteins derived from the plastid or cytosolic ribosomes (Adams et al. 2002b; Mower and Bonen 2009; Kubo and Arimura 2010).

9.4.1.8 A Putative Protein Transporter

Another gene present among all land plants sequenced to date is the somewhat enigmatic gene *mttB*, originally termed *orfX* (Knoop et al. 1991). Homologs (sometimes called *tatC* or *ymf16*) are also present in the mitochondrial genomes of many green and red algae. Given its widespread distribution, it clearly encodes for a functional product, which is homologous to a bacterial membrane transport protein that functions independently of the Sec translocation pathway (Bogsch et al. 1998). Although the role of this protein is uncertain, it presumably continues to function as a protein translocase, perhaps shuttling proteins from the matrix into the intermembrane space (Bogsch et al. 1998).

9.4.1.9 Intron-Encoded Proteins

All land plant mitochondrial genomes sequenced to date contain introns (discussed in Sect. 9.4.3) and some of these introns encode ORFs that are involved in the evolutionary spread and/or splicing of introns. Group I introns may contain homing endonucleases (Fig. 9.1), which play a role in the insertion of replicated introns into new genomic locations and in splicing. Three apparently functional endonucleases are present in the *Marchantia* mitochondrial genome (Oda et al. 1992), and probably in *Pleurozia* as well, given the high degree of intron similarity between the two liverworts (Wang et al. 2009). An endonuclease is sporadically present and potentially functional in many of the horizontally transferred *cox1* introns of various angiosperms (Sanchez-Puerta et al. 2008). All charophyte and many chlorophyte genomes also contain endonuclease genes (Pombert et al. 2006; Turmel et al. 2007; Smith and Lee 2009; Smith et al. 2010b), but few of these ORFs are orthologous to those from land plants. In contrast, no intact endonucleases have been identified in the group I introns of other land plant groups.

Group II introns may encode maturases (Fig. 9.1) that aid in intron splicing and, at least ancestrally, have reverse transcriptase activity to facilitate intron propagation. The most notable of these maturases is encoded by the *matR* gene located in *nad1* intron *nad1*i728, which is typically present in vascular plants but less common in bryophytes (Dombrovska and Qiu 2004). Among seed plants, this is the only functional maturase identified in the mitochondrial genome. Although the mitochondrial genome of the liverwort *Marchantia* does not contain *matR*, it does have nine other maturase genes, all of which are presumably functional (Oda et al. 1992). One of these is orthologous to the sole maturase gene in the lycophyte *Isoetes*, which is found in the intron *atp9*i87 (Grewe et al. 2011). *Physcomitrella* encodes two maturases that are unique to this genome (Terasawa et al. 2007). Maturase genes were not annotated in the mitochondrial genomes of the liverwort *Pleurozia* or the two hornworts, but BLAST searches indicate that they are present and probably functional (Wenhu Guo and JPM, unpublished results). Some green and red algae also have mitochondrial maturase genes, but none have orthologs in land plants.

9.4.1.10 Chimeric Genes

Chimeric ORFs, composed of fragments of one or more characterized mitochondrial genes, are present in the mitochondrial genomes of most, if not all, angiosperms. In some cases, these genes serve no apparent function and may simply be rearranged pseudogenes. In other cases, however, they can cause cytoplasmic male sterility, i.e., a failure to produce viable pollen in some hermaphrodites, leading to the formation of gynodioecious populations (reviewed in Hanson and Bentolila 2004). The first CMS gene to be characterized, *T-urf13* from a maize line with CMS-T type cytoplasm, is composed of a fragment of mitochondrial LSU rRNA gene as well as intergenic DNA flanking the LSU gene (Dewey et al. 1986). Many additional CMS genes have now been characterized from different crops. Most appear to have originated independently, and almost all share the common theme of chimerism involving known mitochondrial genes, often ATP synthase subunits (Hanson and Bentolila 2004).

Several studies have taken a comparative genomics approach to attempt to identify mitochondrial genes that cause CMS (Satoh et al. 2004; Allen et al. 2007; Fujii et al. 2010). In a comparison of sugar beet fertile and CMS lines (Satoh et al. 2004), several genes exhibited nucleotide or transcriptional differences between the lines, and several chimeric ORFs were found that were unique to either the fertile line or to the CMS line. Although most of these differences are likely due to simple evolutionary divergence between the lines, the line-specific ORFs were considered candidates for causing (or preventing) CMS, and further characterization indicated that one of these ORFs was a likely CMS gene in sugar beet (Yamamoto et al. 2005). In a comparison of two fertile and three CMS maize lines, the CMS-T and CMS-S genomes were shown to contain unique chimeric ORFs, which in fact were previously identified to be causative agents of CMS in these lines (Allen et al. 2007). However, for the genome of CMS-C, whose CMS gene has yet to be identified, no obvious candidates emerged from the comparative analysis. Similar comparative analysis of two rice CMS lines with published fertile lines uncovered unique chimeric ORFs (Fujii et al. 2010), but further characterization of these genes is necessary to determine if any of their products cause CMS.

Although chimeric genes are commonplace in angiosperm mitochondrial genomes, their presence and potential functionality in other plant groups has not been explored. However, pseudogene fragments of mitochondrial genes have been described in the lycophyte *Isoetes*, the hornwort *Megaceros*, and the liverwort *Pleurozia* (Grewe et al. 2009; Li et al. 2009; Wang et al. 2009), which suggests that they could constitute new chimeric ORFs. Whether any are expressed and evolutionarily important is not known.

9.4.1.11 Other ORFs

Given the large amount of intergenic DNA in plant mitochondrial genomes, it is not surprising to find an abundance of additional ORFs in each genome. These are often annotated but rarely show conservation among species. Hence, they are generally considered to be spurious predictions. However, some ORFs do show limited conservation among the more poorly sampled lineages, such as bryophytes (Li et al. 2009; Wang et al. 2009). Indeed, one of these ORFs of limited conservation, *orf-bryo1*, was recently identified as the *rpl10* gene present in many streptophytes (Mower and Bonen 2009; Kubo and Arimura 2010). Other ORFs bear resemblance to DNA or RNA polymerases derived from mitochondrial plasmids and viruses, to reverse transcriptases derived from insertions of nuclear retrotransposons, or to plastid genes present in integrated fragments of plastid DNA (see Sect. 9.5.2.1). Some unidentified ORFs are even transcribed (Giegé and Brennicke 1999; Notsu et al. 2002; Handa 2003; Satoh et al. 2004), but whether they encode functional proteins will require deeper genome sampling, especially from bryophytes and charophytes, and further characterization of their putative products.

9.4.2 RNA Genes

9.4.2.1 Ribosomal RNAs

In addition to the LSU and SSU rRNA genes universally present in eukaryotic mitochondrial genomes, nearly all land plants encode a 5S rRNA subunit (Fig. 9.4), although this gene was not found in the *Selaginella* genome (Hecht et al. 2011). All three genes have also been found in all charophytes and the sole glaucophyte sequenced to date, but the 5S subunit has been lost from the mitochondrial genomes of several groups of chlorophytes and rhodophytes. All three genes are clustered in the genomes of liverworts, mosses, glaucophytes, and most charophytes, suggesting this was the ancestral condition. In most other land plants, however, the LSU gene is separated from the other two, whereas in *Isoetes* and *Mesostigma*, LSU and SSU are clustered to the exclusion of 5S. In many of the reduced chlorophyte genomes such as *Chlamydomonas*, the LSU and/or SSU genes are separated into gene fragments scattered across the genome.

9.4.2.2 Transfer RNAs

Nearly all completely sequenced land plant mitochondrial genomes contain tRNA genes, but none contains a complete set (Fig. 9.4), and the mitochondrial genome of *Selaginella* is reported to have none at all (Hecht et al. 2011). To compensate for this deficiency, plant mitochondria import cytosolic tRNAs encoded in the nucleus (Maréchal-Drouard et al. 1988). This is also probably true for the many red and green algae that have incomplete suites of mitochondrion-encoded tRNAs. Some green algae, on the other hand, appear to have a full set of mitochondrial tRNAs (Turmel et al. 1999; Pombert et al. 2004; Robbens et al. 2007). Across land plants, tRNAs with the same anticodon are generally orthologous. Sequence homology in some *Isoetes* and *Cycas* tRNAs suggest that RNA editing is required to restore conserved anticodons (Fig. 9.4). Although the predicted anticodon edits in *Isoetes* and *Cycas* will require experimental confirmation, three of the *Isoetes* edits were already confirmed by direct tRNA sequencing, and in fact many other sites are also edited in *Isoetes* tRNAs (Grewe et al. 2011). Other mitochondrial tRNAs are actually derived from integrated fragments of plastid DNA (Maréchal-Drouard et al. 1990). This has been observed for all seed plant genomes sequenced to date (see Fig. 9.1 in Sloan et al. 2010a) but not reported for other plant groups (Li et al. 2009). Finally, a few tRNAs (*Beta*, *Citrullus*, and *Vigna trnC*-GCA; *Phaeoceros trnV*-UAC; *Isoetes trnS*-GCU; *Cycas trnN*-GUU) are not obviously orthologous to their counterparts in other land plants and, in some cases, show stronger similarity to bacterial tRNAs (Kubo et al. 2000; DBS, unpublished results). Whether these tRNAs have experienced substantial sequence divergence, have undergone codon reassignment, or have been acquired horizontally from a bacterial donor is currently unclear.

9.4.3 Introns

The mitochondrial genomes of land plants contain a large and variable number of introns (Fig. 9.5), from only 19 in the

Fig. 9.4 Distribution of mitochondrial RNA genes in sequenced land plants. Present (*black*), absent (*white*), pseudogene (*gray*). tRNAs whose anticodons are known ("E") or predicted ("e") to be altered by RNA editing are classified according to the edited form

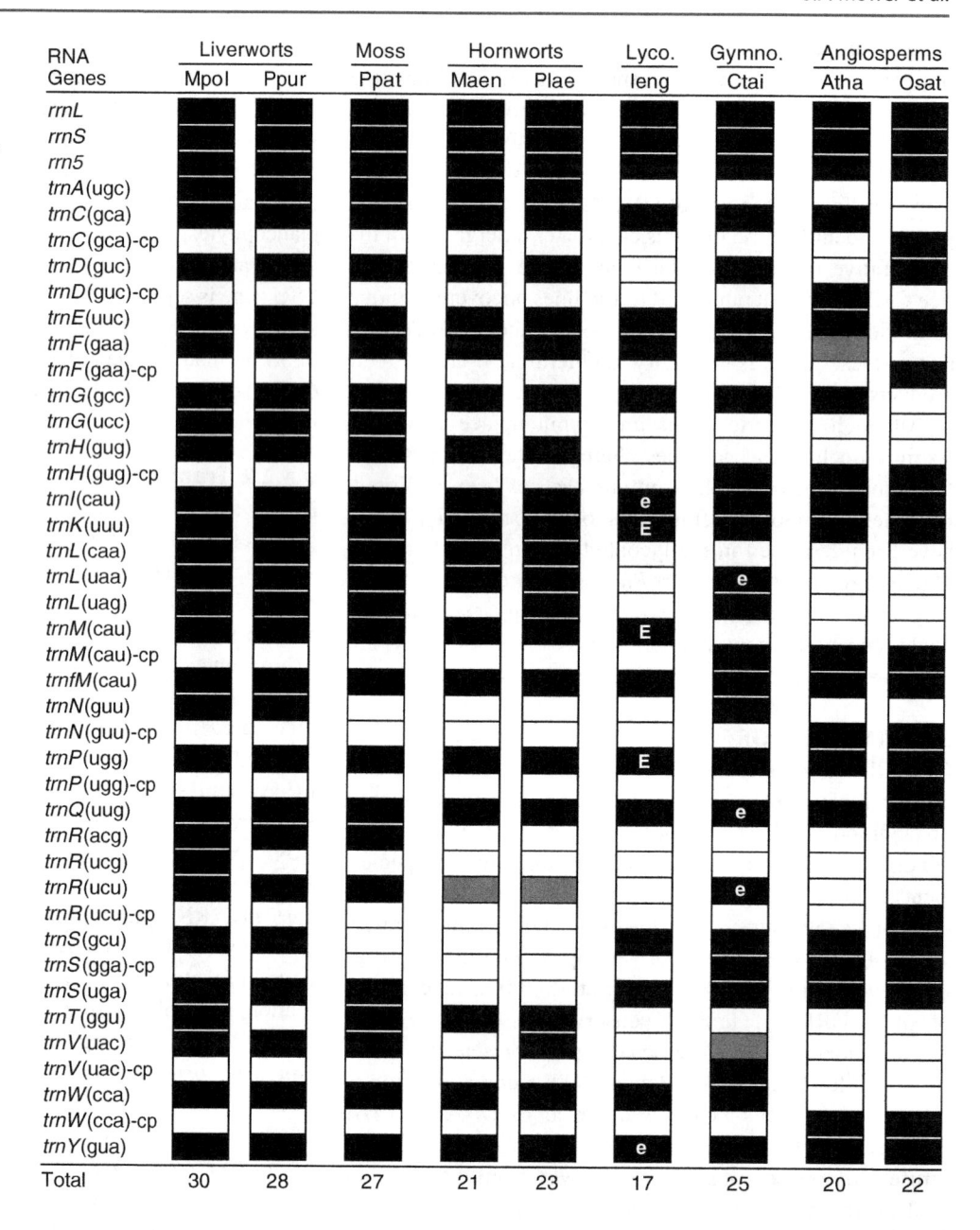

angiosperm *Silene latifolia* to 34 in the hornwort *Phaeoceros laevis* (Sloan et al. 2010a; Xue et al. 2010), although the recently published genome of the lycophyte *Selaginella* is again exceptional in that it contains 37 (Hecht et al. 2011). The introns are of two types, group I and group II, which differ in their secondary structures and splicing mechanisms. Group II introns predominate, although most lineages also have one to several group I introns. Intron content is relatively conserved within each of the major land plant lineages for which multiple genome sequences are available: 31 out of 32 introns are shared between the two liverworts, 30 of 34 between the two hornworts, and 25 of 26 between the gymnosperm *Cycas* and one or more angiosperms (including the *rps10* intron, which is also found in numerous angiosperms

but absent from those included in Fig. 9.5). In contrast, the set of introns shared among the different major lineages is much lower, such that not a single intron is shared by all lineages. In fact, each lineage contains a significant number of unique introns: 20 for liverworts, 6 for mosses, 12 for hornworts, 9 for lycophytes, and 6 for seed plants, indicating a high rate of both gain and loss of introns early in land plant evolution that subsequently stabilized in descendent lineages.

Among algal relatives of land plants, mitochondrial introns are highly variable in position and frequency. All four sequenced genomes from charophytes contain multiple group I and group II introns, a few of which are orthologous to liverwort or moss introns (Turmel et al. 2007). *Chara* contains the most with 27 (14 group I and 13 groups II), and *Mesostigma* contains the fewest with 7 (4 group I, 3 group II). For chlorophytes, most sequenced species have several group I introns, and *Dunaliella salina* has 18 (Smith et al. 2010b). Group II introns are generally absent from most chlorophytes, and species with compact genomes (such as *Polytomella, Pycnococcus,* and *Ostreococcus*) are often entirely devoid of any introns (Robbens et al. 2007; Smith et al. 2010a; Turmel et al. 2010). When present, chlorophyte introns are most often located in the *cob, cox1,* and rRNA genes, but further work is needed to determine if they are orthologous to land plant introns. Among red algae, group I introns are unknown and group II introns are generally rare, with no more than two introns present in any sequenced mitochondrial genome (Leblanc et al. 1995; Ohta et al. 1998; Burger et al. 1999). The glaucophyte *Glaucocystis* contains no mitochondrial introns.

In vascular plants, several different introns have evolved a split structure that requires *trans*-splicing, including the *cox1* intron from lycophytes and introns in the *nad1, nad2,* and *nad5* genes from seed plants (Fig. 9.5), as well as introns in *atp9, cob,* and *cox2* from *Selaginella* (Hecht et al. 2011). Although no fern mitochondrial genomes have yet been sequenced, it is likely that some of their introns will also require *trans*-splicing given the presence of this

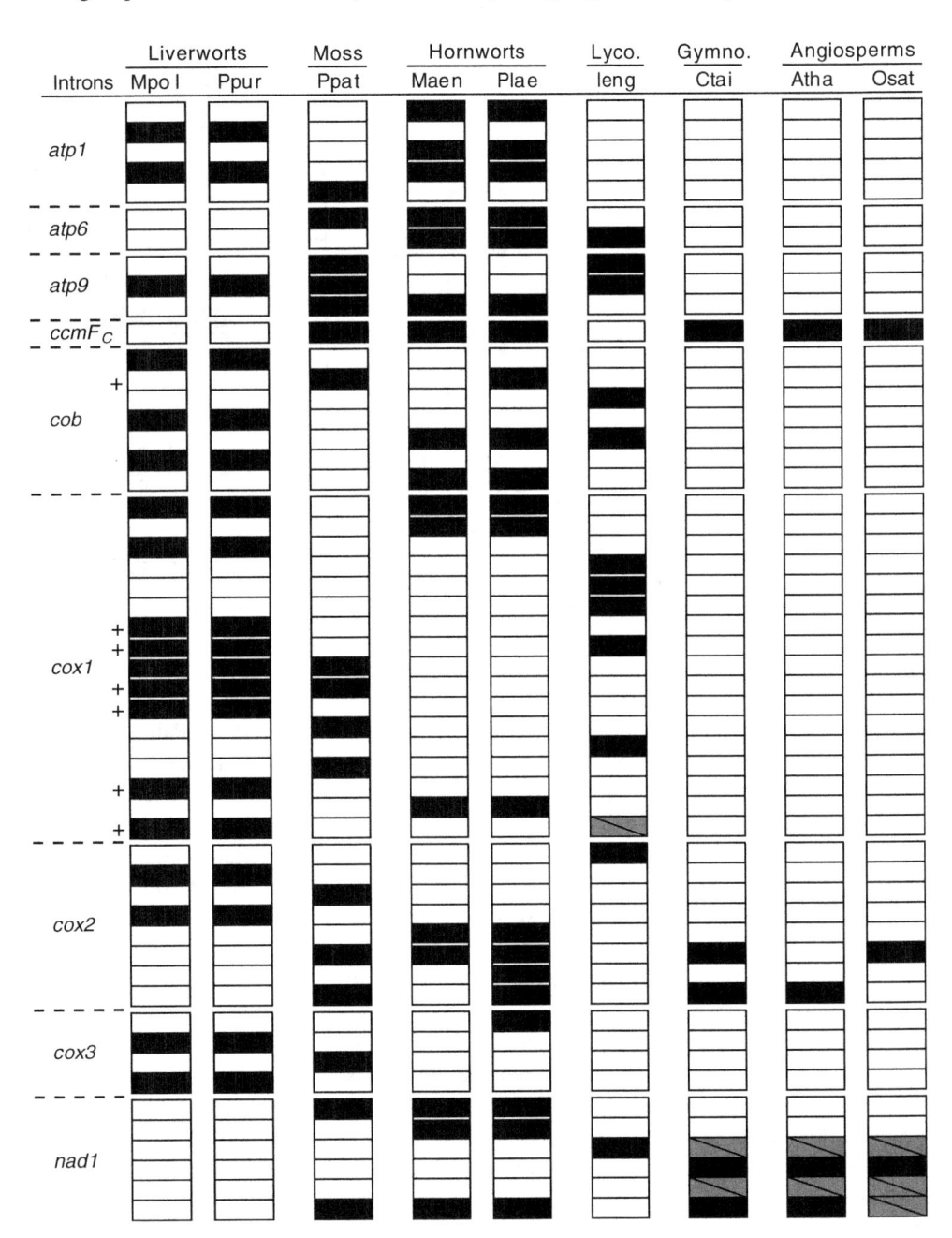

Fig. 9.5 (continued)

Fig. 9.5 Distribution of
mitochondrial introns in
sequenced land plants. *Cis*-pliced
(*black*), *trans*-pliced (*gray* with
crossatch), absent (*white*).
Group I introns are marked ("+");
the remaining introns are group
II. Homology assignment was
based on position in the gene and
by BLAST analysis

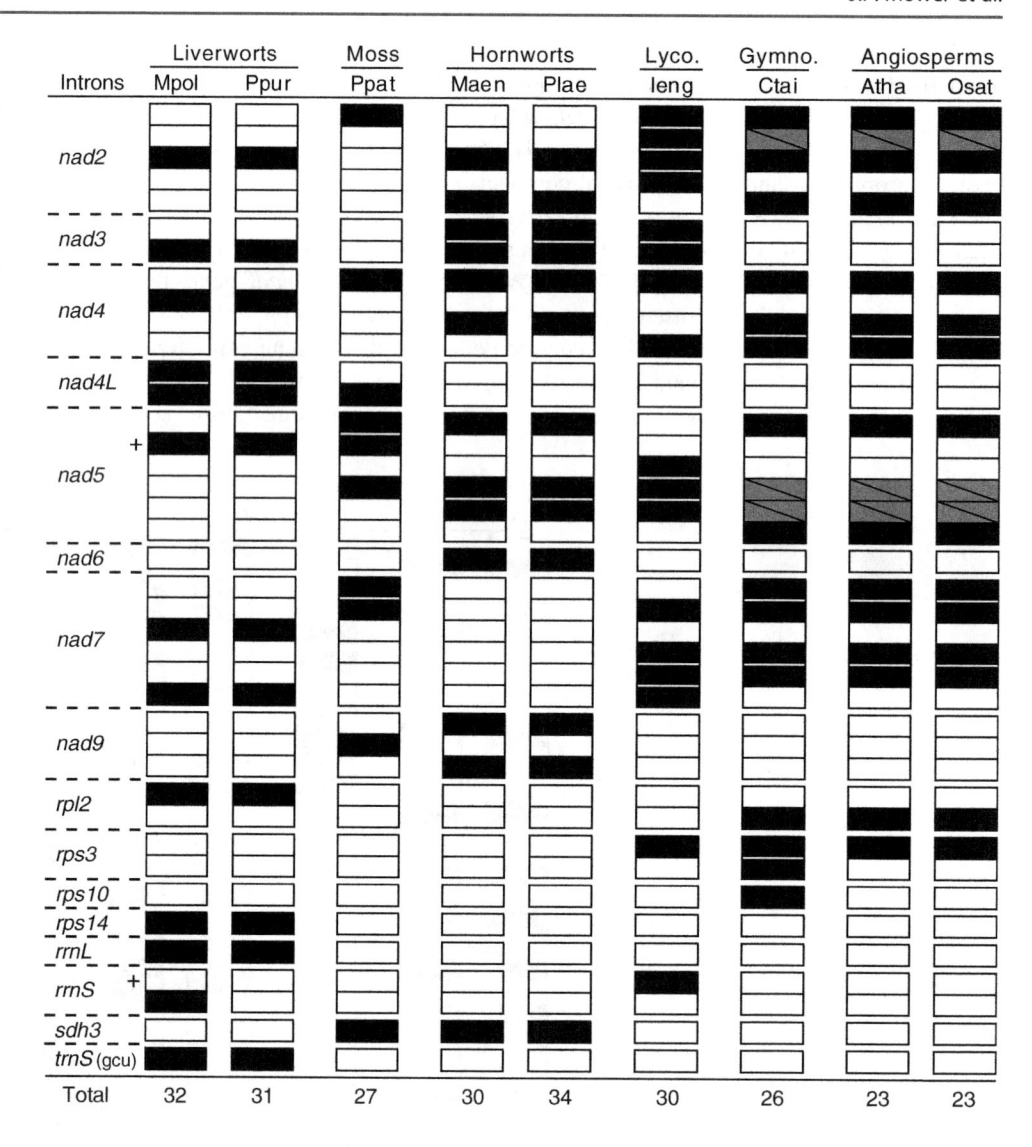

9.5 Plant Mitochondrial Intergenic Content

Although plant mitochondrial genomes differ substantially in gene and intron content, the vast majority of the variance in genome size can be explained by differences in the size of intergenic regions. The mitochondrial genomes of plants, particularly angiosperms, were already well known for the presence of sizeable fractions of repetitive and foreign DNA before any complete genomes were available. Repetitive DNA was shown to constitute 5–10% of the genomes from several different plants (Ward et al. 1981). Large repeats (>1 kb) were first mapped in *Brassica* and maize (Lonsdale

phenomenon in all other vascular plant groups. *Trans*-splicing has also evolved in two introns from the green alga *Mesostigma viride* (Turmel et al. 2002b), but has not been observed in other green or red algae.

et al. 1984; Palmer and Shields 1984), and short (<100 bp) to intermediately sized repeats (100–1,000 bp) were subsequently identified in a broad range of plants (reviewed in André et al. 1992). A surprising finding in angiosperm mitochondrial genomes was the presence of DNA originating from the plastid (Stern and Lonsdale 1982), followed by discoveries of retrotransposon sequences in mitochondria that were acquired from the nuclear genome (Schuster and Brennicke 1987; Knoop et al. 1996). Even more surprising was the identification of mitochondrial genes and introns from entirely different species gained via horizontal transfer (Vaughn et al. 1995; Bergthorsson et al. 2003). However, these additional contributions to the plant mitochondrial genome can only partially account for their massive sizes relative to non-plant genomes, much of which is species-specific. Exactly where all of the "extra" intergenic DNA came from and why it is so common are important unanswered questions.

9.5.1 Repeats

Repeats are sequences present in more than one copy in the genome. Although researchers have used different ways to identify repeats and set different length or percent identity thresholds in defining them, it is nevertheless clear that a large fraction of the overall content in fully sequenced seed plant mitochondrial genomes is repetitive. Like intergenic sequences, repeats are poorly conserved across species, with some genomes containing thousands of short dispersed repeats (e.g., *Cucurbita*) and others containing small numbers of large segmental duplications (e.g., the *Beta* CMS cytotype) (Alverson et al. 2011). The largest genome sequenced to date, from *Cucurbita*, also contains the largest amount of repetitive DNA, both in raw volume (371 kb) and in percentage of the total genome size (38%) (Alverson et al. 2010). Genome size is by no means a perfect indicator of repeat content, however. The *Vitis* mitochondrial genome has only 7% repetitive DNA despite a size of nearly 773 kb (Goremykin et al. 2009), and the moderately sized (401 kb) *Vigna* genome has fewer and smaller repeats than in all other fully sequenced plant mitochondrial genomes, including the much smaller genomes of *Brassica* (222 kb) and *Silene* (253 kb) (Alverson et al. 2011). Repeat content can also differ dramatically over short time frames. For example, the roughly 25% variation in genome size among five genetic lines of maize owes almost exclusively to differences in repeat content (Allen et al. 2007). The sole gymnosperm sequence, from *Cycas*, is noteworthy in that 15% of its genome is repetitive, and most of this coverage is by a short, apparently self-replicating "Bpu element" (Chaw et al. 2008).

Repeats are also usually present but less common in the mitochondrial genomes of other land plants and algae. Bryophyte repetitive DNA content ranges from only 1% in the moss *Physcomitrella* to nearly 7% in the liverwort *Marchantia* (Turmel et al. 2007). Repeat content in charophytes ranges from essentially none (0.1%) in the compact genomes of *Mesostigma* and *Chaetosphaeridium* to nearly 8% in the much larger *Chlorokybus* genome (Turmel et al. 2007). Among chlorophytes, short repeats are present but widely variable in frequency, ranging from almost none in some species, such as *Nephroselmis olivacea* and *C. reinhardtii*, to >50% in *Volvox carteri* (Pombert et al. 2006; Smith and Lee 2009). Larger repeats are not found in any chlorophyte except for an unusual 10 kb inverted repeat present in the prasinophyte *Ostreococcus tauri* (Robbens et al. 2007) and the terminal inverted telomeric repeats of the linear chromosomes from *Chlamydomonas* and *Polytomella* (Smith et al. 2010a). Repeats in rhodophyte mitochondrial DNA are rare to nonexistent, covering <1% of the currently sequenced genomes (Leblanc et al. 1995; Ohta et al. 1998; Burger et al. 1999).

9.5.2 Foreign DNA

9.5.2.1 Integrated Plastid DNA

Insertions of plastid DNA have been found in the mitochondrial genome of every fully sequenced vascular plant genome, but not from any red or green algal genomes. Among angiosperms, the amount of plastid DNA varies quite widely, from only 2–3 kb (1%) in *Arabidopsis*, *Silene* and *Vigna* up to 113 kb (12%) in *Cucurbita*, although the majority of species that have been examined contain 3–6%. The gymnosperm *Cycas* is also within this range, with 4.4% of its genome composed of plastid DNA (Chaw et al. 2008). In lycophytes, a single 1.2 kb segment of plastid DNA was identified in *Isoetes* (Grewe et al. 2009). No information is available from ferns, but given that all other vascular plants contain some level of plastid DNA in their mitochondria, it is likely that ferns will as well. In many cases, these plastid-to-mitochondrion transfers have resulted in the insertion of plastid genes into the mitochondrial genome, most of which clearly are nonfunctional. Evidence for functional transfer of a plastid gene into the mitochondrial genome exists only for tRNAs (Maréchal-Drouard et al. 1990). However, some nonfunctional plastid integrants can occasionally effect changes in the mitochondrial genome. Short segments of the mitochondrial *atp1* and SSU rRNA genes have been replaced by homologous segments of plastid genes, apparently through gene conversion of the mitochondrial genes by their plastid homologs that have been integrated into the mitochondrial genome (Hao and Palmer 2009; Sloan et al. 2010a).

9.5.2.2 Integrated Nuclear DNA

In addition to integrating sequences from the plastid genome, plant mitochondrial genomes also contain varying amounts of sequences derived from the nuclear genome. Poor conservation (i.e., rapid turnover) of intergenic mitochondrial sequences across species combined with the still relatively few fully sequenced plant nuclear genomes makes the detection of nuclear transfers quite difficult. As a result, nucleus-derived sequences are almost certainly underreported. In addition, because of the large amounts of non-coding and seemingly featureless DNA in both the mitochondrial and nuclear genomes of plants, it is also difficult to polarize the direction of sequence transfer (nucleus-to-mitochondrion or vice-versa) of DNA shared between the two genomes (Notsu et al. 2002). Most identifiably nucleus-derived sequences therefore carry an unambiguous nuclear signature, such as similarity to a nuclear transposable element or gene.

Indeed, nuclear transposable elements have been reported from virtually all sequenced angiosperm mitochondrial genomes, dating back to the first sequenced *Arabidopsis* genome, which contains >5% coverage by fragments of *copia-, gypsy-*, and LINE-like retrotransposons (Knoop et al. 1996). Retrotransposon-like sequences have been reported from virtually all other eudicot mitochondrial genomes including *Brassica* (<1%, Handa 2003), *Vigna* (1.6%, Alverson et al. 2011), *Citrullus* (5.5%, Alverson et al. 2010), *Cucurbita* (1.8%, Alverson et al. 2010), *Nicotiana* (Sugiyama et al. 2005), and *Silene* (Sloan et al. 2010a). Nuclear insertions comprise some 18% of the unique regions distinguishing the mitochondrial genomes of fertile and CMS cytotypes of *Beta* (Satoh et al. 2004). Among monocots, few if any transposable-element-like sequences were found in the mitochondrial genomes of maize (Clifton et al. 2004; Allen et al. 2007) and wheat (0.2%, Ogihara et al. 2005), whereas as much as 13% of the rice mitochondrial genome has similarity with the nuclear genome (Notsu et al. 2002). In addition to transposable elements, several angiosperm mitochondrial genomes contain nuclear pseudogenes as well (Satoh et al. 2006; Alverson et al. 2010, 2011).

Outside of angiosperms, the gymnosperm *Cycas* and the lycophyte *Isoetes* are the only other green plants with identifiably nucleus-derived DNA (Grewe et al. 2009; AJA, unpublished results). Finally, all three *Volvox* genomes contain the same short and apparently self-replicating repetitive element, though which genome the element originated in and the pattern of colonization of the other genetic compartments is unclear (Smith and Lee 2009).

9.5.2.3 Integrated DNA From Other Species

One of the more recent—and contentious—findings in the plant mitochondrial field is the discovery that their genomes can integrate genetic material from entirely different species through the process of horizontal transfer. Originally reported in 1995 after phylogenetic examination of the unusual *cox1* group I intron from *Peperomia* (Vaughn et al. 1995), phylogenetic analyses have subsequently uncovered numerous additional cases of horizontal transfer between many different plants (reviewed in Richardson and Palmer 2007). In many instances, the transferred material appears nonfunctional and thus provides no obvious adaptive value. Occasionally, however, this process can add to genetic diversity by replacing genes previously lost from the mitochondrial genome or by generating chimeric genes via gene conversion between native and foreign copies (Bergthorsson et al. 2003; Hao et al. 2010; Mower et al. 2010).

Few of the fully sequenced plant mitochondrial genomes provide clear evidence of horizontal transfer, though it should be pointed out that unambiguous examples may lie undiscovered due to the generally very low substitution rate in plant mitochondria and the highly biased sampling of primarily crop plants. The best example comes from *Citrullus*, the only fully sequenced flowering plant to contain the horizontally acquired *cox1* intron (Alverson et al. 2010).

This intron is absent from the mitochondrial genomes of the other 29 sequenced flowering plants (and most other land plants and green algae as well), underscoring its patchy distribution. This disjunct distribution, plus strong phylogenetic incongruence of the intron sequences compared to known organismal phylogeny, provides much of the evidence that the intron was horizontally transmitted from a (most likely) fungal donor and subsequently transmitted horizontally (and vertically) among angiosperms (Vaughn et al. 1995; Sanchez-Puerta et al. 2008). This interpretation is not universally accepted, however (Cusimano et al. 2008).

Although genome sequences are not yet available, comprehensive PCR-based surveys in two species, *Amborella trichopoda* and *Plantago coronopus*, have uncovered widespread horizontal transfer of multiple mitochondrial genes acquired from one or more plant donors (Bergthorsson et al. 2004; Mower et al. 2010). This PCR-based approach is not without its problems because of the possibility of amplifying a paralogous, nucleus-encoded copy of a mitochondrial gene that may masquerade as a horizontally derived sequence in phylogenetic analysis (Goremykin et al. 2009). Complete mitochondrial genome sequencing for both *Amborella* and *Plantago* are currently underway (Richardson and Palmer 2007; Mower et al. 2010), which will provide greater power to differentiate between horizontal transfer and alternative hypotheses.

9.5.3   The Rest of the Genome...

Based on the current set of sequenced mitochondrial genomes, it appears that the expansion of intergenic regions began in the common ancestor of land plants. In bryophytes, intergenic DNA constitutes a substantial portion (>100 kb in some cases) of the total sequence in their moderately sized (105–209 kb) genomes (Fig. 9.2). Although the identity of bryophyte intergenic sequences is not well characterized, the mitochondrial genome of the lycophyte *Isoetes* contains chloroplast- and nucleus-derived pseudogenes, both of which are hallmarks of the greatly expanded (Fig. 9.2) seed plant mitochondrial genomes. Seed plant mitochondrial genomes contain other kinds of foreign sequence. For example, linear mitochondrial plasmids (typically <10 kb in length) have been integrated, in whole or in part, into the main mitochondrial chromosome of several angiosperm species (reviewed in McDermott et al. 2008). The mitochondrial genomes of *Vitis* and *Vigna* contain small amounts of viral DNA (Goremykin et al. 2009; Alverson et al. 2011). Still, all of these features together comprise a relatively small fraction of the intergenic DNA in most seed plant mitochondrial genomes. As a result, a common and sometimes frustrating theme in plant mitochondrial genomics is that most of the intergenic DNA—which exceeds hundreds of kilobases of sequence in most cases—is seemingly featureless and with no apparent similarity to previously sequenced DNA of any kind. Anonymous sequences comprise >80% of the DNA in the >500 kb mitochondrial genome of maize, for example (Clifton et al. 2004). Moreover, intergenic sequences are generally poorly conserved across species. Illustrative of this, in the Brassicales 79 kb of the 222 kb *Brassica* mitochondrial genome had no similarity to sequences in the mitochondrial genome of *Arabidopsis* (Handa 2003). The high rate of non-coding sequence turnover is also evident in the mitochondrial genomes of two different cytoplasmic lines of the same species, *Beta vulgaris* (sugar beet). The mitochondrial genome of the CMS cytoplasmic line contains >68 kb of sequence (14% of the genome) not found in the fertile cytoplasmic line (Satoh et al. 2006). Although some of these sequences could be traced to mitochondrial plasmids and recent chloroplast- or nucleus-derived insertions, 70% of the unique sequence was unidentifiable (Satoh et al. 2006). These unidentified regions were rich in repetitive DNA, supporting the hypothesis that many of the unique sequences in plant mitochondrial genomes are generated intrinsically through a process of recombinationally driven sequence duplication and continuous reshuffling (Bendich 1985; Lilly and Havey 2001; Satoh et al. 2006).

A major outstanding challenge in plant mitochondrial genomics is to precisely identify the intrinsic and extrinsic sources of the vast amounts of their unique and seemingly anonymous sequences. The current exponential growth of mitochondrial DNA sequences should provide significant insight into the source of currently unidentifiable sequences.

9.6 Plant Mitochondrial Genome Structure

The conventional depiction of genome "maps" often betrays the complex reality of genomic structure and organization. This is particularly true for plant mitochondrial genomes, in which recombinational processes often result in a fluid and heterogeneous population of genomes within a single individual. Mitochondrial genomes also differ greatly in structure and complexity within and among the major lineages of green algae and land plants. The sequenced mitochondrial genomes of green algae generally map as circular molecules, although the short and sometimes fragmented linear genomes of multiple species within the Chlamydomonadales stand out as an exceptions in this group (Smith et al. 2010a). Similarly, all five sequenced bryophyte mitochondrial genomes have circular topologies, and the availability of two complete genome sequences within both liverworts and hornworts has illustrated a high degree of structural and syntenic conservation in each of these groups (Oda et al. 1992; Li et al. 2009; Wang et al. 2009; Xue et al. 2010). In contrast, the evolution of mitochondrial genomes

in vascular plants is characterized by active recombination and rearrangement associated with repeated sequences. This section reviews the consequences of these recombinational processes on the evolution of mitochondrial genome structure in vascular plants, placing particular emphasis on angiosperms.

9.6.1 Inter- and Intraspecific Divergence in Genome Structure

Angiosperm mitochondrial genomes generally contain a diverse set of repeats, ranging from large (>1 kb) repeats that mediate high-frequency homologous recombination to short (<100 nt) repeats that recombine rarely (reviewed in André et al. 1992; Maréchal and Brisson 2010). Although a handful of cistrons (e.g., *rpl5–rps14* and *rps3–rpl16*) have been conserved from their much larger progenitor eubacterial operons, recombination has led to widespread scrambling in gene order among and even within species (Satoh et al. 2006; Allen et al. 2007). For example, a series of sequence insertions and deletions and at least four inversions associated with short (9–376 nt) repeats distinguish the mitochondrial genomes from fertile and CMS lines of *Beta* (Satoh et al. 2006). A more elaborate series of both inversions and tandem sequence duplications followed by random deletion of duplicates is necessary to reconcile the five maize mitochondrial genomes (Darracq et al. 2010). Likewise, experimental observations show that differences in the structure and sequence content of mitochondrial genomes of three *Arabidopsis* ecotypes reflect substoichiometric shifting; that is, the three different genomic configurations are thought to co-occur in each of the three ecotypes, with dramatic changes in relative frequency having led to a different predominant form in each one (Arrieta-Montiel et al. 2009).

9.6.2 Large Repeats and Multipartite Mitochondrial Genomes

In addition to driving mitochondrial genome divergence within and between species, recombinational processes are also responsible for generating structural variation at the individual level. Long before the first complete angiosperm mitochondrial genome sequences were available, restriction mapping studies established that these genomes often co-exist in alternative forms that are capable of interconversion by recombination across large (>1 kb) repeated sequences (Lonsdale et al. 1984; Palmer and Shields 1984). In the extreme, recombination is frequent enough to maintain alternative genome conformations at essentially equal frequencies. The existence of large recombining repeats has proven to be common (but not universal, see Alverson et al. 2010) in angiosperm mitochondrial genomes, with the complexity of genome organization increasing with the number of recombinationally active repeats (Sloan et al. 2010a and references therein).

The limited data available outside angiosperms suggest that frequent intragenomic recombination might be an equally common characteristic of mitochondria in other vascular plants. Complete sequencing of the mitochondrial genome from the lycophyte *Isoetes* detected numerous recombination breakpoints, precluding the identification of any single genome sequence (circular or otherwise) (Grewe et al. 2009). The only available gymnosperm mitochondrial genome sequence contains multiple pairs of repeats >1 kb in length, but these have not yet been assessed for recombinational activity (Chaw et al. 2008).

An important uncertainty in the field of plant mitochondrial genomics involves the question of how the results from mapping and genome assembly projects relate to the in vivo molecular structure of the genome. Attempts to directly observe angiosperm mitochondrial genome structure have generally failed to recover a large population of genome-sized circular molecules, instead finding a complex assemblage of subgenomic and multimeric linear and circular molecules as well as highly branched and sigma-like structures (Oldenburg and Bendich 1996).

9.6.3 Low Frequency Rearrangements and Substoichiometric Shifting

In contrast to their larger counterparts, small repeat sequences of a few 100 nt or less exhibit only rare recombinational activity in angiosperm mitochondrial genomes (Arrieta-Montiel et al. 2009). However, this rare activity has important genomic consequences, generating alternative, low frequency genome conformations ("sublimons"), which undergo occasional, rapid changes in frequency in a process known as substoichiometric shifting (reviewed in Maréchal and Brisson 2010).

Several families of nucleus-encoded proteins have now been shown to control mitochondrial recombination, and a common theme from this growing body of work is that these proteins stabilize the genome by actively suppressing "runaway" recombination across short (<1 kb) repeats (reviewed in Maréchal and Brisson 2010). Several of these genes have clear eubacterial homologs and so likely trace their origins to endosymbiotic gene transfers from early mitochondrial and plastid genomes. The *Arabidopsis* nuclear genome contains at least three organelle-targeted *RecA* homologs that participate in double-strand break repair (Odahara et al. 2007) and suppression of recombination across short repeats (Shedge et al. 2007; Odahara et al. 2009). Likewise, disruption of a

bacterial *MutS* homolog (MSH1) leads to active recombination across virtually all mitochondrial repeats >100 nt in length in *Arabidopsis* (Shedge et al. 2007; Arrieta-Montiel et al. 2009). The asymmetric accumulation of recombination products suggests an important role for MSH1 in the process of substoichiometric shifting of mitochondrial genotypes (Arrieta-Montiel et al. 2009). Disruption of a different, plant-specific DNA binding protein (OSB1) has also been reported to lead to substoichiometric shifting as a consequence of increased recombination across short mitochondrial repeats in *Arabidopsis* (Zaegel et al. 2006).

9.7 Plant Mitochondrial RNA Editing

Early DNA sequence data from angiosperm mitochondria revealed unique patterns of divergence relative to other eukaryotes. For example, observations of frequent replacements of CGG (Arg) codons for highly conserved TGG (Trp) codons in protein genes led to the hypothesis that plant mitochondria may utilize an alternate genetic code (Fox and Leaver 1981). Motivated by these observations, three labs independently discovered that plant mitochondria instead use a novel mechanism to modify their RNA by converting cytidines to uridines (C-to-U editing), restoring the conserved protein sequence (recounted by Gray 2009). Subsequent research showed that some plant mitochondrial transcripts also experience so-called "reverse" U-to-C editing and that similar editing also occurs in plastids. Further studies documenting patterns of mitochondrial RNA editing across diverse plant lineages have found a variable and often high frequency of editing, making plant mitochondrial genomes one of the primary models for investigating the evolution and maintenance of RNA processing mechanisms.

9.7.1 Mechanisms and Functional Importance of RNA Editing

Shortly after the discovery of RNA editing in plant mitochondria, it was determined that the C-to-U conversion is achieved by direct chemical modification (deamination) as opposed to complete excision and replacement (Yu and Schuster 1995). Although the deaminase that mediates this process has not been identified, it has become clear in recent years that the pentatricopeptide repeat (PPR) family of RNA binding proteins includes numerous members that are essential for determining editing site specificity (reviewed in Schmitz-Linneweber and Small 2008). The apparent function of RNA editing is to "correct" the RNA sequence. In protein genes, this generally means restoring codons for conserved amino acids and generating start/stop codons (or often removing internal stop codons in the case of U-to-C editing). In non-coding sequences, editing has been shown to restore conserved base-pairings required for the tRNA folding and intron splicing (Castandet et al. 2010; Grewe et al. 2011).

9.7.2 Variation in the Frequency of RNA Editing

The frequency of RNA editing in plant mitochondrial genomes varies substantially both within and among the major plant lineages. RNA editing has not been found in the mitochondria of green algae, but it occurs in all land plants examined to date with the exception of the marchantiid liverworts (Steinhauser et al. 1999). Because mitochondrial RNA editing occurs in other liverwort lineages, it is believed that it originated in a common ancestor of land plants and was secondarily lost in the marchantiids.

The most extensive study of mitochondrial RNA editing has been conducted in angiosperms with genome wide surveys now available for eight genera: *Arabidopsis* (Giegé and Brennicke 1999), *Oryza* (Notsu et al. 2002), *Brassica* (Handa 2003), *Beta* (Mower and Palmer 2006), *Vitis* (Picardi et al. 2010), *Silene* (Sloan et al. 2010b), *Citrullus* and *Cucurbita* (Alverson et al. 2010). Unpublished surveys have also been conducted in *Nicotiana*, *Liriodendron*, and *Amborella* (Mower 2009). These studies have identified between 189 (*Silene noctiflora*) and 835 (*Amborella trichopoda*) C-to-U editing sites per species, but these values are likely underestimates given the difficulty in identifying sites that experience only partial or tissue-specific editing (Mower and Palmer 2006; Picardi et al. 2010). The vast majority of angiosperm editing sites has been found in the exons of protein genes, affecting roughly 1–2% of all coding sequence. Although mitochondrial editing sites have also been identified in tRNA genes, introns, and UTRs in angiosperms (Giegé and Brennicke 1999; Castandet et al. 2010; Picardi et al. 2010), the full extent of editing in non-coding regions is not known because many studies have restricted their focus solely to protein genes.

Data are more limited outside of the angiosperms, but it is clear that there are substantial differences in the frequency of mitochondrial RNA editing in other plant lineages. In addition to the marchantiid liverworts, which appear to have lost editing altogether, the moss *Physcomitrella patens* experiences minimal mitochondrial editing with only 11 sites identified in the entire genome (Rüdinger et al. 2009). At the other extreme, the complete mitochondrial genome of the lycophyte *Isoetes engelmannii* exhibits a much higher rate of editing than angiosperms with a total of 1,782 sites identified in coding sequences alone (Grewe et al. 2011), and the lycophyte *Selaginella moellendorffii* is even more

impressive with over 2,100 sites (Hecht et al. 2011). There is also a high rate of editing in the non-coding sequence of the *Isoetes* mitochondrial genome, including tRNA genes and introns. Genome-wide quantifications of RNA editing frequency are not available in other plant lineages, but data from individual gene sequences and predictive algorithms indicate that rates of editing in hornworts, ferns and gymnosperms also exceed the levels observed in most angiosperms (Malek et al. 1996; Vangerow et al. 1999; Duff 2006; Salmans et al. 2010; Xue et al. 2010).

Shortly after the discovery of C-to-U RNA editing in angiosperm mitochondria, examples of U-to-C editing were also identified, indicating the co-existence of distinct biochemical pathways involved in the modification of plant mitochondrial RNA sequence. However, whole-genome analyses have indicated that such "reverse" editing is quite rare in seed plant mitochondria, particularly in coding sequence (Picardi et al. 2010; Salmans et al. 2010). Likewise, there has been little evidence of U-to-C editing in liverworts and mosses (Malek et al. 1996). In contrast, U-to-C changes appear to constitute a significant fraction of editing events in lycophytes (e.g., 12% of editing sites in *Isoetes*), hornworts, and ferns (Vangerow et al. 1999; Duff 2006; Grewe et al. 2011).

An outstanding challenge in the field of plant mitochondrial genomics is to distinguish between the suite of competing adaptive and non-adaptive hypotheses (reviewed in Sloan et al. 2010b) regarding the origin and maintenance of RNA editing and the variation observed across land plants. Genome-wide surveys have begun to build on earlier studies in pointing to specific mechanisms that affect the evolution of RNA editing, including the rate of sequence evolution and the frequency of retroprocessing events, in which spliced and edited mRNA sequence is incorporated back into the genome via a reverse transcribed intermediate (Sloan et al. 2010b; Grewe et al. 2011). In some cases, the evolution of RNA editing may interact with other dimensions of genome architecture. For example, it has been proposed that the loss of RNA editing in marchantiid liverworts may have been caused by an increased rate of retroprocessing resulting from the abnormally large number of intron-encoded ORFs with reverse transcriptase activity in this lineage (Grewe et al. 2011).

9.8 Future Prospects

The past 5 years have seen enormous growth in the number of complete genome sequences for plant mitochondria. Considering that sequencing began in the 1980s, it could be argued that the plant mitochondrial genomics revolution has been slow in coming, especially in comparison to the thousands of animal mitochondrial genomes and hundreds of plastid genomes now available in the sequence databases. Of course, this lag is not surprising given the incredible diversity in size, structure, and content of plant mitochondrial genomes relative to their animal mitochondrial and plastid counterparts. In fact, because of the additional complexities that must be dealt with in sequencing, assembly, and downstream analysis of plant mitochondrial genomes, it could be argued that we are much further along than might otherwise be expected.

This genomics revolution has been instrumental in delimiting the origins and evolutionary diversity for many interesting and complex genomic features of plant mitochondria, most of which were discovered prior to any genome sequencing. For instance, there is now a clear, general pattern of expanding genome size from green algae to bryophytes to seed plants, but there is also a wide range of sizes within each of these diverse groups. In addition, we have a firm grasp not only of the full complement of genes likely to have been encoded in the ancestral green plant mitochondrion, but also the subset of genes that possibly are transferred to the nucleus in any particular lineage. The origins and variable extents for many of the other complexities, such as genomic rearrangements, intron *trans*-splicing, and RNA editing, can also be mapped confidently onto the plant tree of life.

Despite this progress, further genomic sequencing is necessary to fully delineate the evolutionary diversity across plants. Our understanding is heavily biased towards angiosperms, which is not surprising given the number of economically important species from this lineage. Few genomes are available from each of the remaining land plant lineages, and ferns have no representative at all. At deeper taxonomic depths, the currently sparse sampling of charophytes, prasinophytes, rhodophytes, and glaucophytes limits our understanding of the early evolution of green plant mitochondria.

More importantly, future sequencing efforts need to move beyond descriptive analyses of genomic diversity in plants and focus on uncovering the evolutionary forces that have shaped this diversity. Perhaps the most glaring question relates to the in vivo structure of plant mitochondrial genomes. While most genome sequence projects display a circular map as proof of sequencing completion, these maps fail to represent the broad genomic diversity within any individual plant: the various linear and circular forms, the different configurations in recombinational equilibrium, and the numerous substoichiometric rearrangements. Certainly, most authors readily acknowledge this fact, and some even provide evidence for the alternative genomic configurations, either by mapping incongruent read-pairs, PCR amplification across recombinationally active repeats, Southern blot hybridization, or direct sequencing of cloned genomic segments. However, future sequencing efforts need to fully quantify the relative contribution of these different genomic configurations as a first step to establishing the evolutionary significance, if any, of such configurations.

Another largely unanswered question centers on the large fluctuations in genome size among plant mitochondrial genomes. What fraction of these sequences derives from a process of recurrent duplication and rearrangement of existing sequences, reflecting the dynamic nature of the plant mitochondrial genome? To what extent do these sequences reflect the propensity of mitochondrial genomes to integrate and maintain foreign DNA? For example, many of these sequences might represent acquisitions (of varying age) from the large plant nuclear genome, for which we still have very few sequences. Are these "missing" sequences in fact missing, or does much of it reside in the largely uncharacterized substoichiometric milieu present in the mitochondria of flowering plants (and probably all plants with actively recombining genomes)? To address this issue, more thorough within-species sequencing is needed, specifically from a species with a completely sequenced nuclear genome and from individual lines whose relationships are clearly established in order to identify the origin and trace the subsequent evolution of extra-genetic DNA in a phylogenetic context.

The plant mitochondrial genomics revolution is now in full swing and has provided enormous detail about genome content, structure and function, although it must be stressed again that much of this information was already established prior to any genomic sequencing. Additional sequencing has the potential to reveal further insights, but simply obtaining new genome sequences from more plants is unlikely to progress our understanding of plant mitochondrial biology. Thus, it is important that future projects focus on species that can expand our understanding of genomic diversity and specifically address the major unresolved questions.

Acknowledgements We thank Alan Christensen, Volker Knoop, Sally Mackenzie, David Smith, Jonathan Wendel and Paul Wolf for providing comments and corrections to improve the manuscript and Wenhu Guo for help in compiling the intron distribution information in Fig. 9.5. Research in the Mower Lab on plant mitochondria is funded by the National Science Foundation (IOS-1027529 and MCB-1125386) and the University of Nebraska-Lincoln. AJA was funded by a National Institutes of Health Ruth L. Kirschstein NRSA Postdoctoral Fellowship (1F32GM080079-01A1).

References

Adams KL, Qiu YL, Stoutemyer M, Palmer JD (2002a) Punctuated evolution of mitochondrial gene content: high and variable rates of mitochondrial gene loss and transfer to the nucleus during angiosperm evolution. Proc Natl Acad Sci USA 99:9905–9912

Adams KL, Daley DO, Whelan J, Palmer JD (2002b) Genes for two mitochondrial ribosomal proteins in flowering plants are derived from their chloroplast or cytosolic counterparts. Plant Cell 14:931–943

Allen JO, Fauron CM, Minx P, Roark L, Oddiraju S, Lin GN, Meyer L, Sun H, Kim K, Wang C, Du F, Xu D, Gibson M, Cifrese J, Clifton SW, Newton KJ (2007) Comparisons among two fertile and three male-sterile mitochondrial genomes of maize. Genetics 177:1173–1192

Alverson AJ, Wei X, Rice DW, Stern DB, Barry K, Palmer JD (2010) Insights into the evolution of mitochondrial genome size from complete sequences of *Citrullus lanatus* and *Cucurbita pepo* (Cucurbitaceae). Mol Biol Evol 27:1436–1448

Alverson AJ, Zhuo S, Rice DW, Sloan DB, Palmer JD (2011) The mitochondrial genome of the legume *Vigna radiata* and the analysis of recombination across short mitochondrial repeats. PLoS One 6: e16404

André C, Levy A, Walbot V (1992) Small repeated sequences and the structure of plant mitochondrial genomes. Trends Genet 8:128–132

Arrieta-Montiel MP, Shedge V, Davila J, Christensen AC, Mackenzie SA (2009) Diversity of the *Arabidopsis* mitochondrial genome occurs via nuclear-controlled recombination activity. Genetics 183:1261–1268

Bendich AJ (1985) Plant mitochondrial DNA: Unusual variation on a common theme. In: Hohn B, Dennis ES (eds) Genetic flux in plants. Springer, Wien/New York, pp 111–138

Bergthorsson U, Adams KL, Thomason B, Palmer JD (2003) Widespread horizontal transfer of mitochondrial genes in flowering plants. Nature 424:197–201

Bergthorsson U, Richardson AO, Young GJ, Goertzen LR, Palmer JD (2004) Massive horizontal transfer of mitochondrial genes from diverse land plant donors to the basal angiosperm *Amborella*. Proc Natl Acad Sci USA 101:17747–17752

Boer PH, Bonen L, Lee RW, Gray MW (1985) Genes for respiratory chain proteins and ribosomal RNAs are present on a 16-kilobase-pair DNA species from Chlamydomonas reinhardtii mitochondria. Proc Natl Acad Sci USA 82:3340–3344

Bogsch EG, Sargent F, Stanley NR, Berks BC, Robinson C, Palmer T (1998) An essential component of a novel bacterial protein export system with homologues in plastids and mitochondria. J Biol Chem 273:18003–18006

Bonen L, Gray MW (1980) Organization and expression of the mitochondrial genome of plants I. The genes for wheat mitochondrial ribosomal and transfer RNA: evidence for an unusual arrangement. Nucleic Acids Res 8:319–335

Boxma B, de Graaf RM, van der Staay GW, van Alen TA, Ricard G, Gabaldon T, van Hoek AH, Moon-van der Staay SY, Koopman WJ, van Hellemond JJ, Tielens AG, Friedrich T, Veenhuis M, Huynen MA, Hackstein JH (2005) An anaerobic mitochondrion that produces hydrogen. Nature 434:74–79

Braun H-P, Kruft V, Schmitz UK (1994) Molecular identification of the ten subunits of cytochrome-*c* reductase from potato mitochondria. Planta 193:99–106

Burger G, Saint-Louis D, Gray MW, Lang BF (1999) Complete sequence of the mitochondrial DNA of the red alga *Porphyra purpurea*. Cyanobacterial introns and shared ancestry of red and green algae. Plant Cell 11:1675–1694

Castandet B, Choury D, Begu D, Jordana X, Araya A (2010) Intron RNA editing is essential for splicing in plant mitochondria. Nucleic Acids Res 38:7112–7121

Chapdelaine Y, Bonen L (1991) The wheat mitochondrial gene for subunit I of the NADH dehydrogenase complex: a *trans*-splicing model for this gene-in-pieces. Cell 65:465–472

Chaw SM, Shih AC, Wang D, Wu YW, Liu SM, Chou TY (2008) The mitochondrial genome of the gymnosperm *Cycas taitungensis* contains a novel family of short interspersed elements, Bpu sequences, and abundant RNA editing sites. Mol Biol Evol 25:603–615

Clifton SW, Minx P, Fauron CM, Gibson M, Allen JO, Sun H, Thompson M, Barbazuk WB, Kanuganti S, Tayloe C, Meyer L, Wilson RK, Newton KJ (2004) Sequence and comparative analysis of the maize NB mitochondrial genome. Plant Physiol 136:3486–3503

Covello PS, Gray MW (1992) Silent mitochondrial and active nuclear genes for subunit 2 of cytochrome c oxidase (*cox2*) in soybean: evidence for RNA-mediated gene transfer. EMBO J 11:3815–3820

Cusimano N, Zhang L-B, Renner SS (2008) Reevaluation of the *cox1* group I intron in Araceae and angiosperms indicates a history dominated by loss rather than horizontal transfer. Mol Biol Evol 25:265–276

Darracq A, Varre JS, Touzet P (2010) A scenario of mitochondrial genome evolution in maize based on rearrangement events. BMC Genomics 11:233

Dewey RE, Levings CS 3rd, Timothy DH (1986) Novel recombinations in the maize mitochondrial genome produce a unique transcriptional unit in the Texas male-sterile cytoplasm. Cell 44:439–449

Dombrovska O, Qiu Y-L (2004) Distribution of introns in the mitochondrial gene *nad1* in land plants: phylogenetic and molecular evolutionary implications. Mol Phylogenet Evol 32:246–263

Duff RJ (2006) Divergent RNA editing frequencies in hornwort mitochondrial *nad5* sequences. Gene 366:285–291

Fox TD, Leaver CJ (1981) The *Zea mays* mitochondrial gene coding cytochrome oxidase subunit II has an intervening sequence and does not contain TGA codons. Cell 26:315–323

Fujii S, Kazama T, Yamada M, Toriyama K (2010) Discovery of global genomic re-organization based on comparison of two newly sequenced rice mitochondrial genomes with cytoplasmic male sterility-related genes. BMC Genomics 11:209

Giegé P, Brennicke A (1999) RNA editing in *Arabidopsis* mitochondria effects 441 C to U changes in ORFs. Proc Natl Acad Sci USA 96:15324–15329

Giegé P, Grienenberger JM, Bonnard G (2008) Cytochrome *c* biogenesis in mitochondria. Mitochondrion 8:61–73

Goremykin VV, Salamini F, Velasco R, Viola R (2009) Mitochondrial DNA of *Vitis vinifera* and the issue of rampant horizontal gene transfer. Mol Biol Evol 26:99–110

Gray MW (2009) RNA editing in plant mitochondria: 20 years later. IUBMB Life 61:1101–1104

Gray MW, Lang BF, Burger G (2004) Mitochondria of protists. Annu Rev Genet 38:477–524

Grewe F, Viehoever P, Weisshaar B, Knoop V (2009) A *trans*-splicing group I intron and tRNA-hyperediting in the mitochondrial genome of the lycophyte *Isoetes engelmannii*. Nucleic Acids Res 37:5093–5104

Grewe F, Herres S, Viehöver P, Polsakiewicz M, Weisshaar B, Knoop V (2011) A unique transcriptome: 1782 positions of RNA editing alter 1406 codon identities in mitochondrial mRNAs of the lycophyte *Isoetes engelmannii*. Nucleic Acids Res 39:2890–2902

Handa H (2003) The complete nucleotide sequence and RNA editing content of the mitochondrial genome of rapeseed (*Brassica napus* L.): comparative analysis of the mitochondrial genomes of rapeseed and *Arabidopsis thaliana*. Nucleic Acids Res 31:5907–5916

Hanson MR, Bentolila S (2004) Interactions of mitochondrial and nuclear genes that affect male gametophyte development. Plant Cell 16(Suppl):S154–S169

Hao W, Palmer JD (2009) Fine-scale mergers of chloroplast and mitochondrial genes create functional, transcompartmentally chimeric mitochondrial genes. Proc Natl Acad Sci USA 106:16728–16733

Hao W, Richardson AO, Zheng Y, Palmer JD (2010) Gorgeous mosaic of mitochondrial genes created by horizontal transfer and gene conversion. Proc Natl Acad Sci USA 107:21576–21581

Heazlewood JL, Whelan J, Millar AH (2003) The products of the mitochondrial *orf25* and *orfB* genes are F_O components in the plant F_1F_O ATP synthase. FEBS Lett 540:201–205

Hecht J, Grewe F, Knoop V (2011) Extreme RNA editing in coding islands and abundant microsatellites in repeat sequences of *Selaginella moellendorffii* mitochondria: the root of frequent plant mtDNA recombination in early tracheophytes. Genome Biol Evol 3:344–358

Klodmann J, Sunderhaus S, Nimtz M, Jansch L, Braun HP (2010) Internal architecture of mitochondrial complex I from *Arabidopsis thaliana*. Plant Cell 22:797–810

Knoop V, Schuster W, Wissinger B, Brennicke A (1991) *Trans* splicing integrates an exon of 22 nucleotides into the *nad5* mRNA in higher plant mitochondria. EMBO J 10:3483–3493

Knoop V, Unseld M, Marienfeld J, Brandt P, Sunkel S, Ullrich H, Brennicke A (1996) *copia*-, *gypsy*- and LINE-like retrotransposon fragments in the mitochondrial genome of *Arabidopsis thaliana*. Genetics 142:579–585

Kobayashi Y, Knoop V, Fukuzawa H, Brennicke A, Ohyama K (1997) Interorganellar gene transfer in bryophytes: the functional *nad7* gene is nuclear encoded in *Marchantia polymorpha*. Mol Gen Genet 256:589–592

Kubo N, Arimura S (2010) Discovery of the *rpl10* gene in diverse plant mitochondrial genomes and its probable replacement by the nuclear gene for chloroplast RPL10 in two lineages of angiosperms. DNA Res 17:1–9

Kubo T, Nishizawa S, Sugawara A, Itchoda N, Estiati A, Mikami T (2000) The complete nucleotide sequence of the mitochondrial genome of sugar beet (*Beta vulgaris* L.) reveals a novel gene for tRNA^{Cys}(GCA). Nucleic Acids Res 28:2571–2576

Lang BF, Gray MW, Burger G (1999) Mitochondrial genome evolution and the origin of eukaryotes. Annu Rev Genet 33:351–397

Leaver CJ, Gray MW (1982) Mitochondrial genome organization and expression in higher plants. Annu Rev Plant Physiol 33:373–402

Leblanc C, Boyen C, Richard O, Bonnard G, Grienenberger JM, Kloareg B (1995) Complete sequence of the mitochondrial DNA of the rhodophyte *Chondrus crispus* (Gigartinales). Gene content and genome organization. J Mol Biol 250:484–495

Li L, Wang B, Liu Y, Qiu YL (2009) The complete mitochondrial genome sequence of the hornwort *Megaceros aenigmaticus* shows a mixed mode of conservative yet dynamic evolution in early land plant mitochondrial genomes. J Mol Evol 68:665–678

Lilly JW, Havey MJ (2001) Small, repetitive DNAs contribute significantly to the expanded mitochondrial genome of cucumber. Genetics 159:317–328

Lonsdale DM, Hodge TP, Fauron CM (1984) The physical map and organisation of the mitochondrial genome from the fertile cytoplasm of maize. Nucleic Acids Res 12:9249–9261

Malek O, Lättig K, Hiesel R, Brennicke A, Knoop V (1996) RNA editing in bryophytes and a molecular phylogeny of land plants. EMBO J 15:1403–1411

Maréchal A, Brisson N (2010) Recombination and the maintenance of plant organelle genome stability. New Phytol 186:299–317

Maréchal-Drouard L, Weil J-H, Guillemaut P (1988) Import of several tRNAs from the cytoplasm into the mitochondria in bean *Phaseolus vulgaris*. Nucleic Acids Res 16:4777–4788

Maréchal-Drouard L, Guillemaut P, Cosset A, Arbogast M, Weber F, Weil J-H, Dietrich A (1990) Transfer RNAs of potato (*Solanum tuberosum*) mitochondria have different genetic origins. Nucleic Acids Res 18:3689–3696

McDermott P, Connolly V, Kavanagh TA (2008) The mitochondrial genome of a cytoplasmic male sterile line of perennial ryegrass (*Lolium perenne* L.) contains an integrated linear plasmid-like element. Theor Appl Genet 117:459–470

Millar AH, Eubel H, Jansch L, Kruft V, Heazlewood JL, Braun HP (2004) Mitochondrial cytochrome *c* oxidase and succinate dehydrogenase complexes contain plant specific subunits. Plant Mol Biol 56:77–90

Mower JP (2009) The PREP suite: predictive RNA editors for plant mitochondrial genes, chloroplast genes and user-defined alignments. Nucleic Acids Res 37(Web Server issue):W253–W259

Mower JP, Bonen L (2009) Ribosomal protein L10 is encoded in the mitochondrial genome of many land plants and green algae. BMC Evol Biol 9:265

Mower JP, Palmer JD (2006) Patterns of partial RNA editing in mitochondrial genes of *Beta vulgaris*. Mol Genet Genomics 276:285–293

Mower JP, Touzet P, Gummow JS, Delph LF, Palmer JD (2007) Extensive variation in synonymous substitution rates in mitochondrial genes of seed plants. BMC Evol Biol 7:135

Mower JP, Stefanović S, Hao W, Gummow JS, Jain K, Ahmed D, Palmer JD (2010) Horizontal acquisition of multiple mitochondrial genes from a parasitic plant followed by gene conversion with host mitochondrial genes. BMC Biol 8:150

Notsu Y, Masood S, Nishikawa T, Kubo N, Akiduki G, Nakazono M, Hirai A, Kadowaki K (2002) The complete sequence of the rice (*Oryza sativa* L.) mitochondrial genome: frequent DNA sequence acquisition and loss during the evolution of flowering plants. Mol Genet Genomics 268:434–445

Nugent JM, Palmer JD (1991) RNA-mediated transfer of the gene *coxII* from the mitochondrion to the nucleus during flowering plant evolution. Cell 66:473–481

Nunes-Nesi A, Fernie AR (2007) Mitochondrial metabolism. In: Logan DC (ed) Plant mitochondria. Blackwell, Oxford, UK, pp 212–277

Oda K, Yamato K, Ohta E, Nakamura Y, Takemura M, Nozato N, Akashi K, Kanegae T, Ogura Y, Kohchi T et al (1992) Gene organization deduced from the complete sequence of liverwort *Marchantia polymorpha* mitochondrial DNA. A primitive form of plant mitochondrial genome. J Mol Biol 223:1–7

Odahara M, Inouye T, Fujita T, Hasebe M, Sekine Y (2007) Involvement of mitochondrial-targeted RecA in the repair of mitochondrial DNA in the moss, *Physcomitrella patens*. Genes Genet Syst 82:43–51

Odahara M, Kuroiwa H, Kuroiwa T, Sekine Y (2009) Suppression of repeat-mediated gross mitochondrial genome rearrangements by RecA in the moss *Physcomitrella patens*. Plant Cell 21:1182–1194

Ogihara Y, Yamazaki Y, Murai K, Kanno A, Terachi T, Shiina T, Miyashita N, Nasuda S, Nakamura C, Mori N, Takumi S, Murata M, Futo S, Tsunewaki K (2005) Structural dynamics of cereal mitochondrial genomes as revealed by complete nucleotide sequencing of the wheat mitochondrial genome. Nucleic Acids Res 33:6235–6250

Ohta N, Sato N, Kuroiwa T (1998) Structure and organization of the mitochondrial genome of the unicellular red alga *Cyanidioschyzon merolae* deduced from the complete nucleotide sequence. Nucleic Acids Res 26:5190–5198

Oldenburg DJ, Bendich AJ (1996) Size and structure of replicating mitochondrial DNA in cultured tobacco cells. Plant Cell 8:447–461

Palmer JD, Herbon LA (1988) Plant mitochondrial DNA evolves rapidly in structure, but slowly in sequence. J Mol Evol 28:87–97

Palmer JD, Shields CR (1984) Tripartite structure of the *Brassica campestris* mitochondrial genome. Nature 307:437–440

Picardi E, Horner DS, Chiara M, Schiavon R, Valle G, Pesole G (2010) Large-scale detection and analysis of RNA editing in grape mtDNA by RNA deep-sequencing. Nucleic Acids Res 38:4755–4767

Pombert JF, Otis C, Lemieux C, Turmel M (2004) The complete mitochondrial DNA sequence of the green alga *Pseudendoclonium akinetum* (Ulvophyceae) highlights distinctive evolutionary trends in the chlorophyta and suggests a sister-group relationship between the Ulvophyceae and Chlorophyceae. Mol Biol Evol 21:922–935

Pombert JF, Beauchamp P, Otis C, Lemieux C, Turmel M (2006) The complete mitochondrial DNA sequence of the green alga *Oltmannsiellopsis viridis*: evolutionary trends of the mitochondrial genome in the Ulvophyceae. Curr Genet 50:137–147

Richardson AO, Palmer JD (2007) Horizontal gene transfer in plants. J Exp Bot 58:1–9

Robbens S, Derelle E, Ferraz C, Wuyts J, Moreau H, Van de Peer Y (2007) The complete chloroplast and mitochondrial DNA sequence of *Ostreococcus tauri*: organelle genomes of the smallest eukaryote are examples of compaction. Mol Biol Evol 24:956–968

Rüdinger M, Funk HT, Rensing SA, Maier UG, Knoop V (2009) RNA editing: only eleven sites are present in the *Physcomitrella patens* mitochondrial transcriptome and a universal nomenclature proposal. Mol Genet Genomics 281:473–481

Ryan R, Grant D, Chiang KS, Swift H (1978) Isolation and characterization of mitochondrial DNA from *Chlamydomonas reinhardtii*. Proc Natl Acad Sci USA 75:3268–3272

Salmans ML, Chaw SM, Lin CP, Shih AC, Wu YW, Mulligan RM (2010) Editing site analysis in a gymnosperm mitochondrial genome reveals similarities with angiosperm mitochondrial genomes. Curr Genet 56:439–446

Sanchez-Puerta MV, Cho Y, Mower JP, Alverson AJ, Palmer JD (2008) Frequent, phylogenetically local horizontal transfer of the *coxI* group I Intron in flowering plant mitochondria. Mol Biol Evol 25:1762–1777

Satoh M, Kubo T, Nishizawa S, Estiati A, Itchoda N, Mikami T (2004) The cytoplasmic male-sterile type and normal type mitochondrial genomes of sugar beet share the same complement of genes of known function but differ in the content of expressed ORFs. Mol Genet Genomics 272:247–256

Satoh M, Kubo T, Mikami T (2006) The Owen mitochondrial genome in sugar beet (*Beta vulgaris* L.): possible mechanisms of extensive rearrangements and the origin of the mitotype-unique regions. Theor Appl Genet 113:477–484

Schmitz-Linneweber C, Small I (2008) Pentatricopeptide repeat proteins: a socket set for organelle gene expression. Trends Plant Sci 13:663–670

Schuster W, Brennicke A (1987) Plastid, nuclear and reverse transcriptase sequences in the mitochondrial genome of *Oenothera*: is genetic information transferred between organelles via RNA? EMBO J 6:2857–2863

Shedge V, Arrieta-Montiel M, Christensen AC, Mackenzie SA (2007) Plant mitochondrial recombination surveillance requires unusual *RecA* and *MutS* homologs. Plant Cell 19:1251–1264

Sloan DB, Oxelman B, Rautenberg A, Taylor DR (2009) Phylogenetic analysis of mitochondrial substitution rate variation in the angiosperm tribe Sileneae. BMC Evol Biol 9:260

Sloan DB, Alverson AJ, Storchova H, Palmer JD, Taylor DR (2010a) Extensive loss of translational genes in the structurally dynamic mitochondrial genome of the angiosperm *Silene latifolia*. BMC Evol Biol 10:274

Sloan DB, MacQueen AH, Alverson AJ, Palmer JD, Taylor DR (2010b) Extensive loss of RNA editing sites in rapidly evolving *Silene* mitochondrial genomes: selection vs. retroprocessing as the driving force. Genetics 185:1369–1380

Smith DR, Lee RW (2009) The mitochondrial and plastid genomes of *Volvox carteri*: bloated molecules rich in repetitive DNA. BMC Genomics 10:132

Smith DR, Hua J, Lee RW (2010a) Evolution of linear mitochondrial DNA in three known lineages of *Polytomella*. Curr Genet 56:427–438

Smith DR, Lee RW, Cushman JC, Magnuson JK, Tran D, Polle JE (2010b) The *Dunaliella salina* organelle genomes: large sequences, inflated with intronic and intergenic DNA. BMC Plant Biol 10:83

Stechmann A, Hamblin K, Perez-Brocal V, Gaston D, Richmond GS, van der Giezen M, Clark CG, Roger AJ (2008) Organelles in *Blastocystis* that blur the distinction between mitochondria and hydrogenosomes. Curr Biol 18:580–585

Steinhauser S, Beckert S, Capesius I, Malek O, Knoop V (1999) Plant mitochondrial RNA editing. J Mol Evol 48:303–312

Stern DB, Lonsdale DM (1982) Mitochondrial and chloroplast genomes of maize have a 12-kilobase DNA sequence in common. Nature 299:698–702

Sugiyama Y, Watase Y, Nagase M, Makita N, Yagura S, Hirai A, Sugiura M (2005) The complete nucleotide sequence and multipartite organization of the tobacco mitochondrial genome: comparative analysis of mitochondrial genomes in higher plants. Mol Genet Genomics 272:603–615

Terasawa K, Odahara M, Kabeya Y, Kikugawa T, Sekine Y, Fujiwara M, Sato N (2007) The mitochondrial genome of the moss *Physcomitrella patens* sheds new light on mitochondrial evolution in land plants. Mol Biol Evol 24:699–709

Tian X, Zheng J, Hu S, Yu J (2006) The rice mitochondrial genomes and their variations. Plant Physiol 140:401–410

Tovar J, Fischer A, Clark CG (1999) The mitosome, a novel organelle related to mitochondria in the amitochondrial parasite *Entamoeba histolytica*. Mol Microbiol 32:1013–1021

Turmel M, Lemieux C, Burger G, Lang BF, Otis C, Plante I, Gray MW (1999) The complete mitochondrial DNA sequences of *Nephroselmis olivacea* and *Pedinomonas minor*. Two radically different evolutionary patterns within green algae. Plant Cell 11:1717–1730

Turmel M, Otis C, Lemieux C (2002a) The chloroplast and mitochondrial genome sequences of the charophyte *Chaetosphaeridium globosum*: insights into the timing of the events that restructured organelle DNAs within the green algal lineage that led to land plants. Proc Natl Acad Sci USA 99:11275–11280

Turmel M, Otis C, Lemieux C (2002b) The complete mitochondrial DNA sequence of *Mesostigma viride* identifies this green alga as the earliest green plant divergence and predicts a highly compact mitochondrial genome in the ancestor of all green plants. Mol Biol Evol 19:24–38

Turmel M, Otis C, Lemieux C (2003) The mitochondrial genome of *Chara vulgaris*: insights into the mitochondrial DNA architecture of the last common ancestor of green algae and land plants. Plant Cell 15:1888–1903

Turmel M, Otis C, Lemieux C (2007) An unexpectedly large and loosely packed mitochondrial genome in the charophycean green alga *Chlorokybus atmophyticus*. BMC Genomics 8:137

Turmel M, Otis C, Lemieux C (2010) A deviant genetic code in the reduced mitochondrial genome of the picoplanktonic green alga *Pycnococcus provasolii*. J Mol Evol 70:203–214

Unseld M, Marienfeld JR, Brandt P, Brennicke A (1997) The mitochondrial genome of *Arabidopsis thaliana* contains 57 genes in 366,924 nucleotides. Nat Genet 15:57–61

Vangerow S, Teerkorn T, Knoop V (1999) Phylogenetic information in the mitochondrial *nad5* gene of pteridophytes: RNA editing and intron sequences. Plant Biol 1:235–243

Vaughn JC, Mason MT, Sper-Whitis GL, Kuhlman P, Palmer JD (1995) Fungal origin by horizontal transfer of a plant mitochondrial group I intron in the chimeric *CoxI* gene of *Peperomia*. J Mol Evol 41:563–572

Wang B, Xue J, Li L, Liu Y, Qiu YL (2009) The complete mitochondrial genome sequence of the liverwort *Pleurozia purpurea* reveals extremely conservative mitochondrial genome evolution in liverworts. Curr Genet 55:601–609

Ward BL, Anderson RS, Bendich AJ (1981) The mitochondrial genome is large and variable in a family of plants (Cucurbitaceae). Cell 25:793–803

Wissinger B, Schuster W, Brennicke A (1991) *Trans* splicing in *Oenothera* mitochondria: *nad1* mRNAs are edited in exon and *trans*-splicing group II intron sequences. Cell 65:473–482

Wolfe KH, Li WH, Sharp PM (1987) Rates of nucleotide substitution vary greatly among plant mitochondrial, chloroplast, and nuclear DNAs. Proc Natl Acad Sci USA 84:9054–9058

Xue JY, Liu Y, Li L, Wang B, Qiu YL (2010) The complete mitochondrial genome sequence of the hornwort *Phaeoceros laevis*: retention of many ancient pseudogenes and conservative evolution of mitochondrial genomes in hornworts. Curr Genet 56:53–61

Yamamoto MP, Kubo T, Mikami T (2005) The 5′-leader sequence of sugar beet mitochondrial *atp6* encodes a novel polypeptide that is characteristic of Owen cytoplasmic male sterility. Mol Genet Genomics 273:342–349

Yu W, Schuster W (1995) Evidence for a site-specific cytidine deamination reaction involved in C to U RNA editing of plant mitochondria. J Biol Chem 270:18227–18233

Zaegel V, Guermann B, Le Ret M, Andres C, Meyer D, Erhardt M, Canaday J, Gualberto JM, Imbault P (2006) The plant-specific ssDNA binding protein OSB1 is involved in the stoichiometric transmission of mitochondrial DNA in *Arabidopsis*. Plant Cell 18:3548–3563

Plastid Genome Diversity

10

Paul G. Wolf

Contents

10.1 Introduction

Most living plant cells contain plastids, which harbour their own DNA: the plastome. Plastomes are significantly less diverse than nuclear genomes, but this lower diversity has the advantage that comparisons can be made across all clades of green plants. Large blocks of synteny can be found among plastomes from different algal lineages and land plants. In this chapter, I review the nature of plastome diversity, starting with some history of research and what is known about packaging, replication, and inheritance of plastid DNA. I review the composition of plastomes, the current status of available data (and future needs), and then examine plastome origins, structural evolution, patterns of nucleotide substitution, gene expression (but not translation or post-translational stages), and plastome biotechnology. The goal is a broad overview, but this is at the expense of important detail to which I defer to cited literature.

10.2 A Brief History of Research on Plastids and Their Genomes

Early descriptions of subcellular structures were made by Leeuwenhoek in the seventeenth century. However, specific descriptions and coining of the terms plastid and chloroplast is usually attributed to Franz Wilhelm Schimper (Schimper 1883, 1885), who also noted the resemblance between chloroplasts and cyanobacteria (See Wayne 2009 for more details). Later, this relationship was further developed into the endosymbiont theory of plastid origins by Mereschkowski (Martin and Kowallik 1999; Mereschkowski 1905). Evidence continues to accumulate that organelles have prokaryotic origins, and the current picture argues strongly for a single origin of primary plastids in plants, with several secondary and tertiary origins throughout eukaryotic lineages (Cavalier-Smith 2002; Keeling 2010). In addition, early in the twentieth century, experimental evidence led to a theory of plastid

P.G. Wolf (✉)
Department of Biology, College of Science, UMC 5305, Utah State University, 5305 Old Main Hill, Logan, UT 84322-5305, USA
e-mail: wolf@biology.usu.edu; paul.wolf@usu.edu

J.F. Wendel et al. (eds.), *Plant Genome Diversity Volume 1*,
DOI 10.1007/978-3-7091-1130-7_10, © Springer-Verlag Wien 2012

inheritance (Hagemann 2000), initially developed by Erwin Baur (Baur 1909) and further refined by Otto Renner (Renner 1929). The endosymbiont and plastid inheritance theories were developed over a 100 years ago and they both predicted that plastids should contain hereditary material. This was demonstrated cytologically in the 1950s (for example Chiba 1951; see Sugiura 2003 for further review). Plastid DNA was first isolated in the early 1960s from *Chlamydomonas* (Sager and Ishida 1963; Sugiura 2003). With the dawn of the age of DNA analysis, plastid DNA was soon examined in detail, starting with a physical map of the *Zea* plastome based on restriction sites (Bedbrook and Kolodner 1979). Subsequently, plastomes of many species were mapped (reviewed by Palmer 1985), verifying that in most (but not all) lineages, plastomes map to a circle with a large single copy region (LSC) and a small single copy region (SSC) separated by two copies of an inverted repeat (IR) which included the ribosomal RNA genes (Palmer 1985). Mapping studies also indicated that within plant cells, the plastome exists in two orientations (Palmer 1983), a pattern that is maintained by a form of recombination between the two copies of the IR (Stein et al. 1986).

10.3 Plastome Packaging, Inheritance and Replication

Most green plant plastomes map to a circle of about 150 kb and no more than 218 kb, with the largest reported at a little over 500 kb (Brouard et al. 2010). However, most plant cells contain many copies of the plastome. Even plants with a single plastid (e.g., *Chlamydomonas*) can contain many copies of the plastome. At the other extreme, wheat cells have more than 50 plastids per cell and more than 300 plastome copies per plastid, and therefore, over 15,000 plastome copies per cell (Boffey and Leech 1982). Thus, although the plastome is a small genome compared to its nuclear counterpart, plastid DNA makes up a significant proportion of total cellular DNA, as much as 20% in some species (Boffey and Leech 1982).

Plastid DNA is neither assembled into chromosomes nor does it reside in the plastid as a series of independent circles. Rather, several plastomes are organized, with proteins and RNA, into structures known as nucleoids (Sato et al. 2003). Nucleoids of green plants differ from those of cyanobacteria in the types of DNA-binding proteins involved. Green plant nucleoids are bound with a sulphate reductase, which has also been linked to repression of transcription (Sekine et al. 2007). Nucleoids are approximately 0.2 μm in diameter, and several are found in each plastid (Sato et al. 2003). Most are attached to the envelope membrane, but mature chloroplasts can also have nucleoids associated with the thylakoid membrane (Sato et al. 2003). It is likely that nucleoid structure plays an important role in plastome replication,

transcription, and post-transcriptional modification. However, the general relationships between plastome packaging and these processes remain poorly understood (Bock 2007).

Although plastomes are typically depicted as circles, most plastid DNA is not in this form in a living plant cell (Bendich 2004; Bock 2007). Studies have not only found linear plastomes, but also concatenated pieces representing multiple plastomes, sometimes circular (Bendich 2004), and even branched forms (Oldenburg and Bendich 2004a, b). This variety of possible conformations is likely to be a function of both phylogenetic divergence and stage of plastome replication. The plastome replication process itself is also poorly understood (Bock 2007), and alternative mechanisms have been proposed. Early models involved bidirectional replication similar to that in bacteria, resulting in displacement (D) loops (Kolodner and Tewari 1975a). Rolling circle amplification (RCA) could also be used to achieve additional replication (Kolodner and Tewari 1975b). A double D-loop mechanism has also been proposed (Kunnimalaiyaan and Nielsen 1997a). However, these models were challenged by Bendich, based on the degree of linear DNA observed (Bendich 2004), and a recombination-dependent mechanism was instead proposed (Oldenburg and Bendich 2004b). The challenge of studying replication is making observations during the actual process. Alternatively, researchers can examine evidence left at the scene, such as changes in nucleotide composition. Genomes can accumulate adenine-to-guanine deaminations during the single-stranded phase of replication. Thus, gradients in A/G composition, especially for non-coding DNA, can reveal origins and directions of replication, as a function of total amount of time spent in the single-stranded phase. This approach was used recently used to show that A/G composition gradients are most consistent with the earlier models (bidirectional and RCA) across a wide range of published green plant plastomes (Krishnan and Rao 2009). Direct testing of these models is now needed. Meanwhile, evidence continues to accumulate for a role of recombination-dependent replication in *Arabidopsis*, especially as a repair process for maintaining plastome integrity (Rowan et al. 2010). Clearly, the evidence suggests that more than one replication process appears to be operating.

One further piece in the replication puzzle comes from analyses of origins of replication. Experiments from the capture of replication forks in pea and tobacco have located the actual origins in plastomes (Kunnimalaiyaan and Nielsen 1997b). Two origins were located (oriA and oriB), but each has two copies because they are in the inverted repeat. However, removal of both copies oriA does not prevent replication (Mühlbauer et al. 2002). More recently it was shown that neither origin is essential (Scharff and Koop 2007), although reduced plant growth was observed. This further supports the idea that multiple replication processes are possible. One possibility is that some processes are involved in replication during a growth phase and others during repair of DNA in mature plastids.

Plastid DNA not only replicates within cells, but then is distributed among daughter cells following cytokinesis. Furthermore, plastids and their genomes can be inherited during sexual reproduction in different ways (see Bock 2007 for a thorough recent review). Although most plants have maternal inheritance of plastid DNA (and this is thought to be the ancestral mode), this is not universal; paternal and biparental patterns are also known. Moreover, several alternative mechanisms can result in maternal inheritance. For example, in the algal model system *Chlamydomonas*, paternal plastid DNA is selectively degraded. In most angiosperms, plastids themselves are either excluded from, or degraded in, the generative cell of the developing pollen grain. In some grasses, plastid exclusion can occur during fertilization. A few lineages (such as *Pelargonium* in the geranium family) have biparental inheritance; plastids and their DNA are transmitted in the pollen. Paternal inheritance is found in many conifers, in which plastids and DNA are transmitted in pollen, and maternal plastids are degraded in the egg cells. As for many aspects of plastid genetics, there is considerable variation in patterns and mechanisms.

10.4 Composition of Plastid Genes and DNA

The earliest studies noted that nuclear and plastid DNA differ in base composition and buoyant density under centrifugation in CsCl (Kolodner and Tewari 1972; Sager and Ishida 1963). This forms the basis of many techniques for isolating plastid DNA (Jansen et al. 2005). The overall organization of the plastome has been reviewed extensively (Palmer 1985). Gene content includes a typical set of about 80 genes in land plant plastomes. These include genes for genetic and photosynthetic systems (Bock 2007; Wicke et al. 2011). The genetic system genes include those for transfer and ribosomal RNA, small and large ribosomal proteins, DNA-dependent RNA polymerase, a maturase (for intron splicing), and a translation initiation factor. Photosynthetic genes include those for light-independent pathways (cys, chl, rbcL, and accD) and light-dependent pathways (photosystem I and II, ATP-synthetase, cytochrome b6/f complex, and NAD(P)H dehydrogenase complex). However, certain genes have been lost in some lineages (see below).

10.5 Complete Sequences of Plastomes

Much of our early understanding of plastomes comes from studies of the 1970s and 1980s. These analyses involved restriction-site and gene mapping via Southern hybridization

Table 10.1 Complete plastome sequence data

Clade	Genomes in GenBank	Projects in progress
Viridiplantae (green plants)	156	
Chlorophyta	18	
Streptophyta	138	
Embryophyta (land plants)	132	(379)
Liverworts	2	8
Hornworts	1	1
Mosses	2	0?
Vascular plants	(128)	(370)
Lycophytes	3	1
Ferns	5	4
Seed plants	(120)	(365)
Gymnosperms	13	200
Angiosperms	107	165

to produce coarse-scale information. However, the finer details of plastome structure and function have come from complete plastome sequencing projects. The first two plastome sequences were both completed in 1986: for tobacco (Shinozaki et al. 1986) and the liverwort *Marchantia* (Ohyama et al. 1986). As of August 2010, the NCBI web site on organelle genome resources (http://www.ncbi.nlm.nih.gov/genomes/GenomesGroup.cgi?taxid=2759 &opt=plastid) lists 156 plastome sequences of Viridiplantae (green plants) of which 18 are from Chlorophyta and 138 from Streptophyta. Of the latter, 132 are from Embryophyta (land plants), most (120) of which are Spermatophyta (seed plants). Additional context, including estimates of sequencing projects in progress is provided in Table 10.1. This information illustrates the range of currently available and accessible plastome sequencing projects. Estimates of the number of plastome sequences in progress is not exhaustive and comes from information provided by colleagues involved in large-scale projects. From these data, it is evident that there is a heavy emphasis on seed plants, particularly angiosperms, whereas some older and more critical lineages of algae and bryophytes are lacking. Furthermore, it is clear that there are probably more projects in progress than there are published plastome sequences. This illustrates the recent acceleration in acquisition on all aspects of genomic data. In the next few years many previously unsampled clades will have representatives with complete plastome sequences, and this should significantly improve our understanding of plastome diversity.

10.6 Diversity and Evolution of Plastome Organization

Plastomes, especially those of land plants, are unusually conserved in their overall structure across lineages that have been diverging for hundreds of millions of years (Palmer

1985; Palmer and Stein 1986). This is especially noticeable when compared to the rapid and drastic changes that characterize nuclear genomes (Wei et al. 2009). But nuclear genomes are larger and it is not yet clear if their rate of synteny loss is different from that of plastomes when considered at the same scale of nucleotide length. Nevertheless, diversity among plastomes is low enough to allow comparisons across green plants, yet sufficient to observe variation useful at a wide range of evolutionary levels. The acquisition of complete plastome sequences has greatly enhanced the comparative study of plastomes. Here, I review some of the best-documented changes among plastomes in the green tree of life.

The earliest evolutionary changes in plastomes occurred after their free-living cyanobacterial ancestors were first absorbed by eukaryotic cells about half a billion years ago (Cavalier-Smith 2002). Early changes to plastomes have been described by comparing modern plastomes to various prokaryotic genomes in a phylogenetic context (Martin et al. 2002). The most significant series of evolutionary events is the transfer of genomic regions from the plastomes to nuclear genomes (Martin 2003), so that only about 5–10% of the genes from the cyanobacterial ancestor still reside in plastomes (Martin et al. 2002). The transfer process also means that many plant nuclear genes are of cyanobacterial origin. Thus, it has been estimated that about 18% of the nuclear protein-encoding genes of *Arabidopsis* have cyanobacterial ancestry (Martin et al. 2002). It would seem that the primary endosymbiotic event represents the most significant example of horizontal gene transfer in evolutionary history. The subsequent transfer of genes to the nucleus appears to have occurred it sporadic pulses (Martin and Herrmann 1998; Martin 2003). As more cyanobacterial and plastid genomes are sequenced a clearer picture of gene transfer events should emerge. Meanwhile, several independent assessments have been made. In one of the broadest analyses (Martin et al. 2002), over 600 transfer events are hypothesized, of which 54 are unique events, and distributed among all major branches of plant phylogeny. Within the green plants, finer detail has been achieved with the completion of plastome sequences from additional critical clades of algae (Turmel et al. 2006), again revealing several transfer events on major branches of the tree (see Fig. 10.1), even on more recent branches within land plants. In addition to the large number of events transferring DNA to the nucleus, there is also evidence that plastid genes have moved to the mitochondrial genome (Wang et al. 2007).

Because 90–95% of the original cyanobacterial genome has been transferred to the nucleus, it seems reasonable to ask why are any genes retained in the plastome? One possibility is that we are indeed on the pathway to complete loss of the plastome, a state that has already been attained for the genome-less hydroxysome, which is thought to have evolved

from a mitochondrion (Bui et al. 1996). There are clear advantages to transferring genes to the nucleus, most notably the release from Muller's ratchet which causes mutations to accumulate faster in asexual lineages, such as plastomes (Race et al. 1999). However, mapping of gene losses onto the green tree of life (Martin et al. 2002) suggests that there are clear patterns of parallel loss of some genes, and selective retention of others. Two main ideas have been proposed to explain this gene retention. Early ideas argued that the retained genes encode proteins that are too hydrophobic to enter the double membrane-bound organelle (von Heijne 1986). More recently the CORR (*Co*-location for *R*edox *R*egulation) hypothesis (Allen 1992; Allen 2003; Race et al. 1999) argues that electron transport processes in organelles must be tightly regulated because of the possible rapid damage from photooxidation. This rapid-response regulation can only occur if the proteins involved in redox balance across membranes are encoded, transcribed, and translated proximally to those membranes. Thus several factors likely are involved in selecting which genes that are transferred and those that are retained within plastomes.

Additional changes in gene content can occur when the "normal" selective environment for a plant changes, such as when parasitism evolves. Specifically, holoparasitic plants have reduced photosynthetic ability, and some have none. This means that they no longer need most genes that encode photosynthetic proteins. The result is that holoparasites have highly reduced plastomes (Krause 2008). The smallest known land plant plastome is that of the holoparasite *Epiphagus virginiana* with only 42 genes and a total size of 70 kb (Wolfe et al. 1992), about half the size of a typical plastome. In addition to loss of many photosynthetic genes, some holoparasite plastomes have lost genes that encode subunits of RNA polymerase. Thus, the remaining plastid genes must be transcribed by nuclear-encoded polymerases. Other holoparasites have undergone fewer losses, either as a function of a reduced level of holoparasitism, or a more recent switch to this life history (Krause 2008). Interestingly, *rbcL* (the gene encoding the large subunit of Rubisco) is present in many holoparasites, raising the possibility that some genes may have multiple functions. Indeed, evidence exists that Rubisco is involved in lipid biosynthesis (Schwender et al. 2004). This illustrates how useful parasitic plants are to the study of plastid gene function (Krause 2008).

In addition to the array of gene transfer events, plastomes in some lineages have also become restructured, most often by genome inversions. Several of the key inversions are noted on Fig. 10.1. Most notable among those within land plants include a 30 kb inversion in the ancestor of all extant vascular plants except lycophytes (Raubeson and Jansen 1992a), a 71 kb inversion incorporating most of the LSC in moss subclass Funariidae (Goffinet et al. 2007), an inversion

Fig. 10.1 Green plant phylogeny with hypothesized plastid gene losses and plastome inversions marked on branches

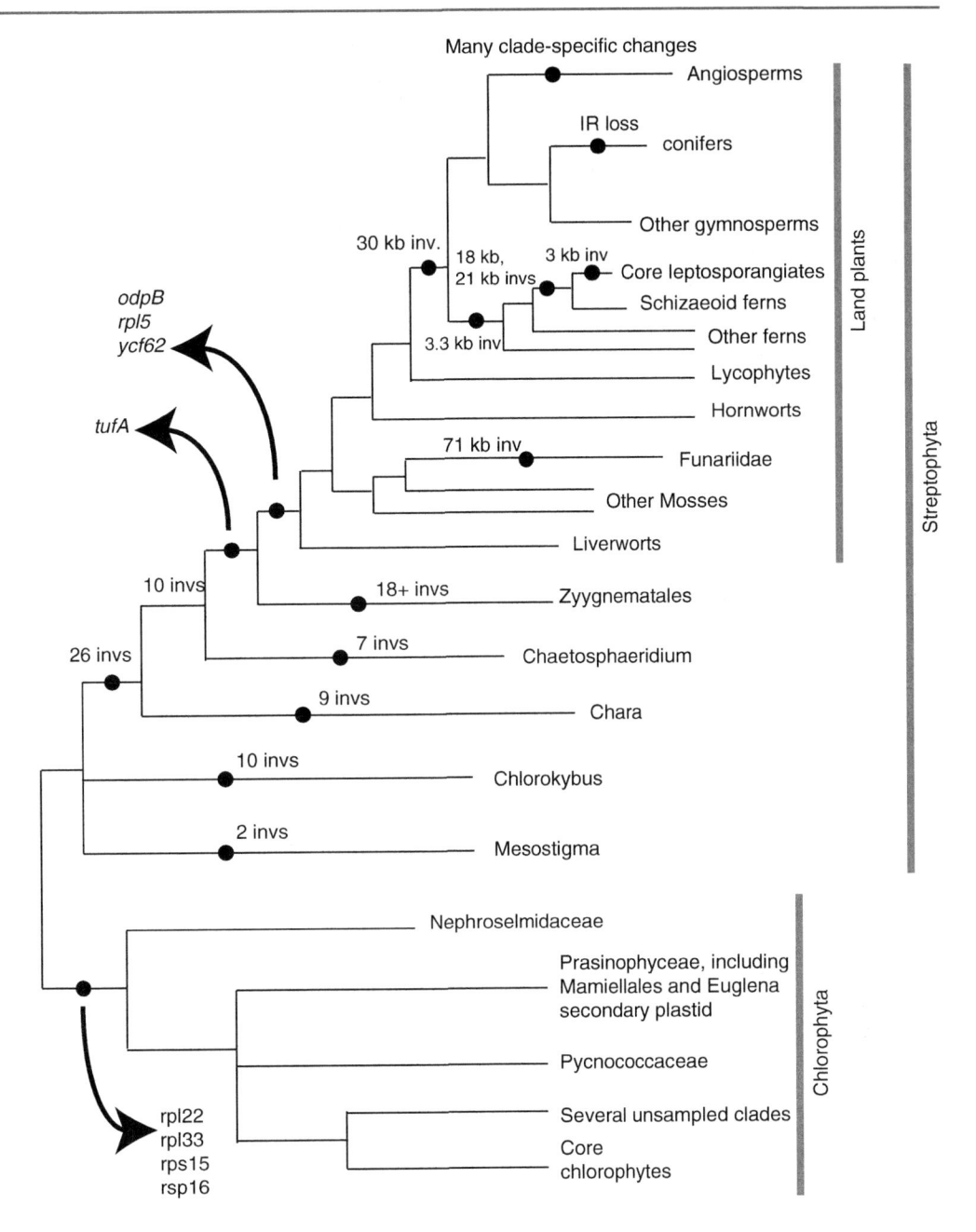

that unites the ferns, (including horsetails, Karol et al. 2010), several inversions within the ferns (Wolf et al. 2010), reduction of the IR and several inversions in conifers (Hirao et al. 2008; Raubeson and Jansen 1992b), several key inversions specific to angiosperm clades (Greiner et al. 2008; Howe et al. 1988; Kim et al. 2005), as well as multiple rearrangements within lineages with destabilized plastomes (Cai et al. 2008; Chumley et al. 2006; Guisinger et al. 2010a; Haberle et al. 2008; Milligan et al. 1989).

Inferring the processes that resulted in ancient plastome inversions is inherently challenging. One approach is to examine the endpoints of inferred inversions for unique features. This has been done for both large (many kb) and

small (less than 200 bp) inversions. Because small inversions tend to be more recent it is likely that they retain signatures of their causes (Catalano et al. 2009; Kim and Lee 2005). Several studies have found that small inversions are flanked by small inverted repeats of 8-50 bp. Recombination between identical (or even similar) genomic regions can then result in an inversion (Palmer 1991). The homologous regions can form stable hairpin structures (with IRs as stems and the inverted region as a loop), similar to the stems and loops in tRNAs. Kelchner and Wendel (1996) hypothesize that the inversion formation is not merely a function of the length of an IR but also is related to its estimated free energy of formation. Hairpins that form most easily can result in independent inversions in different lineages (Catalano et al. 2009), i.e., homoplasy. The larger inversions that have been observed may have been formed by similar processes as the small ones. The *Pelargonium* plastome has undergone a minimum of six inversions as well as several other reorganizational events (Chumley et al. 2006). The *Pelargonium* inversions are also associated with repetitive DNA. But other mechanisms for the inversion process are considered by Chumley et al., including invasion of transposable elements, which has been reported in *Chlamydomonas* (Fan et al. 1995). Another highly rearranged angiosperm plastome is that of *Trachelium* in Campanulaceae (Haberle et al. 2008). Many events are inferred for this plastome, resulting in 18 syntenous blocks. Presumably the total number of inversions relative to an ancestral organization is difficult to determine. Both large and small inversions in *Trachelium* are associated with short repeats or tRNA genes. However, unlike the small inversions described earlier, most of the inversion endpoints are characterized by direct repeats. Also in *Trachelium* some genes (e.g., *psbJ*) are found as three copies, implying more complex mechanisms such as transposition (Haberle et al. 2008). The 15 inversions that have been hypothesized for the *Cryptomeria* plastome are also associated with both direct and inverted repeats (Hirao et al. 2008).

Because inversions vary in size and the nature of flanking sequences it is very likely that they occur by more than one process. Although a mechanism is proposed for inversions flanked by inverted repeats, the process for direct repeats is not clear. One possibility is that these inversions could be moved via a retroposon that uses such a direct repeat as a target for reverse transcriptase. Although there is no evidence of retroposon sequences in land plant plastomes, they have been found in the plastome of *Chlamydomonas* (Fan et al. 1995). Transposition has also been suggested as a mechanism for intron loss in general (Vinogradov 2002) and for loss of one of the *clpP* introns in *Oenothera* (Greiner et al. 2008). Transposition is also able to produce repeats that could then be used for recombination after the retroposon itself is lost. Part of the general problem is that currently we only have evidence of an association between repeats (both direct and inverted) and inversion end-points, yet no evidence of causation. It is possible that the repeats are scar tissue that resulted from DNA repair after the inversion occurred. Further research is needed from lineages that undergo rapid and extensive inversions, such as those that are well-documented in angiosperms. However, it is likely that additional destabilized lineages remain unsampled. Despite gallant efforts by Turmel, Lemieux, and colleagues, plastome diversity remains unknown within most groups of green algae. Such unsampled plastomes may well hold the key to a better understanding of the mechanisms of plastome reorganization.

10.7 Diversity in Nucleotide Substitution Rates

Information on nucleotide substitution rates can provide clues to mutation rates under different conditions, and also enable researchers to choose appropriate genes for phylogenetic analyses. Nucleotide substitution rates vary across genome compartments, across orthologous nucleotide positions, across genes, and across lineages (Muse and Gaut 1997; Muse 2000). Among genome compartments, early comparisons examined average synonymous substitution rates. One of the first studies demonstrated that the rates for plant mitochondrial: chloroplast: nuclear genes are approximately 1:3:6–12 for seed plants (Wolfe et al. 1987). Although data at the time were particularly lacking for nuclear genes, the early studies already detected rate differences among lineages (Gaut et al. 1993; Wolfe et al. 1987). A more recent study, with a larger sample of taxa and genes, modified the ratio slightly to 1:3:10 (Drouin et al. 2008). The higher substitution rate in nuclear than plastid DNA is somewhat counterintuitive. Plastomes are effectively asexual, with little or no recombination among divergent genotypes, thus they are expected to accumulate mutations via Muller's ratchet. This contrasts with sexual systems where recombination can break up linkages and eliminate via selection deleterious to slightly deleterious mutations (Muller 1964). A recent explanation for this paradox argues that mutation accumulation is avoided via the high degree of polyploidy in plastid DNA, in combination with a correction process involving gene conversion (Khakhlova and Bock 2006).

Comparisons across genome compartments can obscure the variation among genes within compartments. Substitution rates vary considerably among chloroplast genes (Gaut et al. 1993). For synonymous substitutions, it appears that *rpl32* evolves about 20 times faster than *rps7* in the Poaceae, but only about five times faster in other angiosperm lineages. Similar variation in such extremes is seen for nonsynonymous rates (Guisinger et al. 2010b). Generally, the faster-evolving genes (and non-coding regions) are more useful for studies among recently diverged groups (such as species), and the more slowly evolving genes can be more effectively used to infer deeper phylogenetic events. However, such inferences can be confounded by variation in substitution rates among lineages. For example, substitution rates for plastid genes are faster in angiosperms than in gymnosperms (Drouin et al. 2008), faster in the branch leading to grasses than in other angiosperms (Gaut et al. 1993; Guisinger et al. 2010b), and faster in ferns than in seed plants (Wolf et al. 2011). In general, lineage-specific variation is more pronounced for nonsynonymous than synonymous rates (Muse and Gaut 1997; Wolfe et al. 1987). Also, increases in lineage-specific rates tend to correlate across plastid genes, consistent with the idea that there are "lineage effects". These effects could be a function of elevated mutation rates, possibly caused by variation in the efficiency of DNA repair mechanisms. It has also been suggested that similar lineage effects are responsible for elevated plastome reorganizations (Guisinger et al. 2010a).

10.8 Expression of Plastid Genes

Most introductory genetics textbooks contrast eukaryotic with prokaryotic gene expression and regulation. Plastid genes fall neatly into neither category; they have many eukaryotic features, retain many aspects of their prokaryotic ancestors, and also have unique components. Much of this complexity stems from the fact that most of the proteins active in plastids are encoded by nuclear genes. Thus, coordination is needed between gene regulation systems in the two compartments. Plastid genes for proteins are transcribed by two different RNA polymerases: a nuclear-encoded RNA polymerase (NEP) and a plastid-encoded RNA polymerase (PEP), each using different promoters (reviewed by Toyoshima et al. 2005). Some genes are transcribed by PEP, some by NEP, and some by both. For genes that can be transcribed by both polymerases, the switch from NEP to PEP is involved in plant cell development. Furthermore, PEP is similar to eubacterial RNA polymerase in that sigma factors are needed for transcription initiation. However, the sigma factors for PEP are nuclear-encoded. Thus, plastid transcription is under tight control from the nucleus (Toyoshima et al. 2005). Moreover, many plastid transcripts resemble prokaryotic transcripts, with multiple products from operons (Palmer 1985; Westhoff and Herrmann 1988). However, post-transcriptional events deviate considerably from those observed in prokaryotes; plastid transcripts are heavily processed in many ways, including several forms of 5′ and 3′ end maturation (reviewed by Stern et al. 2010). The latter can include polyadenylation (Schuster and Stern 2009), which functions in determining RNA stability (Stern et al. 2010). Additional protection from decay can come from 5′ to 3′ capping proteins, including pentatricopeptide repeat (PPR) proteins (Pfalz et al. 2009).

Several plastid genes contain introns. Land plant plastomes typically harbor 15–20 introns, most of which are type II introns, but one that is a type I intron (in *trnL*). One plastid gene (*rps12*) is trans-spliced in most plants. The intron groups are spliced by alternative mechanisms. In plastids, intron splicing is performed by a complex of RNA and proteins (reviewed by Stern et al. 2010). Most of the splicing factors are nuclear-encoded, but one maturase is

encoded by a plastid gene, *matK*. This gene usually sits within the group IIA intron of *trnK*. In general, many introns contain open reading frames like *matK* that encode splicing factors needed for splicing the flanking intron. However, most of these encoded factors appear to have been lost in plastid introns, with the exception of *matK*. Although *matK* is clearly involved in, and preferentially splices, the *trnK* intron (Vogel et al. 1999), it is thought that *matK* is also needed for splicing other plastid group II introns. The evidence for this is that *trnK* and its intron have been lost from several parasitic plant plastomes (Ems et al. 1995) and most ferns (Duffy et al. 2009), but *matK* has remained. More recently direct evidence shows that the MatK protein is closely associated with most plastid group II introns (Zoschke et al. 2010).

In addition to the various aspects of RNA processing that occur, an independent post-transcriptional stage is RNA editing. This process of altering the bases before translation is found throughout eukaryotes, and is especially common in organellar genomes (reviewed by Tillich et al. 2006). The process occurs at fewer than 40 sites in seed plant plastomes, at least ten times that number in ferns and hornworts, but has not been observed in liverworts. In plastid genes, cytosines are edited to uracils, but in hornworts and ferns, additional uracil-to-cytosine edits have been reported (Kugita et al. 2003; Wolf et al. 2004). RNA editing requires both cis- and trans-acting factors, including possible upstream and downstream recognition sequences. However, the latter appear to have no obvious pattern across sites. This might be because the trans-acting factors (nuclear-encoded proteins) are likely to be of several types. To date, over 20 different nuclear factors have been associated with RNA editing in *Arabidopsis* (see Stern et al. 2010), most of which are PPR proteins. High levels of RNA editing can pose a problem for phylogenetic analysis of genomic sequences, especially in hornworts where levels can be so high that the same site can be C to U edited in some taxa and U to C edited in other taxa (Duff and Moore 2005). The reasons for RNA editing are also not clear. It has been argued that the process acts to repair genomic sequences (Jobson and Qiu 2008; Stern et al. 2010), but this seems far less efficient than a simple nucleotide substitution in the DNA. An additional role has been implicated in gene regulation, a phenomenon observed in a few cases in animals, but rarely in plants (Stern et al. 2010). Alternatively, the proteins that are involved in RNA editing could have alternative cellular functions and their editing ability then releases selective constraints for the edited nucleotide positions. Other functions of these proteins are known. For example, in primates the APOBEC family of RNA editing enzymes are cytosine deaminases that act to restrict infection from retroviruses (Bransteitter et al. 2009). Further study of the *Arabidopsis* (and other plant) RNA editing factors is needed to understand better the function and cellular significance of RNA editing.

10.9 Plastome Biotechnology

Since recombinant DNA technology was developed in the 1970s, researchers have sought to convey specific genetic traits to plants. One important class of traits includes agronomic ones such as herbicide resistance, disease resistance, salt tolerance, and drought tolerance. Moreover, plants are ideal factories for the manufacture of vaccines and other biomaterials. However, several problems are associated with transgenic plants, including the possible spread of transgenes into wild relatives (Chandler and Dunwell 2008). Engineering the plastome instead of the nucleus can reduce this risk and provide additional advantages (Grevich and Daniell 2005), since plastome transgenic systems are maternally inherited so transgenes cannot be spread through pollen, the main source of containment failure (Chandler and Dunwell 2008). Also important is the greater copy number of plastomes compared to nuclear genes in a cell. This can result in much higher levels of transgene expression using plastome approaches. Because plastomes are so well characterized, it is relatively simple to target specific uptake sites, which typically are designed between genes to reduce unwanted effects on native genes. Plastid genes are also far less susceptible to gene silencing than are nuclear genes, further improving the success rate. Also, chloroplasts are ideal sites to accumulate substances that can be toxic in other parts of the cell. Finally, energy (from photosynthesis) is abundantly available in chloroplasts for high levels of expression. To date, only about a dozen plant species have been successfully transformed using plastome sequences (Verma et al. 2008), including carrot (Kumar et al. 2004a), potato (Sidorov et al. 1999), and cotton (Kumar et al. 2004b). Although the field is beyond its infancy, more work is needed before major crops with transgenic plastomes will be commercially available (Grevich and Daniell 2005).

Conclusions

Plastomes are a tiny fraction of the size of nuclear genomes, yet they are packed tightly with important genes. Unlike the current small sample of complete plant nuclear genomes, many hundreds of plastome sequences are known from most major clades of plants. This provides a powerful comparative database for studies of evolution, cell function, and applied biotechnology. Plastomes have unique physical features and patterns of gene expression. Their pedigree in the evolutionary analysis of nucleotide sequences, genes, genomes, and

their plant hosts, represents some of the earliest tools in these fields of study. However, despite these advances, our understanding of functional aspects of plastomes lags behind that of purely eukaryotic and prokaryotic systems. More research is needed, especially in the areas of plastome replication and gene expression. Because plastids are the primary structures that this planet uses to harness the sun's energy, it is essential that we understand how their genomes behave.

References

Allen JF (1992) Protein-phosphorylation in regulation of photosynthesis. Biochim Biophys Acta 1098:275–335

Allen JF (2003) Why chloroplasts and mitochondria contain genomes. Comp Funct Genomics 4:31–36

Baur E (1909) Das Wesen und die Erblichkeitsverhältnisse der "Varietates albomarginatae hort." von *Pelargonium zonale*. Ztschr Indukt Abstammungs Vererbungsl 1:330–351

Bedbrook JR, Kolodner R (1979) Structure of chloroplast DNA. Annu Rev Plant Physiol Plant Mol Biol 30:593–620

Bendich AJ (2004) Circular chloroplast chromosomes: the grand illusion. Plant Cell 16:1661–1666

Bock R (2007) Structure, function, and inheritance of plastid genomes. Cell Mol Biol Plast 19:29–63

Boffey SA, Leech RM (1982) Chloroplast DNA levels and the control of chloroplast division in light-grown wheat leaves. Plant Physiol 69:1387–1391

Bransteitter R, Prochnow C, Chen XJS (2009) The current structural and functional understanding of APOBEC deaminases. Cell Mol Life Sci 66:3137–3147

Brouard JS, Otis C, Lemieux C, Turmel M (2010) The exceptionally large chloroplast genome of the green alga *Floydiella terrestris* illuminates the evolutionary history of the Chlorophyceae. Genome Biol Evol 2:240–256

Bui ETN, Bradley PJ, Johnson PJ (1996) A common evolutionary origin for mitochondria and hydrogenosomes. Proc Natl Acad Sci U S A 93:9651–9656

Cai ZQ, Guisinger M, Kim HG, Ruck E, Blazier JC, McMurtry V, Kuehl JV, Boore J, Jansen RK (2008) Extensive reorganization of the plastid genome of *Trifolium subterraneum* (Fabaceae) is associated with numerous repeated sequences and novel DNA insertions. J Mol Evol 67:696–704

Catalano SA, Saidman BO, Vilardi JC (2009) Evolution of small inversions in chloroplast genome: a case study from a recurrent inversion in angiosperms. Cladistics 25:93–104

Cavalier-Smith T (2002) Chloroplast evolution: secondary symbiogenesis and multiple losses. Curr Biol 12:R62–R64

Chandler S, Dunwell JM (2008) Gene flow, risk assessment and the environmental release of transgenic plants. Crit Rev Plant Sci 27:25–49

Chiba Y (1951) Cytochemical studies on chloroplasts. I. Cytologic demonstration of nucleic acids in chloroplasts. Cytologia 16:259–264

Chumley TW, Palmer JD, Mower JP, Fourcade HM, Calie PJ, Boore JL, Jansen RK (2006) The complete chloroplast genome sequence of *Pelargonium × hortorum*: organization and evolution of the largest and most highly rearranged chloroplast genome of land plants. Mol Biol Evol 23:2175–2190

Drouin G, Daoud H, Xia J (2008) Relative rates of synonymous substitutions in the mitochondrial, chloroplast and nuclear genomes of seed plants. Mol Phylogenet Evol 49:827–831

Duff RJ, Moore FBG (2005) Pervasive RNA editing among hornwort *rbcL* transcripts except *Leiosporceros*. J Mol Evol 61:571–578

Duffy AM, Kelchner SA, Wolf PG (2009) Conservation of selection on *matK* following an ancient loss of its flanking intron. Gene 438:17–25

Ems SC, Morden CW, Dixon CK, Wolfe KH, dePamphilis CW, Palmer JD (1995) Transcription, splicing and editing of plastid RNAs in the nonphotosynthetic plant *Epifagus virginiana*. Plant Mol Biol 29:721–733

Fan WH, Woelfle MA, Mosig G (1995) Two copies of a DNA element, 'Wendy', in the chloroplast chromosome of *Chlamydomonas reinhardtii* between rearranged gene clusters. Plant Mol Biol 29:63–80

Gaut BS, Muse SV, Clegg MT (1993) Relative rates of nucleotide substitution in the chloroplast genome. Mol Phylogenet Evol 2:89–96

Goffinet B, Wickett NJ, Werner O, Ros RM, Shaw AJ, Cox CJ (2007) Distribution and phylogenetic significance of the 71-kb inversion in the plastid genome in Funariidae (Bryophyta). Ann Bot 99:747–753

Greiner S, Wang X, Rauwolf U, Silber MV, Mayer K, Meurer J, Haberer G, Herrmann RG (2008) The complete nucleotide sequences of the five genetically distinct plastid genomes of *Oenothera*, subsection *Oenothera*: I. Sequence evaluation and plastome evolution. Nucleic Acids Res 36:2366–2378

Grevich JJ, Daniell H (2005) Chloroplast genetic engineering: recent advances and future perspectives. Crit Rev Plant Sci 24:83–107

Guisinger MM, Chumley TW, Kuehl JV, Boore JL, Jansen RK (2010a) Implications of the plastid genome sequence of *Typha* (Typhaceae, Poales) for understanding genome evolution in Poaceae. J Mol Evol 70:149–166

Guisinger MM, Kuehl JV, Boore JL, Jansen RK (2010b) Extreme reconfiguration of plastid genomes in the angiosperm family Geraniaceae: rearrangements, repeats, and codon usage. Mol Biol Evol 28:583–600

Haberle RC, Fourcade HM, Boore JL, Jansen RK (2008) Extensive rearrangements in the chloroplast genome of *Trachelium caeruleum* are associated with repeats and tRNA genes. J Mol Evol 66:350–361

Hagemann R (2000) Erwin Baur or Carl Correns: who really created the theory of plastid inheritance? J Hered 91:435–440

Hirao T, Watanabe A, Kurita M, Kondo T, Takata K (2008) Complete nucleotide sequence of the *Cryptomeria japonicia* D. Don. chloroplast genome and comparative chloroplast genomics: diversified genomic structure of coniferous species. BMC Plant Biol 8:70

Howe CJ, Barker RF, Bowman CM, Dyer TA (1988) Common features of three inversions in wheat chloroplast DNA. Curr Genet 13:343–349

Jansen RK, Raubeson LA, Boore JL, dePamphilis CW, Chumley TW et al (2005) Methods for obtaining and analyzing whole chloroplast genome sequences. Methods Enzymol 395:348–384

Jobson RW, Qiu YL (2008) Did RNA editing in plant organellar genomes originate under natural selection or through genetic drift? Biol Direct 3:43

Karol KG, Arumuganathan K, Boore JL, Duffy AM, Everett KDE et al (2010) Complete plastome sequences of *Equisetum arvense* and *Isoetes flaccida*: implications for phylogeny and plastid genome evolution of early land plant lineages. BMC Evol Biol 10:321

Kelchner SA, Wendel JF (1996) Hairpins create minute inversions in non-coding regions of chloroplast DNA. Curr Genet 30:259–262

Keeling PJ (2010) The endosymbiotic origin, diversification and fate of plastids. Philos Trans R Soc Lond B 365:729–748

Khakhlova O, Bock R (2006) Elimination of deleterious mutations in plastid genomes by gene conversion. Plant J 46:85–94

Kim KJ, Lee HL (2005) Widespread occurrence of small inversions in the chloroplast genomes of land plants. Mol Cells 19:104–113

Kim KJ, Choi KS, Jansen RK (2005) Two chloroplast DNA inversions originated simultaneously during the early evolution of the sunflower family (Asteraceae). Mol Biol Evol 22:1783–1792

Kolodner R, Tewari KK (1972) Molecular size and conformation of chloroplast deoxyribonucleic acid from pea leaves. J Biol Chem 247:6355–6364

Kolodner R, Tewari KK (1975a) Presence of displacement loops in covalently closed circular chloroplast deoxyribonucleic-acid from higher-plants. J Biol Chem 250:8840–8847

Kolodner R, Tewari KK (1975b) Chloroplast DNA from higher plants replicates by both the Cairns and the rolling circle mechanism. Nature 256:708–711

Krause K (2008) From chloroplasts to "cryptic" plastids: evolution of plastid genomes in parasitic plants. Curr Genet 54:111–121

Krishnan NM, Rao BJ (2009) A comparative approach to elucidate chloroplast genome replication. BMC Genomics 10:237

Kugita M, Yamamoto Y, Fujikawa T, Matsumoto T, Yoshinaga K (2003) RNA editing in hornwort chloroplasts makes more than half the genes functional. Nucleic Acids Res 31:2417–2423

Kumar S, Dhingra A, Daniell H (2004a) Plastid-expressed betaine aldehyde dehydrogenase gene in carrot cultured cells, roots, and leaves confers enhanced salt tolerance. Plant Physiol 136:2843–2854

Kumar S, Dhingra A, Daniell H (2004b) Stable transformation of the cotton plastid genome and maternal inheritance of transgenes. Plant Mol Biol 56:203–216

Kunnimalaiyaan M, Nielsen BL (1997a) Chloroplast DNA replication: mechanism, enzymes and replication origins. J Plant Biochem Biot 6:1–7

Kunnimalaiyaan M, Nielsen BL (1997b) Fine mapping of replication origins (oriA and oriB) in *Nicotiana tabacum* chloroplast DNA. Nucleic Acids Res 25:3681–3686

Martin W (2003) Gene transfer from organelles to the nucleus: frequent and in big chunks. Proc Natl Acad Sci U S A 100:8612–8614

Martin W, Herrmann RG (1998) Gene transfer from organelles to the nucleus: how much, what happens, and why? Plant Physiol 118:9–17

Martin W, Kowallik KV (1999) Annotated English translation of Mereschkowsky's 1905 paper "Über Natur und Ursprung der Chromatophoren im Pflanzenreiche". Eur J Phycol 34:287–295

Martin W, Rujan T, Richly E, Hansen A, Cornelsen S et al (2002) Evolutionary analysis of *Arabidopsis*, cyanobacterial, and chloroplast genomes reveals plastid phylogeny and thousands of cyanobacterial genes in the nucleus. Proc Natl Acad Sci USA 99:12246–12251

Mereschkowski KS (1905) Über Natur und Ursprung der Chromatophoren im Pflanzenreiche. Biol Centralbl 25(593–604):689–691

Milligan BG, Hmpton JN, Palmer JD (1989) Dispersed repeats and structural reorganization in subclover chloroplast DNA. Mol Biol Evol 6:355–368

Mühlbauer SK, Lössl A, Tzekova L, Zou ZR, Koop HU (2002) Functional analysis of plastid DNA replication origins in tobacco by targeted inactivation. Plant J 32:175–184

Muller HJ (1964) The relation of recombination to mutational advance. Mutat Res 1:2–9

Muse SV (2000) Examining rates and patterns of nucleotide substitution in plants. Plant Mol Biol 42:25–43

Muse SV, Gaut BS (1997) Comparing patterns of nucleotide substitution rates among chloroplast loci using the relative ratio test. Genetics 146:393–399

Ohyama K, Fukuzawa H, Kohchi T, Shirai H, Sano T et al (1986) Chloroplast gene organization deduced from complete sequence of liverwort *Marchantia polymorpha* chloroplast DNA. Nature 322:572–574

Oldenburg DJ, Bendich AJ (2004a) Changes in the structure of DNA molecules and the amount of DNA per plastid during chloroplast development in maize. J Mol Biol 344:1311–1330

Oldenburg DJ, Bendich AJ (2004b) Most chloroplast DNA of maize seedlings in linear molecules with defined ends and branched forms. J Mol Biol 335:953–970

Palmer JD (1983) Chloroplast DNA exists in two orientations. Nature 301:92–93

Palmer JD (1985) Comparative organization of chloroplast genomes. Annu Rev Genet 19:325–354

Palmer JD (1991) Plastid chromosomes: structure and evolution. In: Bogorad L, Vasil IK (eds) Cell culture and somatic cell genetics of plants, vol 7A, Molecular biology of plastids. Academic, San Diego, pp 5–53

Palmer JD, Stein DB (1986) Conservation of chloroplast genome structure among vascular plants. Curr Genet 10:823–833

Pfalz J, Bayraktar OA, Prikryl J, Barkan A (2009) Site-specific binding of a PPR protein defines and stabilizes 5′ and 3′ mRNA termini in chloroplasts. EMBO J 28:2042–2052

Race HL, Herrmann RG, Martin W (1999) Why have organelles retained genomes? Trends Genet 15:364–370

Raubeson LA, Jansen RK (1992a) Chloroplast DNA evidence on the ancient evolutionary split in vascular land plants. Science 255:1697–1699

Raubeson LA, Jansen RK (1992b) A rare chloroplast-DNA structural mutation is shared by all conifers. Biochem Syst Ecol 20:17–24

Renner O (1929) Artbastarde bei Pflanzen. Bornträger, Berlin

Rowan BA, Oldenburg DJ, Bendich AJ (2010) RecA maintains the integrity of chloroplast DNA molecules in Arabidopsis. J Exp Bot 61:2575–2588

Sager R, Ishida MR (1963) Chloroplast DNA in Chlamydomonas. Proc Natl Acad Sci U S A 50:725–730

Sato N, Terasawa K, Miyajima K, Kabeya Y (2003) Organization, developmental dynamics, and evolution of plastid nucleoids. Int Rev Cytol 232:217–262

Scharff LB, Koop HU (2007) Targeted inactivation of the tobacco plastome origins of replication A and B. Plant J 50:782–794

Schimper AFW (1883) Ueber die Entwickelung der Chlorophyllköner und Farbkörper. Bot Zeit 41:105–114

Schimper AFW (1885) Untersuchungen über die Chlorophyllköner und die ihnen homologen Gebilde. Jahrb Wiss Bot 16:1–247

Schuster G, Stern D (2009) RNA polyadenylation and decay in mitochondria and chloroplasts. Prog Mol Biol Transl Sci 85:393–422

Schwender J, Goffman F, Ohlrogge JB, Shachar-Hill Y (2004) Rubisco without the Calvin cycle improves the carbon efficiency of developing green seeds. Nature 432:779–782

Sekine K, Fujiwara M, Nakayama M, Takao T, Hase T, Sato N (2007) DNA binding and partial nucleoid localization of the chloroplast stromal enzyme ferredoxin: sulfite reductase. FEBS J 274:2054–2069

Shinozaki K, Ohme M, Tanaka M, Wakasugi T, Hayashida N et al (1986) The complete nucleotide sequence of tobacco chloroplast genome: its gene organization and expression. EMBO J 5:2043–2049

Sidorov VA, Kasten D, Pang SZ, Hajdukiewicz PTJ, Staub JM, Nehra NS (1999) Stable chloroplast transformation in potato: use of green fluorescent protein as a plastid marker. Plant J 19:209–216

Stein DB, Palmer JD, Thompson WF (1986) Structural evolution and flip-flop recombination of chloroplast DNA in the fern genus Osmunda. Curr Genet 10:835–841

Stern DB, Goldschmidt-Clermont M, Hanson MR (2010) Chloroplast RNA metabolism. Annu Rev Plant Biol 61:125–155

Sugiura M (2003) History of chloroplast genomics. Photosynth Res 76:371–377

Tillich M, Lehwark P, Morton BR, Maier UG (2006) The evolution of chloroplast RNA editing. Mol Biol Evol 23:1912–1921

Toyoshima Y, Onda Y, Shiina T, Nakahira Y (2005) Plastid transcription in higher plants. Crit Rev Plant Sci 24:59–81

Turmel M, Otis C, Lemieux C (2006) The chloroplast genome sequence of Chara vulgaris sheds new light into the closest green algal relatives of land plants. Mol Biol Evol 23:1324–1338

Verma D, Samson NP, Koya V, Daniell H (2008) A protocol for expression of foreign genes in chloroplasts. Nat Protoc 3:739–758

Vinogradov AE (2002) Growth and decline of introns. Trends Genet 18:232–236

Vogel J, Borner T, Hess WR (1999) Comparative analysis of splicing of the complete set of chloroplast group II introns in three higher plant mutants. Nucleic Acids Res 27:3866–3874

von Heijne G (1986) Why mitochondria need a genome. FEBS Lett 198:1–4

Wang D, Wu YW, Shih ACC, Wu CS, Wang YN, Chaw SM (2007) Transfer of chloroplast genomic DNA to mitochondrial genome occurred at least 300 MYA. Mol Biol Evol 24:2040–2048

Wayne R (2009) Plant cell biology. Elsevier/Academic, San Diego

Wei FS, Stein JC, Liang CZ, Zhang JW, Fulton RS et al (2009) Detailed analysis of a contiguous 22-Mb region of the maize genome. PLoS Genet 5:e1000728

Westhoff P, Herrmann RG (1988) Complex RNA maturation in chloroplasts. The psbB operon from spinach. Eur J Biochem 171:551–564

Wicke S, Schneeweiss GM, dePamphilis C, Müller KF, Quandt D (2011) The evolution of the plastid chromosome in land plants: gene content, gene order, gene function. Plant Mol Biol 76:273–297

Wolf PG, Rowe CA, Hasebe M (2004) High levels of RNA editing in a vascular plant chloroplast genome: analysis of transcripts from the fern Adiantum capillus-veneris. Gene 339:89–97

Wolf PG, Roper JM, Duffy AM (2010) The evolution of chloroplast genome structure in ferns. Genome 53:731–738

Wolf PG, Der JP, Duffy AM, Jacobson JB, Grusz AL, Pryer KM (2011) The evolution of chloroplast genes and genomes in ferns. Plant Mol Biol 76(3–5):251–261

Wolfe KH, Li W-H, Sharp PM (1987) Rates of nucleotide substitution vary greatly among plant mitochondrial, chloroplast, and nuclear DNAs. Proc Natl Acad Sci U S A 84:9054–9058

Wolfe KH, Morden CW, Palmer JD (1992) Functions and evolution of a minimal plastid genome from a nonphotosynthetic parasitic plant. Proc Natl Acad Sci U S A 89:10648–10652

Zoschke R, Nakamura M, Liere K, Sugiura M, Börner T, Schmitz-Linneweber C (2010) An organellar maturase associates with multiple group II introns. Proc Natl Acad Sci USA 107:3245–3250

Duplications and Turnover in Plant Genomes

11

Michael S. Barker, Gregory J. Baute, and Shao-Lun Liu

Contents

11.1 Introduction

Cytologists have long documented differences in chromosome number and organization among plants, but the truly dynamic nature of plant genome evolution is only becoming apparent with fully sequenced and assembled genomes. A major result of these new data is that duplication—of single genes, chromosomes, and whole genomes—is a major force in the evolution of plant genome structure and content. For example, genomic comparisons among divergent animals are able to recover significant signatures of synteny (Hiller et al. 2004), but less divergent flowering plant genomes often demonstrate relatively lower large-scale collinearity because of cycles of polyploidy and diploidization (Tang et al. 2008; Salse et al. 2009). Among individuals and closely related species, copy number variation and changes in gene family size are now recognized as critical sources of genetic variation (Lynch 2007). Duplication and subsequent resolution have yielded a continually changing genome whose elements are constantly turning-over. Although gene and genome duplication has long garnered attention as a potentially important source of evolutionary novelty (Haldane 1933; Stebbins 1950; Ohno 1970), the perspective of a dynamic plant genome fueled by duplication and loss stands in contrast to classical concepts of a largely stable genome. In this chapter we provide an overview of how duplication-driven genomic turn-over has influenced the evolution and diversity of plant genomes.

11.2 Frequency and Sources of Duplications in Plant Genomes

Plant and other eukaryotic genomes contain an abundance of duplications with a variety of origins. The power of these duplications to influence evolution is attenuated by the rate and nature of gene birth and death. In most eukaryotes, including plants, plots of the age of all gene duplications in

M.S. Barker (✉)
Department of Ecology & Evolutionary Biology, University of Arizona, Tucson, AZ, USA
e-mail: msbarker@email.arizona.edu

J.F. Wendel et al. (eds.), *Plant Genome Diversity Volume 1*,
DOI 10.1007/978-3-7091-1130-7_11, © Springer-Verlag Wien 2012

Fig. 11.1 Age distribution of gene duplications from multiple nuclear gene family phylogenies. (**a**) *Selaginella moellendorffii* (Banks et al. 2011), the only sequenced plant genome without an ancient polyploidy. (**b**) *Helianthus argophyllus* (Barker et al. 2008) with a paleopolyploidy visible as a large peak of gene duplications

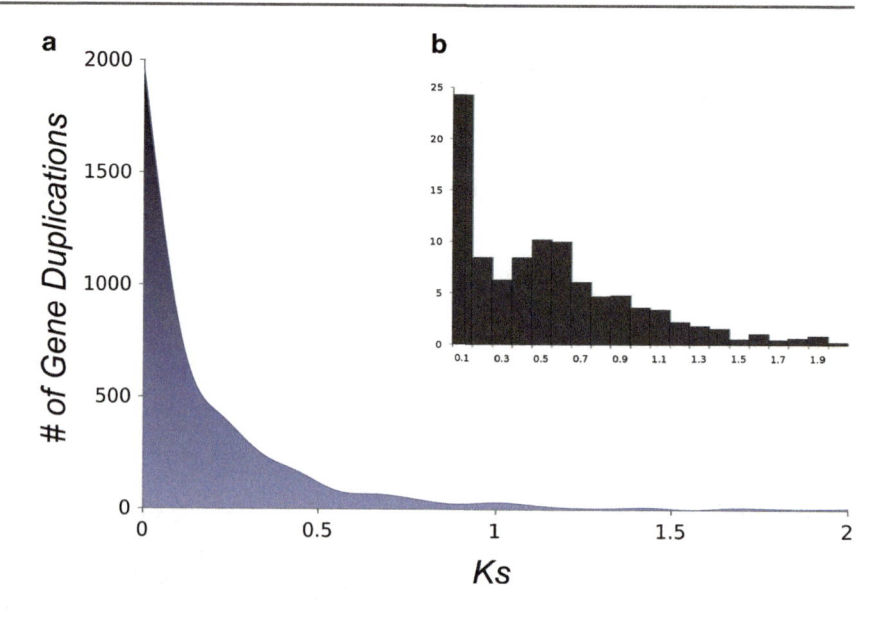

a single genome demonstrate an exponential or power-law distribution, generally with a steep curve indicative of high birth and death rates (Fig. 11.1). An immediate result apparent from these types of plots is that most duplicated genes do not survive very long. However, polyploidy, or whole genome duplication, yields a large and enduring signature in such plots (Fig. 11.1b). Although these polyploid peaks are observable even after many millions of years because of the scale of gene birth, the signal likely persists longer than expected because some gene classes are known to be strongly retained in duplicate following polyploidy (Maere et al. 2005; Doyle et al. 2008; Edger and Pires 2009). Thus, the nature of gene birth has a significant impact on the ultimate fate of duplicates.

How often does a gene duplicate in the average plant genome? Lynch and Conery (2003) measured the birth rate (*B*) of genes in *Arabidopsis thaliana* to be 0.0032 with a death rate (*D*) of 0.033 on a time scale of 1% synonymous divergence per gene for a single individual. Using an alternative approach, Maere et al. (2005) arrived at a nearly identical *B* for *Arabidopsis*—0.03 per 10% synonymous divergence. Although these estimates are based on the initial peak of duplications and should not be confounded by paleopolyploidy, it is possible that the presence of paleopolyploid duplicates influences the birth and death rates of these young, ongoing duplications in plant genomes. To evaluate this possibility, we used Lynch and Connery's (2003) method to estimate birth and death rates for the recently sequenced genome of *Selaginella moellendorffii* (Banks et al. 2011), the only sequenced plant without evidence of ancient polyploidy. We estimate $B = 0.00883$ and $D = 0.0999$ for *S. moellendorffii* on a scale of 1% synonymous divergence. Although these rates are higher than

Arabidopsis, the *B*/*D* ratios (*Arabidopsis B/D* = 0.0970; *Selaginella B/D* = 0.0883) are similar, suggesting that paleopolyploidy likely has a negligible impact on the birth and death rates of ongoing gene duplication in plant genomes. Notably, these *B/D* ratios are the highest estimated rates among sampled eukaryotes, but still of the same order of magnitude (Lynch and Conery 2003). Overall birth rates in most plant genomes will be much higher than these estimates from the initial slope of the birth-death curve because of ancient polyploidy.

To put these rates of gene birth and death into perspective, it is useful to compare them to other evolutionary events. For example, the average gene birth rate for eukaryotes is approximately 40% of the rate of mutation per nucleotide site (Lynch 2007). *Arabidopsis* and *Selaginella* are within the reported range with rates of gene duplication—excluding polyploidy—of nearly 32% and 88% the rate of mutation per nucleotide site. If a whole genome duplication occurs at least once per 100% synonymous divergence—a reasonable estimate for most angiosperms (Blanc and Wolfe 2004a, b; Pfeil et al. 2005; Cui et al. 2006; Barker et al. 2008; Tang et al. 2008; Shi et al. 2010)—then the rate of duplication for plant genes will often exceed the mutation rate over the time scale it takes to achieve mutational saturation. Considering that some whole genome duplications are actually triplications (Jaillon et al. 2007) or possibly higher multiples, and that many plant genomes have experienced more than one round of polyploidy on a time scale of 100% synonymous divergence (e.g., Cui et al. 2006; Barker et al. 2008; Shi et al. 2010), the genome-wide rate of gene duplication is substantially higher than the rate of mutation per nucleotide in plant genomes on a long time scale.

11.2.1 Small-Scale Duplications: Segmental Duplication

Both large and small-scale duplication events produce new genes in plant genomes. Although large-scale duplication events such as whole genome duplications have received significant attention among plant biologists, much more frequent but small-scale duplication events have contributed similar quantities of paralogs. Nearly 25% of the genes in *Arabidopsis* are the product of ancient whole genome duplications (Blanc et al. 2003), but nearly 16% of the genes occur as tandem duplicates (Rizzon et al. 2006), just one class of small-scale duplications. Lynch (2007) describes four mechanisms for the origin of gene-sized duplications in eukaryotic genomes that are collectively referred to as segmental duplications. These include tandemly duplicated genes derived from replication errors such as strand slippage or unequal crossing-over, the replication and ectopic insertion of genes downstream from sloppily transcribed non-LTR retrotransposons, the capture and incorporation of new genes from DNA inserts during repair of double-strand breaks, and duplication by ectopic exchange when the ends of a double-strand break invade a non-homologous region and initiates a recombination event that copies a gene(s) by strand extension before repair of the broken chromosome.

Analyses of plant genomes indicate that these mechanisms actively and variously contribute to the repertoire of gene families across the plant phylogeny. For example, unequal crossing over appears to have been responsible for a large fraction of tandemly arrayed genes in *Arabidopsis thaliana* and *Oryza sativa* (Rizzon et al. 2006). Consistent with duplication by unequal crossing over, Rizzon and colleagues (2006) found nearly 80% and 88% of the tandemly arrayed genes in rice and *Arabidopsis*, respectively, are in direct orientation. Other mechanisms, such as intrachromosomal recombination between repeats, generally produce tandem duplicates that are oriented in opposite directions (Schuermann et al. 2005). Thus, unequal crossing over appears to be responsible for ~14% and ~11% of the genes in the *Arabidopsis* and rice genomes, respectively.

Other mechanisms of segmental duplication have also produced abundant duplications in plant genomes. In particular, repetitive DNA constitutes a large fraction of plant genomes and is an important mechanism of genome size evolution (Leitch et al. 2005). Much of this repetitive DNA is non-coding and derived from a variety of mobile genetic elements. However, many genes or at least gene fragments may be duplicated along with mobile elements. For example, *mutator-like* elements (MULEs) are associated with the transduplication of more than 1,300 genes or gene fragments in rice (Jiang et al. 2004; Juretic et al. 2005). Although most transduplicated gene fragments are pseudogenes (Juretic

et al. 2005), some transduplicated genes may be active and contribute to phenotypic evolution (Jiang et al. 2004). An outstanding case is the retrotransposon mediated gene duplication of a *SUN* gene in a cultivar of *Solanum lycopersicum*. *SUN* is a major gene involved in the control of fruit shape, particularly length, in tomato. By mapping and positionally cloning the *SUN* locus, Xiao and colleagues (2008) discovered that the region, which encoded a member of a IQ67 domain-containing family, arose by duplication of a 24.7 kb fragment mediated by the *Rider* long terminal repeat retrotransposon. Critically, duplication and movement of this gene led to increased *SUN* expression that significantly altered the phenotype. Considering the frequency of repetitive elements in plant genomes, their potential contributions to both gene and gene expression evolution may be significant.

Functionally transduplicated gene families may also expand selfishly and contribute little to the host organism. Hoen and colleagues (2006) recently found evidence for functional transduplication and expansion of a family of *ULP*-like genes in *Arabidopsis* associated with MULEs. Unlike most functional cellular gene families, these *ULP*-like genes were targeted by small RNAs and silenced in most tissues but with elevated expression in the pollen. Further analyses indicate that these genes may actually encode for proteins that disrupt host transposon-silencing mechanisms. Thus, this gene family may be selfish rather than contributing to cellular processes.

Perhaps more significant for the origin of new functional paralogs and synteny evolution is the discovery of single-gene transposition in plant genomes. Strikingly, Freeling and colleagues (2008) found evidence that 25–75% of all genes in *Arabidopsis* have moved location by single-gene transposition since divergence from *Carica*. Recent analyses in maize (Woodhouse et al. 2010) have uncovered evidence that flanking repeats are associated with many of these transposed genes as well as membership in tandem arrays, suggesting that intrachromosomal recombination between tandem arrays may be responsible for the duplication and movement of these sequences. Evidence for a similar process has been found among other plant lineages, particularly the nucleotide-binding site (NBS)-Leucine-rich repeat (LRR) gene family in *Medicago*. Composed of 400–500 members, the NBS-LRR gene family in *Medicago* appears to have expanded through a combination of tandem duplication and transduplication (Ameline-Torregrosa et al. 2008). Additional high quality genome sequences, especially from relatively closely related plants where mutation and turnover has not completely erased the genomic footprints of duplication mechanisms, are needed to further resolve the contribution of various molecular processes and ultimately evolutionary forces that drive segmental duplications in plant genomes.

11.2.2 Large-Scale Duplications: Polyploidy

Perhaps one of the largest changes ushered in by analyses of plant genomes is a paradigm shift on the importance of polyploidy, or whole genome duplication, in the evolution of plant genome organization and diversity. Prior to genomic analyses of the *Arabidopsis* genome that uncovered evidence of ancient whole genome duplications (Vision et al. 2000), some researchers had come to regard polyploid species as nothing more than evolutionary "dead-ends" (e.g., Wagner 1970). Despite accounting for an abundance of extant plant diversity—nearly 35% of vascular plants are recent polyploids (Wood et al. 2009)—some botanists surmised that polyploidy had, for various reasons, not replaced diploidy as the predominant genetic system and thus polyploids must largely go extinct (Wagner 1970; Stebbins 1971). Although analyses of recent polyploids show that most polyploid species go extinct at much higher rates than related diploids (Mayrose et al. 2011), a fraction of the polyploid species formed over the history of plant evolution have survived and their legacies are evident in plant genomes. Analyses of the *Arabidopsis* genome revealed that a plant long-considered a model diploid species was in fact an ancient polyploid that had experienced at least three rounds of paleopolyploidy that were obscured by gene fractionation and genetic diploidization over millions of years (Vision et al. 2000; Blanc et al. 2003; Bowers et al. 2003). Additional analyses of plant genomic data have uncovered evidence of widespread paleopolyploidy across the plant phylogeny (Blanc and Wolfe 2004a; Pfeil et al. 2005; Cui et al. 2006; Barker et al. 2008, 2009; Tang et al. 2008; Shi et al. 2010), including duplications before the origin of seed plants and angiosperms (Jiao et al. 2011). Far from being an evolutionary dead-end, polyploidy has occurred repeatedly throughout the evolutionary history of plants.

Unlike ongoing small-scale segmental duplications that often produce dead-on-arrival genes or gene fragments, polyploidy punctually creates functional copies of the entire genome. This scale of duplication means that although polyploidies occur at a much lower rate than for example, tandem duplications, they still have a large impact on gene content. The fraction of genes in contemporary "diploid" genomes directly derived from paleopolyploidy varies, but is generally substantial. Near the low end of the range for plants is *Arabidopsis thaliana* with nearly 25% of its genes derived from paleopolyploidy (Blanc et al. 2003). Other plants generally have higher fractions of genes retained from ancient polyploidy: 50% in rice (Wang et al. 2005), 67% in soybean (Schmutz et al. 2010), and ~30% in *Populus* (Tuskan et al. 2006). The time since polyploidy and the number of duplications does not appear to entirely explain the different contributions of polyploidy to plant genomes. For example, *Arabidopsis* genome has been multiplied at

least 12 times (Tang et al. 2008) yet contains fewer genes and chromosomes than *Vitis* (Jaillon et al. 2007) whose genome has been multiplied only six times (Tang et al. 2008). Considering that *Arabidopsis* has also experienced two rounds of whole genome duplications since its divergence from *Vitis* (Tang et al. 2008; Barker et al. 2009), it is indicative that variation in rates of gene fractionation and diploidization are just as important as new duplications for shaping the content of plant genomes.

It is important to note that although polyploidy has directly produced a significant fraction of the genes present in many plant genomes, the current complement of genes does not appear to be entirely dependent on polyploidy. A single sequenced plant genome, *Selaginella moellendorffii*, demonstrates no evidence of paleopolyploidy (Banks et al. 2011). However, its genome contains 22,285 annotated genes, nearly 85% of the annotated gene count in *Arabidopsis thaliana* (Haas et al. 2005). *Selaginella's* relatively small genome size and low gene count—the lowest among sequenced land plant genomes—is probably partially attributable to its lack of paleopolyploidy. Although polyploidy has clearly been a predominant force in creating new genes in many plant genomes, the fact that the *Selaginella* genome contains nearly as many genes as an angiosperm whose genome has been multiplied at least 12 times underscores that polyploid history does not, by itself, predict the current complement of genes in plant genomes.

11.3 Differential Fates of Duplicated Genes in Plant Genomes

11.3.1 Introduction and Nonfunctionalization

Newly duplicated genes face several possible fates, but most frequently one member of a newly formed gene pair is lost (Walsh 1995). This is because the rate of deleterious mutation is much higher than that of beneficial mutation (Kimura 1983), and if duplicated genes act redundantly one copy may eventually accumulate deleterious mutations and undergo pseudogenization. Computational simulation supports the view that the rate of gene loss after duplication is at least an order of magnitude higher than gene divergence (e.g., Ohta 1987; Walsh 1995). One need only look at the steepness of the Ks curve for duplicate genes in any eukaryotic genome to see that there are many young duplicates but that relatively few persist. Although the reversion to single copy after gene duplication is expected, there are still numerous duplicated genes retained in all eukaryotic genomes studied. Several different models have been proposed that attempt to explain why and how duplicates are retained (Fig. 11.2). Here we attempt to only provide a more detailed introduction for three predominant models: (1) Neofunctionalization;

Fig. 11.2 Alternative fates of duplicated genes in plant genomes. Panels (**a–d**) *upper* gene model depicts preduplication ancestor, *below* are the duplicates. (**a**) Protein subfunctionalization. (**b**) Expression subfunctionalization. (**c**) Protein neofunctionalization. (**d**) Expression neofunctionalization. Panels (**e–g**) portray dosage-sensitive genes and macromolecular complexes produced from their products. (**e**) A dosage sensitive gene contributing a component to a macromolecular product. (**f**) Small scale duplication, gene dosage is unbalanced. (**g**) Large scale duplication resulting in duplication of all genes in the pathway and balanced gene dosage

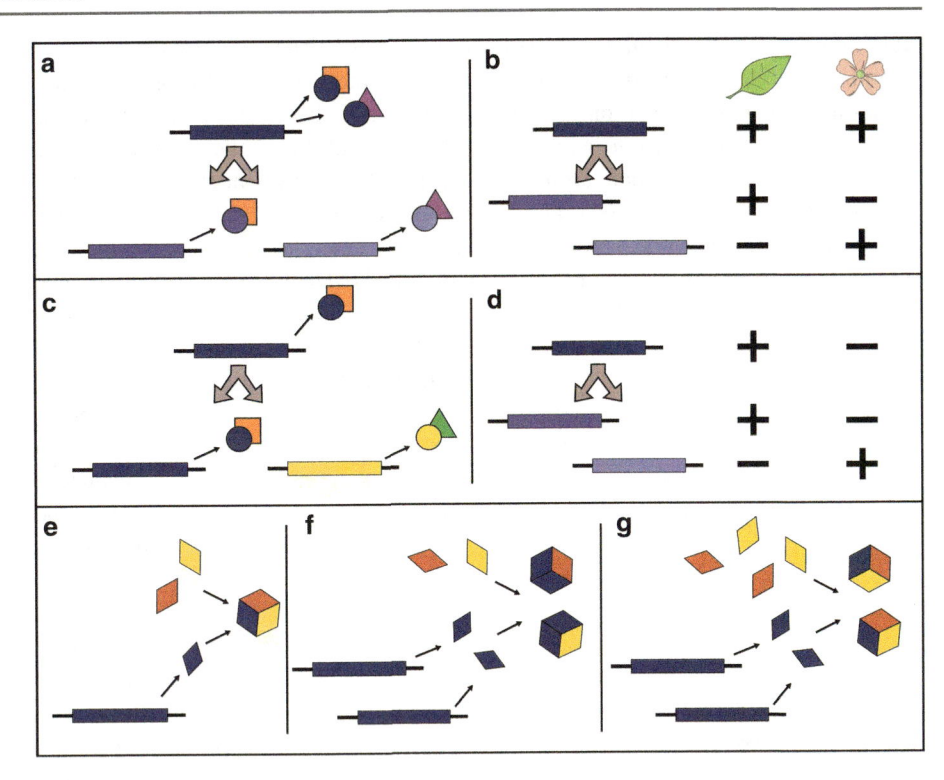

(2) Subfunctionalization; and (3) Gene dosage balance. More complete discussion of all the proposed routes to duplicate gene retention can be found in recent reviews, e.g., Sémon and Wolfe (2007), Conant and Wolfe (2008), Hahn (2009), and Innan and Kondrashov (2010).

11.3.2 Neofunctionalization

Long hypothesized to be an important source of new variation (Harland 1936; Stephens 1951; Ohno 1970), the neofunctionalization model anticipates the gain of a new function by one gene of a duplicated pair. In this model, duplicated genes are expected to have redundant function so one copy is free to accumulate mutations without affecting the gene's ancestral role. Either rapidly or over a long period of time, this copy may gain a new function. As discussed below with various examples, this process may happen by completely neutral processes (Dykhuizen–Hartl effect), as an adaptive process (positive Darwinian selection), or by some combination of the two.

11.3.2.1 Dykhuizen–Hartl Effect (Neutral Process)

In this scenario, a new function arises as a by-product of the accumulation of chance mutations in the cis-regulatory or protein coding regions. Critically, these neutral mutations become prevalent in the population by genetic drift, a dominant process in eukaryotes with small population sizes. Although initially increased in frequency by non-adaptive

mechanisms these mutations might later become adaptive, especially if the ecological or biochemical environment changes, the so called Dykhuizen–Hartl effect (Kimura 1983). Thus, there may be no need for positive Darwinian selection to drive the acquisition of a new function. Liu and Adams (2010) provide a potential example that shows neofunctionalization of a duplicated gene by neutral process. They studied a pair of duplicated genes, *Short Suspensor* (*SSP*) and *Brassinosteroid Kinase 1* (*BSK1*), derived from the most recent Brassicaceae-wide paleopolyploidization, the At-α duplication (Tang et al. 2008; Barker et al. 2009). After gene duplication, *SSP* underwent both expression regulatory and protein sequence changes, acquiring a pollen-specific expression pattern and the loss of a protein kinase domain. Changes to both *cis*-regulatory and protein coding regions allowed *SSP* to gain a new function, transitioning from being a component of the brassinosteroid signaling pathway to a paternal regulator of early embryogenesis. Although *SSP* experienced accelerated amino acid changes, a codon-based statistical method for testing positive selection supported a relaxation of purifying selection instead of positive Darwinian selection, suggesting that the neofunctionalization of *SSP* may be the result of neutral processes (Liu and Adams 2010).

11.3.2.2 Positive Darwinian Selection (Adaptive Process)

The second scenario is neofunctionalization by an adaptive process. Positive Darwinian selection may play an important role in promoting the fixation of beneficial mutations in one member of a new duplicate pair, facilitating neofunctionalization following duplication. Under this model, positive selection drives one copy to gain a new function and thus preserve it—subsequent deleterious mutations to this copy would carry a fitness cost. Several examples of neofunctionalization have been documented in plants. One example of neofunctionalization by positive selection is *MEDEA*, which is a gene duplicated by the most ancient whole genome duplication in Brassicaceae. *MEDEA*, a *SET*-domain *Polycomp* group protein, is a paternally imprinted gene that regulates the development of endosperm in *Arabidopsis*. After duplication, *MEDEA*'s duplicated partner retained the ancestral function whereas *MEDEA* underwent positive Darwinian selection due to parental conflict (Spillane et al. 2007), with a larger magnitude of positive selection in outcrossing lineages in comparison to selfing lineages (Spillane et al. 2007; Miyake et al. 2009). Although it is controversial whether positive selection or balancing selection act on the new function of *MEDEA* (Kawabe et al. 2007), a recent study suggested that both types of selection contributed to the gain of new function (Miyake et al. 2009).

Another example of neofunctionalization driven by positive selection is a pair of functionally redundant cytochrome P450 genes (*CYP98A8* and *CYP98A9*), duplicated by retroposition in Brassicaceae (Matsuno et al. 2009). Their ancestral gene, *CYP98A3*, is involved in the formation of lignin monomer. The ancestral gene is expressed in many organ types but not in pollen, whereas *CYP98A8* and *CYP98A9* are highly expressed in pollen. Their new function has been showed to be involved in a novel phenolic pathway for pollen development. Codon-based methods found that several codons in these two genes have undergone repeated amino acid changes, suggesting that positive Darwinian selection has contributed to the acquisition of new function in *CYP98A8* and *CYP98A9*. Similarly, other cytochrome P450 genes (CYP79s and CYP83s), duplicates from ancient WGDs, have evolved distinct functions in novel biochemical defense pathways (Schranz et al. 2011). However, whether the evolution of these pathways was an adaptive process is still unclear.

11.3.3 Subfunctionalization

An alternative to the long-popular neofunctionalization model is subfuntionalization, which explains duplicate gene retention by the partitioning of multiple ancestral functions or expression patterns among the new paralogs (Hughes 1994; Force et al. 1999). Assuming that a gene has multiple functions or expression patterns before duplication, subfunctionalization would predict that these ancestral functions are simply partitioned among the new paralogs.

It should be noted that expression subfunctionalization and protein subfunctionalization will result in two different outcomes. Duplicated genes may share the same function for the subfunctionalization of expression patterns, but both are needed to maintain proper spatial or temporal expression—the loss of one copy will be detrimental to fitness because of mis-expression or loss of expression. For protein subfunctionalization, the ancestral gene must have multiple functions that can be partitioned among the new paralogs. Thus, the loss of a protein subfunctionalized paralog will be deleterious because retention of both copies is essential to fully complement the ancestral function. Such a dichotomous scenario, however, oversimplifies the complexity of possible combinations between expression and protein subfunctionalization and it is possible that the retention of some duplicated genes might be driven by both expression and protein subfunctionalization.

Based on population genetic simulation, subfunctionalization may initially be a neutral stage of duplicate gene evolution, as each copy accumulates mutations that may be reciprocally deleterious without interrupting the total function of their ancestral state. Thus, it has been argued that subfunctionalization is likely more important than neofunctionalization, especially in organisms with small effective population sizes, because it is easier to lose an existing function by mutation than gain a new one (Lynch and Force 2000). As we discuss in more detail below, this does not necessarily mean that positive selection will not act on subfunctionalized gene duplicates for long-term preservation. Subfunctionalization may be further classified into two different models: (1) duplication, degeneration, and complementation (DDC) model and (2) escape from adaptive conflict (EAC) model.

11.3.3.1 DDC Model (Neutral Process)

When Force et al. (1999) initially proposed the DDC model they focused on regulatory aspects of duplicate gene evolution. According to this model, both copies will neutrally accumulate deleterious mutations on their cis-regulatory regions. After sufficient mutation accumulation significantly impairs function, each copy will retain only a fraction of the ancestral phenotype (i.e., subfunction) and complement each other to cover the full spectrum of their ancestral expression pattern. Thus, this kind of subfunctionalization has been termed as duplication-degeneration-complement (DDC) model. It has been shown that the probability of DDC subfunctionalization is much higher than that of regulatory neofunctionalization, especially in organisms with small effective population sizes where positive selection is less efficient and genetic drift is more predominant (Force et al. 1999). A nice example of the DDC model in plants was illustrated by Force et al. (1999). A pair of MADS-box transcription factor genes (*ZAG1* and *ZMM2*) in maize, which originated from an allopolyploidization event, showed reciprocal expression pattern where *ZAG1* is highly expressed throughout carpel development but weakly expressed in stamen, and *ZMM2* is highly expressed in stamen (Mena et al. 1996). In comparison to their putative ancestral expression pattern from a single orthologous gene in *Arabidopsis (AGAMOUS)* and *Antirrhinum (PLENA)* that are highly expressed in both carpels and stamen (Yanofsky et al. 1990; Coen and Meyerowitz 1991; Bradley et al. 1993), it is reasonable to infer that *ZAG1* and *ZMM2* have subfunctionalized (i.e. DDC) since their duplication. Another example consistent with the DDC model is an organ-specific reciprocal gene silencing pattern for a pair of homoeologous alcohol dehydrogenase genes (*AdhA*) in allopolyploid cotton (Adams et al. 2003). In this case, one copy is only expressed in petals and stamen whereas the other copy is only expressed in carpels, indicative of regulatory subfunctionalization across different organ types. Because the polyploidization event in cotton has been inferred to occur about 1–2 million years ago (Cronn et al. 2002; Senchina et al. 2003) and polyploid plants likely have small effective population sizes at the onset of polyploidization, the *AdhA* example supports the perspective that the DDC model is important for the retention of duplicated genes in organisms where genetic drift predominates their population genetic landscape.

11.3.3.2 EAC Model (Adaptive Process)

Under the Escape from Adaptive Conflict (EAC) model, a new function arises in the ancestral gene but it reduces the performance of the original function. This creates an adaptive conflict where improving either function comes at a cost to the other. One solution to resolve such adaptive conflicts is by gene duplication. After duplication, one copy is free to improve the new function whereas the other copy may improve the original function. Discriminating subfunctionalization by EAC and neofunctionalization by adaptive process relies on these improvements because both processes will leave signatures of positive Darwinian selection. However, both copies will experience adaptive changes under the EAC model whereas only one copy will show adaptive change if neofunctionalization occurs. Further, the ancestral function will be improved if EAC took place but not under neofunctionalization. Des Marais and Rausher (2008) report a nice example of the EAC model in tandemly duplicates of the anthocyanin biosynthesis pathway gene dihydroflavonol-4-reductase (*DFR*) in morning glories (*Ipomea*). The major function of *DFR* is the regulation of flower color by via the anthocyanin biosynthesis pathway. After the first gene duplication event, one copy (*DFR-A/C* clade) underwent repeated positive Darwinian selection based on codon-based models of sequence evolution while the copy (*DFR-B* clade) showed an improvement of its ancestral function by increasing its

enzyme activity on dihydrokaempferol (DHK), dihydroquercetin (DHQ) and dihydromyricetin (DHM). In addition, the copy in *DFR-A/C* clade has been demonstrated to lose its ancestral function, suggesting that a new function has been acquired after gene duplication although this new function remains uncharacterized.

11.3.4 Frequency of Subfunctionalization and Neofunctionalization on a Genome-Wide Scale

A number of examples demonstrate that both neofunctionalization and subfunctionalization contribute to the evolution of new genes in plant genomes. However, the relative contribution of these two processes to duplicate gene retention is a major question in the evolution of plant genome content. Large-scale genomic studies, as outlined below, are being used to investigate expression pattern and protein sequence divergence to make inferences about the importance of sub- and neofunctionalization.

11.3.4.1 Expression Pattern Divergence

Genome-wide studies of expression divergence between duplicated genes have largely focused on rice, *Arabidopsis*, and cotton. Blanc and Wolfe (2004b) and Haberer et al. (2004) investigated the expression divergence between duplicated genes in *Arabidopsis*. Overall, they found that the majority of duplicated genes have significantly divergent expression patterns. For example, 57% of *At-α* whole genome duplicated genes and 73% of older *At-β* or *At-γ* whole genome duplicated genes have diverged in their expression pattern from 62 different developmental stages and organ types, indicating that many paralogs may functionally diversify after duplication even though protein divergence between most duplicated genes is small (Blanc and Wolfe 2004a, b; Haberer et al. 2004). In rice, 88% of paralogs had divergent expression (Throude et al. 2008). The correlation of expression divergence and protein divergence is indeed observed in later studies where Casneuf et al. (2006) and Ganko et al. (2007) found that older paralogs showed more functional divergence than younger gene duplicates. A similar trend was also found in rice duplicated genes (Li et al. 2009). Further, studies by Casneuf et al. (2006) and Ganko et al. (2007) demonstrated that expression pattern in small-scale duplications (i.e., tandem duplication or dispersed duplication) are less correlated than those in large-scale duplications (i.e., chromosomal or whole genome duplication), suggesting that small-scale duplications potentially have a higher chance to be neofunctionalized than large-scale duplications. Although these studies demonstrate that paralogs frequently have different expression patterns, they are unable to infer regulatory subfunctionalization or neofunctionalization due to the lack of information about their ancestral expression pattern.

The first attempt to ascertain the relative contribution of regulatory sub- or neofunctionalization for the retention of duplicated genes in plants on a genome-wide scale was recently conducted by Zou and colleagues (2009). They used the stress response of paralogs to gain a better understanding of the relative contributions of regulatory sub- and neofunctionalization in duplicate evolution. Using a maximum likelihood probabilistic framework with gene family phylogenies, they employed a Bayesian method to reconstruct the putative most recent common ancestral expression pattern for every duplicated gene pair in *Arabidopsis*. From their analysis, ~61% duplicated genes retained ancestral responsiveness to stress, ~30% showed the loss of ancestral responsiveness to stress, ~6% experienced the gain of a new response to stress, and only ~2% experienced a functional switch that changed their response to the opposite of their ancestral state. For the ~30% of paralogs where the ancestral expression pattern was lost, the ancestral functions were partitioned between gene duplicates in most cases, suggesting that subfunctionalization plays an important role in paralog retention in response to stresses. Their analyses also found that older duplicates tended to be more neofunctionalized than younger paralogs, which were more often subfunctionalized. Although Zou et al. (2009) found that subfunctionalization is more frequent than neofunctionalization, it should be noted that they only analyzed the gene expression pattern under different stress conditions. It is reasonable to assume that the relative contribution of subfunctionalization and neofunctionalization will differ when someone survey different dataset (e.g., developmental stages, cell types, organ types) or more conditions (e.g., temporal and spatial). For example, Duarte et al. (2006) found that 85% of 280 *Arabidopsis* duplicated pairs showed significant gene by organ effect across six different organs based on an analysis of variance approach, indicative of potentially regulatory subfunctionalization or regulatory neofunctionalization.

How fast can the regulatory sub-/neofunctionalization arise after gene duplication? To answer this question, Adams et al. (2003) and Chaudhary et al. (2009) surveyed, in total, 103 duplicated genes in various tissues/organ types in recently formed polyploid cotton, *Gossypium hirsutum*. Interestingly, eight cases of regulatory subfunctionalization (~8%) and 15 cases of probable regulatory neofunctionalization (~15%) were observed in their analyses. Their studies provide striking examples that regulatory subfunctionalization and neofunctionalization can quickly arise soon after genome duplication, suggesting that both mechanisms are important for the retention of duplicated genes over evolution.

11.3.4.2 Protein Sequence Divergence

Neofunctionalization predicts that gene duplicates will evolve asymmetrically. In other words, one copy will experience relaxed purifying selection, accumulate more nonsynonymous substitution or positive selected sites, whereas the other copy will experience strong purifying selection and accumulate mutations without evidence of significant amino acid change or positive selection. Based on these assumptions, Blanc and Wolfe (2004a, b) surveyed the asymmetric rate of sequence evolution between duplicated genes in *Arabidopsis* to assess the relative contribution of neofunctionalization in the retention of duplicated genes in plants. Nearly 21% of paralogs derived from the At-α whole genome duplication showed asymmetric sequence rate evolution, indicative of potential neofunctionalization. Their observation suggests that neofunctionalization may play an important role in paralog retention in plants. However, it should be noted that using asymmetric sequence rate analysis is not a perfect test for neofunctionalization. As mentioned previously, asymmetric sequence rate evolution is also possible under the EAC model and it is difficult to use difference in paralog rate evolution as a method to distinguish between subfunctionalization and neofunctionalization.

To resolve this issue, better knowledge of the ancestral protein function or structure, pre-duplication, will be important. However, characterization of gene function is a tedious and daunting task and plant biologists often must apply a case-by-case approach. It is therefore difficult to apply a rigorous, genome-wide analysis in this regard. Analyses of protein structure may be high-throughput, and those related to protein subcellular relocalization (PSR) offer a promising approach to evaluate post-duplication fates. PSR is often associated with the 5′ N-terminal signal peptide (detailed review in Byun-McKay and Geeta 2007). By comparing this region with other closely related species, it is possible to infer if the change of subcellular localization results from subfunctionalization or neofunctionalization. An unpublished analysis of differential subcellular localization found that the N-terminal peptide in *Arabidopsis* duplicated genes evolved rapidly relative to the rest of the protein region, implying that PSR might greatly contribute to functional diversification between duplicated genes in plants (Byun-McKay et al. 2009). So far, studies that aim to understand the relative importance of subfunctionalization and neofunctionalization after gene duplication at the protein level on a genome-wide scale remain rare. In addition to rice and *Arabidopsis*, more genomic resources from closely related plant species will help us to understand the relative contribution of protein subfunctionalization and neofunctionalization for the retention of duplicated genes in plants.

11.3.5 Gene Dosage Balance

The concept of dosage balance arose from studies of chromosome segregation in plants during the early twentieth century (e.g., Blakeslee and Avery 1919; Blakeslee et al. 1920; Blakeslee 1921). Pioneering plant geneticists found that unbalanced chromosome numbers such as trisomy or monosomy often caused significant phenotypic changes. Some of these phenotypic changes were strongly deleterious and severely impacted plant growth. Based on these observations, plant geneticists proposed that imbalance in the products of genetic materials carried by the chromosomes were responsible for these phenotypic changes (Blakeslee 1921). Support for this hypothesis was found in examples of chromosomal changes that restored proper dosage. For example, a secondary isochromosome (fusion of two sister chromatids) in a trisomic plant would alleviate the phenotypic changes of a pure trisomic line (Belling and Blakeslee 1924). In this case, the secondary isochromosome yields four identical chromosomal arms, which consequently produced proper gene dosage balance and ameliorate deleterious phenotypic changes. These concepts have been updated recently by researchers utilizing molecular genetic approaches to study the effects of gene copy number, or dosage, on phenotype and gene duplication in yeast (Veitia 2002, 2003; Papp et al. 2003), fruitfly, and plants (Birchler et al. 2001). Significantly, this recent work has found that the products of genes involved in regulatory roles (e.g., transcriptional factors and signal transductions) appear to require maintenance of their stoichiometric relationships, or dosage balance, likely because they participate in the formation of macromolecular complexes (see review by Birchler and Veitia 2007 and references therein). These concepts have been synthesized into the gene dosage balance hypothesis, whereby genes involved in more interaction, so called "connected" genes, are more dosage-sensitive (see introduction by Freeling and Thomas 2006).

How does gene dosage influence the retention of duplicated genes in plants? After gene duplication, gene dosage balance is commonly invoked to explain that either gene loss or retention is required to maintain the proper amount of protein product stoichiometry, or dosage, in genomes (e.g., Birchler et al. 2001; Veitia 2002, 2003; Papp et al. 2003). Based on this model, duplicated genes that have large pleotropic effects (or more interactions with other proteins) are expected to be retained more often after a large scale duplication event (i.e., whole genome duplication) than a small scale duplication event (i.e., tandem duplication, dispersed duplication, and retroposition). This is because other interacting partners will also be duplicated in large-scale events and retention of all genes is necessary to maintain stoichiometry, whereas not all interacting loci will be duplicated in a small-scale event and loss of the new paralog is required for proper dosage (see review by Freeling 2009). Specifically, the dosage balance hypothesis predicts that genes such as transcription factors or involved in signal transduction are expected to be retained after whole genome duplication such that entire genetic network still has proper amount of upstream regulator to regulate the doubled downstream pathway. If such highly connected genes are created by small scale events they would disrupt the non-duplicated downstream elements. In contrast, genes involved in the terminals of genetic networks are not prevented from being retained following small-scale duplications because additional copies of these genes is less likely to cause significant imbalance on the genetic network.

As expected, genes involved in transcriptional factors and signaling transduction often are overrepresented in duplicated genes from whole genome duplication while genes involved in abiotic or biotic stresses are overrepresented in those from tandem duplication in *Arabidopsis*, rice, and kiwifruit (Blanc and Wolfe 2004a, b; Tian et al. 2005; Shi et al. 2010). Further evidence supporting gene dosage balance and the retention of duplicated genes was recently reported by Liang et al. (2008). They investigated the relationship between protein under-wrapping (i.e., a molecular quantifier of the reliance of the protein on binding partnerships to maintain its protein structure stability or, simply the degrees of interaction with other molecules) and gene duplicability across various organisms, including *E. coli*, yeast, nematodes, fruit fly, mammals, and plants. A strong negative correlation between protein under-wrapping and gene duplicability was found, suggesting that dosage-sensitive genes are unlikely to be duplicated by small-scale duplication events. Their analysis supports the argument that proteins with highly interacted partners are more sensitive to dosage imbalance and be less like to be retained. It should be noted that not all analyses of ancient polyploidy in plants have found results consistent with dosage balance hypothesis. In the Compositae, Barker et al. (2008) found the opposite pattern predicted by the dosage balance hypothesis; signaling genes and transcription factors were significantly lost across whole genome duplications in multiple species, whereas metabolic and structural genes were strongly retained in duplicate (Fig. 11.3). It may be that the dosage-sensitivity of gene families varies across the plant phylogeny, and further study in model systems outside of the Brassicales would be valuable to better understand the evolution of plant genomes in these disparate lineages.

One intriguing possibility is that the shrinkage or expansion of plant gene families across the phylogeny may be largely attributed to the degree of dosage sensitivity. Cannon et al. (2004) investigated the relative contribution of large-scale and tandem gene duplication on the evolution of 50 different large gene families in *Arabidopsis*. As predicted by gene dosage balance hypothesis, gene families that function

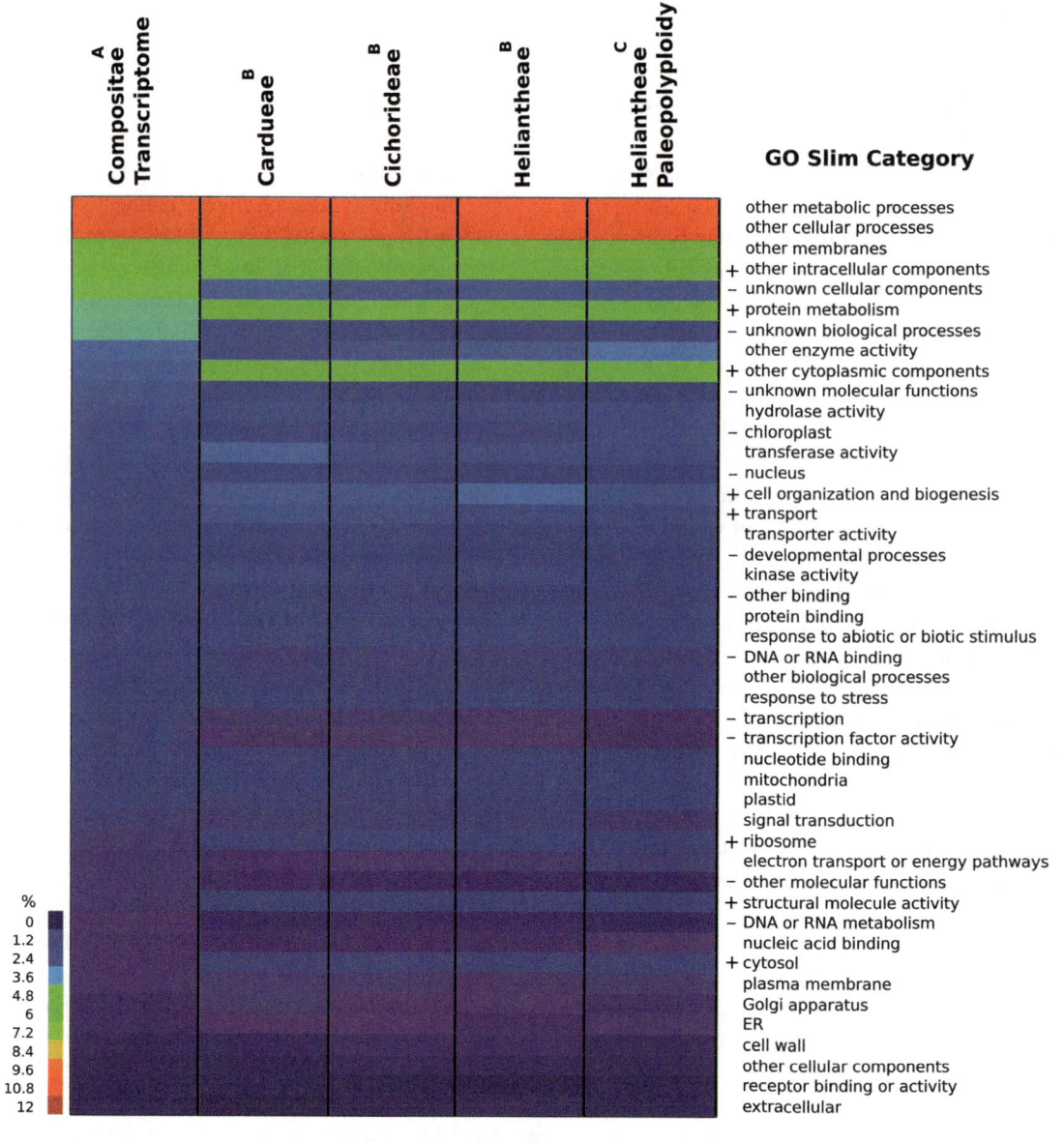

Fig. 11.3 Cell plot of gene ontology frequencies demonstrating biased gene retention following ancient polyploidy in the Compositae (Barker et al. 2008)

as transcription factor (e.g., MYB transcription factor) and signaling molecules (e.g., GTP binding protein and calmodulin) contained a higher proportion of genes derived from large-scale duplications, whereas gene families that function as pathogen defenses (e.g., major latex protein and chlorophyll a-b binding) were composed of a high proportion of tandemly duplicated genes. Such an observation further supports the importance of gene dosage balance on the retention of duplicated genes, whereby more "connected" genes (i.e., upstream regulator of genetic network such as transcription factor and signaling) would be likely retained after large scale duplication and less "connected" genes (i.e., terminal nodes of genetic network such as genes in response to biotic or abiotic stresses) should be likely retained by

small scale duplication. This reciprocal pattern of retention (i.e. WGD versus tandem duplicates) is supported by the distribution of duplicates across the entire *Arabidopsis* metabolic network (Bekaert et al. 2011). The retained duplicates from ancient WGDs were significantly more clustered (i.e. "connected") and involved in high flux reactions across the entire network compared to tandem duplicates. Similarly, Freeling (2009) observed a reciprocal pattern in gene retention following WGDs compared to tandem duplications across various GO terms.

Under the gene dosage balance hypothesis, what will be the evolutionary driving force operating on the retention of duplicated genes? Veitia (2005) illustrated that mutations for genes encoding multidomain proteins can produce dominant negative phenotypes, thereby inactivating the macromolecular complexes. For example, within the context of a dimer AA, one mutant allele in a tetraploid will result in only 56% of the AA dimers being active. Assuming that the decrease of active macromolecular complexes reduces fitness, such dominant negative effects may serve as strong purifying selection for the retention and maintenance of duplicated dosage-sensitive genes (see detail illustrations and description in Veitia 2005). In contrast, the retention of numerous copies of dosage-insensitive genes could be driven by positive Darwinian selection. For example, Hanikenne et al. (2008) provided a nice example about the expansion of small-scale duplicated genes in related to abiotic stress where the tolerance of high heavy metal concentration in *Arabidopsis halleri* was due to the triplication of three tandemly arrayed heavy metal aptase 4 (*HMA4*) genes. The amplification of copy number for HMA4 allows *A. halleri* to colonize heavy-metal polluted soil, suggesting that the retention of small-scale duplicated genes were largely driven by positive selection. In addition, the authors showed that enhanced expression level of HMA4 by changes of cis-element region also contribute to the tolerance of heavy metal in *A. halleri*. Thus, the increase of protein product itself, rather than maintenance of stoichiometry, may be beneficial for fitness in some circumstances.

Which model best explains the patterns of retention and loss of duplicated genes in plant genomes? Although all models likely act in plant genomes, gene dosage balance has become favored by many researchers because it makes explicit predictions about retention and loss. Neofunctionalization and subfunctionalization models explain how and why duplicate genes may be retained, but they are unable to generate predictions with regards to which genes will be retained after different types of duplication. Only gene dosage balance model can predict the reciprocal gene function retention pattern between whole genome duplicates and tandem duplicates in several plants such as *Arabidopsis*, rice, and kiwifruit (Freeling 2009; Edger and Pires 2009; Shi et al. 2010). However, these models are not mutually exclusive. For example, it is likely that the retention of dosage-insensitive gene duplicates from small-scale duplications, which are essentially dosage-neutral, may often be best explained by the sub- and neofunctionalization models. Thus, it is likely that all of these models contribute to the content of plant genomes with their efficacies varying across gene families and duplication type.

11.4 Duplication, Turnover, and Re-organization of Plant Genomes

All plant species have at least undergone one round of whole genome duplication over their evolutionary history (Blanc and Wolfe 2004a, b; Pfeil et al. 2005; Cui et al. 2006; Barker

et al. 2008; Tang et al. 2008; Shi et al. 2010; Jiao et al. 2011). However, most of these species demonstrate diploid chromosomal behavior and disomic inheritance restored through a process termed diploidization (Wolfe 2001). Diploidization is a dynamic process of fractionation, shuffling, and divergence of duplicated portions of a genome. The scale of events that contribute to diploidization range from loss of single genes to entire chromosomes. Various aspects of the process have been studied at time scales ranging from early generation polyploids to fully diploidized paleopolyploids. Diploidization results in the large scale re-organization of plant genomes and parallels the process of gene duplication, gene loss and functional divergence (Fig. 11.4). In this section, we present recent work on the effect of whole genome duplication on the dynamics of genomic reorganization from early generation polyploids to fully diploidized paleopolyploids in plants. These examples

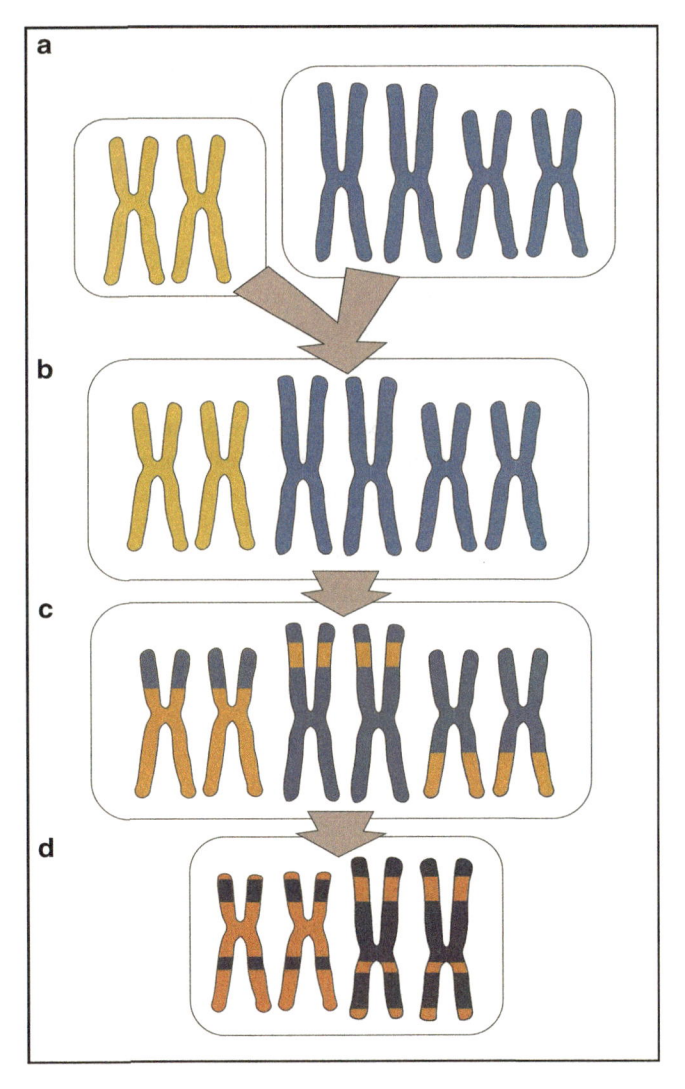

Fig. 11.4 Schematic of progressive stages of genomic diploidization following an ancient whole genome duplication depicting homoeologous pairing and chromosomal loss

will provide some insight on our current understanding of turnover and genome evolution in plants.

11.4.1 Re-organization of Polyploid Plant Genomes

Diploidization may begin as early as the first generation following genome duplication. Chromosomal rearrangements have been observed in many early generation polyploids. For example, in 50 synthetic allopolyploid lines of *Brassica napus* genomic changes were present but rare in the first generation (Lukens et al. 2006) and by the fifth generation every line had evidence for recombination between homologous chromosomes. These recombination events resulted in the loss of alleles from one of the parents and a doubling of the other. A recent analysis of synthetic *B. napa* allopolyploid lines found variation in chromosome number as early as the first generation (Xiong et al. 2011). Further, Xiong and colleagues (2011) discovered that although there was extensive loss and shuffling of chromosomes, the total number was maintained close to the tetraploid count with apparent compensatory replacement of homoeologous sets, putatively to maintain dosage balance. The rate of these early generation changes appears to vary significantly across lineages, even within the genome of a species. Allohaploid lines from a diverse panel of *B. napus* were found to have markedly different rates of homoeologous recombination and suggest different levels of crossover suppression between homoeologs (Cifuentes et al. 2010). Similar early generation changes have also been found in natural polyploids. A striking example is the *Tragopogon* allopolyploids, which formed less than 80 years ago (Soltis et al. 2004), but which have experienced a variety of genomic changes following polyploidy, including chromosomal rearrangements and loss (Lim et al. 2008). However, not all polyploids demonstrate significant changes in genome organization. The genomes of nearly cotton allopolyploids demonstrate very little genomic changes (Liu et al. 2001). It is not clear why some plants, such as cotton, do not experience rapid changes in chromosomal rearrangement after polyploidization. At larger time scales, the rate of synteny evolution is clearly variable across plants (Jaillon et al. 2007), and it may be a similar or the same process at work here. Future work associating the rates of turnover and synteny evolution with life history traits, generation time, and effective population sizes may be fruitful avenues for advancing our understanding of these patterns of plant genome evolution.

The process of chromosomal rearrangements and sequence divergence is continuous following polyploid formation. Allopolyploid tobacco species with similar parentage but a range of times since formation have been observed

to follow this expectation. In a recent study, young polyploids were found to have fewer rearrangements and less sequence divergence than older ones and the most ancient allopolyploid of the group (~5 million years old) experienced too much sequence divergence for the technique employed (Genomic In Situ Hybridization) to identify the parental alleles within its genome (Lim et al. 2007). Australian relatives of *Arabidopsis* experienced a lineage-specific whole genome duplication event since the well documented *At-α* WGD, 6–9 million years ago. Strikingly, the three species studied have a relatively low chromosome counts of $n = 4$, 5, and 6 with the genetic behavior of diploids. Reconstruction of their polyploid ancestor suggests there would have been 16 chromosomes in their ancestral genomes. Mandáková and colleagues (2010) found that most segments of their genomes still exist in duplicate although extensive and independent rearrangement has occurred in these different species. The large amount of chromosome number reduction would have required multiple chromosome loss and fusion events as well as the loss of several centromeres (Mandáková et al. 2010). Most rearrangements in plant genomes probably involve high repeat regions, telomeres and centromeres (Lysak et al. 2006), and in this case it seems that substantial amounts of end-to-end chromosomal fusion and unequal reciprocal translocations were the main mechanisms of chromosomal reduction. Whether or not such cases are outstanding or routine among plant lineages remains to be seen.

An interesting observation made when comparing syntenic blocks of ancient whole genome duplicates is a bias in which of the blocks undergo fractionation or loss of duplicates. The majority of syntenic blocks in *Arabidopsis* remaining from the *At-α* WGD have biased fractionation where clusters of genes were retained within blocks on the same homoeolog, but lost from the other (Thomas et al. 2006). Surprisingly, Thomas and colleagues (2006) found that ~85% of the *Arabidopsis* genome experienced this biased fractionation, most often involving genes in the same genetic network consistent with the dosage-balance hypothesis. A similar study in maize, which had the benefit of having an extant pre-duplication outgroup for comparison, also found evidence of biased fractionation. The authors propose a mechanism of the deletion of homologs by small intrachromosomal recombination (Woodhouse et al. 2010). It is possible that some of these differences existed in the ancestral parental genomes before the genome merger and duplication event; however, research with younger polyploids suggest that the biased gene content in the different blocks is due to biased fractionation following polyploidy. For example, the young (<80 years) allotetraploid *Tragopogon mirus* also exhibits biased loss towards one parent (Koh et al. 2010). Significantly, synthetic F1 crosses between the two parental species demonstrated only additive patterns of both locus presence and expression. These results suggest that the biased patterns of homoeolog loss observed in natural polyploid populations are the result of later generation changes and are not the immediate result of hybridization. Biased fractionation following WGD has also been documented in protozoan, fungal, and animal lineages (e.g. Paramecium, yeast, teleost; Sankoff et al. 2010). Further studies on patterns of gene expression, homoeolog loss, cytonuclear interactions, and epigenetic phenomena in related allo- and autopolyploids are needed to disentangle the effects of hybridization and polyploidy on biased fractionation in plant genomes.

Conclusions

Duplication and loss are fundamental forces in the evolution of plant genomes. Polyploidy and smaller scale duplications, while operating at different scales and frequencies, have provided the raw material and set the pace for the evolution of plant gene families as well as genome organization. Differential loss and retention of paralogs has led to significantly different gene family content across the vascular plants (Banks et al. 2011) that were likely important for the evolution of the defining phenotypes of many lineages of land plants (Banks et al. 2011; Jiao et al. 2011) and their evolutionary success. The relatively high cycles of duplication and loss experienced by plants have also driven substantial changes in genome organization, yielding more rapid changes among the genomes of related plants compared to animals (Salse et al. 2009). Resolving the relationships among repetitive element content, rearrangements, chromosomal loss and gain, dosage balance, and life history features promises to be an exciting synthesis of plant genome evolution over the next decade. Although much progress has been made in the last decade, the growing genomic data from across plants will refine our knowledge of plant genomes and population and clade-scale genomics will provide the ultimate resources to test the numerous hypotheses proposed by plant geneticists over the past century.

References

Adams KL, Cronn R, Percifield R, Wendel J (2003) Genes duplicated by polyploidy show unequal contributions to the transcriptome and organ-specific reciprocal silencing. Proc Natl Acad Sci U S A 100:4649–4654

Ameline-Torregrosa C, Wang B-B, O'Bleness MS, Deshpande S, Zhu H, Roe B, Young ND, Cannon SB (2008) Identification and characterization of nucleotide-binding site-leucine-rich repeat genes in the model plant Medicago truncatula. Plant Physiol 146:5–21

Banks JA et al (2011) The Selaginella genome identifies genetic changes associated with the evolution of vascular plants. Science 332:960–963

Barker MS, Kane NC, Matvienko M, Kozik A, Michelmore RW, Knapp SJ, Rieseberg LH (2008) Multiple paleopolyploidizations during the evolution of the Compositae reveal parallel patterns of duplicate gene retention after millions of years. Mol Biol Evol 25:2445–2455

Barker MS, Vogel H, Schranz ME (2009) Paleopolyploidy in the Brassicales: analyses of the Cleome transcriptome elucidate the history of genome duplications in Arabidopsis and other Brassicales. Genome Biol Evol 1:391–399

Bekaert M, Edger PP, Pires JC, Conant GC (2011) Two phase resolution of polyploidy in Arabidopsis metabolic network gives rise to relative and absolute dosage constraints. Plant Cell 23:1719–1728

Belling J, Blakeslee AF (1924) The configurations and sizes of the chromosomes in trivalents of 25-chromosome Daturas. Proc Natl Acad Sci U S A 10:116–120

Birchler JA, Veitia RA (2007) The gene balance hypothesis: from classical genetics to modern genomics. Plant Cell 19:395–402

Birchler JA, Bhadra MP, Auger DL (2001) Dosage-dependent gene regulation in multicellular eukaryotes: implications for dosage compensation, aneuploid syndromes, and quantitative traits. Dev Biol 234:275–288

Blakeslee AF (1921) Types of mutations and their possible significance in evolution. Am Nat 5:254–267

Blakeslee AF, Avery BT (1919) Mutations in the Jimson weed. J Hered 10:111–120

Blakeslee AF, Belling J, Farnham ME (1920) Chromosomal duplication and Mendelian phenomena in Datura mutants. Science 52:388–390

Blanc G, Wolfe KH (2004a) Widespread paleopolyploidy in model plant species inferred from age distributions of duplicate genes. Plant Cell 16:1667–1678

Blanc G, Wolfe KH (2004b) Functional divergence of duplicated genes formed by polyploidy during Arabidopsis evolution. Plant Cell 16:1679–1691

Blanc G, Hokamp K, Wolfe KH (2003) A recent polyploidy superimposed on older large-scale duplications in the Arabidopsis genome. Genome Res 13:137–144

Bowers JE, Chapman BA, Rong JK, Paterson AH (2003) Unravelling angiosperm genome evolution by phylogenetic analysis of chromosomal duplication events. Nature 422:433–438

Bradley D, Carpenter R, Sommer H, Hatley N, Coen E (1993) Complementary floral homeotic phenotypes result from opposite orientations of a transposon at the plena locus of Antirrhinum. Cell 72:85–95

Byun-McKay SA, Geeta R (2007) Protein subcellular relocalization: a new perspective on the origin of novel genes. Trends Ecol Evol 22:338–344

Byun-McKay SA, Geeta R, Duggan R, Carroll B, McKay SJ (2009) Missing the subcellular target: a mechanism of eukaryotic gene evolution. In: Pontarotti P (ed) Evolutionary biology: concept, modeling, and application. Springer, Berlin/Heidelberg/New York, pp 175–183

Cannon SB, Mitra A, Baumgarten A, Young ND, May G (2004) The roles of segmental and tandem gene duplication in the evolution of large gene families in Arabidopsis thaliana. BMC Plant Biol 4:10

Casneuf T, De Bolt S, Raes J, Maere S, Van de Peer Y (2006) Nonrandom divergence of gene expression following gene and genome duplications in the flowering plant Arabidopsis thaliana. Genome Biol 7:R13

Chaudhary B, Flagel L, Stupar RM, Udall JA, Verma N, Springer NM, Wendel J (2009) Reciprocal silencing, transcriptional bias and functional divergence of homoeologs in polyploid cotton (Gossypium). Genetics 182:503–517

Cifuentes M, Eber F, Lucas M-O, Lode M, Chèvre A-M, Jenczewski E (2010) Repeated polyploidy drove different levels of crossover suppression between homoeologous chromosomes in *Brassica napus* allohaploids. Plant Cell 22:2265–2276

Coen ES, Meyerowitz EM (1991) The war of the whorls: genetic interactions controlling flower development. Nature 353:31–37

Conant GC, Wolfe KH (2008) Turning a hobby into a job: how duplicated genes find new functions. Nat Rev Genet 9:938–950

Cronn RC, Small RL, Haselkorn T, Wendel J (2002) Rapid diversification of the cotton genus *(Gossypium:* Malvaceae) revealed by analysis of sixteen nuclear and chloroplast genes. Am J Bot 89:707–725

Cui LY, Wall PK, Leebens-Mack JH, Lindsay BG, Soltis DE, Doyle JJ, Soltis PS, Carlson JE, Arumuganathan K, Barakat A, Albert VA, Ma H, dePamphilis CW (2006) Widespread genome duplications throughout the history of flowering plants. Genome Res 16:738–749

Des Marais DL, Rausher MD (2008) Escape from adaptive conflict after duplication in an anthocyanin pathway gene. Nature 454:762–765

Doyle JJ, Flagel LE, Paterson AH, Rapp RA, Soltis DE, Soltis PS, Wendel JF (2008) Evolutionary genetics of genome merger and doubling in plants. Annu Rev Genet 42:443–461

Duarte JM, Cui L, Wall PK, Zhang Q, Zhang X, Leebens-Mack J, Ma H, Altman N, dePamphilis CW (2006) Expression pattern shifts following duplication indicative of subfunctionalization and neofunctionalization in regulatory genes of *Arabidopsis*. Mol Biol Evol 23:467–478

Edger PP, Pires JC (2009) Gene and genome duplications: the impact of dosage-sensitivity on the fate of nuclear genes. Chromosome Res 17:699–717

Force A, Lynch M, Pickett FB, Amores A, Yan YL, Postlethwait J (1999) Preservation of duplicate genes by complementary, degenerative mutations. Genetics 151:1531–1545

Freeling M (2009) Bias in plant gene content following different sorts of duplication: tandem, whole-genome, segmental, or by transposition. Annu Rev Plant Biol 60:433–453

Freeling M, Thomas BC (2006) Gene-balanced duplications, like tetraploidy, provide predictable drive to increase morphological complexity. Genome Res 16:805–814

Freeling M, Lyons E, Pederson B, Alam M, Ming R, Lisch D (2008) Many or most genes in *Arabidopsis* transposed after the origin of the order Brassicales. Genome Res 18:1924–1937

Ganko EW, Meyers BC, Vision TJ (2007) Divergence in expression between duplicated genes in *Arabidopsis*. Mol Biol Evol 24:2298–2309

Haas BJ et al (2005) Complete reannotation of the *Arabidopsis* genome: methods, tools, protocols and the final release. BMC Biol 3:7

Haberer G, Hindemitt T, Meyers BC, Mayer KFX (2004) Transcriptional similarities, dissimilarities, and conservation of cis-elements in duplicated genes of *Arabidopsis*. Plant Physiol 136:3009–3022

Hahn MW (2009) Distinguishing among evolutionary models for the maintenance of gene duplicates. J Hered 100:605–617

Haldane JBS (1933) The part played by recurrent mutation in evolution. Am Nat 67:5–9

Hanikenne M, Talke IN, Haydon MJ, Lanz C, Nolte A, Motte P, Kroymann J, Weigel D, Kramer U (2008) Evolution of metal hyperaccumulation required cis-regulatory changes and triplication of HMA4. Nature 453:391–395

Harland SC (1936) The genetical conception of the species. Camb Philos Soc Biol Rev 11:83–112

Hiller LDW et al (2004) Sequence and comparative analysis of the chicken genome provide unique perspectives on vertebrate evolution. Nature 432:695–716

Hoen DR, Park KC, Elrouby N, Yu Z, Mohabir N, Cowan RK, Bureau TE (2006) Transposon-mediated expansion and diversification of a family of *ULP*-like genes. Mol Biol Evol 23:1254–1268

Hughes AL (1994) The evolution of functionally novel proteins after gene duplication. Proc R Soc Lond B Biol Sci 256:119–124

Innan H, Kondrashov F (2010) The evolution of gene duplications: classifying and distinguishing between models. Nat Rev Genet 11:97–108

Jaillon O et al (2007) The grapevine genome sequence suggests ancestral hexaploidization in major angiosperm phyla. Nature 449:463–467

Jiang N, Bao Z, Zhang X, Eddy SR, Wessler SR (2004) Pack-MULE transposable elements mediate gene evolution in plants. Nature 431:569–573

Jiao Y, Wickett NJ, Ayyampalayam S, Chanderball AS, Landherr L, Ralph PE, Tomsho LP, Hu Y, Llang H, Soltis PS, Soltis DE, Clifton SW, Schlarbaum SE, Schuster SC, Ma H, Leebens-Mack J, dePamphilis CW (2011) Ancestral polyploidy in seed plants and angiosperms. Nature 473:97–100

Juretic N, Hoen DR, Huynh ML, Harrison PM, Bureau TE (2005) The evolutionary fate of MULE-mediated duplications of host gene fragments in rice. Genome Res 15:1292–1297

Kawabe A, Fujimoto R, Charlesworth D (2007) High diversity due to balancing selection in the promoter region of the Medea gene in *Arabidopsis lyrata*. Curr Biol 17:1885–1889

Kimura M (1983) The neutral theory of molecular evolution. Cambridge University Press, Cambridge

Koh J, Soltis PS, Soltis DE (2010) Homeolog loss and expression changes in natural populations of the recently and repeatedly formed allotetraploid *Tragopogon mirus* (Asteraceae). BMC Genomics 11:97

Leitch IJ, Soltis DE, Soltis PS, Bennett MD (2005) Evolution of DNA amounts across land plants (Embryophyta). Ann Bot 95:207–217

Li Z, Zhang H, Ge S, Gu X, Gao G, Luo J (2009) Expression pattern divergence of duplicated gene in rice. BMC Bioinformatics 10:S8

Liang H, Plazonic KR, Chen J, Li WH, Fernández A (2008) Protein under-wrapping causes dosage sensitivity and decreases gene duplicability. PLoS Genet 4:e11

Lim KY, Kovarik A, Matyasek R, Chase MW, Clarkson JJ, Grandbastien MA, Leitch AR (2007) Sequence of events leading to near-complete genome turnover in allopolyploid *Nicotiana* within five million years. New Phytol 175:756–763

Lim KY, Soltis DE, Soltis PS, Tate J, Matyasek R, Srubarova H, Kovarik A, Pires JC, Xiong Z, Leitch AR (2008) Rapid chromosome evolution in recently formed polyploids in *Tragopogon* (Asteraceae). PLoS One 3:e3353

Liu S-L, Adams KL (2010) Dramatic change in function and expression pattern of a gene duplicated by polyploidy created a paternal effect gene in the Brassicaceae. Mol Biol Evol 27:2817–2828

Liu B, Brubaker CL, Mergeai G, Cronn RC, Wendel JF (2001) Polyploid formation in cotton is not accompanied by rapid genomic changes. Genome 44:321–330

Lukens LN, Pires JC, Leon E, Vogelzang R, Oslach L, Osborn T (2006) Patterns of sequence loss and cytosine methylation within a population of newly resynthesized *Brassica napus*. Plant Physiol 140:336–348

Lynch M (2007) The origins of genome architecture. Sinnauer, Sunderland

Lynch M, Conery JS (2003) The evolutionary demography of duplicate genes. J Struct Funct Genomics 3:35–44

Lynch M, Force AG (2000) The origin of interspecific genomic incompatibility via gene duplication. Am Nat 156:590–605

Lysak MA, Berr A, Pecinka A, Schmidt R, McBreen K, Schubert I (2006) Mechanisms of chromosome number reduction in *Arabidopsis thaliana* and related Brassicaceae species. Proc Natl Acad Sci U S A 103:5224–5229

Maere S, De Bodt S, Raes J, Casneuf T, Montagu M, Kuiper M, Van de Peer Y (2005) Modeling gene and genome duplications in eukaryotes. Proc Natl Acad Sci U S A 102:5454–5459

Mandáková T, Joly S, Krzywinksi M, Mummenhoff K, Lysak MA (2010) Fast diploidization in close mesopolyploid relatives of *Arabidopsis*. Plant Cell 22:2277–2290

Matsuno M, Compagnon V, Schoch GA, Schmitt M, Debayle D, Bassard J-E, Pollet B, Hehn A, Heintz D, Ullmann P, Lapierre C, Bernier F, Ehlting J, Werck-Reichhart D (2009) Evolution of a novel phenolic pathway for pollen development. Science 325:1688–1692

Mayrose I, Zhan SH, Rothfels CJ, Magnuson-Ford K, Barker MS, Rieseberg LH, Otto SP (2011) Recently formed polyploid plants diversify at lower rates. Science 333:1257

Mena M, Ambrose BA, Meeley RB, Briggs SP, Yanofsky MF, Schmidt RJ (1996) Diversification of C function activity in maize flower development. Science 274:1537–1540

Miyake T, Takebayashi N, Wolf DE (2009) Possible diversifying selection in the imprinted gene, *MEDEA*, in *Arabidopsis*. Mol Biol Evol 26:843–857

Ohno S (1970) Evolution by gene duplication. Springer, Berlin/Heidelberg/New York

Ohta T (1987) Simulating evolution by gene duplication. Genetics 115:207–213

Papp I, Mette MF, Aufsatz W, Daxinger L, Schauer SE, Ray A, van der Winden J, Matzke M, Matzke AJM (2003) Evidence for nuclear processing of plant micro RNA and short interfering RNA precursors. Plant Physiol 132:1382–1390

Pfeil BE, Schlueter JA, Shoemaker RC, Doyle JJ (2005) Placing paleopolyploidy in relation to taxon divergence: a phylogenetic analysis in legumes using 39 gene families. Syst Biol 54:441–454

Rizzon C, Ponger L, Gaut BS (2006) Striking similarities in the genomic distribution of tandemly arrayed genes in *Arabidopsis* and rice. PLoS Comput Biol 2:e115

Salse J, Abrouk M, Bolot S, Guilhot N, Courcelle E, Faraut T, Waugh R, Close TJ, Messing J, Feuillet C (2009) Reconstruction of monocotyledonous proto-chromosomes reveals faster evolution in plants than in animals. Proc Natl Acad Sci U S A 106:14908–14913

Sankoff D, Zheng C, Zhu Q (2010) The collapse of gene complement following whole genome duplication. BMC Genomics 11:313

Schmutz J et al (2010) Genome sequence of the paleopolyploid soybean. Nature 463:178–183

Schranz ME, Edger PP, Pires JC, van Dam N, Wheat CW (2011) Comparative genomics in the Brassicales: ancient genome duplications, glucosinolate diversification and Pierinae herbivore radiation. In: Edwards D, Batley J, Parkin I, Kole C (eds) Oilseeds: genetics, genomics, and breeding of oilseed brassicas. Science, Boca Raton

Schuermann D, Molinier J, Fritsch O, Hohn B (2005) The dual nature of homologous recombination in plants. Trends Genet 21:172–181

Sémon M, Wolfe KH (2007) Consequences of genome duplication. Curr Opin Genet Dev 17:505–512

Senchina DS, Alvarez I, Cronn RC, Liu B, Rong J, Noyes RD, Paterson AH, Wing RA, Wilkind TA, Wendel J (2003) Rate variation among nuclear genes and the age of polyploidy in *Gossypium*. Mol Biol Evol 20:633–643

Shi T, Huang H, Barker MS (2010) Ancient genome duplications during the evolution of kiwifruit (*Actinidia*) and related Ericales. Ann Bot 106:497–504

Soltis DE, Soltis PS, Pires JC, Kovarik A, Tate JA, Mavrodiev E (2004) Recent and recurrent polyploidy in *Tragopogon* (Asteraceae): cytogenetic, genomic and genetic comparisons. Biol J Linn Soc 82:485–501

Spillane C, Schmid KJ, Laoueillé-Duprat S, Pien S, Escobar-Restrepo J-M, Baroux C, Gagliardini V, Page DR, Wolfe KH, Gossniklaus U (2007) Positive Darwinian selection at the imprinted *MEDEA* locus in plants. Nature 448:349–352

Stebbins GL (1950) Variation and evolution in plants. Columbia University Press, New York

Stebbins GL (1971) Chromosomal evolution in higher plants. Addison-Wesley, Reading

Stephens SG (1951) Possible significance of duplication in evolution. Adv Genet 4:247–265

Tang HB, Bowers JE, Wang XY, Ming R, Alam M, Paterson AH (2008) Synteny and collinearity in plant genomes. Science 320:486–488

Thomas BC, Pedersen B, Freeling M (2006) Following tetraploidy in an *Arabidopsis* ancestor, genes were removed preferentially from one homeolog leaving clusters enriched in dose-sensitive genes. Genome Res 16:934–946

Throude M, Bolot S, Bosio M, Pont C, Sarda X, Quraishi UM, Bourgis F, Lessard P, Rogowsky P, Ghesquiere A, Murigneux A, Charmet G, Perez P, Salse J (2008) Structure and expression analysis of rice paleo duplications. Nucleic Acids Res 37:1248–1259

Tian C, Xiong Y, Liu T, Sun S, Chan L, Chen M (2005) Evidence for an ancient whole-genome duplication event in rice and other cereals. Acta Genet Sin 32:519–527

Tuskan GA et al (2006) The genome of black cottonwood, *Populus trichocarpa* (Torr. & Gray). Science 313:1596–1604

Veitia RA (2002) Exploring the etiology of haploinsufficiency. Bioessays 24:175–184

Veitia RA (2003) Nonlinear effects in macromolecular assembly and dosage sensitivity. J Theor Biol 220:19–25

Veitia RA (2005) Paralogs in polyploids: one for all and all for one? Plant Cell 17:4–11

Vision TJ, Brown DG, Tanksley SD (2000) The origins of genomic duplications in *Arabidopsis*. Science 290:2114–2117

Wagner WH (1970) Biosystematics and evolutionary noise. Taxon 19:146–151

Walsh JB (1995) How often do duplicated genes evolve new functions? Genetics 139:421–428

Wang X, Shi X, Hao B, Ge S, Luo J (2005) Duplication and DNA segmental loss in the rice genome: implications for diploidization. New Phytol 165:937–946

Wolfe KH (2001) Yesterday's polyploids and the mystery of diploidization. Nat Rev Genet 2:333–341

Wood TE, Takebayashi N, Barker MS, Mayrose I, Greenspoon PB, Rieseberg LH (2009) The frequency of polyploid speciation in vascular plants. Proc Natl Acad Sci U S A 106:13875–13879

Woodhouse MR, Schnable JC, Pedersen BS, Lyons E, Lisch D, Subramaniam S, Freeling M (2010) Following tetraploidy in maize, a short deletion mechanism removed genes preferentially from one of the two homeologs. PLoS Biol 8(6):e1000409

Xiao H, Jiang N, Schaffner E, Stockinger EJ, van der Knaap E (2008) A retrotransposon-mediated gene duplication underlies morphological variation of tomato fruit. Science 319:1527–1530

Xiong Z, Gaeta RT, Pires JC (2011) Homoeologous shuffling and chromosome compensation maintain genome balance in resynthesized allopolyploid Brassica napus. Proc Natl Acad Sci U S A 108:7908–7913

Yanofsky MF, Ma H, Bowman JL, Drews GN, Feldmann KA, Meyerowitz EM (1990) The protein encoded by the *Arabidopsis* homeotic gene *agamous* resembles transcription factors. Nature 346:35–39

Zou C, Lehti-Shiu MD, Thomashow M, Shiu S-H (2009) Evolution of stress-regulated gene expression in duplicate genes of *Arabidopsis thaliana*. PLoS Genet 5:e1000581

Concerted Evolution of Multigene Families and Homoeologous Recombination

12

Gonzalo Nieto Feliner and Josep A. Rosselló

Contents

12.1 Introduction

The dynamism of genomes is one of the most thoroughly documented paradigms in the genomic era, envisaged by the cytogenetic school during middle decades of the twentieth century. Such dynamism refers not just to the evolutionary changes that take place across deep time but also to the myriad changes at different levels (from SNPs to large structural rearrangements) that shape and adjust genomes over a smaller time scale. This chapter reviews two of the many forces that provide genome dynamism in plants. These two forces, concerted evolution of multigene families and homoeologous recombination of hybridized genomes, in principle contribute to shape the plant genomes through opposite effects but, in fact, both represent some of the most important manifestations of non-independent evolution of DNA sequences.

Multigene families are sets of genes descended by duplication and variation from some ancestral gene. They may be clustered together on the same chromosome or dispersed on different chromosomes. Examples of multigene families in plants include those that encode the hemoglobins, immunoglobulins, actins, tubulins, keratins, heat shock proteins, cuticle proteins, and phaseolins. But those known as reiterated genes, i.e., histones and ribosomal RNA, are characterized by being composed of hundreds of identical genes present in a tandem array (King and Stansfield 1990). Underlying the functionality of multigene families is the principle of genetic redundancy for production of higher amounts of a single product, specific proteins, ribosomal RNAs, etc., ultimately resulting in a higher individual fitness. Concerted evolution of these multigene families is likely to be a correction mechanism counteracting the undesired effects of mutations. Homogenization of sequences across these regions provides stability by minimizing deviations from single products that are required to be present uniformly in large quantities.

G. Nieto Feliner (✉)
Real Jardín Botánico, (CSIC), Plaza de Murillo 2, Madrid 28014, Spain
e-mail: nieto@rjb.csic.es

J.F. Wendel et al. (eds.), *Plant Genome Diversity Volume 1*,
DOI 10.1007/978-3-7091-1130-7_12, © Springer-Verlag Wien 2012

On the other side, recombination—specifically meiotic homologous recombination—generates new allelic combinations by exchange of sequence parts from homologous genes. This shuffling of genes is considered to be a major engine of genetic variation and thus to play a fundamental role in evolution. In this chapter, we focus on homoeologous recombination, that is, exchange between duplicated genes and chromosomes that have diverged. In plants, homoeologous recombination is an alternative mechanism that generates genetic diversity. However, recombination in general has other possible functions, operating not only in meiotic division but also in somatic cells, like repairing damaged DNA. In addition, recombination may produce different outcomes not all involving reciprocal exchange, e.g., gene conversion, in which DNA sequence information from one DNA helix is transferred to another whose sequence is modified, while the former DNA helix remains unchanged. The consequence of gene conversion is sequence uniformity along a stretch of two interacting DNA molecules, and in fact gene conversion is also one of the proposed mechanisms for concerted evolution. Therefore, depending on the outcome of recombination, this phenomenon plays a role similar to concerted evolution or represents an opposite force that generates diversity.

The review has a clear evolutionary focus. Since both mechanisms have implications not only at the individual organism level but also at higher levels, attention is paid to methods for detection and interpretation of patterns resulting from those mechanisms. This is very much in the spirit of Sydney Brenner's Challenge of reconstructing the past from what we now know of the genomes (e.g., http://www.alanmacfarlane.com/DO/filmshow/brenner2_fast.htm).

12.2 Concerted Evolution of Multigene Families

12.2.1 Discovery of the Phenomenon

The description of the phenomenon of concerted evolution goes back to when cloning facilities and DNA sequencing were not available (Nei and Rooney 2005). The unexpected finding by Brown et al. (1972), who used DNA and RNA hybridization techniques in *Xenopus*, was that the nucleotide sequences of the non-coding regions in the 45S ribosomal genes, both the transcribed (ITS and ETS) and the non-transcribed (NTS) spacers, were very similar within individuals and within a species but differed as much as 10% between two closely related species. Such observations did not fit with the prevailing model of divergent molecular evolution that would predict changes to accumulate in sequences even within the same repeat. Brown et al. concluded that a correction mechanism should operate faster

than the pace at which new changes arise in these genes. Such a phenomenon, described as 'horizontal evolution', was later named 'concerted evolution' by Zimmer et al. (1980), referring to the idea that repeats in an array evolve "in concert" as a unit with the same or a very similar sequence (Fig. 12.1).

Thus, concerted evolution basically means that members within a repetitive family do not evolve independently of each other. It is a biologically universal phenomenon, which following the introduction of restriction enzyme and sequencing techniques, has been recorded in most living groups from bacteria to mammals, and for which a large amount of data has been generated (reviews in Ohta 1980; Arnheim 1983; Dover et al. 1993; Elder and Turner 1995; Liao 1999). Lack of independence in the evolution of multigene or repetitive DNA families refers not only to members within an individual genome, in the same or in different loci, but also within a species and ultimately a reproductive group, either within the same species or involving different species (e.g., Fuertes Aguilar et al. 1999b).

Much of the research undertaken since the original reports by Brown et al. has been devoted to unravelling the mechanisms responsible for concerted evolution, in the process revealing a wide range of situations concerning homogenization of repeats. A primary focus has been on ribosomal DNA, because it plays a crucial cellular role and because it is highly reiterated and hence subject to concerted evolutionary forces. Because of these two facts, ribosomal DNA has turned out to be the model gene family for research on concerted evolution. However, concerted evolution also occurs in low copy nuclear genes (e.g., Clegg et al. 1997).

12.2.2 Repetitive DNA

The nuclear genome of most plants is largely composed of different classes of repetitive DNA elements, either in tandem arrays or dispersed many hundreds or thousands of times throughout the genome. Large-scale genome sequencing projects have confirmed what earlier works anticipated based on RFLPs, Southern analysis, and DNA reassociation and molecular cytogenetic techniques, i.e., that stretches of nucleotide sequences that occur once or in low-copy number represent only a small fraction of the plant nuclear genome, whereas repetitive DNA constitutes the majority of the genome (up to 90% or more of all DNA, Kubis et al. 1998; Heslop-Harrison 2000; Macas et al. 2007). These repetitive DNA elements consist of core sequence motifs ranging in size from dinucleotides to more than 10 kb, whose degree of sequence similarity and patterns of intragenomic homogenisation is usually species-specific, as is their genomic distribution and organization and copy number (Kubis et al. 1998).

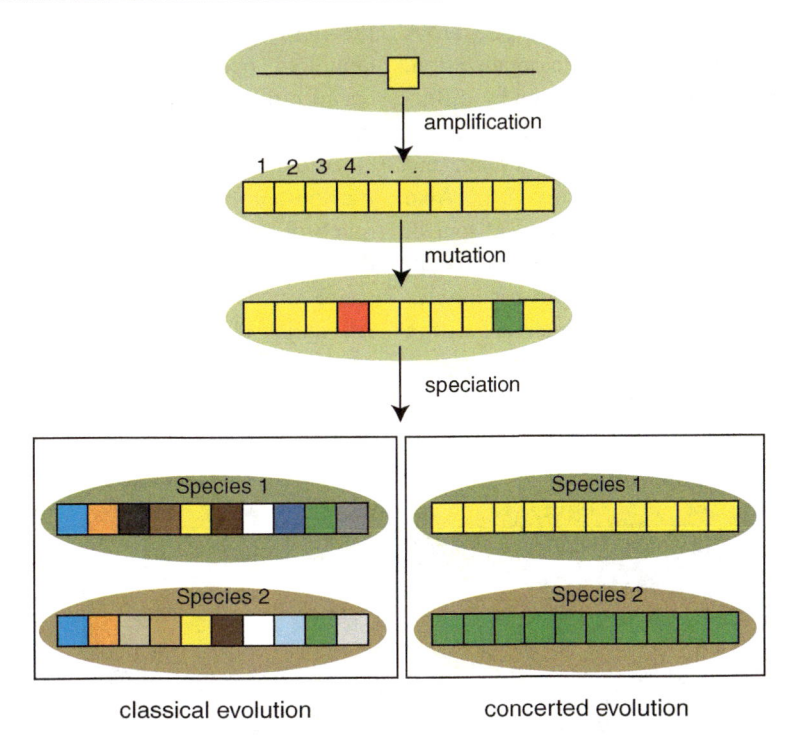

Fig. 12.1 Patterns of concerted versus classical evolution in multigene families. Repeats (*individual boxes*) in an array are formed by gene amplification events. The repeats accumulate mutations (each represented by a different color) through time. After speciation events, under no internally driven homogenizing mechanism (classical evolution), mutations accumulate and each repeat maintains orthologous relationships with the corresponding repeat in the sister species. Repeats may remain unchanged, as the first from the *left* (in *blue*) or diverge, as the first from the *right* (*light* and *dark grey*); but their relationships can be traced to their most common recent ancestor. Under a concerted evolution model, arrays are rapidly homogenized either to maintain the original repeat or to fix a new variant across the array (as in species 2) but homogenization sweeps the orthologous relationships between repeats of the same position in the two sister species (Adapted from Ganley and Kobayashi 2007)

Two of the most widespread repetitive sequences are located on defined chromosomal sites, namely, the telomeres and centromeres (see Chap. 4 by Hirsch and Jiang, this volume). In contrast, ribosomal rDNA arrays apparently do not show any evolutionary constraint regarding genomic location in plants.

Satellite DNA consists of highly repetitive, short, non-coding DNA sequences organized in long arrays, composed of thousands to millions of tandemly arranged units, contributing to the formation of the constitutive heterochromatin of chromosomes (Ugarkovic and Plohl 2002). Satellite DNA usually shows tertiary structures, due to sequence-induced DNA curvature, that could be important for packing of DNA and proteins in heterochromatin, and hence are hypothesized to be under selective pressure and evolutionary constraints (Ugarkovic and Plohl 2002). Usually, the structure, organization, sequence, and genomic distribution of satellite DNA are species-specific, even between closely related lineages (Suárez-Santiago et al. 2007).

The telomeric region is composed of long arrays of short minisatellite, tandemly-repeated GC-rich sequences that stabilize chromosome ends (Zakian 1995). Telomeres are highly conserved in green plants (Fuchs et al. 1995), both in sequence (but see Weiss and Scherthan 2002 and Sýkorová et al. 2003 for divergent types) and location, being primarily located at the chromosome termini. The finding of interstitial telomeric sequences, in addition to the terminal ones, may indicate karyotype rearrangements, involving chromosome fusion or inversions (Fuchs et al. 1995).

The cytologically defined primary constriction of chromosomes, or centromere, is essential for proper chromosome function during mitosis and meiosis (see Chap. 4 by Hirsch and Jiang). Usually, centromeres are embedded within megabases of complex regions (including retrotransposon families; Jiang et al. 2003) of long arrays of tandem DNA repeats, with a head-to-tail orientation, whose basic repeat units show little variation in length (120–300 bp or multiples; Henikoff et al. 2001). Centromeric DNA repeats are rapidly evolving sequences and little similarity between species, or even ecotypes, has been usually detected (Hall et al. 2004).

The most abundant tandemly-repeated DNA in plants belongs to two ribosomal gene families that show a conserved genic role in ribosome assembly and nucleolus formation. rDNA genes include the multigenic 45S cistron (with the 18S, 5.8S, and 25S genes, internal transcribed spacers 'ITS' between genes, and long intergenic spacers

'IGS' separating the tandem repeats) and the 5S rDNA genes. These rDNA families differ in genomic organization, regulatory elements of transcription, transcription by different RNA polymerases, size of the repeat unit, coding regions and intergenic spacers, number of loci, copy number, and degree of intragenomic homogenization (Fig. 12.2). Also both rDNA families typically occur as independent loci in vascular plant species (but see García et al. 2009), but the 5S genes are linked to the 45S cistron in some algae, bryophytes and ferns. In these cases, the 5S genes are included within the intergenic spacer separating the 25S and 18S coding regions. The finding that in certain species of liverworts (Orzechowska et al. 2010) some 5S loci could be linked to

the 45S rDNA family while other 5S loci appear as independent loci, opens new windows for a reassessment of the organization and molecular evolution of rDNA genes in green plants.

Another class of repetitive DNA, collectively comprising transposable elements, is non-structural, dispersed over much of the chromosome's length, and is one of the most dynamic and rapidly evolving components of the genome (see Kejnovsky et al. 2012; Slotkin et al. 2012; this volume). All major classes of transposable elements appear to be present in all plant species, but their quantitative and qualitative contributions are enormously variable even between closely related species (Bennetzen 2000). Although only a

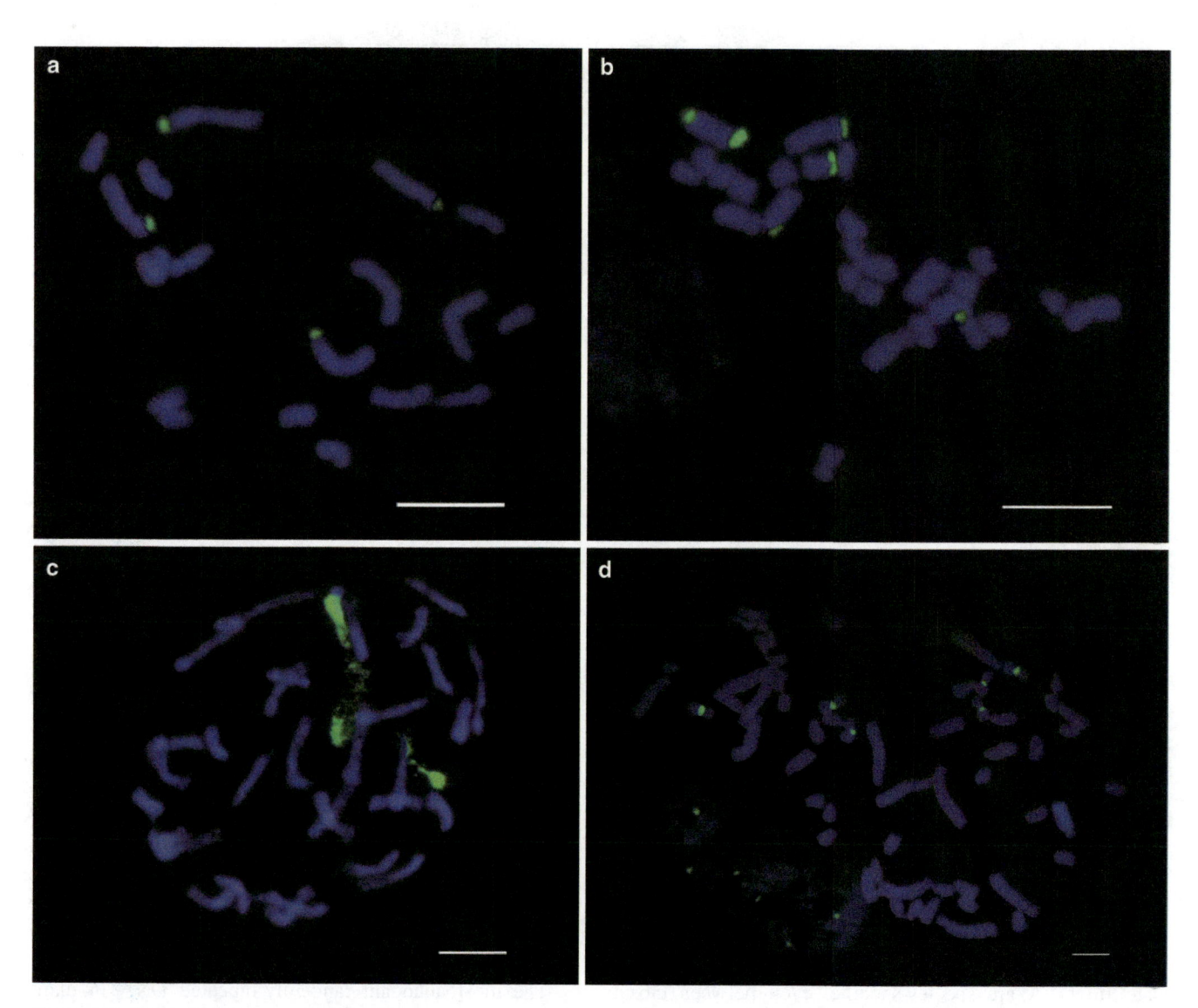

Fig. 12.2 Physical location of 45S rDNA genes on mitotic chromosomes by fluorescent in situ hybridization (FISH). Chromosomes are counterstained with DAPI (*blue fluorescence*); *green* signals are digoxigenin-labelled 45S rDNA probe detected by FITC fluorescence. (**a**) *Limonium minutum*, a diploid species (2n = 18) with two 45S rDNA loci. (**b**) *Limonium artruchium*, a triploid apomictic

species (2n = 27) showing six 45S rDNA sites; one chromosome, apparently lacking its homolog, has two terminal 45S rDNA signals. (**c**) *Medicago arborea*, a tetraploid species (2n = 32) with only a single 45S rDNA locus. (**d**) A hexaploid cytotype of *Urginea maritima* (2n = 60) with three 45S rDNA loci; an interphase nucleus showing six ribosomal sites can be also seen at the *left*. *Scale bars*: 10 μm few elements –or even none—may be active within plant genomes at a given time, the genomic alterations they cause may result in major outcomes for species evolution (Raskina et al. 2004).

Given the variety of structure and function of repetitive DNA in plant genomes, it is not surprising that rates and mechanisms of sequence homogenization do not apply equally to all genomic elements and that they are not well understood in all of them. For instance, sequencing of the chromosome 9S in rice has shown that the telomere repeat TTTAGGG array is interspersed by the deviant TTTTAGGG and TTAGGG variants (Fujisawa et al. 2006). In addition, there are many studies reporting that in the plant nuclear ribosomal 5S family, intragenomic sequence polymorphisms in the gene and associated non-transcribed spacers are the rule rather than the exception, both within and between arrays. For example, Kellogg and Appels (1995) found that, for 28 diploid species of the wheat tribe (Triticeae), nucleotide diversity within an array was up to 6.2% in the genes, not significantly different from that of the nontranscribed spacers, suggesting that rates of concerted evolution were insufficient to homogenize the entire 5S array.

Homogenization is not well understood in transposable elements either, and patterns vary. It has been proposed that elements that are still active in retrotransposition have a higher degree of sequence similarity, as a result of recent amplification. However, gene conversion has been reported for MITE elements in *Arabidopsis* (Le et al. 2000), so that there are active mechanisms promoting sequence similarity. Features subject to strong selection, like the location of repetitive elements in a sex chromosome (in which recombination is precluded or minimized), do not impose a single pattern either. For instance, in *Silene latifolia* Kejnovsky et al. (2007) report a high frequency of homogenization in gypsy-like retrotransposons in the Y chromosome, which may be achieved through gene conversion. Conversely, in another dioecious plant presenting two Y chromosomes, *Rumex acetosa*, the variability another repetitive family, satellite DNA, was shown to be higher in Y-chromosome satellites than in autosomal ones (Navajas-Perez et al. 2005).

12.2.3 Molecular Mechanisms for Concerted Evolution

The concept of concerted evolution was suggested (Brown et al. 1972) and coined (Zimmer et al. 1980) in the early stages of the refinement and diversification of molecular techniques based on a few organisms. Nonetheless, the simple model proposed to account for cohesive evolution in tandemly arranged sequences has turned out to be widely applicable (Dover 1982). Indeed, nearly four decades later, a majority of studies have confirmed that most tandemly arrayed multigene families and repetitive DNA sequences evolve in a concerted manner rather than independently.

During the 1990s, the standard model of concerted evolution was questioned by theoretical and empirical studies in some multigene families and an alternative model, called birth-and-death, was proposed (Nei and Hughes 1992). Controversy about the pros and cons of each model to explain the patterns and processes of sequence evolution, even in universal gene families like those involving rDNA (e.g. Rooney and Ward 2005; Ganley and Kobayashi 2007) is by no means solved and will likely increase as the number of sequenced complex genomes increases (see below). As summarized by Nei and Rooney (2005), distinguishing between the two models may be difficult, particularly when sequence differences are small. Due to the very long size of the tandem-repeat arrays and the high levels of similarity of these arrays, DNA sequencing and contig assembling challenges are frequently faced when performing large scale sequencing genome projects. This usually leads to gaps of uncharted regions of repetitive DNA, and ultimately to challenges in assessing the extent of sequence uniformity in any particularly large tandem array.

A final point of caution with respect to mechanisms and models is the non-random selection of plant case studies. Economic and practical constraints (interest to humans, genome size, chromosome number, life cycle and culture versatility) have determined the choice of species that have been subjected to human agronomic selection and that have ended up being our model organisms. Those constraints, as well as the domestication process itself, have surely influenced the evolutionary fate of multigene families in those organisms. The extent to which studies of sequence homogenization of repetitive DNA in these few species mirror the dynamics of what happens in the genomes of nearly 300,000 species of land plant species is both an intriguing question and a challenge for future studies.

Two main mechanisms have been proposed as responsible for concerted evolution in multigene families: unequal crossing over and gene conversion. Initially, homogenization was thought to be exclusively driven by random and recurrent unequal crossing-over between repeat units from homologous loci. This mechanism theoretically could lead to sequence homogenization, either favoring the spread of new mutations through the entire gene family or deleting them by recurrent unequal recombination events along chromosomal lineages. Whereas sequence homogenization within arrays of multigene families is predicted by this model, variation in the number of tandem units can also be generated (Graur and Li 2000). Thus, selection should counterbalance random change in copy number, maintaining a critical threshold of the number of gene copies for selectively favored functionality.

Unequal crossing-over mechanisms necessarily imply that meiosis, and hence sexual reproduction, is occurring in the organisms where concerted evolution operates. However, the analysis of rDNA variation in parthenogenetic (i. e., sexually non-reproducing) lizards (*Heteronotia binoei*) of hybrid origin showed that concerted evolution of rDNA occurred and was strongly biased toward one of two parental sequences (Hillis et al. 1991). Since meiosis is suppressed in these organisms, the authors suggested that homogenization of rDNA arrays across all nucleolar organizer regions was due to biased gene conversion as the operative mechanism (Hillis et al. 1991). Gene conversion was predicted by theoretical modeling as the alternative mechanism to concerted evolution (Ohta and Dover 1983) and it has been defined as a non-reciprocal transfer of genetic information between similar sequences leading to homogenization of multigene families (Liao et al. 1997; Liao 1999). Gene conversion is a wide concept that encompasses independent, poorly known molecular mechanisms that could be involved in sequence homogenization. These include DNA repair, replication mechanisms involving DNA breakage, unequal sister-chromatid exchange, mitotic non-homologous DNA exchange, organization of DNA interphase, and chromatin and DNA modifications (see Sect. 12.3 of this chapter). However, the frequency, extent, and efficacy of these mechanisms in generating sequence homogenization are currently unknown, and experimental evidence regarding how widespread is gene conversion in plants is limited (Salmon et al. 2010).

Although gene conversion and unequal crossing over are the most likely mechanisms that explain concerted evolution (Liao 1999, 2000), other molecular mechanisms are involved in the production and maintenance of homogeneity within repeated DNA families (Dover 1982; Elder and Turner 1995). These include transposition and subsequent reamplification of a DNA sequence, slippage-replication and RNA-mediated exchanges. In addition, the presence of a single type of parental rDNA sequence in plants of hybrid origin may be also due to rapid and extensive gene rearrangements after hybridization events (including allopolyploidy), involving sequence deletion and loss of entire loci of either parental species, which could mirror concerted evolution.

Finally, as mentioned above, there is a model that explains the patterns of certain gene families not by constricting mutations through the spread of copies (homogenization) but by gene duplication and strong purifying selection. The birth-and-death model of multigene families implies that whereas some duplicated genes are maintained in the genomes for a long time, others are deleted or inactivated through deleterious mutations (Nei and Rooney 2005). The presence of pseudogenes in a multigene family supports the involvement of the birth-and-death model in its molecular evolution (Rooney and Ward 2005). It has been suggested that the disease resistance loci and MADS-box gene families have followed this model of evolution in plants (Michelmore and Meyers 1998; Nam et al. 2004), and that the 5S rDNA multigene family in filamentous fungi is subject to birth-and-death evolution under strong purifying selection (Rooney and Ward 2005), a hypothesis that may also apply to plant species.

As has been suggested by Nei and Rooney (2005), it is possible that the models of concerted evolution and birth-and-death evolution operate within the same multigene family shaping differentially its uniformity (and variability). Specifically, these authors state that the high conservation of the coding regions of ribosomal genes is primarily caused by purifying selection rather than by concerted evolution. This would conciliate the observations that escapes of sequence homogenization are likely the exception, not the rule, in some plant multigene families, that pseudogenes are present within single loci of 5S and 45S rDNA families, and that a gradient of sequence homogenization appears to be operating in the latter family.

12.2.4 Tempo and Degree of Concerted Evolution

During the 1990s, cases were increasingly reported in which homogenization was inferred to be absent or less efficient, resulting in variability in sequences and copy numbers from a multigene family within a species or even within an individual. Modeling by computer simulations indicated that the rate of spread of a variant through an array should be faster that the mutation rate in order to result in homogeneous arrays (reviewed in Elder and Turner 1995). Particularly insightful cases were those in which different copies of the gene family in question had been merged together in one genome via hybridization. Provided that some time frame could be estimated, those cases allowed assessing both the direction of the homogenization, either uni- or biparental, and the rate at which sequence homogenization was operating (see below)

Patterns of departure from full homogenization in gene families subject to concerted evolution vary. Degree of homogenization may be related to the gene family itself. For instance, homogenization has been found to be very efficient in ribosomal DNA from five species of fungi using whole-genome shotgun sequence data (Ganley and Kobayashi 2007). But different patterns are also found for the same gene family in different organisms. Studies in plant groups reported contrasting degrees of homogenization in ribosomal DNA. In some cases, fast homogenization was found in natural hybrids from *Cardamine* (Franzke and Mummenhoff 1999), *Tragopogon* (Kovarik et al. 2005), *Senecio* (Abbott and Lowe 2004), as well as in artificial hybrids in *Armeria*. In the latter case, clear signs of homogenization were found already in the F_2 offspring in the ITS, which suggests gene conversion as the underlying mechanism (Fuertes Aguilar et al. 1999a). In contrast, some allopolyploids maintain polymorphic parental rDNAs (e.g., *Krigia*, Kim and Jansen 1994; *Arabidopsis*, O'Kane et al. 1996; *Brassica*, Waters and Schaal 1996; *Paeonia*, Zhang and Sang 1999; *Glycine*, Rauscher et al. 2004) as do some agamospermic species (e.g., *Amelanchier*, Campbell et al. 1997; *Limonium*, Palacios et al. 2000) and species groups with frequent hybridization (*Coprosma*, Wichman et al. 2002). For some genera, like *Nicotiana*, contrasting situations have been reported, depending on the species (Kovarik et al. 2004).

Several causes have been hypothesized to account for intragenomic heterogeneity. On theoretical grounds, it was suggested early that the number of loci, the number of copies at each locus, and the chromosomal location of the loci could differently modulate the tempo of concerted evolution of multigene families (Arnheim 1983; Slatkin 1986). Also in the 1980s, the total number of copies in the repetitive DNA or multigene family, together with the conversion rate within each chromosome, were suggested to be the most critical factors affecting rates of sequence homogenization (Ohta and Dover 1983). In this theoretical study, the authors indicated that the presence of paralogous loci on several chromosomes should have a relatively small effect on homogenization across loci, unless the conversion rate between genes on nonhomologous chromosomes was extremely low, or unless the number of nonhomologous chromosomes on which gene members are dispersed was large (Ohta and Dover 1983). In fact, Dover (1986), with his "molecular drive" concept, stressed the idea that mutations were able to spread through a multigene family (homogenization) and through a population (fixation) via a variety of mechanisms of nonreciprocal DNA transfer within and between chromosomes (gene conversion, unequal crossing-over, transposition, slippage-replication and RNA-mediated exchanges).This position was in contrast to ideas that chromosomal location (at least for rDNA loci) had a more substantial impact than the number of loci on the tempo of concerted evolution through either unequal crossing-over or gene conversion (Zhang and Sang 1999). The importance attributed to genomic and chromosomal location involved in such a view was consistent with findings in *Drosophila* that intrachromosomal rates of homogenization are faster than interchromosomal (Schlötterer and Tautz 1994). In the end, these factors are likely to play a role in homogenization and are thus valuable guidelines for further research. But, their overall predictive value has yet to be determined due to the virtual lack of case studies where all key parameters potentially affecting rates of sequence homogenization are adequately known.

Many species have more than one chromosomal locus containing rDNA (up to 10 loci in several gymnosperm groups; Liu et al. 2003). However, all rDNA loci are not necessarily involved in rRNA transcription and genesis of nucleoli. Thus, plant genomes can show the joint presence of active (NOR) as well as permanently silenced rDNA loci (Cuñado et al. 2000), with the latter usually occurring in higher numbers than the former. It seems reasonable that active rDNA loci are more selectively constrained than silenced paralogous loci and thus concerted evolution mechanisms are stronger in active rDNA loci, both within and between, than in silenced paralogous loci, where sequence homogenization could be largely relaxed. This association between relative invulnerability to homogenization and epigenetic silencing in rDNA has been found in *Nicotiana* allopolyploids (Lim et al. 2000; Kovarik et al. 2008). Retardation or failure of concerted evolution between functionally different paralogous loci could lead to progressive sequence divergence, reduction of copy number units, and ultimately to the formation of pseudogenes. Moreover, the presence of rDNA loci on supernumerary chromosomes (B chromosomes) could lead to a failure in sequence homogenization in these multigene families since B chromosomes behave as singular elements and do not pair with the other chromosomes in meiosis or mitosis (Camacho et al. 2000).

Other authors have argued that plant biological features, especially reproductive ones, could be also involved in retarding sequence homogenization. Sang et al. (1995) detected additivity of ribosomal ITS sequences in 14 diploid and tetraploid *Paeonia* species affecting nucleotide sites that are variable in the remaining 12 species of section *Paeonia*, so that they postulated reticulate evolution through homoploid hybridization from currently allopatric parentals. These authors suggested that lack of sequence homogenization could be related to the long generation time of *Paeonia* species (vegetative multiplication by rhizomes is frequent in the genus) which might reduce rates of concerted evolution. This hypothesis, however, does not explain why three allotetraploid species (*P. arietina*, *P. officinalis*, and *P. parnassica*) showing similar life cycles, underground rhizomes, and accordingly, similar generation times present completely homogenized ITS sequences. Long generation time has been also invoked to account for low nrDNA intra- and inter-genomic homogeneity in tree species although in gymnosperm groups additional factors may be involved such as ITS-region length, high copy number and number of loci (e.g. *Pinus*, Gernandt et al. 2001; *Larix*, Wei et al. 2003).

12.2.5 Alternative Mechanisms for Homogeneous Gene or Protein Products

Concerted evolution is not the only way to achieve homogeneous amounts of the same gene or protein product. There are independent molecular processes that could achieve similar effects in the absence of sequence homogenization. Epigenetic modifications leading to gene silencing through several processes (e.g., by methylation) are the main alternative, leading first to transcriptional inactivation and ultimately to pseudogene formation of duplicate loci. One of the first recorded and best studied phenomena of gene silencing is nucleolar dominance. This is an epigenetic process occurring in hybrids and allopolyploids leading to transcriptional repression of one parental set of potentially active 45S rDNA genes (Navashin 1934). Cytological manifestations of this NOR repression are the selective suppression of both secondary constrictions (and hence the satellites) and nucleoli formation contributed by one of the two parents (Pikaard 1999, 2000). Nucleolar dominance is a reversible phenomenon in which the repressed rDNA genes of a particular genome in a hybrid species could become the transcriptionally active ones when they are put in another genome environment (i.e., in hybrids involving other species). Why these divergent methods (homogenization by concerted evolution vs epigenetic silencing) leading to the same transcriptional uniformity are diversely operating in plants is not known.

12.2.6 Biological Implications of Concerted Evolution

It is of interest to consider the different mechanisms that might maintain homogeneity of gene copies within multigene families. High metabolic demands for rRNA exist in growing cells and, unlike protein-coding RNA in which expression of a primary product can be amplified, in structural molecules such as rRNA, translation is not available and transcription of a single gene is not enough (Liao 1999). A high number of copies guarantees the rDNA needed for the cellular machinery, although as discussed not all of the multiple copies need be functional, so that the relationship between copy number and rRNA production is not straightforward. Yeasts, for instance, tend to alter rates of transcription of active rDNA genes rather than activating additional genes under increased requirements for rRNA (Reeder 1999). Also, in recently formed allopolyploid species of *Tragopogon*, rDNA alleles from one parent that are present in much reduced numbers relative to those from the other parent are expressed preferentially (Matyásek et al. 2007).

Ribosomal RNA molecules, required in large quantities, function only when assembled into ribosomes, presumably with sufficient structural constraint that relative homogeneity of rRNAs is needed. Under this view, a possible biological function of concerted evolution is to maintain homogenous gene copies within such families so that homogeneous gene products are produced. Thus, concerted evolution is seen as a quality control process (Liao 1999, 2003).

Further, it can be speculated that the maintenance of sequence homogeneity within populations and species may accelerate population differentiation and species isolation (Liao 2003) as illustrated by alpha-satellite of the centromeric heterochromatin exhibiting rapid evolution and within-species homogenization (Schueler et al. 2001). In contrast, in some multigene families, such as the major histocompatibility genes (MHC) in vertebrates, there are apparently selective pressures for maintaining different alleles. It is not surprising from a functional point of view that the evolutionary pattern in this family, and others like the inmunoglobulin genes, are more consistent with a birth-and-death model of evolution than with a homogenization model via concerted evolution (Nei et al. 1997; see above).

A number of studies have focused on the functional significance of the variation in rDNA, specifically in copy number and length of the intergenic spacer (IGS) (reviewed in Weider et al. 2005). Such variation is necessarily related to concerted evolution or to the alternative models underlying changes in rDNA. In plants, these studies can be grouped into four categories. A first category includes surveys of rDNA variation in wild versus cultivated varieties or species. In *Oryza*, for instance, significant differences in IGS length were found between cultivated (no variation) and wild species, which might suggest purging/purifying of IGS length variation in cultivated species (Cordesse et al. 1990). A second group of studies have detected significant associations between rDNA structure and important fitness traits. In *Avena* cultivars/strains selected for higher yields exhibited consistently larger IGS than wild strains under controlled growing conditions (Polanco and Pérez de la Vega 1997). In *Hordeum*, relevant fitness traits that have been subjected to artificial selection, such as yield, were significantly associated with rDNA alleles (Powell et al. 1992).

A third group of studies have found associations between environmental parameters and rDNA variation in wild and/or cultivated species. Among cultivated species, several studies in *Hordeum* have found significant links between IGS length variants and up to nine environmental factors (humidity, rainfall, edaphic factors, etc.) suggesting that selection is operating on the genotypes (Saghai-Maroof et al. 1990; Zhang et al. 1990; Sharma et al. 2004). Similar associations have been found in wild species, for instance in two species of *Picea*, where the copy number of distinct alleles (repeat types) varied across a latitudinal gradient (Bobola et al. 1992) and in *Pinus*, where fire-related environmental stress was correlated with low copy number of rDNA (Govindaraju and Cullis 1992). Still, a fourth group of studies add another component to the functional significance of rDNA variation by relating rDNA variation, associated to environmental parameters, and differentiation, either ongoing or complete. In western American populations of the introduced *Avena barbata*, seven different rDNA genotypes were detected that were associated with different habitats following a pattern that was coincident both with allozyme data and morphology. Thus, ecological diversification that occurred over a short time span after the introduction of the species in the American continent likely involved ecological selection affecting rDNA (Cluster and Allard 1995).

In sum, there is evidence that variation in rDNA can be associated with fitness-related traits and/or to ecological factors, implying that such variation has ecological significance and may have responded to artificial or natural selection. In most if not all of these cases cited above, however, alternative explanations involving genetic bottlenecks, history, or linkage drag have not been completely excluded.

But what might be the role of concerted evolution in this view of functional significance of rDNA variation? If selection is involved in the predominance of rDNA variants (either copy number, IGS length and structure or allele composition) in specific environments, concerted evolution may contribute to spread and eventually fix the favored rDNA variants within populations subjected to those selective pressures (Elder and Turner 1995). This combined action of selection pressures and concerted evolution would require that no internal constraint occurred against favored, i.e. adaptive, rDNA variants, in which case concerted evolution would counteract selection. Some of the examples mentioned above from the literature are consistent with the dual functional role of concerted evolution hypothesized by Liao (2003): the maintenance of homogeneity among groups that interbreed as opposed to other such groups may have enhanced population differentiation.

Hybridization merges different rDNA sequences within the same genome and thus allows following the effects of concerted evolution –early effects if hybrids are recently formed. Therefore case studies examining rDNA variation in hybrids allow some considerations on the occurrence of selection and/or bias even if not necessarily simplifying the detection of selection operating on the different rDNA alleles, which is crucial to assessing their biological significance. For instance, different fates in different individuals of the same hybrid or allopolyploid concerning the direction and degree of homogenization indicates no intrinsic selection against any of the parental rDNA types, e.g. in *Arabis* (Koch et al. 2003), *Tragopogon* (Kovarik et al. 2005) or *Oryza* (Bao et al. 2010). These situations could be due to environmental-dependent selection or to the possibility that concerted evolution is primarily a stochastic phenomenon only directed towards achieving homogeneous arrays. If on the contrary a bias occurs, i.e., the rDNA types of one of the parents predominate in the hybrid, it would be suggestive of intrinsic selection. No matter if the bias was in the direction of the mother progenitor as in *Cardamine* (Franzke and Mummenhoff 1999) or to specific repeats independently of the mother progenitor as in artificial hybrids of *Armeria* (Fuertes Aguilar et al. 1999a), environmentally-driven selection would be less likely unless the hybrid's habitats were extremely uniform.

An intriguing case, also in the genus *Armeria* suggests that biased concerted evolution has been driven by selection although some aspects of the system are not well understood. The southern Spanish *Armeria villosa* is diversified so that several subspecies that have defined geographic ranges can be recognized on the basis of both their morphology as well as their ecological requirements. A study revealed that the rDNA (ITS) sequence found in each sample of *A. villosa* depended, not on the subspecies, but on its geographical location and that the same ITS sequence was found also in other sympatric species and subspecies (Fuertes Aguilar et al. 1999b). The most likely explanation for this geographic pattern of variation independent of taxonomy is that there had been extensive hybridization with congeners, which is frequent in the genus, and the same bias in the homogenization within the same area. The system fits the classical scheme of species colonizing novel habitats through introgression with adapted congeners leading to differentiation. But, in the context of biological implications of concerted evolution, what is noteworthy from this case is that if rDNA types (ribotypes) are linked to ecological settings and areas, it is likely that those ribotypes have been selected and that the homogenization process that has lead to such pattern has some adaptive significance (Fuertes Aguilar et al. 1999a; Nieto Feliner et al. 2001). In the end, the Gordian knot of understanding the biological significance of concerted evolution lies on unravelling how it interacts with selection because in theory both mechanisms can either be synergic or opposed.

Another line of research requiring further attention aims at searching for functional consequences of rDNA variation in a very different level following the perspective of biological stoichiometry (which studies the balance of energy and chemical elements in living systems). This is based on the 'growth rate hypothesis' (Elser et al. 1996), which states that variation in organismal C:N:P ratios reflects differences in growth rate because of differential allocation to P-rich rRNA (RNA is c. 9.6% P by mass) that is needed to meet the protein synthesis demands of growth. This hypothesis was extended to focus on its genetic basis, postulating that variations in rDNA IGS length and copy number underpin variation in growth and therefore RNA allocation, and thus P content and C:N:P ratios (Elser et al. 2000).

12.2.7 Concerted Evolution and Phylogenetic Inference

The relevance of concerted evolution for plant phylogenetic inference is intimately connected to the enormous success that rDNA and specifically the ITS regions have had in the botanical community as the most widely used molecular marker (Baldwin et al. 1995; Hughes et al. 2006). The usefulness of this marker is based on a number of advantages. But problems with its use have been highlighted as well (Alvarez and Wendel 2003).

Advantages of the ribosomal DNA and specifically ITS are firstly derived from their universality. There are sets of primers that work for a wide variety of living organisms (White et al. 1990; Gardes and Bruns 1993). Also, their multicopy structure facilitates amplification even from herbarium samples in which DNA quality is usually reduced. The moderate size of the ITS sequences, with notable exceptions in gymnosperms (Gernandt et al. 2001), normally allows amplification and sequencing without internal primers, which is very convenient given the increasingly higher demands for thorough samplings in all kinds of sequence-based biodiversity studies no matter if they are focused at deep or shallow phylogenetic levels. The fourth major advantage relates to the information content derived from the level of variation exhibited by these regions. This is frequently enough for phylogenetic inference at the species level to the point that this marker was explored as a possible candidate for barcoding identification of plants (Kress et al. 2005). Additional factors have also contributed to the wide use of ITS in plant studies such as the scarcity of markers within the three genomes providing variability at the species level for phylogenetic inference and the fact that, unlike plastid and mitochondrial markers, ITS are biparentally inherited and thus, ideally may provide information on maternal and paternal genealogies.

Other features of these markers affecting phylogenetic inference are directly related to the speed and bias of concerted evolution and are considered advantageous or disadvantageous depending on the objective of the study. Intragenomic and interpopulational homogeneity resulting from active concerted evolution and interbreeding, respectively, were deemed useful for accurate reconstruction of species relationships (Baldwin et al. 1995). This proposal assumed that concerted evolution would tend to fix novel mutations within species thus converting them in synapomorphies while reducing the need for intraspecific sampling since conspecific individuals would tend to have homogeneous sequences. On the other end, for studies aiming to document hybridization or introgression, incomplete or failed homogenization is a desirable property since it may allow identifying both progenitors whenever species-specific ITS sequences are available.

But there are a number of factors that represent potential pitfalls for the use of these markers in phylogeny reconstruction independently of the specific objectives of the study. The multicopy nature of ribosomal DNA itself is a source of problems since it complicates the assumption that orthologous sequences are being compared. Phylogeny reconstruction, as other disciplines aiming to understand biodiversity, follows a comparative approach so that establishing the terms of comparison is fundamental to the meaning of the results obtained. In principle, distinct but homologous loci coexisting in a single individual are paralogous, i.e., arisen from gene duplications (Doyle and Davis 1998) whereas if one is interested in relationships among species she must look for orthologous genes, that is, those arisen due to divergence of organismal lineages.

The worst effect of concerted evolution in phylogenetic inference is eliminating the phylogenetic distinction between orthologous and paralogous loci. Homology relationships among concertedly evolving gene families are complex to the point that they received a specific name, 'plerology' (Patterson 1988; Doyle and Davis 1998). In practical terms, if concerted evolution is complete, any member of the gene family can be used for reconstructing organism phylogeny, making complete concerted evolution in general a useful attribute of a gene family.

However, intermediate levels of concerted evolution always present problems in the reconstruction of gene trees. Hybrids and, in particular, allopolyploids exemplify this question since they allow examining the fate of divergent rDNA repeats merged together in one genome. Three situations are expected in this respect. First, divergent rDNA copies may evolve independently without interacting with each other via recombination or homogenization, and thus be retained (O'Kane et al. 1996; Popp and Oxelman 2001; Zhang et al. 2002; Bao et al. 2010). A second possible situation is that one repeat becomes dominant within the new genome and the other is lost via concerted evolution (Sang et al. 1995; Wendel et al. 1995; Volkov et al. 1999; Kotseruba et al. 2003; Kim et al. 2008; Bao et al. 2010). A third possible situation is the formation of mosaic repeats differing from both progenitor rDNA types (Buckler et al. 1997; Nieto Feliner et al. 2004), which could become dominant through homogenization (Volkov et al. 2007)

There are also intermediate situations where the pattern as to degree of homogenization and bias is not consistent across individuals or populations of the same hybrid or allopolyploid. This has been found, for instance, in individuals and populations of the recently formed allopolyploids *Tragopogon mirus* and *T. miscellus*, although a trend toward elimination of repeats from one (*T. dubius*) of the two parental diploid genomes has been noted (Kovarik et al. 2005; Matyasek et al. 2007; Lim et al. 2008).

An additional concern, also derived from the multicopy nature of ribosomal DNA, is that sequences retrieved from a PCR product may derive from several arrays on different chromosomes. Therefore, a consensus ITS sequence obtained by direct amplicon sequencing cannot reveal either the number of ribosomal loci nor the presence of allelic variants. Cloning often unveils divergent intragenomic copies, which may have various origins that cannot be deduced without other comparative evidence. Different copies may come from a single locus that has not fully homogenized, or from duplicated arrays from some ancestral duplication process. Amplification of pre-existing arrays by ectopic recombination between terminal chromosomal regions is a possibility (Pedrosa-Harand et al. 2006) as is dispersion of NORs by other processes (Castro et al. 2001). Different ITS sequences retrieved after cloning may also belong to homoeologous (xenologous) loci incorporated into the nuclear genome through hybridization, either with or without polyploidy. In addition, significantly different sequences may be the result of functional gene loss leading to pseudogenes. Finally, divergent sequences may also represent true allelic variants of a homologous locus.

These potential pitfalls of rDNA prompted Álvarez and Wendel (2003) to recommend abandoning these markers in favour of low copy nuclear genes (LCNGs) for sequence-based phylogenetic reconstruction. However, previous characterization of gene family composition is highly recommended with LCNGs due to the need to compare orthologous genes and this may represent considerable effort, particularly in allopolyploids (Hughes et al. 2006). These and other challenges with LCNGs support an alternative view which argues that the systematic and phylogenetic plant community is "not yet ripe for abandoning ITS as an important and easy-to-work, albeit not *the only*, nuclear marker" (Nieto Feliner and Rosselló 2007). Instead, we proposed that pitfalls could be overcome in many cases by following three general recommendations: representative samplings following prospective pilot studies, careful lab protocols, and mindful analysis. To help extract phylogenetic signal from ITS sequences, we proposed guidelines in the form of flow-charts, aiming to improve the quality of ribosomal data and their interpretation in the context of organism phylogenies of plant groups (Figs. 12.3, 12.4; Nieto Feliner and Rosselló 2007).

12.3 Homoeologous Recombination in Hybridized Genomes

12.3.1 Homologous Recombination

Recombination in a broad sense is a wide-ranging phenomenon about which much remains to be learned, both mechanistically and with respect to evolutionary implications. The best known aspects of recombination are those related to the homologous recombination, originally from cytogenetic approaches in the last half of the twentieth century (e.g. Sybenga 1975) and more recently using diverse molecular and genetic approaches. Operationally, homologous recombination is the process by which DNA loci of nearly identical nucleotide sequences interact and exchange DNA structure and sequence information (Cromie and Smith 2007).

Two major consequences are widely acknowledged for recombination. One is its central role in meiosis in sexually reproducing eukaryotic organisms, ensuring the proper segregation of chromosomes and creating new linkages that ultimately are transmitted into the haploid gametes. The random combinations of alleles generated by recombination break down linkage disequilibrium, can be subjected to natural selection and therefore represent a fundamental process in evolutionary genetics. The second function is to provide a universally important mechanism for repairing damaged DNA, which occurs also in somatic tissue. DNA double-strand breaks (DSBs), caused by external or internal factors, occur frequently; in plants and vertebrates, most of these breaks are repaired by nonhomologous end joining (NHEJ). But this is an error-prone process, while homologous recombination enables a precise repair of those breaks, although in somatic tissue it may change the copy number of genes (Schuermann et al. 2005).

Homologous recombination events can be classified as crossovers (COs) or non-crossovers (NCOs). In COs, the two strands of each homologous duplex are reciprocally broken and joined, so that alleles flanking the break site are exchanged. In NCOs, there is no such exchange and therefore flanking alleles maintain their original linkage. In both COs and NCOs, localized non-reciprocal 'donation' of sequence information from one homologous locus to another can also take place; the donor remaining unchanged. This is termed 'gene conversion' (Cromie and Smith 2007), sometimes considered equal to NCOs (e.g., Gaeta and Pires 2010).

Recombination is initiated by the formation of DNA double-strand breaks (DSBs). Double-stranded DNA is resected (degraded by nucleases) in the 5′ to 3′ direction so that 3′ single-stranded ends are produced. From here on, to explain how homologous recombination proceeds to repair DSBs, two primary pathways currently are accepted. However, there have been numerous variants proposed during the last 50 years to accommodate new cytological as well molecular evidence (see Haber 2008 for a review). The first is the Double Strand DNA Break Repair model, DSBR (Szostak et al. 1983; Fig. 12.5), which has played the most influential role over the years in interpreting meiotic recombination. However, recent studies in different model organisms suggest that the Szostak model is not universal for meiotic recombination and points out mechanistic differences between meiotic recombination in different organisms. The second model is the synthesis-dependent strand annealing (SDSA) model, which unlike the DBSR model, does not involve

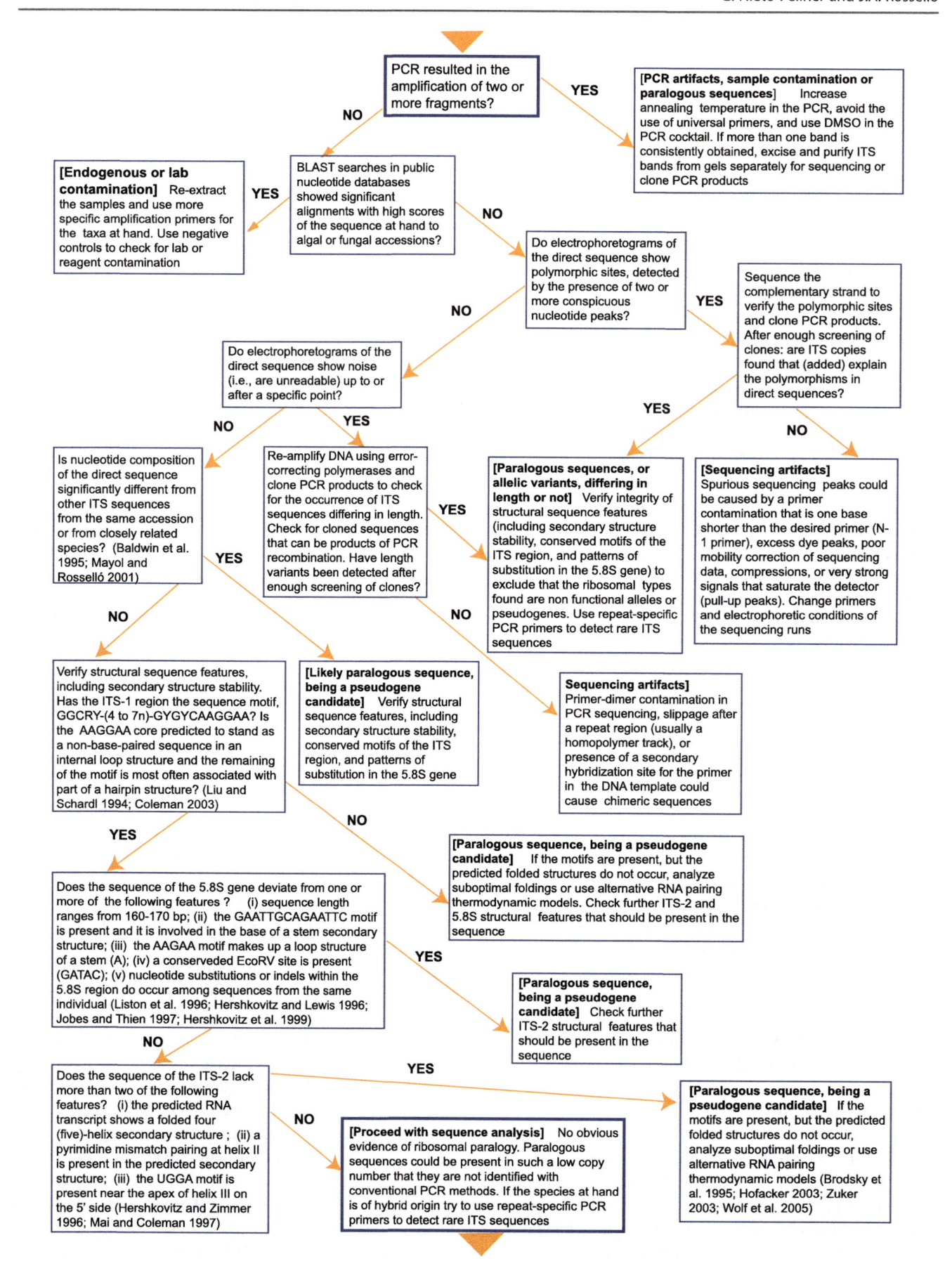

Fig. 12.3 Guidelines for obtaining ITS sequences in plants. This chart is intended to be used to help solve problems of amplification, detection of pseudogenes and paralogs, contamination and sequence artifacts (Adapted from Nieto Feliner and Rosselló 2007)

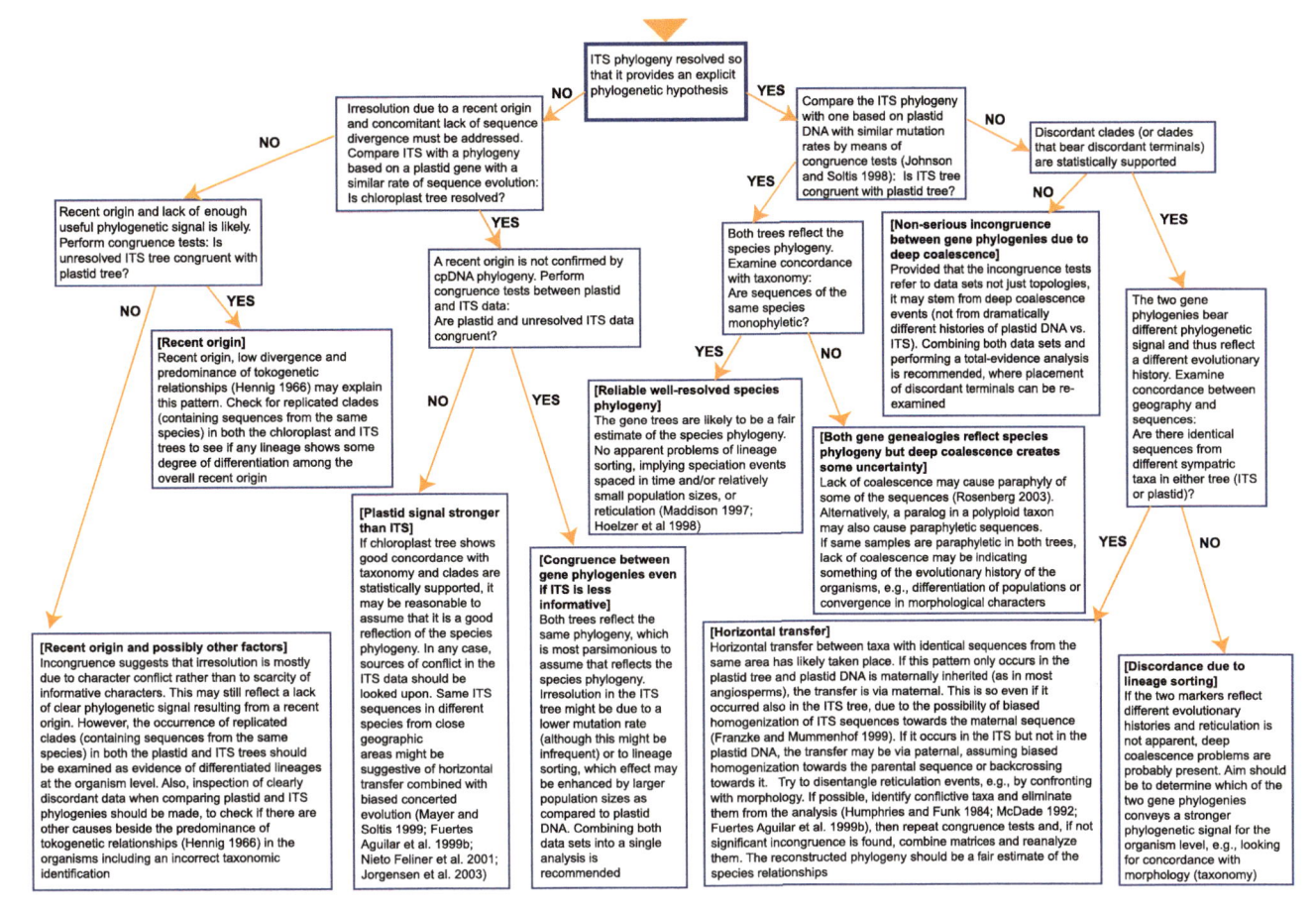

Fig. 12.4 Guidelines for analyzing ITS sequences as estimators of species-level phylogenies. This chart is intended to help discover causes for unresolved clades, integrating gene phylogenies, detecting horizontal transfer and lineage sorting, and revealing if a phylogenetic hypothesis derived from ITS is a reliable estimate of organism phylogeny. The term *resolved* in the entrance to the chart is intended to be understood in the context of a gene tree that seeks to represent species relationships; therefore, a terminal polytomy composed of sequences retrieved from the same species is not meant as absence of resolution. Also, a largely resolved tree with a few small terminal unresolved clades containing sequences from more than one species is not considered, in this context, unresolved either (Adapted from Nieto Feliner and Rosselló 2007)

Holliday junctions (Nassif et al. 1994; Cromie and Smith 2007, Fig. 12.5). It has been suggested, based on a study in yeast (Terasawa et al. 2007), that this model is valid for NCO while the DSBR model, modified, is essentially valid for CO recombination (Cromie and Smith 2007).

A major process by which genomic diversity is generated in plants is structural evolution, caused by polyploidy, localized gene duplication or deletion, and other rearrangements. Recently, the role of recombination in genome change and in fostering genome evolution has been revitalized (Gaut et al. 2007). Even if the frequency of recombination-related mechanisms that generate mutations is unknown, there is clear evidence of the types of mutations than can be generated. Unequal crossing over between sister chromatids—or between homologous chromosomes—can increase or decrease the copy number of a repetitive element (one of the proposed mechanisms for concerted evolution) and this occurs at a non-negligible rate in *Arabidopsis*

(Jelesko et al. 2004). Also, recombination, not between homologous chromosomes or chromatids but between direct or inverted repeats of the same chromatid, can generate sequence deletions, inversions and gene conversion (Molinier et al. 2004; Schuermann et al. 2005; Lysak et al. 2006; Gaut et al. 2007). In addition, it is thought that unequal recombination slows the growth in genome size caused by retrotransposon amplification, with illegitimate recombination (that is between non-homologous sequences) being an important mechanism for the removal of non-essential DNA from genomes (Bennetzen 2002; Devos et al. 2002). In fact, recombination between sequences at different genomic locations (ectopic exchange) may potentially lead to large chromosomal rearrangements (Lysak et al. 2006; Gaut et al. 2007).

In addition to having a role in generating mutations, recombination facilitates linkage of advantageous ones, and thus may play a role in selection and adaptation (Gaut et al. 2007). Since recombination does not occur randomly

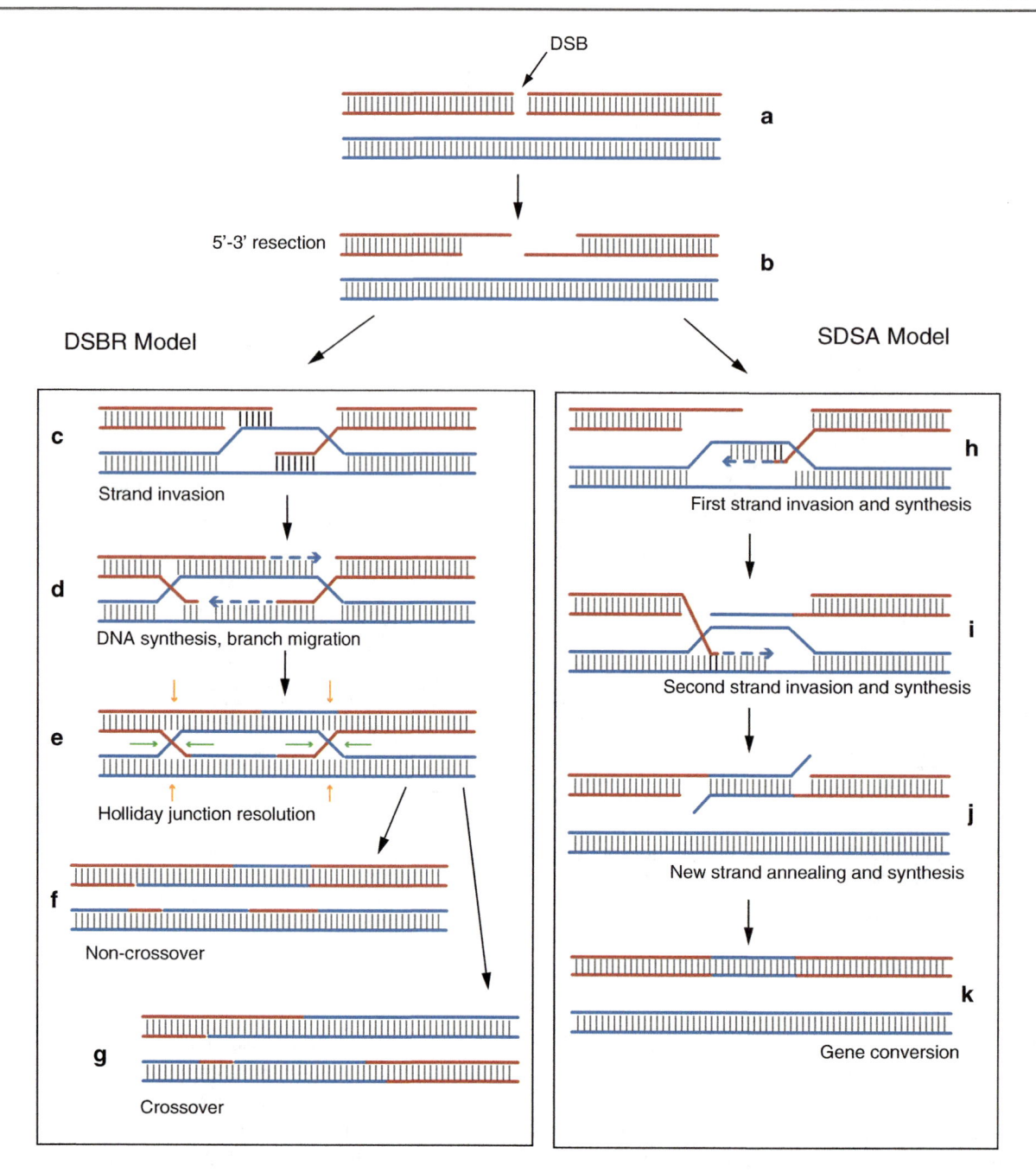

Fig. 12.5 Basic features in the two primary models for homologous recombination. Both models start with a double strand break (DSB) in one of the two homologous chromosomes represented in *red* and *blue* (**a**); the DSB is then resected (degraded by nucleases) in the 5′ to 3′ direction (**b**) resulting in two 3′ single-stranded ends (**b**). In the Double Strand DNA Break Repair model, **DSBR** (**c–g**), one 3′ end (*red*) invades the homologous chromosome (*blue*) and creates a displacement-loop structure to which the second 3′ end can anneal (**c**). This structure has one Holliday junction (HJ) and one half HJ, but branch migration (extension of the D-loop) allows the formation of a second complete HJ (**d**). New DNA synthesis fills the gaps, forming a fully ligated, double HJ structure (**e**). Resolution of the HJs may lead to non-crossover of flanking regions (**f**) when cleavages occur in the same orientation (either the *orange* or the *green arrows*); when cleavages involves both orientations (*orange* and *green*), crossover of flanking markers takes place (**g**). In the synthesis-dependent strand annealing model, **SDSA** (**h–k**), one of the 3′ single-stranded ends (*red*) invades the homologous duplex and is extended by DNA synthesis using the homologous (*blue*) chromosome as template (**h**). The other 3′ end can also invade the homologous chromosome and copy the homologous (*blue*) chromosome by DNA synthesis (**i**). This strand is displaced from the donor chromosome (**j**), pairs with the other 3′ end and the repair is completed with further DNA synthesis and ligation, resulting in gene conversion, in which the donor chromosome remains unchanged (**k**)

along chromosomes (Schnable et al. 1998; Anderson et al. 2001), it has been proposed that spatial organization of genomes (and disruption of synteny) may be related to recombination. Evidence evaluating this suggestion is inconclusive but some observations are consistent with it, including the correlation between gene density and recombination in maize (Anderson et al. 2006) and rice (International Rice Genome Sequencing Project 2005), and the finding that recombination takes place preferentially within genes in maize (Civardi et al. 1994).

12.3.2 Homoeologous Recombination and Evolution of Genomes

This section focuses on homoeologous recombination, that is, on exchange of sequences and genetic structure between duplicated chromosomes and loci that diverged independently in their respective ancestral lineages but became merged in the same genome through hybridization (in a wide sense, i.e., involving populations distinguishable on the basis of one or more heritable characters; Harrison 1990). In line with the long held position that natural hybridization may play a creative role in plant evolution (Stebbins 1959; Arnold 1997; Soltis and Soltis 2009), homoeologous recombination in hybridized genomes generates structural and gene diversity, and thus may be important for the evolution of plant genomes. However, since it operates on genetic material from different backgrounds, it presumably generates a high number of deleterious or unfit rearrangements.

Most of what we know from homoeologous exchanges comes from the study of allopolyploid model organisms like *Brassica, Nicotiana, Gossypium* or *Tragopogon*. A number of genetic, epigenetic, chromosomal and genomic changes produced in recently formed allopolyploids have been recorded (Wendel 2000; Doyle et al. 2008; Leitch and Leitch 2008), perhaps in response to the "genomic shock" resulting from the merging of genomes from different genetic backgrounds into a single cell (McClintock 1984). But the mechanisms leading to those changes are poorly understood, as are the factors that control them. Responses to hybridization and polyploidy include loss of DNA (e.g., Ozkan et al. 2001; Shaked et al. 2001; Tate et al. 2006; Gaeta et al. 2007), chromosomal rearrangements (e.g., Pires et al. 2004; Pontes et al. 2004; Udall et al. 2005), changes in rDNA loci (e.g., Lim et al. 2000, 2008; Joly et al. 2004; Pontes et al. 2004), gene conversion (e.g., Wendel et al. 1995; Kovarik et al. 2004, 2005; Salmon et al. 2010), repatterning of epigenetic marks (e.g., Salmon et al. 2005; Lukens et al. 2006; Chen 2007), transposon activation (e.g., Kashkush et al. 2003; Madlung et al. 2005; Parisod et al. 2010) and biased gene expression (e.g.,

Hegarty et al. 2006; Tate et al. 2006; Adams 2007; Flagel et al. 2008; Rapp et al. 2009).

Meiotic recombination occurs predominantly between homologous chromosomes (Naranjo and Corredor 2008), but in allopolyploids that have not acquired diploidization recombination may take place ectopically between paralogous or homoeologous sequences. And these types of non-homologous recombination are believed to generate structural changes that play a role in the evolution of these hybridized genomes. Therefore, some of the changes reported in allopolyploids are likely due to homoeologous recombination and gene conversion. Early generations of allopolyploid organisms face the challenge of achieving homologous pairing in meiosis and avoiding non-homologous pairing. Otherwise the disruption of disomic inheritance may lead to chromosomal rearrangements and aneuploidy, which compromise fertility (Gillies 1989). Fertile allopolyploids are likely to solve this problem by strict genetic control of homologous pairing (Griffiths et al. 2006), either pre-existing or evolving after their formation. But it is also possible that genomic rearrangements occurring in newly formed polyploids may contribute to divergence among homoeologous chromosomes, thereby reducing the probability of homoeologous pairing and thus facilitating regular meiosis and fertility (Gaeta and Pires 2010).

A major difficulty with assessing the frequency and evolutionary impact of homoeologous recombination is its detection. The panoply of molecular markers that have become available in the last two decades have facilitated diagnosis, but reconstructing the evolutionary origin and timing of exchanges has remained challenging. In this respect, resynthesized allopolyploids have provided crucial experimental complements to searches for changes in natural allopolyploids. This has made possible the detection of homoeologous recombination, as well as allowed their effects to be followed in subsequent progenies. Particularly illustrative in this regard is the study of Szadkowski et al. (2010), who used molecular and cytological approaches to investigate the first meiosis in allopolyploid *Brassica napus*. They demonstrated rearrangements involving homoeologous chromosomes, possibly through recombination, that promote restructuring in future generations and thus, that homoeologous exchanges may play an important role in the subsequent evolutionary fate of allopolyploids .

In fact, the amphidiploid *Brassica napus* and resynthesized allopolyploids from its parental species *B. oleracea* and *B. rapa* probably have contributed more than other systems to the study of homoeologous recombination as a causal mechanism for genetic changes in allopolyploids (reviewed in Gaeta and Pires 2010). For instance, RFLP markers have revealed loss and duplications at homoeologous loci (Osborn et al. 2003; Pires et al. 2004; Udall et al. 2005; Gaeta et al. 2007), described as homoeologous nonreciprocal transpositions (HNRTs) since they are detected as the loss of one marker coinciding with the duplication of the other (although they might in fact be reciprocal; Nicolas et al. 2008). The consequences of allopolyploidy have been studied in up to six generations in 50 resynthesyzed *Brassica napus* allopolyploid lines obtained independently (Lukens et al. 2006; Gaeta et al. 2007). Interestingly, chromosomal rearrangements derived from homoeologous recombination have been found in natural *B. napus* individuals (Udall et al. 2005), although at a much lower frequency than in the resynthesized lines (Gaeta et al. 2007), which suggests a role for selection in early stages of allopolyploid formation. As mentioned above, it is likely that extensive genome rearrangements, directly deleterious or non-adaptive, have been selected against in natural individuals (Gaeta and Pires 2010).

In the recently formed allopolyploids *Tragopogon miscellus* and *T. mirus*, elimination of parental loci, possibly due to homoeologous rearrangements, has been detected using various types of evidence, including FISH analysis (Lim et al. 2008). Recombination in a wide sense has been also reported as a likely mechanism for changes in rDNA genes, specifically losses, translocations and amplifications. For instance, in *Nicotiana* allopolyploids, rDNA translocation, losses and amplifications may have been caused by homoeologous recombination and gene conversion, associated with epigenetic mechanisms like methylation (Lim et al. 2000; Kovarik et al. 2008). In tetraploid *Arabidopsis suecica*, a possible mechanism for the loss or translocation of rDNA loci with respect to the parental species is transposable element-mediated chromosome breakage and recombination (Pontes et al. 2004).

In contrast to other polyploid model systems, homoeologous recombination has not been detected in resynthesized allopolyploids of wheat and cotton. In synthetic wheat, this finding coincides with the presence of the *Ph1* gene (pairing homoeologous 1) which restricts meiotic pairing to homologous chromosomes. In cotton, few genomic changes been detected in early generations of resynthesized allopolyplods (Liu et al. 2001; Rong et al. 2004), and this constrasts with reports of homoeologous interactions between rDNA from the two diploid genomes (A, D) involved in natural allopolyploid *Gossypium* (Wendel et al. 1995). However, Salmon et al. (2010) recently assessed the extent of nonreciprocal homoeologous recombination (NRHR) in allopolyploid cottons using a finer scale marker system and single- or low-copy genes. They searched for single nucleotide polymorphisms (SNPs) in expressed sequenced tags (EST) from the two diploid progenitor genomes of allopolyploid cotton and from multiple cDNA libraries derived from diverse tissues of allopolyploid cotton (Udall et al. 2006). They found that NRHR had modified 1.8% of the cotton transcriptome. Since homoeologous pairing and recombination

have rarely been seen in this genus (Rong et al. 2004), the authors suggest that gene conversion events are responsible for those NRHRs (Salmon et al. 2010). They also incorporated a phylogenetic context and assessed the tempo for the generation of those NRHRs and conclude that homoeologous recombination events accumulate gradually rather that episodically along the 1.5-million-year history of the allopolyploid reticulate lineage in *Gossypium*. This is consistent with the idea that homoeologous pairing during meiosis is suspected to be avoided in the first generations following polyploid formation (Ramsey and Schemske 2002), because it can lead to infertility as a result of unbalanced gamete formation following meiosis. It also fits the expectation that a significant proportion of interactions between different genomes gathered in a hybrid or polyploidy is likely to result in unfit individuals and thus the viable resulting changes accumulate sporadically at a low frequency.

Homoeologous recombination has been also found to contribute positively to polyploids originating several times from the same progenitors. In *Glycine tabacina*, recombination among the different independently originated polyploid genomes, which could provide a wide genetic base and possible an increased potential for colonization, was documented for a single copy gene (Doyle et al. 2002).

12.3.3 Recombination and Phylogenetic Inference

In phylogenetic studies, recombination represents a violation of the assumption that there is a single genealogy underlying the evolution of the sequences under study. When recombinant sequences are included in phylogenetic studies, different phylogenetic histories are mixed or, in other words, regions of sequences that possess different most recent common ancestors are put together (Doyle 1996). Consequently, results of phylogenetic analysis may be compromised by recombination (Arenas and Posada 2010). Gene trees constructed including recombinant sequences will increase homoplasy and create topological anomalies, and consequently sequences subject to recombination may more adequately be represented by networks instead of bifurcating trees. Homoplasy introduced by recombination does not result from incorrect homology assessments but is a kind of mixed-signal homoplasy (Doyle 1996).

For phylogenetic purposes, three types of recombination can be distinguished. First, recombination may be interallelic, that is, recombinant sequences may result from alleles that exchange sequence parts with other such alleles from the same locus. If this phenomenon takes place between alleles of a single species that are monophyletic with respect to other species, i.e., if alleles coalesce within species, recombination does not affect reconstruction of species phylogenies from the gene tree in question although it does have implications for phylogenetic analysis. But alleles—specially low-copy nuclear alleles—may coalesce deeper that the speciation events, due to large effective population sizes, in which case trans-species polymorphism occurs and even recombination within species may create some confusion.

However, the effects in reconstructing organismal phylogenies based on gene geneaologies are even more distortive when recombination occurs between paralogous sequences than between orthologous sequences. Since recombination is a similarity-driven phenomenon that involves a 'homology search' (e.g., Pawlowski et al. 2003), it is also likely that paralogous sequences that have diverged more recently have a higher probability to be involved in this kind of recombination (Small et al. 2004). Likewise, for mechanistic reasons paralogous loci that are more closely located in the genome are more likely to suffer recombination. Yet on average paralogous sequences will have diverged more than alleles from a single locus, and so the effects of recombination on phylogenies are expected to be more profoundly distorting.

A third type of phylogeny-distorting recombination-related mechanism is gene conversion, already mentioned above. Its effects on phylogeny reconstruction are the worse of the three because non-reciprocal 'donation' of sequence information not only erases signal in the transformed sequence but also adds a false signal to it.

As explained above, homoeologous recombinant sequences may be related to hybridization, allopolyploidy, or horizontal transfer. Therefore, the occurrence of recombinant sequences may actually provide a hint or unsuspected reticulation (e.g., Kelly et al. 2010). In fact, it may even help identify progenitors whenever alleles involved in the recombinant sequence are species-specific or lineage-specific (Gehrke et al. 2010). Along this line, although recombination in organellar genomes is considered to be rare (Palmer et al. 1988; Rokas et al. 2003), it has been reported in plants for mitochondrial DNA (Jaramillo-Correa and Bousquet 2005) and plastid DNA (Städler and Delph 2002; Wolfe and Randle 2002; Ansell et al. 2007) and has been associated in some of those cases with hybridization or hybrid zones. Organellar recombination implies the existence of—at least transiently—heteroplasmy, i.e., a mixture of more than one type of an organellar genome within a cell or individual. In both organelles, heteroplasmy could be due to leakage of paternal or maternal organellar DNA (the alternative progenitor to the one through which organellar DNA is inherited in the group in question, e.g., paternal for cpDNA in most angiosperms). Such leakage is common in animal groups (e. g., Hoarau et al. 2002), and has also been found in plants (Wagner et al. 1991; Hattori et al. 2002).

In addition to hybridization, recombinant DNA sequences could be produced in vitro by PCR amplification of non-identical templates (Cronn et al. 2002). This phenomenon, which produces chimeric sequences, is more frequent in repetitive sequences such as rDNA due to the existence of different templates in the same genome whenever homogenization is not complete (Bradley and Hillis 1997; Alvarez and Wendel 2003) although it may also occur in low copy genes (Small et al. 2004). Recommendations to cope with this potential problem include designing repeat-specific primers (Rauscher et al. 2002) or using a combination of primers specific to each of the two putatively in vitro recombined paralogous loci to amplify genomic DNA (Sang et al. 1997). Following this procedure, if no amplicon is obtained it can be deduced that the recombined sequence obtained in a previous amplification was a PCR artifact.

Because of the potentially distorting effect of recombination in phylogenetic analysis it is convenient to identify recombinant sequences and remove them from the data set prior to analysis. But identifying recombinant sequences is not easy, particularly if sequence divergence is below 5%. There are many statistical methods to detect recombinant sequences based on different criteria, and there also are several studies that evaluate the performance of different methods on empirical or simulated data (Posada and Crandall 2001; Wiuf et al. 2001; Posada 2002). Methods can be classified as distance methods that look for inversions of distance patterns among the sequences, phylogenetic methods that look for adjacent sequences yielding different branching patterns, compatibility methods that partition phylogenetic incongruence site by site, and substitution distribution methods that look for a significant clustering of substitutions (Posada and Crandall 2001). Because the performance of the different methods depends on the data in question, there are particularly helpful packages that use different methods to attempt to identify recombinant sequences and recombination breakpoints (Martin et al. 2005, 2010).

12.4 Concluding Remarks

Almost four decades after the discovery of the phenomenon of concerted evolution a basic fact remains intact: members of some multigene families and some single-copy genes do not evolve independently. Concerted evolution remains a phenomenon controlled by a suite of molecular mechanisms operating at the genic or genomic level. Concerted evolution is a pattern (Dover 1986), generated by different mechanisms that converge to produce either homogenization, in the case of fully "converted" members of multigene families, or novel alleles resulting from partial gene conversion of homologous or paralogous alleles. This view accommodates different homogenizing mechanisms into a common conceptual platform but also is compatible with alternative models of evolution (birth-and-death, Nei and Hughes 1992) of multigene families.

The phenomenon treated in this chapter termed homoeologous recombination or in a wider sense non-homologous recombination, comprises part of the family of phenomena under the umbrella term "concerted evolution". HR is also far from simple mechanistically, since it encompasses several phenomena, some insufficiently understood. As noted, the outcomes of non-homologous recombination may vary both in magnitude and direction, operating to eliminate variation or create novel variants. Less is understood about the possible functional and evolutionary significance, yet there is ample reason to believe that it has importance with respect to genome evolution. It also is clear that homoeologous and non-homologous recombination underlie many of the genome rearrangements that occur in hybridized genomes.

A combination of approaches, some of which are already being used by some groups, will be needed to fill gaps in the knowledge of both concerted evolution and homoeologous and non-homologous recombination and to improve our understanding of the diverse underlying mechanisms involved and how they interact. First, the use of molecular cytogenetic methods provides the necessary link between two crucial levels of analysis (classical cytogenetic studies and molecular genetics). For instance, methods specifically designed to stretch chromosomes before applying FISH (e.g. fiber-FISH) have increased resolution (up to 3.12 kb/μm) and allowed for finer-scale mapping of sequences than has been possible using conventional FISH on mitotic or meiotic chromosomes (Figueroa and Bass 2010). Second, obtaining comprehensive and comparative sequence data from multigene families as well as single-copy genes is necessary for inferring the underlying model of evolution and mechanisms. Next-generation DNA sequencing techniques will undoubtedly play an important role in this respect. Third, analyzing synthetic hybrids or allopolyploids and their descendants provides an evolutionary perspective on the changes caused by both concerted evolution and homoeologous recombination, particularly when compared to stabilized natural derivatives that are much older. Fourth, to enlarge our understanding of the diversity of these phenomena across plant groups, other species besides model organisms should be studied. An example of this involves neopolyploidy in North-American *Tragopogon* and the pioneering studies by Doug and Pamela Soltis (Soltis and Soltis 1989; Soltis et al. 2004; Kovarik et al. 2005; Tate et al. 2006; Matyasek et al. 2007).

Finally, additional detailed studies of other repetitive families (e.g., satellite, centromeric and telomeric regions

and transposable elements) as well as single-copy genes need to be performed, focusing on the patterns of similarity and on the responsible mechanisms and models. These studies will also serve as the necessary prelude to investigations of functional and ecological impact, which of course is of paramount importance for developing a full appreciation of evolutionary significance.

Acknowledgements We are grateful to Marcela Rosato for providing photographs for Fig. 12.2 and to Inés Álvarez, Myriam Heuertz and Marcela Rosato for helpful comments. Our work has been supported by the Spanish Ministry of Science and Innovation through grants CGL2007-66516 to (GNF) and CGL2007-60550/BOS, CGL2010-22347-C02-01, 2009 SGR 608 (to JAR).

References

Abbott RJ, Lowe AJ (2004) Origins, establishment and evolution of new polyploid species: *Senecio cambrensis* and *S. eboracensis* in the British Isles. Biol J Linn Soc 82:467–474

Adams KL (2007) Evolution of duplicate gene expression in polyploid and hybrid plants. J Hered 98:136–141

Álvarez I, Wendel JF (2003) Ribosomal ITS sequences and plant phylogenetic inference. Mol Phylogenet Evol 29:417–434

Anderson LK, Hooker KD, Stack SM (2001) The distribution of early recombination nodules on zygotene bivalents from plants. Genetics 159:1259–1269

Anderson LK, Lai A, Stack SM, Rizzon C, Gaut BS (2006) Uneven distribution of expressed sequence tag loci on maize pachytene chromosomes. Genome Res 16:115–122

Ansell SW, Schneider H, Pedersen N, Grundmann M (2007) Recombination diversifies chloroplast trnF pseudogenes in *Arabidopsis lyrata*. J Evol Biol 20:2400–2411

Arenas M, Posada D (2010) The effect of recombination on the reconstruction of ancestral sequences. Genetics 184:1133–1139

Arnheim N (1983) Concerted evolution of multigene families. In: Nei M, Koehn RK (eds) Evolution of genes and proteins. Sinauer, Sunderland, pp 38–61

Arnold ML (1997) Natural hybridization and evolution. Oxford University Press, New York

Baldwin BG, Sanderson MJ, Porter JM, Wojciechowski MF, Campbell CS, Donoghue MJ (1995) The ITS region of nuclear ribosomal DNA: a valuable source of evidence on angiosperm phylogeny. Ann Mo Bot Gard 82:247–277

Bao Y, Wendel JF, Ge S (2010) Multiple patterns of rDNA evolution following polyploidy in *Oryza*. Mol Phylogenet Evol 55:136–142

Bennetzen JL (2000) Transposable element contributions to plant gene and genome evolution. Plant Mol Biol 42:251–269

Bennetzen JL (2002) Mechanisms and rates of genome expansion and contraction in flowering plants. Genetica 115:29–36

Bobola MS, Eckert RT, Klein AS (1992) Restriction fragment variation in the nuclear ribosomal DNA repeat unit within and between *Picea rubens* and *Picea mariana*. Can J Forest Res 22:255–263

Bradley RD, Hillis DA (1997) Recombinant DNA sequences generated by PCR amplification. Mol Biol Evol 14:592–593

Brodsky LI, Ivanov VV, Kalai YL, Leontovich AM, Nikolaev VK, Feranchuk SI, Drachev VA (1995) GeneBee-NET: Internet-based server for analyzing biopolymers structure. Biochemistry 60:923–928

Brown DD, Wensink PC, Jordan E (1972) A comparison of the ribosomal DNA's of *Xenopus laevis* and *Xenopus mulleri*: the evolution of tandem genes. J Mol Biol 63:57–73

Buckler ESI, Ippolito A, Holtsford TP (1997) The evolution of ribosomal DNA: divergent paralogs and phylogenetic implications. Genetics 145:821–832

Camacho JPM, Sharbel TF, Beukeboom LW (2000) B-chromosome evolution. Philos Trans R Soc Lond B 355:163–178

Campbell CS, Wojciechowski MF, Baldwin BG, Alice LA, Donoghue MJ (1997) Persistent nuclear ribosomal DNA sequence polymorphism in the *Amelanchier* agamic complex (Rosaceae). Mol Biol Evol 14:81–90

Castro J, Rodríguez S, Pardo BG, Sánchez L, Martínez P (2001) Population analysis of an unusual NOR-site polymorphism in brown trout (*Salmo trutta* L.). Heredity 86:291–302

Chen ZJ (2007) Genetic and epigenetic mechanisms for gene expression and phenotypic variation in plant polyploids. Annu Rev Plant Biol 58:377–406

Civardi L, Xia Y, Edwards KJ, Schnable PS, Nikolau BJ (1994) The relationship between genetic and physical distances in the cloned *a1-sh2* interval of the *Zea mays* L. genome. Proc Natl Acad Sci U S A 91:8268–8272

Clegg MT, Cummings MP, Durbin ML (1997) The evolution of plant nuclear genes. Proc Natl Acad Sci U S A 94:7791–7798

Cluster PD, Allard RW (1995) Evolution of ribosomal DNA (rDNA) genetic structure in colonial populations of *Avena barbata*. Genetics 139:941–954

Coleman A (2003) ITS2 is a double-edged tool for eukaryote evolutionary comparisons. Trends Genet 19:370–375

Cordesse F, Second G, Delseny M (1990) Ribosomal gene spacer length variability in cultivated and wild rice species. Theor Appl Genet 79:81–88

Cromie GA, Smith GR (2007) Branching out: meiotic recombination and its regulation. Trends Cell Biol 17:448–455

Cronn R, Cedroni M, Haselkorn T, Osborne C, Wendel JF (2002) PCR-mediated recombination in amplification products derived from polyploid cotton. Theor Appl Genet 104:482–489

Cuñado N, de la Herrán R, Santos JL, Ruiz Rejón C, Garrido-Ramos MA, Ruiz Rejón M (2000) The evolution of the ribosomal loci in the subgenus *Leopoldia* of the genus *Muscari* (Hyacinthaceae). Plant Syst Evol 221:245–252

Devos KM, Brown JK, Bennetzen JL (2002) Genome size reduction through illegitimate recombination counteracts genome expansion in *Arabidopsis*. Genome Res 12:1075–1079

Dover GA (1982) Molecular drive: a cohesive mode of species evolution. Nature 299:111–116

Dover GA (1986) Molecular drive in multigene families: how biological novelties arise, spread and are assimilated. Trends Genet 2:159–165

Dover GA, Linares AR, Bowen T, Hancock JM (1993) Detection and quantification of concerted evolution and molecular drive. Methods Enzymol 224:525–541

Doyle JJ (1996) Homoplasy connections and disconnections: genes and species, molecules and morphology. In: Sanderson MJ, Hufford L (eds) Homoplasy: the recurrence of similarity in evolution. Academic, San Diego, pp 67–89

Doyle JJ, Davis JI (1998) Homology in molecular phylogenetics: a parsimony perspective. In: Soltis DE, Soltis PS, Doyle JJ (eds) Molecular systematics of plants, 2nd edn. Kluwer, Dordrecht, pp 101–131

Doyle JJ, Doyle JL, Brown AHD, Palmer RG (2002) Genomes, multiple origins, and lineage recombination in the *Glycine tomentella* (Leguminsae) polyploid complex: histone H3-D gene sequences. Evolution 56:1388–1402

Doyle JJ, Flagel LE, Paterson AH, Rapp RA, Soltis DE, Soltis PS, Wendel JF (2008) Evolutionary genetics of genome merger and doubling in plants. Annu Rev Genet 42:443–461

Elder JF, Turner BJ (1995) Concerted evolution of repetitive DNA sequences in eukaryotes. Q Rev Biol 70:297–320

Elser JJ, Dobberfuhl D, MacKay NA, Schampel JH (1996) Organism size, life history, and N:P stoichiometry: towards a unified view of cellular and ecosystem processes. Bioscience 46:674–684

Elser JJ, Sterner RW, Gorokhova E, Fagan WF, Markow TA, Cotner JB, Harrison JF, Hobbie SE, Odell GM, Weider LJ (2000) Biological stoichiometry from genes to ecosystems. Ecol Lett 3:540–550

Figueroa DM, Bass HW (2010) A historical and modern perspective on plant cytogenetics. Brief Funct Genomics 9:95–102

Flagel L, Udall J, Nettleton D, Wendel J (2008) Duplicate gene expression in allopolyploid *Gossypium* reveals two temporally distinct phases of expression evolution. BMC Biol 6:16

Franzke A, Mummenhoff K (1999) Recent hybrid speciation in *Cardamine* (Brassicaceae)—conversion of nuclear ribosomal ITS sequences in statu nascendi. Theor Appl Genet 98:831–834

Fuchs J, Brandes A, Schubert I (1995) Telomere sequence localization and karyotype evolution in higher plants. Plant Syst Evol 196: 227–241

Fuertes Aguilar J, Rosselló JA, Nieto Feliner G (1999a) Nuclear ribosomal DNA (nrDNA) concerted evolution in natural and artificial hybrids of *Armeria* (Plumbaginaceae). Mol Ecol 8:1341–1346

Fuertes Aguilar J, Rosselló JA, Nieto Feliner G (1999b) Molecular evidence for the compilospecies model of reticulate evolution in *Armeria* (Plumbaginaceae). Syst Biol 44:735–754

Fujisawa M, Yamagata H, Kamiya K, Nakamura M, Saji S, Kanamori H, Wu J, Matsumoto T, Sasaki T (2006) Sequence comparison of distal and proximal ribosomal DNA arrays in rice (*Oryza sativa* L.) chromosome 9S and analysis of their flanking regions. Theor Appl Genet 113:419–428

Gaeta RT, Pires JC (2010) Homoeologous recombination in allopolyploids: the polyploid ratchet. New Phytol 186:18–28

Gaeta RT, Pires JC, Iniguez-Luy F, Leon E, Osborn TC (2007) Genomic changes in resynthesized *Brassica napus* and their effect on gene expression and phenotype. Plant Cell 19:3403–3417

Ganley AR, Kobayashi T (2007) Highly efficient concerted evolution in the ribosomal DNA repeats: total rDNA repeat variation revealed by whole-genome shotgun sequence data. Genome Res 17:184–191

García S, Lim KY, Chester M, Garnatje T, Pellicer J, Vallès J, Leitch AR, Kovarik A (2009) Linkage of 35S and 5S rRNA genes in *Artemisia* (family Asteraceae): first evidence from angiosperms. Chromosoma 118:85–97

Gardes M, Bruns TD (1993) ITS primers with enhanced specificity for basidiomycetes: application to the identification of mycorrhizae and rusts. Mol Ecol 2:113–118

Gaut BS, Wright SI, Rizzon C, Dvorak J, Anderson LK (2007) Recombination: an underappreciated factor in the evolution of plant genomes. Nat Rev Genet 8:77–84

Gehrke B, Martín-Bravo S, Muasya M, Luceño M (2010) Monophyly, phylogenetic position and the role of hybridization in *Schoenoxiphium* Nees (Cariceae, Cyperaceae). Mol Phylogenet Evol 56:380–392

Gernandt DS, Liston A, Piñero D (2001) Variation in the nrDNA ITS of *Pinus* subsection *Cembroides*: implications for molecular systematic studies of pine species complexes. Mol Phylogenet Evol 21:449–467

Gillies CB (1989) Fertility and chromosome pairing: recent studies in plants and animals. CRC Press, Boca Raton

Govindaraju DR, Cullis CA (1992) Ribosomal DNA variation among populations of a *Pinus rigida* Mill. (pitch pine) ecosystem: I. Distribution of copy numbers. Heredity 69:133–140

Graur D, Li W-H (2000) Fundamentals of molecular evolution. Sinauer, Sunderland

Griffiths S, Sharp R, Foote TN, Bertin I, Wanous M, Reader S, Colas I, Moore G (2006) Molecular characterization of *Ph1* as a major chromosome pairing locus in polyploid wheat. Nature 439:749–752

Haber JE (2008) Evolution of models of homologous recombination. In: Egel R, Lankenau D-H (eds) Recombination and meiosis, vol 3,

Genome dynamics and stability. Springer, Berlin/Heidelberg/New York, pp 1–64

Hall AE, Keith KC, Hall SE, Copenhaver GP, Preuss D (2004) The rapidly evolving field of plant centromeres. Curr Opin Plant Biol 7:108–114

Harrison RG (1990) Hybrid zones: windows on evolutionary process. Oxf Surv Evol Biol 7:69–128

Hattori N, Kitagawa K, Takumi S, Nakamura C (2002) Mitochondrial DNA heteroplasmy in wheat, *Aegilops* and their nucleus-cytoplasm hybrids. Genetics 160:1619–1630

Hegarty MJ, Barker GL, Wilson ID, Abbott RJ, Edwards KJ, Hiscock SJ (2006) Transcriptome shock after interspecific hybridization in *Senecio* is ameliorated by genome duplication. Curr Biol 16:1652–1659

Henikoff S, Ahmad K, Malik HS (2001) The centromere paradox: stable inheritance with rapidly evolving DNA. Science 293:1098–1102

Hennig W (1966) Phylogenetic systematics. University of Illinois Press, Urbana

Hershkovitz MA, Lewis LA (1996) Deep-level diagnostic value of the rDNA-ITS region. Mol Biol Evol 13:1276–1295

Hershkovitz MA, Zimmer EA (1996) Conservation patterns in angiosperm rDNA ITS2 sequences. Nucleic Acids Res 24:2857–2876

Hershkovitz MA, Zimmer EA, Hahn WJ (1999) Ribosomal DNA sequences and angiosperm systematics. In: Hollingsworth PM, Bateman RM, Gornall RJ (eds) Molecular systematics and plant evolution. Taylor & Francis, London, pp 268–326

Heslop-Harrison JS (2000) Comparative genome organization in plants: from sequence and markers to chromatin and chromosomes. Plant Cell 12:617–636

Hillis DM, Moritz C, Porter CA, Baker RJ (1991) Evidence for biased gene conversion in concerted evolution of ribosomal DNA. Science 251:308–310

Hoarau G, Holla S, Lescasse R, Stam WT, Olsen JL (2002) Heteroplasmy and evidence for recombination in the mitochondrial control region of the flatfish *Platichthys flesus*. Mol Biol Evol 19:2261–2264

Hoelzer GA, Wallman J, Melnick DJ (1998) The effects of social structure, geographical structure and population size on the evolution of mitochondrial DNA. II. Molecular clocks and the lineage sorting period. J Mol Evol 47:21–31

Hofacker IL (2003) Vienna RNA secondary structure server. Nucleic Acids Res 31:3429–3431

Hughes CE, Eastwood RJ, Bailey CD (2006) From famine to feast? Selecting nuclear DNA sequence loci for plant species-level phylogeny reconstruction. Philos Trans R Soc Lond B 361:211–225

Humphries CJ, Funk VA (1984) Cladistics methodology. In: Heywood VH, Moore DM (eds) Current concepts in plant taxonomy. Academic, London, pp 323–362

International Rice Genome Sequencing Project (2005) The map-based sequence of the rice genome. Nature 436:793–800

Jaramillo-Correa JP, Bousquet J (2005) Mitochondrial genome recombination in the zone of contact between two hybridizing conifers. Genetics 171:1951–1962

Jelesko JG, Carter K, Thompson W, Kinoshita Y, Gruissem W (2004) Meiotic recombination between paralogous *RBCSB* genes on sister chromatids of *Arabidopsis thaliana*. Genetics 166:947–957

Jiang J, Birchler JA, Parrott WA, Dawe RK (2003) A molecular view of plant centromeres. Trends Plant Sci 8:570–575

Jobes DV, Thien LB (1997) A conserved motif in the 5.8S Ribosomal RNA (rRNA) gene is a useful diagnostic marker for plant internal transcribed spacer (ITS) sequences. Plant Mol Biol Rep 15:326–334

Johnson LA, Soltis DE (1998) Assessing congruence: empirical examples from molecular data. In: Soltis DE, Soltis PS, Doyle JJ (eds) Molecular systematics of plants II. DNA sequencing. Kluwer, Boston, pp 297–348

Joly S, Rauscher JT, Sherman-Broyles SL, Brown AHD, Doyle JJ (2004) Evolutionary dynamics and preferential expression of homeologous 18S-5.8S-26S nuclear ribosomal genes in natural and artificial *Glycine* allopolyploids. Mol Biol Evol 21:1409–1421

Jorgensen JL, Stehlik I, Brochmann C, Conti E (2003) Implications of ITS sequences and RAPD markers for the taxonomy and biogeography of the *Oxytropis campestris* and *O. arctica* (Fabaceae) complexes in Alaska. Am J Bot 90:1470–1480

Kashkush K, Feldman M, Levy AA (2003) Transcriptional activation of retrotransposons alters the expression of adjacent genes in wheat. Nat Genet 33:102–106

Kejnovsky E, Hobza R, Kubat Z, Widmer A, Marais GAB, Vyskot B (2007) High intrachromosomal similarity of retrotransposon long terminal repeats: evidence for homogenization by gene conversion on plant sex chromosomes? Gene 390:92–97

Kejnovsky E, Hawkins J, Feschotte C (2012) Plant transposable elements: biology and evolution. In: Wendel JF (ed) Plant genome diversity, vol 1, Plant genomes, their residents, and their evolutionary dynamics. Springer, Wien/New York

Kellogg EA, Appels R (1995) Intraspecific and interspecific variation in 5S RNA genes are decoupled in diploid wheat relatives. Genetics 140:325–343

Kelly LJ, Leitch AR, Clarkson JJ, Hunter RB, Knapp S, Chase MW (2010) Intragenic recombination events and evidence for hybrid speciation in *Nicotiana* (Solanaceae). Mol Biol Evol 27:781–799

Kim KJ, Jansen RK (1994) Comparisons of phylogenetic hypotheses among different data sets in dwarf dandelions (*Krigia*): additional information from internal transcribed spacer sequences of nuclear ribosomal DNA. Plant Syst Evol 190:157–159

Kim ST, Sultan SE, Donoghue MJ (2008) Allopolyploid speciation in *Persicaria* (Polygonaceae): insights from a low-copy nuclear region. Proc Natl Acad Sci U S A 105:12370–12375

King RC, Stansfield WD (1990) A dictionary of genetics, 4th edn. Oxford University Press, New York

Koch MA, Dobeš C, Mitchell-Olds T (2003) Multiple hybrid formation in natural populations: concerted evolution of the internal transcribed spacer of nuclear ribosomal DNA (ITS) in North American *Arabis divaricarpa* (Brassicaceae). Mol Biol Evol 20:338–350

Kotseruba V, Gernand D, Meister A, Houben A (2003) Uniparental loss of ribosomal DNA in the allotetraploid grass *Zingeria trichopoda* (2n = 8). Genome 46:156–163

Kovarik A, Matyasek R, Lim KY, Skalicka K, Koukalova B, Knapp S, Chase M, Leitch AR (2004) Concerted evolution of 18–5.8-26S rDNA repeats in *Nicotiana* allotetraploids. Biol J Linn Soc 82:615–625

Kovarik A, Pires JC, Leitch AR, Lim KY, Sherwood AM, Matyasek R, Rocca J, Soltis DE, Soltis PS (2005) Rapid concerted evolution of nuclear ribosomal DNA in two *Tragopogon* allopolyploids of recent and recurrent origin. Genetics 169:931–944

Kovarik A, Dadejova M, Lim YK, Chase MW, Clarkson JJ, Knapp S, Leitch AR (2008) Evolution of rDNA in *Nicotiana* allopolyploids: a potential link between rDNA homogenization and epigenetics. Ann Bot 101:815–823

Kress WJ, Wurdack KJ, Zimmer EA, Weigt LA, Janzen DH (2005) Use of DNA barcodes to identify flowering plants. Proc Natl Acad Sci U S A 102:8369–8374

Kubis S, Schmidt T, Heslop-Harrison JS (1998) Repetitive DNA elements as a major component of plant genomes. Ann Bot 82 (Suppl A):45–55

Le QH, Wright S, Yu Z, Bureau T (2000) Transposon diversity in *Arabidopsis thaliana*. Proc Natl Acad Sci U S A 97:7376–7381

Leitch AR, Leitch IJ (2008) Genomic plasticity and the diversity of polyploid plants. Science 320:481–483

Liao D (1999) Concerted evolution: molecular mechanisms and biological implications. Am J Hum Genet 64:24–30

Liao D (2000) Gene conversion drives within genic sequences: concerted evolution of ribosomal RNA genes in Bacteria and Archaea. J Mol Evol 51:305–317

Liao D (2003) Concerted evolution. In: Encyclopedia of the human genome. Macmillian/Nature Publishing Group, London, pp 1–6

Liao D, Pavelitz T, Kidd JR, Kidd KK, Weiner AM (1997) Concerted evolution of the tandemly repeated genes encoding human U2 sn-RNA (the RNU2 locus) involves rapid intrachromosomal homogenization and rare interchromosomal gene conversion. EMBO J 16:588–598

Lim KY, Kovarik A, Matyasek R, Bezdeek M, Lichtenstein CP, Leitch AR (2000) Gene conversion of ribosomal DNA in Nicotiana tabacum is associated with undermethylated, decondensed and probably active gene units. Chromosoma 109:161–172

Lim KY, Soltis DE, Soltis PS, Tate J, Matyasek R, Srubarova H, Kovarik A, Pires JC, Xiong Z, Leitch AR (2008) Rapid chromosome evolution in recently formed polyploids in Tragopogon (Asteraceae). PLoS One 3:e3353

Liston A, Robinson WA, Oliphant JM, Alvarez Buylla ER (1996) Length variation in the nuclear ribosomal DNA internal transcribed spacer region of non-flowering plants. Syst Bot 21:109–120

Liu J-S, Schardl CL (1994) A conserved sequence in internal transcribed spacer 1 of plant nuclear rRNA genes. Plant Mol Biol 26:775–778

Liu B, Brubaker CL, Mergeai G, Cronn RC, Wendel JF (2001) Polyploid formation in cotton is not accompanied by rapid genomic changes. Genome 44:321–330

Liu Z-L, Zhang D, Hong D-Y, Wang XR (2003) Chromosomal localization of 5S and 18S-5.8S-25S ribosomal DNA sites in five Asian pines using fluorescence in situ hybridization. Theor Appl Genet 106:198–204

Lukens LN, Pires JC, Leon E, Vogelzang R, Oslach L, Osborn T (2006) Patterns of sequence loss and cytosine methylation within a population of newly resynthesized Brassica napus allopolyploids. Plant Physiol 140:336–348

Lysak MA, Berr A, Pecinka A, Schmidt R, McBreen K, Schubert I (2006) Mechanisms of chromosome number reduction in Arabidopsis thaliana and related Brassicaceae species. Proc Natl Acad Sci U S A 103:5224–5229

Macas J, Neumann P, Navrátilova A (2007) Repetitive DNA in the pea (Pisum sativum L.) genome: comprehensive characterization using 454 sequencing and comparison to soybean and Medicago truncatula. BMC Genomics 8:427

Maddison WP (1997) Gene trees in species trees. Syst Biol 46:523–536

Madlung A, Tyagi AP, Watson B, Jiang HM, Kagochi T, Doerge RW, Martienssen R, Comai L (2005) Genomic changes in synthetic Arabidopsis polyploids. Plant J 41:221–230

Mai JV, Coleman A (1997) The internal transcribed spacer 2 exhibits a common secondary structure in green algae and flowering plants. J Mol Evol 44:258–271

Martin DP, Williamson C, Posada D (2005) RDP2: recombination detection and analysis from sequence alignments. Bioinformatics 21:260–262

Martin DP, Lemey P, Lott M, Moulton V, Posada D, Lefeuvre P (2010) RDP3: a flexible and fast computer program for analyzing recombination. Bioinformatics 26:2462–2463

Matyásek R, Tate JA, Lim YK, Srubarová H, Koh J, Leitch AR, Soltis DE, Soltis PS, Kovarík A (2007) Concerted evolution of rDNA in recently formed Tragopogon allotetraploids is typically associated with an inverse correlation between gene copy number and expression. Genetics 176:2509–2519

Mayer MS, Soltis PS (1999) Intraspecific phylogeny analysis using ITS sequences: insights from studies of the Streptanthus glandulosus complex (Cruciferae). Syst Bot 24:47–61

Mayol M, Rosselló JA (2001) Why nuclear ribosomal DNA spacers (ITS) tell different stories in Quercus. Mol Phylogenet Evol 19:167–176

McClintock B (1984) The significance of responses of the genome to challenge. Science 226:792–801

McDade LA (1992) Hybrids and phylogenetic systematics. II. The impact of hybrids on cladistic analysis. Evolution 46:1329–1346

Michelmore RW, Meyers BC (1998) Clusters of resistance genes in plants evolve by divergent selection and a birth-and-death process. Genome Res 8:1323–1344

Molinier J, Stamm ME, Hohn B (2004) SNM-dependent recombinational repair of oxidatively induced DNA damage in Arabidopsis thaliana. EMBO Rep 5:994–999

Nam J, de Pamphilis CW, Ma H, Nei M (2004) Type I MADS-box genes have experienced faster birth-and-death evolution than type II MADS-box genes in angiosperms. Proc Natl Acad Sci U S A 101:1910–1915

Naranjo T, Corredor E (2008) Nuclear architecture and chromosome dynamics in the search of the pairing partner in meiosis in plants. Cytogenet Genome Res 120:320–330

Nassif N, Penney J, Pal S, Engels WR, Gloor GB (1994) Efficient copying of nonhomologous sequences from ectopic sites via P-element-induced gap repair. Mol Cell Biol 14:1613–1625

Navajas-Pérez R, de la Herrán R, Jamilena M, Lozano R, Ruiz Rejón C, Ruiz Rejón M, Garrido-Ramos MA (2005) Reduced rates of sequence evolution of Y-linked satellite DNA in Rumex (Polygonaceae). J Mol Evol 60:391–399

Navashin M (1934) Chromosomal alterations caused by hybridization and their bearing upon certain general genetic problems. Cytologia 5:169–203

Nei M, Hughes AL (1992) Balanced polymorphism and evolution by the birth-and-death process in the MHC loci. In: Tsuji K, Aizawa M, Sasazuki T (eds) 11th Histocompatibility workshop and conference. Oxford University Press, Oxford, pp 27–38

Nei M, Rooney AP (2005) Concerted and birth-and-death evolution of multigene families. Annu Rev Genet 39:121–152

Nei M, Gu X, Sitnikova T (1997) Evolution by the birth-and-death process in multigene families of the vertebrate immune system. Proc Natl Acad Sci U S A 94:7799–7806

Nicolas SD, Leflon M, Liu Z, Eber F, Chelysheva L, Coriton O, Chevre AM, Jenczewski E (2008) Chromosome 'speed dating' during meiosis of polyploid Brassica hybrids and haploids. Cytogenet Genome Res 120:331–338

Nieto Feliner G, Rosselló JA (2007) Better the devil you know? Guidelines for insightful utilization of nrDNA ITS in species-level evolutionary studies in plants. Mol Phylogenet Evol 44:911–919

Nieto Feliner G, Fuertes Aguilar J, Rosselló JA (2001) Can extensive reticulation and concerted evolution result in a cladistically structured molecular data set? Cladistics 17:301–312

Nieto Feliner G, Gutiérrez Larena B, Fuertes Aguilar J (2004) Fine scale geographical structure, intra-individual polymorphism and recombination in nuclear ribosomal internal transcribed spacers in Armeria (Plumbaginaceae). Ann Bot 93:189–200

O'Kane SL, Schaal BA, Al-Shehbaz IA (1996) The origins of Arabidopsis suecica as indicated by nuclear rDNA sequences. Syst Bot 21:559–566

Ohta T (1980) Evolution and variation of multigene families. Springer, Berlin/Heidelberg/New York

Ohta T, Dover GA (1983) Population genetics of multigene families that are dispersed into two or more chromosomes. Proc Natl Acad Sci U S A 80:4079–4083

Orzechowska M, Siwinska D, Maluszynska J (2010) Molecular cytogenetic analyses of haploid and allopolyploid Pellia species. J Bryol 32:113–121

Osborn TC, Butrulle DV, Sharpe AG, Pickering KJ, Parkin IAP, Parker JS, Lydiate DJ (2003) Detection and effects of a homeologous reciprocal transposition in Brassica napus. Genetics 165:1569–1577

Ozkan H, Levy AA, Feldman M (2001) Allopolyploidy-induced rapid genome evolution in the wheat (Aegilops–Triticum) group. Plant Cell 13:1735–1747

Palacios C, Rosselló JA, González-Candelas F (2000) Study of the evolutionary relationships among *Limonium* species (Plumbaginaceae) using nuclear and cytoplasmic molecular markers. Mol Phylogenet Evol 14:232–249

Palmer JD, Jansen RK, Michaels HJ, Chase MW, Manhart JR (1988) Chloroplast DNA variation and plant phylogeny. Ann Mo Bot Gard 75:1180–1206

Parisod C, Alix K, Just J, Petit M, Sarilar V, Mhiri C, Ainouche M, Chalhoub B, Grandbastien MA (2010) Impact of transposable elements on the organization and function of allopolyploid genomes. New Phytol 186:37–45

Patterson C (1988) Homology in classical and molecular biology. Mol Biol Evol 5:603–625

Pawlowski WP, Golubovskaya IN, Cande ZW (2003) Altered nuclear distribution of recombination protein RAD51 in maize mutants suggests the involvement of RAD51 in meiotic homology recognition. Plant Cell 15:1807–1816

Pedrosa-Harand A, Souza de Almeida CC, Mosiolek MM, Blair MW, Schweizer D, Guerra M (2006) Extensive ribosomal DNA amplification during Andean common bean (*Phaseolus vulgaris* L.) evolution. Theor Appl Genet 112:924–933

Pikaard CS (1999) Nucleolar dominance and silencing of transcription. Trends Plant Sci 4:478–483

Pikaard CS (2000) The epigenetics of nucleolar dominance. Trends Genet 16:495–500

Pires JC, Zhao J, Schranz ME, Leon EJ, Quijada PA, Lukens LN, Osborn TC (2004) Flowering time divergence and genomic rearrangements in resynthesized *Brassica* polyploids (Brassicaceae). Biol J Linn Soc 82:675–688

Polanco C, Pérez de la Vega M (1997) Intergenic ribosomal spacer variability in hexaploid oat cultivars and landraces. Heredity 78:115–123

Pontes O, Neves N, Silva M, Lewis MS, Madlung A, Comai L, Viegas W, Pikaard CS (2004) Chromosomal locus rearrangements are a rapid response to formation of the allotetraploid *Arabidopsis suecica* genome. Proc Natl Acad Sci U S A 101:18240–18245

Popp M, Oxelman B (2001) Inferring the history of the polyploid *Silene aegaea* (Caryophyllaceae) using plastid and homoeologous nuclear DNA sequences. Mol Phylogenet Evol 20:474–481

Posada D (2002) Evaluation of methods for detecting recombination from DNA sequences: empirical data. Mol Biol Evol 19:708–717

Posada D, Crandall KA (2001) Evaluation of methods for detecting recombination from DNA sequences: computer simulations. Proc Natl Acad Sci U S A 98:13757–13762

Powell W, Thomas WTB, Thompson DM, Swanston JS, Waugh R (1992) Association between rDNA alleles and quantitative traits in doubled haploid populations of barley. Genetics 130:187–194

Ramsey J, Schemske DW (2002) Neopolyploidy in flowering plants. Annu Rev Ecol Syst 33:589–639

Rapp RA, Udall JA, Wendel JF (2009) Genomic expression dominance in allopolyploids. BMC Biol 7:18

Raskina O, Belyayev A, Nevo E (2004) Quantum speciation in *Aegilops*: molecular cytogenetic evidence from rDNA cluster variability in natural populations. Proc Natl Acad Sci U S A 101:14818–14823

Rauscher JT, Doyle JJ, Brown HD (2002) Internal transcribed spacer repeat-specific primers and the analysis of hybridization in the *Glycine tomentella* (Leguminosae) polyploid complex. Mol Ecol 11:2691–2702

Rauscher JT, Doyle JJ, Brown HD (2004) Multiple origins and nrDNA internal transcribed spacer homeologue evolution in the *Glycine tomentella* (Leguminosae) allopolyploid complex. Genetics 166:987–998

Reeder RH (1999) Regulation of RNA polymerase I transcription in yeast and vertebrates. Prog Nucleic Acid Res Mol Biol 62:293–327

Rokas A, Ladoukakis E, Zouros E (2003) Animal mitochondrial DNA recombination revisited. Trends Ecol Evol 18:411–417

Rong J, Abbey C, Bowers JE, Brubaker CL, Chang C, Chee PW, Delmonte TE, Ding X, Garza JJ, Marler BS, Park C-H, Pierce GJ, Rainey KM, Rastogi VK, Schulze SR, Trolinder NL, Wendel JF, Wilkins TA, Williams-Coplin TD, Wing RA, Wright RJ, Zhao X, Zhu L, Paterson AH (2004) A 3347-locus genetic recombination map of sequence-tagged sites reveals features of genome organization, transmission and evolution of cotton (*Gossypium*). Genetics 166:389–417

Rooney AP, Ward TJ (2005) Evolution of a large ribosomal RNA multigene family in filamentous fungi: birth and death of a concerted evolution paradigm. Proc Natl Acad Sci U S A 102:5084–5089

Rosenberg NA (2003) The shapes of neutral gene genealogies in two species: probabilities of monophyly, paraphyly, and polyphyly in a coalescent model. Evolution 57:1465–1477

Saghai-Maroof MA, Allard RW, Zhang Q (1990) Genetic diversity and ecogeographical differentiation among ribosomal DNA alleles in wild and cultivated barley. Proc Natl Acad Sci U S A 87:8486–8490

Salmon A, Ainouche ML, Wendel JF (2005) Genetic and epigenetic consequences of recent hybridization and polyploidy in *Spartina* (Poaceae). Mol Ecol 14:1163–1175

Salmon A, Flagel L, Ying B, Udall JA, Wendel JF (2010) Homoeologous non-reciprocal recombination in polyploid cotton. New Phytol 186:123–134

Sang T, Crawford DJ, Stuessy TF (1995) Documentation of reticulate evolution in peonies (*Paeonia*) using internal transcribed spacer sequences of nuclear ribosomal DNA: implications for biogeography and concerted evolution. Proc Natl Acad Sci U S A 92:6813–6817

Sang T, Donoghue MJ, Zhang D (1997) Evolution of alcohol dehydrogenase genes in peonies (*Paeonia*): phylogenetic relationships of putative nonhybrid species. Mol Biol Evol 14:994–1007

Schlötterer C, Tautz D (1994) Chromosomal homogeneity of *Drosophila* ribosomal DNA arrays suggests intrachromosomal exchanges drive concerted evolution. Curr Biol 4:777–783

Schnable PS, Hsia AP, Nikolau BJ (1998) Genetic recombination in plants. Curr Opin Plant Biol 1:123–129

Schueler MG, Higgins AW, Rudd MK, Gustashaw K, Willard HF (2001) Genomic and genetic definition of a functional human centromere. Science 294:109–115

Schuermann D, Molinier J, Fritsch O, Hohn B (2005) The dual nature of homologous recombination in plants. Trends Genet 21:172–181

Shaked H, Kashkush K, Ozkan H, Feldman M, Levy AA (2001) Sequence elimination and cytosine methylation are rapid and reproducible responses of the genome to wide hybridization and allopolyploidy in wheat. Plant Cell 13:1749–1759

Sharma S, Beharav A, Balyan HS, Nevo E, Gupta PK (2004) Ribosomal DNA polymorphism and its association with geographical and climatic variables in 27 wild barley populations from Jordan. Plant Sci 166:467–477

Slatkin M (1986) Interchromosomal biased gene conversion, mutation and selection in a multigene family. Genetics 112:681–698

Slotkin R, Nuthikattu S, Jiang N (2012) The impact of transposable elements on gene and genome evolution. In: Wendel JF (ed) Plant genome diversity, vol 1, Plant genomes, their residents, and their evolutionary dynamics. Springer, Wien/New York

Small RL, Cronn R, Wendel JF (2004) Use of nuclear genes for phylogeny reconstruction in plants. Aust Syst Bot 17:145–170

Soltis DE, Soltis PS (1989) Allopolyploid speciation in *Tragopogon*: insights from chloroplast DNA. Am J Bot 76:1119–1124

Soltis PS, Soltis DE (2009) The role of hybridization in plant speciation. Annu Rev Plant Biol 60:561–588

Soltis DE, Soltis PS, Tate JA (2004) Advances in the study of polyploidy since Plant speciation. New Phytol 161:173–191

Städler T, Delph LF (2002) Ancient mitochondrial haplotypes and evidence for intragenic recombination in a gynodioecious plant. Proc Natl Acad Sci U S A 99:11730–11735

Stebbins GL (1959) The role of hybridization in evolution. P Am Philos Soc 103:231–251

Suárez-Santiago VN, Blanca G, Ruiz-Rejón M, Garrido-Ramos MA (2007) Satellite-DNA evolutionary patterns under a complex evolutionary scenario: the case of Acrolophus subgroup (Centaurea L., Compositae) from the western Mediterranean. Gene 404:80–92

Sybenga J (1975) Meiotic configurations. Springer, Berlin/Heidelberg/New York

Sýkorová E, Lim KY, Kunicka Z, Chase MW, Bennett MD, Fajkus J, Leitch AR (2003) Telomere variability in the monocotyledonous plant order Asparagales. Proc R Soc Lond B Bio 270:1893–1904

Szadkowski E, Eber F, Huteau V, Lodé M, Huneau C, Belcram H, Coriton O, Manzanares-Dauleux M, Delourme R, King G, Chalhoub B, Jenczewski E, Chèvre A-M (2010) The first meiosis of resynthesized Brassica napus, a genome blender. New Phytol 186:102–112

Szostak JW, Orr WT, Rothstein RJ, Stahl FW (1983) The double-strand-break repair model for recombination. Cell 33:25–35

Tate JA, Ni ZF, Scheen AC, Koh J, Gilbert CA, Lefkowitz D, Chen ZJ, Soltis PS, Soltis DE (2006) Evolution and expression of homeologous loci in Tragopogon miscellus (Asteraceae), a recent and reciprocally formed allopolyploid. Genetics 173:1599–1611

Terasawa M, Ogawa H, Tsukamoto Y, Shinohara M, Shirahige K, Kleckner N, Ogawa T (2007) Meiotic recombination-related DNA synthesis and its implications for cross-over and non-cross-over recombinant formation. Proc Natl Acad Sci U S A 104:5965–5970

Udall JA, Quijada PA, Osborn TC (2005) Detection of chromosomal rearrangements derived from homeologous recombination in four mapping populations of Brassica napus L. Genetics 169:967–979

Udall JA, Swanson JM, Nettleton D, Percifield RJ, Wendel JF (2006) A novel approach for characterizing expression levels of genes duplicated by polyploidy. Genetics 173:1823–1827

Ugarkovic D, Plohl M (2002) Variation in satellite DNA profiles—causes and effects. EMBO J 21:5955–5959

Volkov RA, Borisjuk NV, Panchuk II, Schweizer D, Hemleben V (1999) Elimination and rearrangement of parental rDNA in the allopolyploid Nicotiana tabacum. Mol Biol Evol 16:311–320

Volkov RA, Komarova NY, Hemleben V (2007) Ribosomal DNA in plant hybrids: inheritance, rearrangement, expression. Syst Biodivers 5:261–276

Wagner DB, Dong J, Carlson MR, Yanchuk AD (1991) Paternal leakage of mitochondrial DNA in Pinus. Theor Appl Genet 82:510–514

Waters ER, Schaal BA (1996) Biased gene conversion is not occurring among rDNA repeats in the Brassica triangle. Genome 39:150–154

Wei X-X, Wang X-Q, Hong D-Y (2003) Marked intragenomic heterogeneity and geographical differentiation of nrDNA ITS in Larix potaninii (Pinaceae). J Mol Evol 57:623–635

Weider LJ, Elser JJ, Crease TJ, Mateos M, Cotner JB, Markow TA (2005) The functional significance of ribosomal (r)DNA variation. Impacts on the evolutionary ecology of organisms. Annu Rev Ecol Evol Syst 36:219–242

Weiss H, Scherthan H (2002) Aloe spp.: plants with vertebrate-like telomeric sequences. Chromosome Res 10:155–164

Wendel JF (2000) Genome evolution in polyploids. Plant Mol Biol 42:225–249

Wendel JF, Schnabel A, Seelanan T (1995) Bidirectional interlocus concerted evolution following allopolyploid speciation in cotton (Gossypium). Proc Natl Acad Sci U S A 92:280–284

White TJ, Bruns T, Lee S, Taylor J (1990) Amplification and direct sequencing of fungal ribosomal RNA genes for phylogenetics. In: Innis M, Gelfand D, Sninsky J, White T (eds) PCR Protocols: a guide to methods and applications. Academic, San Diego, pp 315–322

Wichman SR, Wright SD, Cameron EK, Keeling DJ, Gardner RC (2002) Elevated genetic heterogeneity and Pleistocene climatic instability: inferences from nrDNA in New Zealand Coprosma (Rubiaceae). J Biogeogr 29:943–954

Wiuf C, Christensen T, Hein J (2001) A simulation study of the reliability of recombination detection methods. Mol Biol Evol 18:1929–1939

Wolf M, Achtziger M, Schultz J, Dandekar T, Müller T (2005) Homology modeling revealed more than 20,000 rRNA internal transcribed spacer 2 (ITS2) secondary structures. RNA 11:1616–1623

Wolfe AD, Randle CP (2002) Recombination, heteroplasmy, haplotype polymorphism, and paralogy in plastid genes: implications for plant molecular systematics. Syst Bot 29:1011–1020

Zakian V (1995) Telomeres: beginning to understand the end. Science 270:1601–1607

Zhang D, Sang T (1999) Physical mapping of ribosomal RNA genes in peonies (Paeonia, Paeoniaceae) by fluorescent insitu hybridization: implications for phylogeny and concerted evolution. Am J Bot 86:735–740

Zhang Q, Saghai Maroof MA, Allard RW (1990) Effects of adaptedness of variations in ribosomal DNA copy number in populations of wild barley (Hordeum vulgare ssp. spontaneum). Proc Natl Acad Sci U S A 87:8741–8745

Zhang W, Qu LJ, Gu H, Gao W, Liu M, Chen J, Chen Z (2002) Studies on the origin and evolution of polyploid wheats based on the internal transcribed spacer (ITS) sequences of nuclear ribosomal DNA. Theor Appl Genet 104:1099–1106

Zimmer EA, Martin SL, Beverley SM, Kan YW, Wilson AC (1980) Rapid duplication and loss of genes coding for the chains of hemoglobin. Proc Natl Acad Sci U S A 77:2158–2162

Zuker M (2003) MFOLD web server for nucleic acid folding and hybridization prediction. Nucleic Acids Res 31:3406–3415

Synteny and Genomic Rearrangements

13

A.H. Paterson, X. Wang, H. Tang, and T.H. Lee

Contents

A.H. Paterson (✉)
Plant Genome Mapping Laboratory, University of Georgia,
111 Riverbend Road Rm 228, Athens, GA 30602, USA
e-mail: paterson@plantbio.uga.edu; paterson@dogwood.botany.uga.edu

13.1 Introduction

The concept of synteny, derived from the Greek *syn tene* ('same thread') has come to refer broadly to parallels in gene arrangement in divergent genomes. Gene collinearity, a special case of synteny in which groups of genes occur in largely corresponding orders along the chromosomes of respective genomes, provides valuable information for inferring gene orthology or paralogy. Relative stasis in gene order accompanied by dramatic variation in DNA content (physical size) makes understanding of comparative gene organization of high value in plants. Plant genome sizes vary over about 2,400-fold, from species of *Genlisea* (Lentibulariaceae) with 63.4 million base pairs to more than 150 billion base pairs for *Paris japonica* (Pellicer et al. 2010). The vast majority of this genome size variation is due to the evolution and propagation of repetitive DNA, largely long-terminal repeat retrotransposon-like element families (Peterson et al. 2002), that are allowed to accumulate in gene-poor heterochromatin while being largely purged from gene-rich euchromatin (Bowers et al. 2005; Paterson et al. 2009). In many genomes including those of flowering plants (angiosperms), the content and arrangement of genes is much more static than that of other genomic constituents (Coghlan et al. 2005).

A thorough understanding of comparative genome organization facilitates 'translational genomics', the *in silico* utilization of hard-won functional information from one taxon in better understanding genome structure, function and/or evolution in others. Nearly two decades ago, positional correspondence of major genes and large-effect QTLs contributing to phenotypic variation began to be observed in divergent species that had experienced parallel but independent selection regimes (Lin et al. 1995; Paterson et al. 1995). More recently, the richness of gene functional information in botanical models such as *Arabidopsis thaliana* (mouse-ear cress) and *Oryza sativa* (rice) has made it particularly attractive to understand how alignment of other genomes to

J.F. Wendel et al. (eds.), *Plant Genome Diversity Volume 1*,
DOI 10.1007/978-3-7091-1130-7_13, © Springer-Verlag Wien 2012

Arabidopsis (for example), facilitates inferences about the probable functions of similar genes and gene-like sequences in these other taxa. This often turns out to be more challenging than was envisioned when *Arabidopsis thaliana* was singled out as a facile botanical model (see below).

Plant genomes evolve at highly variable rates, but on average experience more rapid changes in gene order than those of animals, due largely to more frequent episodes of genome duplication and subsequent adaptation of genomes to the duplicated state. Gene order conservation in vertebrates is evident after hundreds of millions of years of divergence (e.g., Consortium MGS 2002; Smith et al. 2002). In contrast, the two major branches of the angiosperms, eudicots and monocots, estimated to have diverged 125–140 My (Davies et al. 2004) to 170–235 mya (Yang et al. 1999), show much more rapid structural evolution (Paterson et al. 1996; Grant et al. 2000; Ku et al. 2000; Lee et al. 2001; Liu et al. 2001; Mayer et al. 2001; Rossberg et al. 2001; Salse et al. 2002; Vandepoele et al. 2002a; Simillion et al. 2004). This difference appears to be due largely to the propensity of angiosperms to experience chromosomal duplication and subsequent gene loss (Coghlan et al. 2005), fragmenting ancestral linkage arrangements across multiple chromosomes (Bowers et al. 2003a; Paterson et al. 2004, 2005a). It has long been suspected that many angiosperms were paleopolyploids (Stebbins 1966). Analyses of genome sequences (Bowers et al. 2003a; Paterson et al. 2004; Tuskan et al. 2006; Jaillon et al. 2007a) and ESTs (Blanc and Wolfe 2004; Pfeil et al. 2005; Cui et al. 2006) now suggest that all angiosperms are paleopolyploids.

Polyploidy profoundly complicates study of gene synteny and genomic rearrangement in plants. While the value of understanding comparative gene arrangements in plant genomes has been appreciated for at least two decades (Bonierbale et al. 1988; Hulbert et al. 1990), only with the complete sequencing of a few genomes have we fully realized the importance of accounting for both neo- and paleo-polyploidy in making accurate deductions about comparative gene arrangement. Much of this chapter emphasizes the increased knowledge of plant comparative genomics that emerges from inclusion of this heretofore under-appreciated factor. A few examples are also provided of how this knowledge has contributed to understanding of angiosperm genome biology and evolutionary history.

13.2 Singular Needs for Robust Genome Comparisons in Angiosperms

There are several limitations of vertebrate or other genome alignment algorithms for plant genome comparisons, due to the paleopolyploid (and sometimes also neopolyploid) architecture of plant genomes (Tang et al. 2008a).

Algorithms (e.g. BLASTZ/CHAINNET pipeline, and LAGAN/SUPERMAP pipeline) commonly used in vertebrate genome alignments focus on identifying orthologous regions while largely ignoring paralogous regions (Kent et al. 2003). A general theme for detection of distant synteny relationships is to use "all vs. all" BLASTP searches as inputs, and model the matches in a homology matrix representation where synteny is uncovered by clustering neighboring matches inside the matrix. Such an approach is central to ADHoRe (Vandepoele et al. 2002b) and DiagHunter (Cannon et al. 2003) and influences other algorithms (Calabrese et al. 2003). Two recent methods DAGchainer (Haas et al. 2004) and CollinearScan (Wang et al. 2006) formulate the problem by dynamic programming and use empirical or statistical strategies which effectively improve sensitivity/specificity of collinearity prediction. However, each method still only predicts pairwise collinearities. If a chromosomal region is preserved in three genomes A, B and C, then pair-wise predictions would yield three inferences about ancestral gene repertoire and order (A-B, B-C, A-C). It would be more useful to combine related pairwise collinear segments into one inferred order which utilizes multiple collinearity (A-B-C). This limitation is especially serious in dicots, and in *Arabidopsis thaliana* specifically, because an ancestral triplication affecting most if not all dicots was followed by two additional whole-genome duplications in the *Arabidopsis thaliana* lineage, potentially creating groups of 12 (!) syntenic segments although in most cases only a subset can still be discerned (Bowers et al. 2003a; Tang et al. 2010).

Early approaches for the computational de-convolution of paleopolyploidy for deduction of ancestral gene orders have been 'bottom-up', in which one attempts to resolve one duplication event at a time, starting with the most recent one. Bottom-up inference is exemplified by studies in *Arabidopsis* and *Paramecium* where the most recently-duplicated segments are merged to generate hypothetical intermediate profiles that are further recursively merged (Bowers et al. 2003a; Aury et al. 2006).

Alternative and often complementary "top-down" algorithms combine related pairwise collinear segments into one inferred order based upon *multiple collinearity* (A-B-C). Collinear segments are picked based on whole genome BLASTP results to produce gene-based alignments of multiple syntenic regions, including cryptic synteny based on transitive homology that has been referred to as "ghost duplications" (Vandepoele et al. 2002b; Vandepoele et al. 2002a, 2003). For a simplistic case, two pairwise segments are identified as A-B and C-D (where A, B, C, D represent subsegments)—if B and C occupy the same chromosomal region, then the final syntenic block would be A-(BC)-D. This is not uncommon—A and D may be duplicated regions which retain few common genes due to reciprocal gene loss, but can be recognized as they both show synteny to BC. Overlapping subsegments (B and C) are then merged to produce mutually distinct regions (A, BC, D) that would be candidates for multiple alignments. Multiple alignments can be evaluated in terms of gene pair densities, sequence similarities and comparison to random occurrences by a permutation test.

13.2.1 Mcscan: A 'top-down' Tool to Compare Gene Orders in Genomes and Subgenomes

A key need, both in revealing consequences of paleopolyploidy within a genome and in providing multiple alignments of multiple paleopolyploid genomes to one another, is to combine related pairwise collinear segments into one inferred order which utilizes multiple collinearity (A-B-C). Multiple collinearity is virtually universal in plant genomes, in which polyploidy forms complicated networks of homologous regions that show multiple-to-multiple matches. Looking only for regions that correspond in a one-to-one manner between genomes would miss much of the comparative information in many plants.

To tackle such problems, we implemented a robust multiple gene-order alignment tool MCscan (Multiple Collinearity scan) (Tang et al. 2008a) and successfully applied it to newly sequenced genomes, including *Carica* (Ming et al. 2008) and *Sorghum* (Paterson et al. 2009). MCscan provides multiple alignments which better reflect the true relationships among angiosperm genomes, in which whole genome duplications are frequently superimposed on speciations (Tang et al. 2008a).

The MCscan algorithm is briefly described as follows. First, all pairwise comparisons (by BLASTP) between genes are performed from all possible pairs of genomes being compared, as well as within each genome. From BLASTP results, we detect pairwise segments by a hybrid approach incorporating features from both DAGchainer and CollinearScan. Any pairwise segment consists of subsegments from two distinct genomic locations with a number of aligned, collinear genes as "anchor points". These pairwise segments are disassembled into subsegments generating a pool of stretches of regions represented by anchor points. We define subsegments as vertices in a graph, and define an edge between any two vertices (subsegments) meeting one of two conditions: (1) show collinearity, or (2) overlap the same genomic region. Then, all related subsegments can be found by looking for connected components within the graph. The second condition is novel, since it infers synteny between seemingly unrelated regions by using a third region as a pivot, as detailed above for "ghost duplications" (Vandepoele et al. 2002a, 2003).

Multiple alignment of gene anchors is equivalent to the "Shortest Common Supersequence" problem and is NP-hard (Garey and Johnson 1979). A heuristic is therefore applied which constructs the multiple alignment progressively by greedily picking one closest-related region at a time. Finally the alignment is evaluated in terms of gene pair densities, sequence similarities and comparison to random occurrences by a permutation test. MCscan is a stand-alone program written in C++ and freely available from our website.

Using a preliminary version of MCScan, *Arabidopsis thaliana*, *Populus trichocarpa*, and *Carica papaya* genomes were compared to each other by BLASTP (e-value 1e-5, top 5 hits). Fig. 13.1 (next page) shows a portion of a representative syntenic block affected by duplication event γ which predates the divergence of the three species (Bowers et al. 2003a). Due to the limitations of conventional genome comparisons (two-way dot plots), γ could previously only be inferred over about 15% of the *Arabidopsis* genome (Bowers et al. 2003a). Using MCscan with additional information from *Populus* and *Carica*, we see evidence of γ over most of the genome. Three distinct 'paleologous' versions of this block co-occur in all species, each doubled in *Populus* by one additional duplication (Tuskan et al. 2006) and quadruplicated in *Arabidopsis* by two (Bowers et al. 2003a).

Comparison to the *Vitis* genome validates the reconstructed order and triplicated structure inferred in our pilot study. *Vitis* is within the *Vitaceae* branch of the core eudicots, outside of the well-supported rosid clade which contains *Arabidopsis-Carica* (Eurosids II) and *Populus* (Eurosids I) (Davies et al. 2004; Soltis et al. 2005): thus *Vitis* provides an independent lineage suitable to test our inferred gene order. Paleo-hexaploidy ('triplication') is discernible over 94.5% of the *Vitis* genome, thanks to a surprisingly high level of retention of triplicated genes, and remarkably little chromosomal rearrangement relative to other angiosperms. When the *Arabidopsis-Carica-Populus* comparison is aligned to *Vitis*, the two triplication patterns correspond closely (Tang et al. 2008a). This validates the MCScan algorithm, revealing triplication that eluded prior detection in both *Arabidopsis* (Bowers et al. 2003a) and *Populus* (Tuskan et al. 2006), and showing that triplication occurred in a common ancestor of *Vitis*, *Arabidopsis*, *Carica*, and *Populus*.

Top-down and bottom-up approaches are often complementary. For example, about ~90% of the *Arabidopsis* genome can be traced to an ancestral, syntenic thread by a top-down approach, but many of the genes in these otherwise-syntenic regions are revealed to be positional exceptions by a bottom-up approach, and these exceptions are strongly biased as to gene family and to preferred genomic location. Further analysis at the intersection of these complementary approaches promises new insights into gene

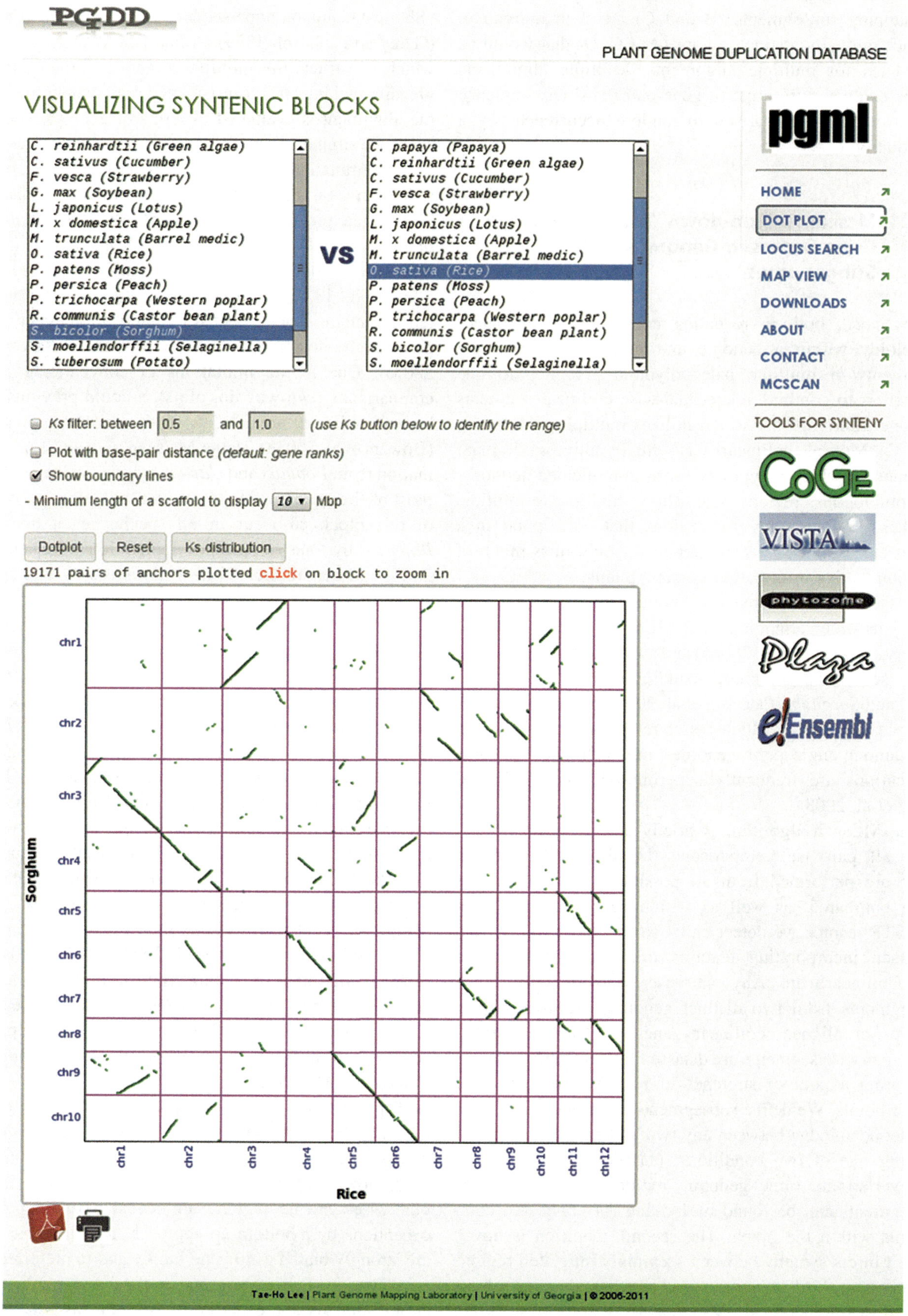

PGDD

VISUALIZING SYNTENIC BLOCKS

Fig. 13.1 Dot-plot visualization of syntenic relationships between genomes. Users can click on any region of the dot-plot to zoom in and view fine structure of syntenic segments. The figure can be recreated at http://chibba.agtec.uga.edu/duplication/index/dotplot

function and regulation on a genome-wide scale (Paterson et al. 2010).

13.3 Recent Findings from Selected Angiosperm Genome Alignments

13.3.1 *Arabidopsis*, One of the Most Facile Angiosperm Models, Has Numerous Limitations as a Foundation for Comparative Genomics

Some reasons for the limitations of *Arabidopsis* as a foundation for angiosperm comparative genomics are exemplified by its comparison to the second fully-sequenced Brassicales genome, *Carica papaya* (Ming et al. 2008). Among the 200 longest papaya scaffolds, 124 show collinearity with *Arabidopsis*, with 26 corresponding to only one *Arabidopsis* segment, 41 to two, 21 to three, 30 to four, and only 3 to more than 4. The fact that so many papaya segments show collinearity with three or four *Arabidopsis* segments, constitutes compelling evidence that *two* genome duplications have affected *Arabidopsis* since its divergence from *Carica*. It was anticipated that the most recent *Arabidopsis* genome duplication, 'alpha' (Bowers et al. 2003a), might affect only a subset of the Brassicales (Schranz and Mitchell-Olds 2006). However, our phylogenomic dating (Bowers et al. 2003a) suggested that the more ancient 'beta' duplication preceded the *Arabidopsis-Carica* divergence. Individual *Arabidopsis* genome segments correspond to only one *Carica* segment, showing that the *Carica* genome has not been duplicated since its divergence from *Arabidopsis*.

Extensive gene loss and fractionation of ancestral gene order in *Arabidopsis* suggests that the *Carica* genome may more closely resemble those of ancestral Brassicales. The *Arabidopsis thaliana* genome has extensively restructured in the few million years since its divergence from *Arabidopsis lyrata* (Kuittinen et al. 2004); *A. lyrata* and related *Capsella rubella*, show near-perfect collinearity while *A. thaliana* differs from each by 9 to 10 rearrangements (Koch and Kiefer 2005; Yogeeswaran et al. 2005). Moreover, much higher DNA content of *A. lyrata* (230 Mb) and *C. rubella* (250 Mb) is inferred to be due to lineage-specific reduction (to ~160 Mb) in *A. thaliana* (Johnston et al. 2005). The *A. lyrata* and *C. rubella* sequences (in progress: www.jgi.doe.gov) will provide, respectively, a close comparator and an appropriate outgroup for analysis of the impact of this size reduction on individual genes and functional groups, perhaps contributing to the limitations of *Arabidopsis* as a model. As a robust foundation for angiosperm comparative genomics we need to include a broader group of angiosperms, sampling more taxonomic diversity and employing new computational approaches to reduce the impact of lineage-specific evolutionary constraints.

13.3.2 Multiple Alignment of Eudicot Genomes Reveals Striking Genome-Wide Variations in Nucleotide Evolutionary Rates

Determination of a consensus gene order among multiple taxa has made it possible to directly compare estimations of the ages of gene duplications based on rates of nucleotide substitution per synonymous site (Ks) between 'paleolog' pairs (syntenic paralogs), filtering out the inevitable influence of single gene duplications which superimpose an L-shaped curve on the relics of whole-genome duplication (Maere et al. 2005; Tang et al. 2008b). The distribution of Ks for paleologs remaining from the gamma triplication within the *Arabidopsis*, *Carica*, *Populus* and *Vitis* genomes differs substantially. The median Ks between gamma-paleologs in *Vitis* (1.27) is much lower than in *Carica* (1.83), with *Populus* being intermediate (1.51), all highly significant differences. Since all these species share the same gamma-triplication, if they evolved at comparable rates then the respective sets of paleologs should show similar Ks values. *Arabidopsis* has a still-faster molecular clock—although the 'beta' duplication occurred after its divergence with *Carica*, median Ks values between *Arabidopsis* beta-duplicates are ~1.8, near that of the much older gamma-duplicates in *Populus*. Indeed, more rapid substitutions occurring at synonymous sites in *Arabidopsis* than in the other three eudicot genomes largely mask the gamma-triplication in Ks distribution plots, only the more recent alpha and beta duplications remaining discernible.

In sum, these four dicots have unexpectedly large variability of nucleotide substitution rates, with the *Arabidopsis* lineage evolving the most quickly of the four at the DNA sequence level. This finding offers two messages. First, at least within the eudicots, there appears to exist no universally applicable molecular clock, consistent with findings suggesting that life history traits (for example) may alter evolutionary rates (Smith and Donoghue 2008). We expect that this will hold true across many taxa. Second, the many merits of *Arabidopsis thaliana* as a botanical model notwithstanding, it is something of an outlier among angiosperm genomes in being very rapidly evolving and also more extensively rearranged than most. This may relate to having experienced at least one triplication and two duplications in its evolutionary history, or may be because *A. thaliana* has experienced 30% genome size reduction and at least nine rearrangements in the short time since its divergence from *A. lyrata* (Kuittinen et al. 2004; Koch and Kiefer 2005; Hansson et al. 2006; Kawabe et al. 2006).

The unusually rapid evolution of *Arabidopsis* genes was largely responsible for erroneous dating of some genome duplication events using a phylogenomic approach (Bowers et al. 2003a; Chapman et al. 2004), errors that were eventually revealed by whole-genome sequences for additional taxa (Ming et al. 2008). This is a nice example of how more data from more plants is likely to continue to improve our understanding of angiosperm evolution.

13.3.3 Rice-Sorghum Genome Comparison Elucidates the Diversification of the Major Cereal Lineages

For more than 20 years the cereals have benefited from a host of comparative genomics studies, facilitated by the fact that so many different cereal species have each been domesticated independently, on different continents and by different societies, for the production of carbohydrate-rich seeds. The sequencing of the genome of sorghum (Paterson et al. 2009), the second cereal to be largely sequenced, permitted careful scrutiny of these two genomes at an unprecedented level of resolution to address several interesting questions.

One natural question regarded the nature and genomic distribution of the much larger quantity of genomic DNA in sorghum (~740 Mb) than rice (~420 Mb). The ~75% larger quantity of DNA in the genome of sorghum than rice is mostly heterochromatin. Alignment to genetic (Bowers et al. 2003b) and cytological maps (Kim et al. 2005) suggested that sorghum and rice have similar quantities of euchromatin (252 and 309 mbp) (Paterson et al. 2009), accounting for 97–98% of recombination (1,025.2 and 1,496.5 cM) and 75.4–94.2% of genes in the respective cereals, with largely collinear gene order (Bowers et al. 2005). In contrast, sorghum heterochromatin occupies at least 460 mbp (62%), far more than in rice (63 mbp, 15%). The ~3x genome expansion in maize since its divergence from sorghum (Swigonova et al. 2004a) has been more dispersed –recombinogenic DNA has grown 4.5× to ~1,382 mbp, much more than can be explained by genome duplication(Swigonova et al. 2004b).

The net size expansion of the sorghum genome relative to rice largely involved LTR-retrotransposons. The sorghum genome contains 55% retrotransposons, intermediate between the larger maize genome (79%) and smaller rice genome (26%). Sorghum more closely resembles rice in having a higher ratio of *gypsy*- to *copia*-like elements (3.7–1 and 4.9–1) than maize (1.6–1: Supplementary Table 10). While recent retroelement activity is widely distributed across the sorghum genome, turnover is rapid (as in other cereals (Swigonova et al. 2005)) with pericentromeric elements persisting longer. Recent LTR-retrotransposon insertions (<0.01 mya) appear randomly distributed along chromosomes, suggesting that they are preferentially eliminated from gene-rich regions (Bowers et al. 2005) but accumulate in gene-poor regions.

A second natural question regards the impact of a whole-genome duplication in an common ancestor of cereals (Paterson et al. 2003; Paterson et al. 2004), on the diversity and divergence among cereal genomes. In other taxa, genome duplication has been implicated as a facilitator of speciation (Werth and Windham 1991; Scannell et al. 2006), and a natural question is whether the pan-cereal duplication may have facilitated cereal diversification. Whole-genome duplication in a common ancestor of cereals is reflected in sorghum and rice gene 'quartets' (Paterson et al. 2009). A total of 19,929 (57.8%) sorghum gene models could be unambiguously assigned to blocks collinear with rice. Following the shared whole-genome duplication, only one copy was retained for 13,667 (68.6%) collinear genes with 13,526 (99%) being orthologous in rice-sorghum, suggesting that most gene losses predate taxon divergence. Both sorghum and rice retained both copies of 4,912 (14.2%) genes, while sorghum lost one copy of 1,070 (3.1%) and rice lost one copy of 634 (1.8%). In partial summary, the vast majority of genomic adaptation to the pan-cereal duplication appears to have taken place long before the divergence of the major cereal lineages (Paterson et al. 2004).

A curious and unexpected finding from sorghum-rice synteny analysis was that one and only one genomic region duplicated in the pan-cereal duplication has been subject to a high level of concerted evolution. While the two members of most paleo-duplicated chromosome pairs have evolved largely independently of one another, one pair is a striking exception (Wang et al. 2009). Rice chromosomes 11 and 12 share a ~3 Mb duplicated DNA segment at the termini of their short arms, dated based on synonymous substitutions to ~5–7 mya based on synonymous substitutions and once suspected to represent a segmental duplication more recent than the pan-grass WGD (Consortia CRTaS 2005; Wang et al. 2005; Yu et al. 2005). Remarkably, the corresponding region(s) of the sorghum genome (S5 and S8, respectively) also contained such an apparently-recent duplication (Paterson et al. 2009). Physical and genetic maps suggest shared terminal segments of the corresponding chromosomes in wheat (4, 5), foxtail millet (VII, VIII), and pearl millet (linkage groups 1, 4) (Singh et al. 2007; Srinivasachary et al. 2007). It would be exceedingly unlikely for segmental duplications to each happen independently at such closely corresponding locations in several reproductively isolated lineages. A much more parsimonious hypothesis is that the R11/12 and S5/8 regions each resulted from the pan-grass duplication 70 mya but have an unusual evolutionary history (Paterson et al. 2009). Independently, it has been shown that gene conversion and illegitimate recombination are more frequent in the rice 11–12 and sorghum 5–8 region than elsewhere in the genome (Wang et al. 2007, 2011).

Finally, sorghum-rice synteny analysis showed that plant genome architecture may reflect euchromatin-specific effects of recombination and selection, superimposed on non-adaptive processes of mutation and drift that may apply to all genomic regions (Lynch and Conery 2003). Patterns of gene and repetitive DNA organization remain correlated in homoeologous chromosomes duplicated 70-mya, despite extensive turnover of specific repetitive elements (Paterson et al. 2009). Synteny is highest and retroelement density is lowest in distal chromosomal regions. More rapid retroelement removal from gene-rich euchromatin that frequently recombines than from heterochromatin that rarely recombines, supports the hypothesis that recombination may preserve gene structure, order, and/ or spacing by exposing new insertions to selection (Bowers et al. 2005). This general pattern appears to pertain to a wide range of monocot and dicot plants based on comparisons of sequences (Schmutz et al. 2010) and physical maps (Lin et al. 2010), although others have suggested alternative underlying mechanisms such as deleterious effects of epigenetic silencing in gene-rich regions (Hollister and Gaut 2009; Hollister et al. 2011). Less euchromatin-heterochromatin polarization in maize, where retrotransposon persistence in euchromatin appears more frequent, may reflect variation in grass genome architecture or perhaps a lingering consequence of more recent genome duplication (Wei et al. 2007).

13.3.4 Multi-Alignments that Account for Paleopolyploidy Improve the Ability to Discern Correspondence Between Cereal and Eudicot Genomes

Similarities between monocot and eudicot genomes resulting from common ancestry have been obscured by many rounds of paleo-polyploidy and numerous genome rearrangements (Liu et al. 2001; Jaillon et al. 2007b). Recurring polyploidy events pose significant challenges when comparing monocot and eudicot genomes because of the degeneration caused by independent gene fractionation (or "diploidization") following several rounds of paleo-polyploidy in each lineage.

To compare monocot and eudicot genomes, a hierarchical clustering approach (Tang et al. 2010) partially circumvented such difficulties, and was effective in identifying synteny across grape and rice. Briefly, the chromosomes were first cut into small segments and comparisons were made between every pair of rice and grape segments. For example, assume we have rice segments *O1* and *O2*, grape

segment *V1*, and comparisons *O1-V1* and *O2-V1* show a significant number of homologs. Based on this information, *O1* and *O2* can be clustered together, because they both match the same grape region(s). In this approach, only the "dense" (syntenic) portions of the whole-genome dot plot are clustered, assembled and interpreted; the "sparse" (non-syntenic) portions are ignored from further analysis.

Based on our clustering approach, duplicated segments retained in grape following the eudicot 'gamma' hexaploidy event (Jaillon et al. 2007b), and homologous segments retained in rice following at least two rounds of duplication ('rho' and 'sigma'), contain 38 "putative ancestral regions" (PAR), that show high densities of homologs. The PARs collectively explain 19.1% of all observed homolog pairs and 31.0% of reciprocal best hits between grape and rice genes, although by chance they should only explain 2.1% for both categories, thus achieving a ~10-fold enrichment. The PARs interleave multiple grape and rice genomic regions collectively covering ~70% of each genome. By consolidating much of the redundancy in each genome, the PARs create syntenic blocks with much less ambiguity.

When a particular PAR is scrutinized, syntenic relationships among the clustered regions are more informative than analyzing any individual pair of syntenic segments that contribute to the PAR. For example, in PAR17, three grape regions resulting from the gamma triplication (gamma-6) (Jaillon et al. 2007b; Tang et al. 2008b) correspond to several regions in rice that match each other, which can be partially explained by rho-1 (and additional duplications unobserved in intra-genomic comparisons). This approach is not only useful for comparing highly divergent genomes but also for improving intra-genomic comparisons—some "ghost duplications" (Van de Peer 2004) in rice that were not identified through intra-genomic comparisons (due to reciprocal gene losses in largely complementary fashion) are much clearer in cross-species comparisons (Tang et al. 2010).

13.4 Toward Intra- and Inter-Genome Angiosperm Genome Alignments Based on Uniform Standards: A Plant Genome Duplication Database (PGDD)

The burgeoning set of angiosperm genome sequences provide the foundation for a host of investigations into the functional and evolutionary consequences of gene and genome duplication, as well as the means to deduce otherwise-cryptic orthology, paralogy, and other relationships among particular genes.

Angiosperms are an ideal group for studying the consequences of genome duplication (GD), an evolutionary process that has profoundly shaped the architecture and function of many higher eukaryotic genomes. Classical views suggest that GD is potentially advantageous as a primary source of genes with new functions (Stephens 1951; Ohno 1970), a view that remains generally supported (albeit somewhat modified) (Lynch and Force 2000; He and Zhang 2005). While new data from microbes such as yeast (Gu et al. 2003; Christoffels et al. 2004; Scannell et al. 2006) and *Paramecium* (Aury et al. 2006) are shedding valuable light on GD and its consequences, population genetic theory predicts that the consequences of GD should be very different in organisms with much smaller effective population sizes such as most higher eukaryotes (Lynch et al. 2001; Lynch 2006). With many independent genome duplications providing 'natural replicates' and also permitting study of the temporal orders and rates at which different events/ mechanisms occur, as well as being well-characterized phylogenetically, often with diploid relatives extant, and with ample genetic, physical, and functional genomic tools available, the angiosperms are an outstanding higher-eukaryote model in which to eludicate consequences of genome duplication and mechanisms by which they are resolved.

Different analytical approaches employed by leading practitioners have sometimes yielded rather different conclusions about genome evolution. For example, early analyses of the rice genome sequence drew markedly different conclusions, one interpretation finding only partial duplication ('ancient aneuploidy': Vandepoele et al. 2003; Simillion et al. 2004) and another (our own) suggesting that the rice duplication involved most if not all of the genome (Paterson et al. 2003, 2004). The fact that our study detected a chromosome 1–5 duplication that had long been known from RFLP mapping (Kishimoto et al. 1994), but was not found in the other studies (Vandepoele et al. 2003; Simillion et al. 2004), raised doubts about the assertion of ancient aneuploidy. An independent study using a different genome assembly and new methods (Wang et al. 2005) drew largely the same conclusions that we had (Paterson et al. 2003, 2004), and a joint publication by the two groups (Paterson et al. 2005b) determined that the differing conclusions were largely attributable to the use of different statistical thresholds for inferring duplication. While this particular incongruity was resolved quickly, such is not always the case.

Improved genome alignments also promise to clarify angiosperm evolutionary history, and provide a firm foundation upon which to base translational genomics, the leveraging of hard-won structural and functional genomic information from leading botanical models to dissect specific evolutionary novelties in other organisms. While our own interest is in the detailed analysis of genome duplication, improved orthology inferences have broader application. Prior strategies to distinguish possible orthologs from paralogs can basically be classified into two groups: *phylogeny-based approaches*, including RIO (Resampled Inference of Orthology) (Zmasek and Eddy 2002) and Orthostrapper/HOPS (Hierarchical grouping of Orthologous and Paralogous Sequences) (Storm and Sonnhammer 2002, 2003) and *BLAST-based approaches*, including RBH (Reciprocal Best Hit), OrthoMCL, COG (Cluster of Orthologous Groups)/KOG, and Inparanoid (Tatusov et al. 2003; O'Brien et al. 2005). Phylogeny-based methods typically exhibit high false negative rates, while BLAST-based methods exhibit high false positive rates (Chen et al. 2006, 2007). The problems with these approaches are often due to 'out-paralogs' produced by duplication prior to two lineages' divergence, and 'in-paralogs' produced in each lineage. The multiple rounds of polyploidy in plant lineages accentuate these problems, as do divergence in evolutionary rates of different lineages (Tang et al. 2008a). As noted, MCScan incorporates both gene phylogeny and gene collinearity, improving power to distinguish between orthology and paralogy.

Intra- and inter-genome alignments from a single resource based on uniform standards are needed to facilitate a host of comparative studies that are founded upon consistent interpretations of genomes. Moreover, as a robust foundation for angiosperm comparative genomics we need to engage a broader group of angiosperms, sampling more taxonomic diversity and employing new computational approaches to reduce the impact of lineage-specific constraints. Toward these needs, we have created the Plant Genome Duplication Database (PGDD: http://chibba.agtec.uga.edu/duplication/), a public database to identify and catalog plant genes in terms of intragenome or cross-genome syntenic relationships. A detailed stand-alone description of PGDD is planned under separate cover, so only a brief introduction is provided here. New genomes are incorporated as they are released for public analysis (i.e. once "Ft Lauderdale" restrictions on whole-genome analyses are lifted by primary publication of the respective sequences). At present, 24 genomes are included. For gene-level synteny comparisons, we retrieve the top five matching genes as possible homologs for a query protein based on BLASTP search between every possible pair of chromosomes in multiple genomes. Then, the homologous matches between genes are clustered into multi-way anchors through a Markov clustering algorithm MCL (Enright et al. 2002), in order to simplify the correspondences among multiple loci. MCscan determines the syntenic regions from the homologous gene pairs and information from the gene clustering. The resulting pairwise and/or multi-aligned syntenic regions are reported in a tabular format and parsed into relational table 'blocks' used as a central resource in the PGDD database.

Basic functionalities of PGDD presently include:

1. Display of macro-scale syntenic blocks. Traditional 'dot-plots' of synteny relationships are provided (Fig. 13.1). Users can directly click on any region of the dot-plot to zoom in on the fine structure of syntenic segments. Syntenic blocks can be filtered with regard to chromosomal positions (to focus on specific rearrangements) and Ks distances between gene pairs (reducing noise from extraneous duplications, to focus on the 'signal' of a specific event as illustrated (Paterson et al. 2004)).

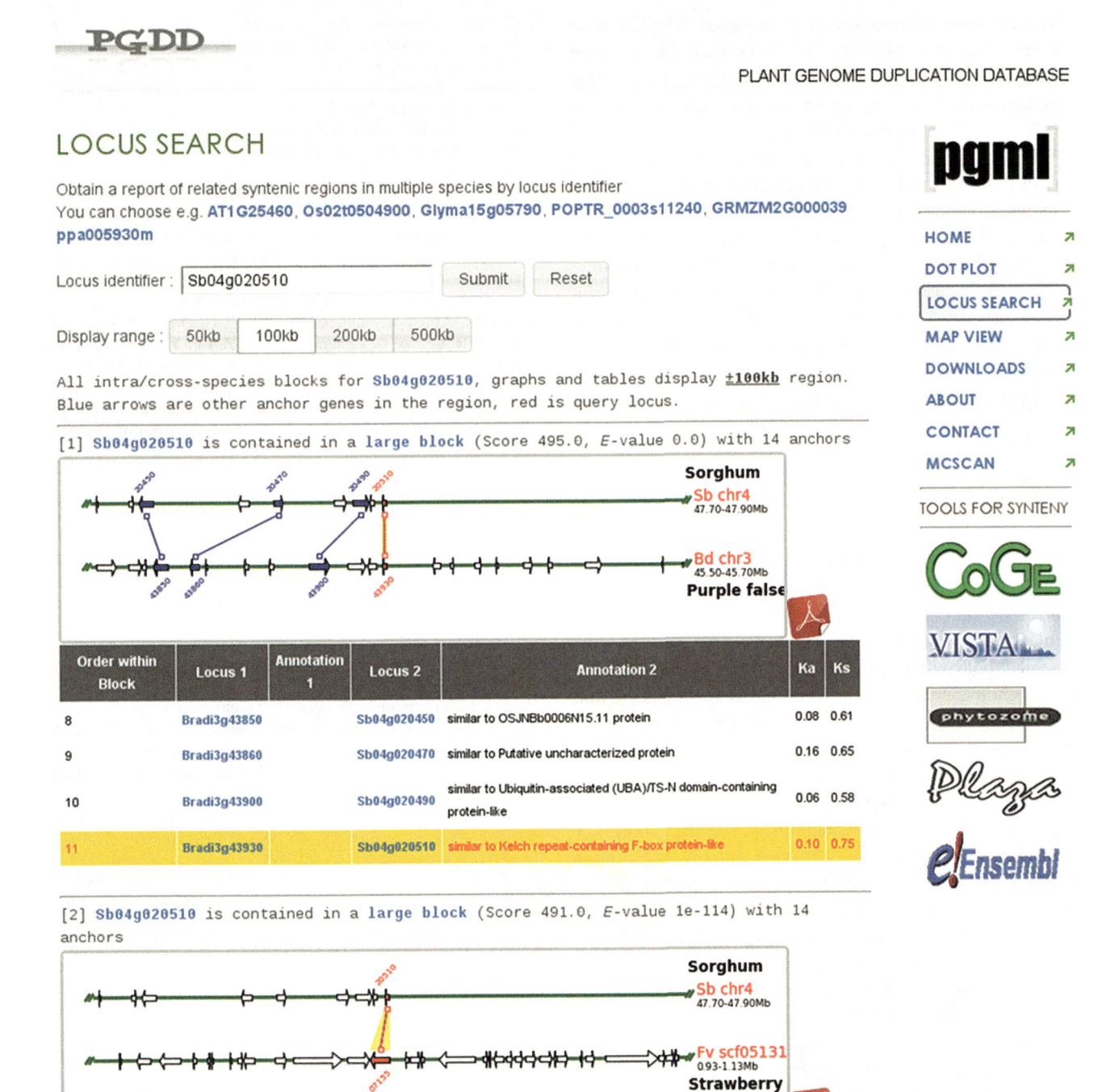

Fig. 13.2 Locus search of a specific gene within- and among-species shows structural changes in homologous chromosomal segments

2. Display of the fine-scale structure of syntenic regions of interest. This is a query service where a user enters a locus ID for a gene model and the server interactively displays the syntenic genes as well as the segments on which they belong (Fig. 13.2).

Numerous improvements are planned or being explored, including:

1. The database schema will be redesigned. The database schema is a foundation of a database system. The growing size and complexity of PGDD, as a result of additional genomes and new opportunities for analysis, require redesigning the database schema.

2. With many new genome sequence data, needs and opportunities for new analysis method or user interfaces increase. To better respond to these needs and opportunities, PGDD will be rewritten in Django, a web framework which increases the productivity of development.

3. We will change the current static synteny browser to a user interactive one. The new browser will display the fine structures of syntenic regions with supporting panning and zooming. Additionally, the new browser will be able to display multiple alignments of synteny regions.

13.5 Synthesis

Synteny-based approaches to clarify comparative genome arrangements, originally based on genetic maps, grow in value as we gain finer-scale information derived from physical maps and whole-genome sequences. Angiosperms have presented both challenges and opportunities for synteny-based genome comparisons. The challenge has been that both neo- and paleo-polyploidy have necessitated novel approaches for genome comparisons, relative to existing methods designed for vertebrates. The combination of bottom-up and top-down multi-alignment approaches has largely met this need (although there remains scope for improvements). Numerous short-term opportunities have already been realized, clarifying specific features of angiosperm genome organization and evolution. Of singular importance is the long-term opportunity to clarify the consequences of genome duplication (GD), an evolutionary event that has shaped the architecture and function of many higher eukaryotic genomes. Toward this end, and also toward reaping the potential benefits of translational genomic approaches to leverage functional genomic knowledge of facile genomes in the study of more complex genomes, intra- and inter-genome alignments from a single resource based on uniform standards are needed. The Plant Genome Duplication Database has been created to facilitate this goal.

References

Aury JM, Jaillon O, Duret L, Noel B, Jubin C, Porcel BM, Segurens B, Daubin V, Anthouard V, Aiach N, Arnaiz O, Billaut A, Beisson J, Blanc I, Bouhouche K, Camara F, Duharcourt S, Guigo R, Gogendeau D, Katinka M, Keller AM, Kissmehl R, Klotz C, Koll F, Le Mouel A, Lepere G, Malinsky S, Nowacki M, Nowak JK, Plattner H, Poulain J, Ruiz F, Serrano V, Zagulski M, Dessen P, Betermier M, Weissenbach J, Scarpelli C, Schachter V, Sperling L, Meyer E, Cohen J, Wincker P (2006) Global trends of whole-genome duplications revealed by the ciliate *Paramecium tetraurelia*. Nature 444:171–178

Blanc G, Wolfe KH (2004) Widespread paleopolyploidy in model plant species inferred from age distributions of duplicate genes. Plant Cell 16:1667–1678

Bonierbale MW, Plaisted RL, Tanksley SD (1988) RFLP maps based on a common set of clones reveal modes of chromosomal evolution in potato and tomato. Genetics 120:1095–1103

Bowers JE, Chapman BA, Rong J, Paterson AH (2003a) Unravelling angiosperm genome evolution by phylogenetic analysis of chromosomal duplication events. Nature 422:433–438

Bowers JE, Abbey C, Anderson S, Chang C, Draye X, Hoppe AH, Jessup R, Lemke C, Lennington J, Li Z, Lin YR, Liu SC, Luo L, Marler BS, Ming R, Mitchell SE, Kresovich S, Schertz KF, Paterson AH (2003b) A high-density genetic recombination map of sequence-tagged sites for sorghum, as a framework for comparative structural and evolutionary genomics of tropical grains and grasses. Genetics 165:367–386

Bowers JE, Arias MA, Asher R, Avise JA, Ball RT, Brewer GA, Buss RW, Chen AH, Edwards TM, Estill JC, Exum HE, Goff VH, Herrick KL, Steele CLJ, Karunakaran S, Lafayette GK, Lemke C, Marler BS, Masters SL, McMillan JM, Nelson LK, Newsome GA, Nwakanma CC, Odeh RN, Phelps CA, Rarick EA, Rogers CJ, Ryan SP, Slaughter KA, Soderlund CA, Tang HB, Wing RA, Paterson AH (2005) Comparative physical mapping links conservation of microsynteny to chromosome structure and recombination in grasses. Proc Natl Acad Sci U S A 102:13206–13211

Calabrese PP, Chakravarty S, Vision TJ (2003) Fast identification and statistical evaluation of segmental homologies in comparative maps. Bioinformatics 19(suppl 1):i74–i80

Cannon SB, Kozik A, Chan B, Michelmore R, Young ND (2003) DiagHunter and GenoPix2D: programs for genomic comparisons, large-scale homology discovery and visualization. Genome Biol 4: R68

Chapman BA, Bowers JE, Schulze SR, Paterson AH (2004) A comparative phylogenetic approach for dating whole genome duplication events. Bioinformatics 20:180–185

Chen F, Mackey AJ, Stoeckert CJ Jr, Roos DS (2006) OrthoMCL-DB: querying a comprehensive multi-species collection of ortholog groups. Nucleic Acids Res 34:D363–D368

Chen F, Mackey AJ, Vermunt JK, Roos DS (2007) Assessing performance of orthology detection strategies applied to eukaryotic genomes. PLoS One 2:e383

Christoffels A, Koh EGL, Chia JM, Brenner S, Aparicio S, Venkatesh B (2004) Fugu genome analysis provides evidence for a whole-genome duplication early during the evolution of ray-finned fishes. Mol Biol Evol 21:1146–1151

Coghlan A, Eichler EE, Oliver SG, Paterson AH, Stein L (2005) Chromosome evolution in eukaryotes: a multi-kingdom perspective. Trends Genet 21:673–682

Consortia TRCaS, (2005) The sequence of rice chromosomes 11 and 12, rich in disease resistance genes and recent gene duplications. BMC Biol 3:20

Consortium MGS (2002) Initial sequencing and comparative analysis of the mouse genome. Nature 420:520–562

Cui LY, Wall PK, Leebens-Mack JH, Lindsay BG, Soltis DE, Doyle JJ, Soltis PS, Carlson JE, Arumuganathan K, Barakat A, Albert VA, Ma H, dePamphilis CW (2006) Widespread genome duplications throughout the history of flowering plants. Genome Res 16:738–749

Davies TJ, Barraclough TG, Chase MW, Soltis PS, Soltis DE, Savolainen V (2004) Darwin's abominable mystery: insights from a supertree of the angiosperms. Proc Natl Acad Sci U S A 101:1904–1909

Enright AJ, John B, Gaul U, Tuschl T, Sander C, Marks DS (2002) An efficient algorithm for large scale detection of protein families. Nucleic Acids Res 30:1575–1584

Garey M, Johnson DS (1979) Computers and intractability: a guide to the theory of NP-completeness. W. H. Freeman, New York

Grant D, Cregan P, Shoemaker RC (2000) Genome organization in dicots: genome duplication in *Arabidopsis* and synteny between soybean and *Arabidopsis*. Proc Natl Acad Sci U S A 97:4168–4173

Gu ZL, Steinmetz LM, Gu X, Scharfe C, Davis RW, Li WH (2003) Role of duplicate genes in genetic robustness against null mutations. Nature 421:63–66

Haas BJ, Delcher AL, Wortman JR, Salzberg SL (2004) DAGchainer: a tool for mining segmental genome duplications and synteny. Bioinformatics 20:3643–3646

Hansson B, Kawabe A, Preuss S, Kuittinen H, Charlesworth D (2006) Comparative gene mapping in *Arabidopsis lyrata* chromosomes 1 and 2 and the corresponding *A. thaliana* chromosome 1: recombination rates, rearrangements and centromere location. Genet Res 87:75–85

He XL, Zhang JZ (2005) Rapid subfunctionalization accompanied by prolonged and substantial neofunctionalization in duplicate gene evolution. Genetics 169:1157–1164

Hollister JD, Gaut BS (2009) Epigenetic silencing of transposable elements: a trade-off between reduced transposition and deleterious effects on neighboring gene expression. Genome Res 19:1419–1428

Hollister JD, Smith LM, Guo YL, Weigel D, Gaut BS (2011) Transposable elements and small RNAs contribute to gene expression divergence between *Arabidopsis thaliana* and *Arabidopsis lyrata*. Proc Natl Acad Sci U S A 108:2322–2327

Hulbert SH, Richter TE, Axtell JD, Bennetzen JL (1990) Genetic mapping and characterization of sorghum and related crops by means of maize DNA probes. Proc Natl Acad Sci U S A 87:4251–4255

Jaillon O, Aury JM et al (2007a) The grapevine genome sequence suggests ancestral hexaploidization in major angiosperm phyla. Nature 449:463–467

Jaillon O, Aury JM, Noel B, Policriti A, Clepet C, Casagrande A, Choisne N, Aubourg S, Vitulo N, Jubin C, Vezzi A, Legeai F, Hugueney P, Dasilva C, Horner D, Mica E, Jublot D, Poulain J, Bruyere C, Billault A, Segurens B, Gouyvenoux M, Ugarte E, Cattonaro F, Anthouard V, Vico V, Del Fabbro C, Alaux M, Di Gaspero G, Dumas V, Felice N, Paillard S, Juman I, Moroldo M, Scalabrin S, Canaguier A, Le Clainche I, Malacrida G, Durand E, Pesole G, Laucou V, Chatelet P, Merdinoglu D, Delledonne M, Pezzotti M, Lecharny A, Scarpelli C, Artiguenave F, Pe ME, Valle G, Morgante M, Caboche M, Adam-Blondon AF, Weissenbach J, Quetier F, Wincker P (2007b) The grapevine genome sequence suggests ancestral hexaploidization in major angiosperm phyla. Nature 449:463–467

Johnston JS, Pepper AE, Hall AE, Chen ZJ, Hodnett G, Drabek J, Lopez R, Price HJ (2005) Evolution of genome size in Brassicaceae. Am J Bot 95:229–235

Kawabe A, Hansson B, Hagenblad J, Forrest A, Charlesworth D (2006) Centromere locations and associated chromosome rearrangements in *Arabidopsis lyrata* and *A. thaliana*. Genetics 173:1613–1619

Kent WJ, Baertsch R, Hinrichs A, Miller W, Haussler D (2003) Evolution's cauldron: duplication, deletion, and rearrangement in the mouse and human genomes. Proc Natl Acad Sci U S A 100:11484–11489

Kim JS, Klein PE, Klein RR, Price HJ, Mullet JE, Stelly DM (2005) Chromosome identification and nomenclature of *Sorghum bicolor*. Genetics 169:1169–1173

Kishimoto N, Higo H, Abe K, Arai S, Saito A, Higo K (1994) Identification of the duplicated segments in rice chromosomes 1 and 5 by linkage analysis of cDNA markers of known functions. Theor Appl Genet 88:722–726

Koch MA, Kiefer M (2005) Genome evolution among cruciferous plants: A lecture from the comparison of the genetic maps of three diploid species—*Capsella rubella*, *Arabidopsis lyrata* ssp. *petraea*, and *A. thaliana*. Am J Bot 92:761–767

Ku HM, Vision T, Liu JP, Tanksley SD (2000) Comparing sequenced segments of the tomato and *Arabidopsis* genomes: large-scale duplication followed by selective gene loss creates a network of synteny. Proc Natl Acad Sci U S A 97:9121–9126

Kuittinen H, de Haan AA, Vogl C, Oikarinen S, Leppala J, Koch M, Mitchell-Olds T, Langley CH, Savolainen O (2004) Comparing the linkage maps of the close relatives *Arabidopsis lyrata* and *A. thaliana*. Genetics 168:1575–1584

Lee JM, Grant D, Vallejos CE, Shoemaker RC (2001) Genome organization in dicots. II. Arabidopsis as a 'bridging species' to resolve genome evolution events among legumes. Theor Appl Genet 103:765–773

Lin Y, Schertz K, Paterson A (1995) Comparative analysis of QTLs affecting plant height and maturity across the Poaceae, in reference to an interspecific sorghum population. Genetics 141:391–411

Lin L, Pierce GJ, Bowers JE, Estill JC, Compton RO, Rainville LK, Kim C, Lemke C, Rong J, Tang H, Wang X, Braidotti M, Chen AH, Chicola K, Collura K, Epps E, Golser W, Grover C, Ingles J, Karunakaran S, Kudrna D, Olive J, Tabassum N, Um E, Wissotski M, Yu Y, Zuccolo A, Rahman MU, Peterson DG, Wing RA, Wendel JF, Paterson AH (2010) A draft physical map of a D-genome cotton species (*Gossypium raimondii*). BMC Genomics 11:395

Liu H, Sachidanandam R, Stein L (2001) Comparative genomics between rice and *Arabidopsis* shows scant collinearity in gene order. Genome Res 11:2020–2026

Lynch M (2006) The origins of eukaryotic gene structure. Mol Biol Evol 23:450–468

Lynch M, Conery JS (2003) The origins of genome complexity. Science 302:1401–1404

Lynch M, Force A (2000) The probability of duplicate gene preservation by subfunctionalization. Genetics 154:459–473

Lynch M, O'Hely M, Walsh B, Force A (2001) The probability of preservation of a newly arisen gene duplicate. Genetics 159:1789–1804

Maere S, De Bodt S, Raes J, Casneuf T, Van Montagu M, Kuiper M, Van de Peer Y (2005) Modeling gene and genome duplications in eukaryotes. Proc Natl Acad Sci U S A 102:5454–5459

Mayer K, Murphy G, Tarchini R, Wanbutt R, Volckaert G, Pohl T, Dusterhoft A, Stiekema W, Entian K-D, Terryn N, Lemcke K, Haase D, Hall CR, van Dodeweerd A-M, Tingey SV, Mewes H-W, Bevan MW, Bancroft I (2001) Conservation of microstructure between a sequenced region of the genome of rice and multiple segments of the genome of *Arabidopsis thaliana*. Genome Res 11:1167–1174

Ming R, Hou S, Feng Y, Yu QY, Dionne-Laporte A, Saw J, Senin P, Wang W, Salzberg SL, Tang H, Lyons E, Rice D, Riley M, Skelton R, Murray J, Chen C, Eustice M, Tong E, Albert H, Paull RE, Wang ML, Zhu Y, Schatz M, Nagarajan N, Agbayani R, Guan P, Blas A, Wang J, Na JK, Michael T, Shakirov EV, Haas B, Thimmapuram J, Nelson D, Wang X, Bowers JE, Suzuki J, Tripathi S, Neupane K,

Wei H, Singh R, Irikura B, Jiang N, Zhang W, Wall K, Presting G, Gschwend A, Li Y, Windsor AJ, Navajas-Perez R, Torres MJ, Feltus FA, Porter B, Paidi M, Luo MC, Liu L, Christopher D, Moore PH, Sugimura T, dePamphilis C, Jiang J, Schuler M, Mitchell-Olds T, Shippen D, Palmer JD, Freeling M, Paterson AH, Gonsalves D, Wang L, Alam M (2008) The draft genome of the transgenic tropical fruit tree papaya (*Carica papaya* Linnaeus). Nature 452:991–997

O'Brien KP, Remm M, Sonnhammer EL (2005) Inparanoid: a comprehensive database of eukaryotic orthologs. Nucleic Acids Res 33: D476–D480

Ohno S (1970) Evolution by gene duplication. Springer, Berlin/ Heidelberg/New York

Paterson AH, Lin YR, Li ZK, Schertz KF, Doebley JF, Pinson SRM, Liu SC, Stansel JW, Irvine JE (1995) Convergent domestication of cereal crops by independent mutations at corresponding genetic loci. Science 269:1714–1718

Paterson AH, Lan TH, Reischmann KP, Chang C, Lin YR, Liu SC, Burow MD, Kowalski SP, Katsar CS, DelMonte TA, Feldmann KA, Schertz KF, Wendel JF (1996) Toward a unified genetic map of higher plants, transcending the monocot-dicot divergence. Nat Genet 14:380–382

Paterson AH, Bowers J, Peterson D, Estill J, Chapman B (2003) Structure and evolution of cereal genomes. Curr Opin Genet Dev 13:644–650

Paterson AH, Bowers JE, Chapman BA (2004) Ancient polyploidization predating divergence of the cereals, and its consequences for comparative genomics. Proc Natl Acad Sci U S A 101:9903–9908

Paterson AH, Freeling M, Sasaki T (2005a) Grains of knowledge: genomics of model cereals. Genome Res 15:1643–1650

Paterson AH, Bowers JE, Vandepoele K, Van de Peer Y (2005b) Ancient duplication of cereal genomes. New Phytol 165:658–661

Paterson AH, Bowers JE, Bruggmann R, Dubchak I, Grimwood J, Gundlach H, Haberer G, Hellsten U, Mitros T, Poliakov A, Schmutz J, Spannagl M, Tang H, Wang X, Wicker T, Bharti AK, Chapman J, Feltus FA, Gowik U, Grigoriev IV, Lyons E, Maher CA, Martis M, Narechania A, Otillar RP, Penning BW, Salamov AA, Wang Y, Zhang L, Carpita NC, Freeling M, Gingle AR, Hash CT, Keller B, Klein P, Kresovich S, McCann MC, Ming R, Peterson DG, Mehboob ur R, Ware D, Westhoff P, Mayer KF, Messing J, Rokhsar DS (2009) The *Sorghum bicolor* genome and the diversification of grasses. Nature 457:551–556

Paterson AH, Freeling M, Tang H, Wang X (2010) Insights from the comparison of plant genome sequences. Annu Rev Plant Biol 61:349–372

Pellicer J, Fay MF, Leitch IJ (2010) The largest eukaryotic genome of them all? Bot J Linn Soc 164:10–15

Peterson DG, Wessler SR, Paterson AH (2002) Efficient capture of unique sequences from eukaryotic genomes. Trends Genet 18:547–550

Pfeil BE, Schlueter JA, Shoemaker RC, Doyle JJ (2005) Placing paleopolyploidy in relation to taxon divergence: a phylogenetic analysis in legumes using 39 gene families. Syst Biol 54:441–454

Rossberg M, Theres K, Acarkan A, Herrero R, Schmitt T, Schumacher K, Schmitz G, Schmidt R (2001) Comparative sequence analysis reveals extensive microcolinearity in the lateral suppressor regions of the tomato, *Arabidopsis*, and *Capsella* genomes. Plant Cell 13:979–988

Salse J, Piegu B, Cooke R, Delseny M (2002) Synteny between *Arabidopsis thaliana* and rice at the genome level: a tool to identify conservation in the ongoing rice genome sequencing project. Nucleic Acids Res 30:2317–2328

Scannell DR, Byrne KP, Gordon JL, Wong S, Wolfe KH (2006) Multiple rounds of speciation associated with reciprocal gene loss in polyploid yeasts. Nature 440:341–345

Schmutz J, Cannon SB, Schlueter J, Ma JX, Mitros T, Nelson W, Hyten DL, Song QJ, Thelen JJ, Cheng JL, Xu D, Hellsten U, May GD, Yu Y, Sakurai T, Umezawa T, Bhattacharyya MK, Sandhu D, Valliyodan B, Lindquist E, Peto M, Grant D, Shu SQ, Goodstein D, Barry K, Futrell-Griggs M, Abernathy B, Du JC, Tian ZX, Zhu LC, Gill N, Joshi T, Libault M, Sethuraman A, Zhang XC, Shinozaki K, Nguyen HT, Wing RA, Cregan P, Specht J, Grimwood J, Rokhsar D, Stacey G, Shoemaker RC, Jackson SA (2010) Genome sequence of the palaeopolyploid soybean. Nature 463:178–183

Schranz ME, Mitchell-Olds T (2006) Independent ancient polyploidy events in the sister families Brassicaceae and Cleomaceae. Plant Cell 18:1152–1165

Simillion C, Vandepoele K, Saeys Y, Van de Peer Y (2004) Building genomic profiles for uncovering segmental homology in the twilight zone. Genome Res 14:1095–1106

Singh NK, Dalal V, Batra K, Singh BK, Chitra G, Singh A, Ghazi IA, Yadav M, Pandit A, Dixit R, Singh PK, Singh H, Koundal KR, Gaikwad K, Mohapatra T, Sharma TR (2007) Single-copy genes define a conserved order between rice and wheat for understanding differences caused by duplication, deletion, and transposition of genes. Funct Integr Genomics 7:17–35

Smith SA, Donoghue MJ (2008) Rates of molecular evolution are linked to life history in flowering plants. Science 322:86–89

Smith SF, Snell P, Gruetzner F, Bench AJ, Haaf T, Metcalfe JA, Green AR, Elgar G (2002) Analyses of the extent of shared synteny and conserved gene orders between the genome of *Fugu rubripes* and human 20q. Genome Res 12:776–784

Soltis DE, Soltis PS, Endress PK, Chase MW (2005) Phylogeny and evolution of angiosperms. Sinauer, Sunderland

Srinivasachary DMM, Gale MD, Devos KM (2007) Comparative analyses reveal high levels of conserved colinearity between the finger millet and rice genomes. Theor Appl Genet 115:489–499

Stebbins G (1966) Chromosomal variation and evolution; polyploidy and chromosome size and number shed light on evolutionary processes in higher plants. Science 152:1463–1469

Stephens S (1951) Possible significance of duplications in evolution. Adv Genet 4:247–265

Storm CE, Sonnhammer EL (2002) Automated ortholog inference from phylogenetic trees and calculation of orthology reliability. Bioinformatics 18:92–99

Storm CE, Sonnhammer EL (2003) Comprehensive analysis of orthologous protein domains using the HOPS database. Genome Res 13:2353–2362

Swigonova Z, Lai J, Ma J, Ramakrishna W, Llaca V, Bennetzen JL, Messing J (2004a) Close split of sorghum and maize genome progenitors. Genome Res 14:1916–1923

Swigonova Z, Lai JS, Ma JX, Ramakrishna W, Llaca M, Bennetzen JL, Messing J (2004b) On the tetraploid origin of the maize genome. Comp Funct Genomics 5:281–284

Swigonova Z, Bennetzen JL, Messing J (2005) Structure and evolution of the r/b chromosomal regions in rice, maize and sorghum. Genetics 169:891–906

Tang H, Bowers JE, Wang X, Ming R, Alam M, Paterson AH (2008a) Synteny and collinearity in plant genomes. Science 320:486–488

Tang H, Wang X, Bowers JE, Ming R, Alam M, Paterson AH (2008b) Unraveling ancient hexaploidy through multiply-aligned angiosperm gene maps. Genome Res 18:1944–1954

Tang H, Bowers JE, Wang X, Paterson AH (2010) Angiosperm genome comparisons reveal early polyploidy in the monocot lineage. Proc Natl Acad Sci U S A 107:472–477

Tatusov RL, Fedorova ND, Jackson JD, Jacobs AR, Kiryutin B, Koonin EV, Krylov DM, Mazumder R, Mekhedov SL, Nikolskaya AN, Rao BS, Smirnov S, Sverdlov AV, Vasudevan S, Wolf YI, Yin JJ, Natale DA (2003) The COG database: an updated version includes eukaryotes. BMC Bioinformatics 4:41

Tuskan GA, DiFazio S, Jansson S, Bohlmann J, Grigoriev I, Hellsten U, Putnam N, Ralph S, Rombauts S, Salamov A, Schein J, Sterck L, Aerts A, Bhalerao RR, Bhalerao RP, Blaudez D, Boerjan W, Brun A, Brunner A, Busov V, Campbell M, Carlson J, Chalot M, Chapman J, Chen GL, Cooper D, Coutinho PM, Couturier J, Covert S, Cronk Q, Cunningham R, Davis J, Degroeve S, Dejardin A, Depamphilis C, Detter J, Dirks B, Dubchak I, Duplessis S, Ehlting J, Ellis B, Gendler K, Goodstein D, Gribskov M, Grimwood J, Groover A, Gunter L, Hamberger B, Heinze B, Helariutta Y, Henrissat B, Holligan D, Holt R, Huang W, Islam-Faridi N, Jones S, Jones-Rhoades M, Jorgensen R, Joshi C, Kangasjarvi J, Karlsson J, Kelleher C, Kirkpatrick R, Kirst M, Kohler A, Kalluri U, Larimer F, Leebens-Mack J, Leple JC, Locascio P, Lou Y, Lucas S, Martin F, Montanini B, Napoli C, Nelson DR, Nelson C, Nieminen K, Nilsson O, Pereda V, Peter G, Philippe R, Pilate G, Poliakov A, Razumovskaya J, Richardson P, Rinaldi C, Ritland K, Rouze P, Ryaboy D, Schmutz J, Schrader J, Segerman B, Shin H, Siddiqui A, Sterky F, Terry A, Tsai CJ, Uberbacher E, Unneberg P, Vahala J, Wall K, Wessler S, Yang G, Yin T, Douglas C, Marra M, Sandberg G, Van de Peer Y, Rokhsar D (2006) The genome of black cottonwood, *Populus trichocarpa* (Torr. & Gray). Science 313:1596–1604

Van de Peer Y (2004) Computational approaches to unveiling ancient genome duplications. Nat Rev Genet 5:752–763

Vandepoele K, Saeys Y, Simillion C, Raes J, Van de Peer Y (2002a) The automatic detection of homologous regions (ADHoRe) and its application to microcolinearity between *Arabidopsis* and rice. Genome Res 12:1792–1801

Vandepoele K, Simillion C, Van de Peer Y (2002b) Detecting the undetectable: uncovering duplicated segments in *Arabidopsis* by comparison with rice. Trends Genet 18:606–608

Vandepoele K, Simillion C, Van de Peer Y (2003) Evidence that rice and other cereals are ancient aneuploids. Plant Cell 15:2192–2202

Wang X, Shi X, Hao B, Ge S, Luo J (2005) Duplication and DNA segmental loss in the rice genome: implications for diploidization. New Phytol 165:937–946

Wang XY, Shi XL, Li Z, Zhu QH, Kong L, Tang W, Ge S, Luo JC (2006) Statistical inference of chromosomal homology based on gene colinearity and applications to *Arabidopsis* and rice. BMC Bioinformatics 7:447

Wang X, Tang H, Bowers JE, Feltus FA, Paterson AH (2007) Extensive concerted evolution of rice paralogs and the road to regaining independence. Genetics 177:1753–1763

Wang X, Tang H, Bowers JE, Paterson AH (2009) Comparative inference of illegitimate recombination between rice and sorghum duplicated genes produced by polyploidization. Genome Res 19:1026–1032

Wang X, Tang H, Paterson AH (2011) Seventy million years of concerted evolution of a homoeologous chromosome pair, in parallel, in major Poaceae lineages. Plant Cell 23:27–37

Wei F, Coe E, Nelson W, Bharti AK, Engler F, Butler E, Kim H, Goicoechea JL, Chen M, Lee S, Fuks G, Sanchez-Villeda H, Schroeder S, Fang Z, McMullen M, Davis G, Bowers JE, Paterson AH, Schaeffer M, Gardiner J, Cone K, Messing J, Soderlund C, Wing RA (2007) Physical and genetic structure of the maize genome reflects its complex evolutionary history. PLoS Genet 3:e123

Werth CR, Windham MD (1991) A model for divergent, allopatric speciation of polyploid pteridophytes resulting from silencing of duplicate-gene expression. Am Natur 137:515–526

Yang YW, Lai KN, Tai PY, Li WH (1999) Rates of nucleotide substitution in angiosperm mitochondrial DNA sequences and dates of divergence between *Brassica* and other angiosperm lineages. J Mol Evol 48:597–604

Yogeeswaran K, Frary A, York TL, Amenta A, Lesser AH, Nasrallah JB, Tanksley SD, Nasrallah ME (2005) Comparative genome analyses of *Arabidopsis* spp.: inferring chromosomal rearrangement events in the evolutionary history of *A. thaliana*. Genome Res 15:505–515

Yu J, Wang J, Lin W, Li SG, Li H, Zhou J, Ni PX, Dong W, Hu SN, Zeng CQ, Zhang JG, Zhang Y, Li RQ, Xu ZY, Li ST, Li XR, Zheng HK, Cong LJ, Lin L, Yin JN, Geng JN, Li GY, Shi JP, Liu J, Lv H, Li J, Deng YJ, Ran LH, Shi XL, Wang XY, Wu QF, Li CF, Ren XY, Wang JQ, Wang XL, Li DW, Liu DY, Zhang XW, Ji ZD, Zhao WM, Sun YQ, Zhang ZP, Bao JY, Han YJ, Dong LL, Ji J, Chen P, Wu SM, Liu JS, Xiao Y, Bu DB, Tan JL, Yang L, Ye C, Zhang JF, Xu JY, Zhou Y, Yu YP, Zhang B, Zhuang SL, Wei HB, Liu B, Lei M, Yu H, Li YZ, Xu H, Wei SL, He XM, Fang LJ, Zhang ZJ, Zhang YZ, Huang XG, Su ZX, Tong W, Li JH, Tong ZZ, Li SL, Ye J, Wang LS, Fang L, Lei TT, Chen C, Chen H, Xu Z, Li HH, Huang HY, Zhang F, Xu HY, Li N, Zhao CF, Dong LJ, Huang YQ, Li L, Xi Y, Qi QH, Li WJ, Hu W, Zhang YL, Tian XJ, Jiao YZ, Liang XH, Jin JA, Gao L, Zheng WM, Hao BL, Liu SQ, Wang W, Yuan LP, Cao ML, McDermott J, Samudrala R, Wong GKS, Yang HM (2005) The genomes of *Oryza sativa*: a history of duplications. PLoS Biol 3:266–281

Zmasek CM, Eddy SR (2002) RIO: analyzing proteomes by automated phylogenomics using resampled inference of orthologs. BMC Bioinformatics 3:14

The Variation of Base Composition in Plant Genomes

14

Petr Šmarda and Petr Bureš

Contents

P. Šmarda (✉)
Department of Botany and Zoology, Masaryk University, Kotlářská 2, CZ-61137 Brno, Czech Republic
e-mail: smardap@sci.muni.cz

14.1 Introduction

The proportions of the four nucleotides in DNA may vary significantly in various genome components and with different selective forces acting on particular sequences. Base composition is conventionally expressed as the percentage of guanine (G) and cytosine (C) bases (GC content) in a given region or for the entire genome (genomic GC content). Based on universal base-pairing rules, the concentration of adenine (A) plus thymine (T) (i.e., the AT content) may be calculated as $1 - GC$, and the concentration of either A or T bases is $((1 - GC)/2)$.

Both the knowledge of mechanisms causing changes in GC content and their ecological consequences are fundamental for understanding genome evolution. The study of GC content has a long tradition in prokaryotic biology and systematics, where it serves as an important criterion in taxa delimitation and predictions of life-styles, such as thermotolerance, growth rate or aerobiosis (Mann and Phoebe-Chen 2010). The GC content of taxa also has been widely discussed in animal genomics, namely, in the consequences of the evolution of the isochore structure of humans and other warm-blooded vertebrates (Bernardi 2000a, b, 2007; Eyre-Walker and Hurst 2001; Constantini and Bernardi 2008a). However, less attention has been paid to the GC content of plant genomes, for which the knowledge of detailed base composition and its meaning in the ecology and evolution of particular taxa is still poor.

Here, we survey existing research concerning GC content of plant nuclear genomes and outline some directions for possible interpretations of the biological meaning of variation in GC content viewed in the context of work done on bacterial and animal genomes. We hope that the examples and proposed explanations will stimulate future testing of their validity in plant genomes. We primarily concentrate on overall genomic base composition. Nevertheless, variation of GC content and its function in particular genomic components is also discussed.

J.F. Wendel et al. (eds.), *Plant Genome Diversity Volume 1*,
DOI 10.1007/978-3-7091-1130-7_14, © Springer-Verlag Wien 2012

14.2 Genomic GC Content Variation in Plants and Other Organisms

The most complete knowledge of GC content comes from bacteria, with >1,000 complete genomes already sequenced. The less complex genome organization of prokaryotes and direct association with their external environment results in a wide range in GC content of prokaryotic genomes, with a GC content as low as 16.6% in the smallest bacterial genome (*Candidatus Carsonella ruddii*; Nakabachi et al. 2006) and as high as 74.9% in the anaerobic bacterium *Anaeromyxobacter dehalogenans* (Sanford et al. 2002). The genomes of eukaryotes generally have a narrower variation in their overall GC content, and a decrease in variation is seen with the incidence of multicellularity and increase in structural and functional complexity (Fig. 14.1).

A complete list of GC content estimates for 214 modern species of plants was recently summarized by Meister and Barow (2007). Over the last several years, our laboratory has completed measurements of the GC contents of >1,000 other plant species (still mostly unpublished), which are considered here in terms of the range of GC contents across plants as a whole. In plants, the highest variation is found in green

algae, where GC content ranges from 39.0% in *Spirogyra* to 63.8% in *Scenedesmus acuminatus* (Serenkov 1962). A high degree of variation is also found in ferns (monilophytes), where the range is from ~38% to 40% in some *Davalliaceae* and *Polypodiaceae* to 48% in *Lygodium* (Bureš et al., unpublished). Within spermatophytes, GC contents as low as 33% have been documented in the *Cyperaceae + Juncaceae + Thurniaceae* clade, and low GC content is also found in some Commelinales (Šmarda unpublished). A very low GC content (34.7%) also distinguishes *Ginkgo biloba* from the other remaining gymnosperms (Barow and Meister 2002). Very high GC contents, usually ranging from 43% to 50%, are typical of grasses (*Poaceae*), distinguish this group from all other angiosperm families (Barow and Meister 2002; Šmarda et al., unpublished). Some of our recent results indicate that such remarkably high GC contents (up to 49%) may also be found in other monocot groups, such as the *Xyridaceae + Eriocaulaceae* clade, some *Liliaceae* and *Dasypogonaceae*. High GC contents have also been detected in the *Cyclanthaceae*, and the entire *Alismatales* clade, with a remarkable increase in the *Araceae* (Šmarda et al. unpubl. res.).

The extreme difference in GC content found between the genomes of *Arabidopsis* and those of model monocots

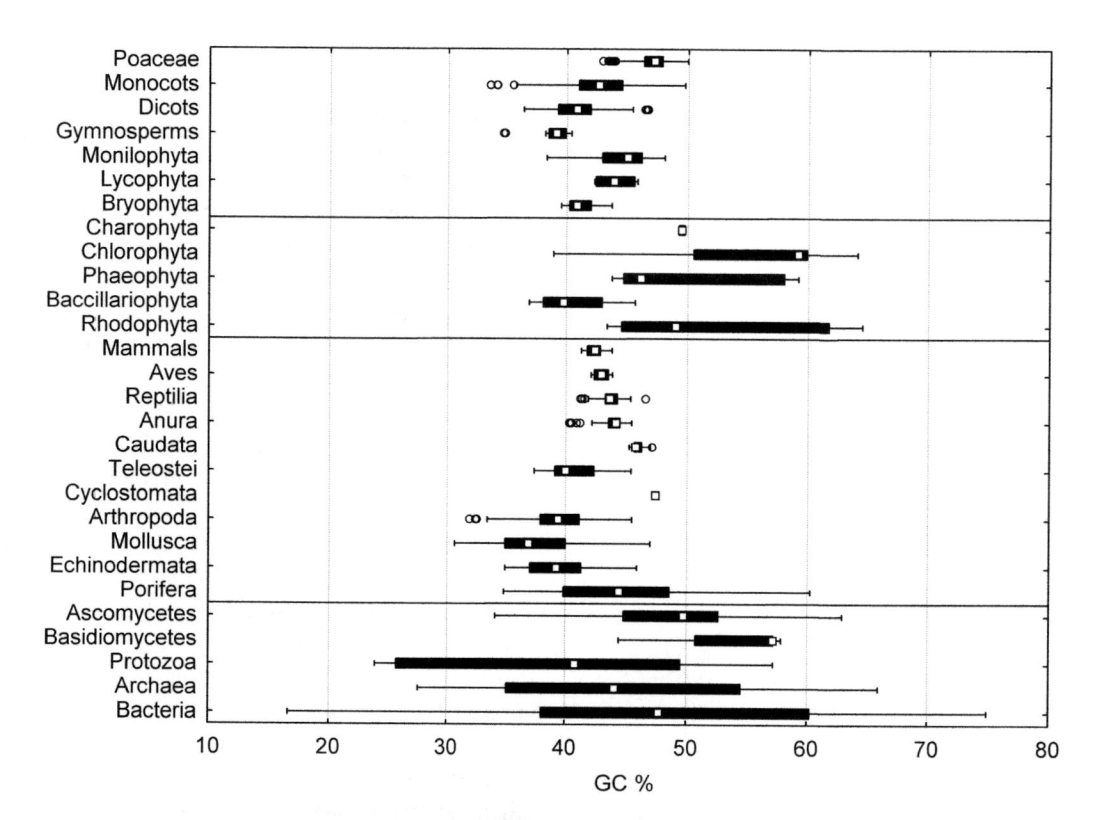

Fig. 14.1 Variation in genomic GC contents of various groups of organisms. *Boxplots* show median (*white squares*) interquartile range (*boxes*) and outliers with extremes (*empty circles*) (This analysis was based on the compilation of historical data by Shapiro (1976; excluding prokaryotes, vertebrates and higher plants), data from the Entrez

Genome project (http://www.ncbi.nlm.nih.gov/genomes/lproks.cgi; prokaryotic genomes), Vinogradov (1998; vertebrates), Meister and Barow (2007; gymnosperms, angiosperms, algae) and unpublished data from our laboratory (land plants))

(represented nearly exclusively by grasses) led to the conclusion that this difference in GC content may be common between monocots and dicots (Carels and Bernardi 2000). Indeed, our preliminary analysis demonstrates that monocots possess statistically greater GC contents than do dicots (Šmarda et al. 2010). Nevertheless, the most apparent difference between both groups is that the GC contents of monocots vary widely (they occasionally contain high GC taxa), while the variation in GC content is comparatively narrow in dicots (cf. Barow and Meister 2002). The genomic GC contents of some common and well-known plants are given in Table 14.1.

At lower taxonomic levels, the variation in genomic GC content is more limited, and genomic GC contents are generally similar in closely related taxa, such as within genera or tribes. Nevertheless, a significant degree of variation may be found within particular groups of plants, perhaps reflecting ongoing evolution related to intensive genome or chromosomal reorganization and the proliferation/removal of repetitive DNA. The combination of GC content and genome size data serves as a simple tool for basic genome characterization and taxa delimitation. In combination with knowledge of species phylogeny, this data also provides a way to track genome-wide associated processes. For example, in Eurasian *Festuca*, the combination of GC contents and genome sizes (average chromosome sizes) enables the distinction of this genus from its close relatives, the demarcation of some sub-generic taxa and the distinction between ancestral and derived evolutionary lineages (Fig. 14.2). Combined with phylogenetic information, the data indicate approximately a twofold increase in the genome size of ancestral lineages and a subsequent reduction in genome size proceeding independently in both major evolutionary lineages that is associated with their massive diversification (Šmarda et al. 2008; Fig. 14.2). It is assumed that such a pattern may result from an ancient amplification of GC-rich retrotransposons and their subsequent lineage-specific elimination (Šmarda et al. 2008). In a similar way, major genome types and genome size changes associated with an amplification of GC-rich retrotransposons also exist in the genus *Oryza* (Šmarda et al. in prep.). A clear separation of phylogenetic groups, in terms of their genome sizes and GC contents, has also been observed in *Cirsium* (Bureš et al. 2004), as well as for many other plant genera in our pilot data.

In comparison to the above-discussed nuclear genomes, chloroplast DNA usually possesses a lower GC content (Liu and Xue 2005). In land plants, chloroplast GC contents range from 28.5% in the moss *Physcomitrella patens* to 42.0% in the fern *Adiantum capillus-veneris*. Extremely GC-rich plastomes are found in the spike moss genus *Selaginella*, with a maximum GC content of 54.8% in the plastomes of *Selaginella uncinata* (Smith 2009). Further extremes might be searched for by comparing GC content of *rbc*L sequences correlating with overall chloroplast GC content to a great extent (cf. Smith 2009). The DNA in plant mitochondria is comparably GC-richer than in chloroplasts, with

Table 14.1 Published genomic GC contents of selected plants

Species	Group	GC (%)	Method[a]
Zea mays	Angiosperms, monocots, Poaceae	47.0	GSeq[b]
Oryza sativa	Angiosperms, monocots, Poaceae	43.6	GSeq[c]
Asparagus officinalis	Angiosperms, monocots, Asparagaceae	39.6	FCM
Allium cepa	Angiosperms, monocots, Amaryllidaceae	34.7	FCM
Lactuca sativa	Angiosperms, dicots, Asteraceae	39.7	FCM
Arabidopsis thaliana	Angiosperms, dicots, Brassicaceae	39.1	FCM
Populus trichocarpa	Angiosperms, dicots, Salicaceae	36.7	GSeq[c]
Beta vulgaris	Angiosperms, dicots, Amaranthaceae	37.3	FCM
Pinus sylvestris	Gymnosperms, Pinaceae	39.1	FCM
Ginkgo biloba	Gymnosperms, Ginkgoaceae	34.7	FCM
Physcomitrella patens	Mosses	38.7	GSeq[c]
Volvox carteri	Multicellular green algae	56.0	GSeq[d]
Chlamydomonas reinhardtii	Unicellular green algae	64.0	GSeq[c]
Ostreococcus tauri	Unicellular green algae	58.0	GSeq[c]
Phaeodactylum tricornutum	Diatoms	48.5	GSeq[c]
Thalassiosira pseudonana	Diatoms	47.0	GSeq[c]
Cyanidioschyzon merolae	Unicellular red algae	55.0	GSeq[c]

[a]*GSeq* data from complete genome sequences, *FCM* data from flow cytometric experiments by Barow and Meister (2002)
[b]Meyers et al. (2001)
[c]After compilation by Lang et al. (2008)
[d]Prochnik et al. (2010)

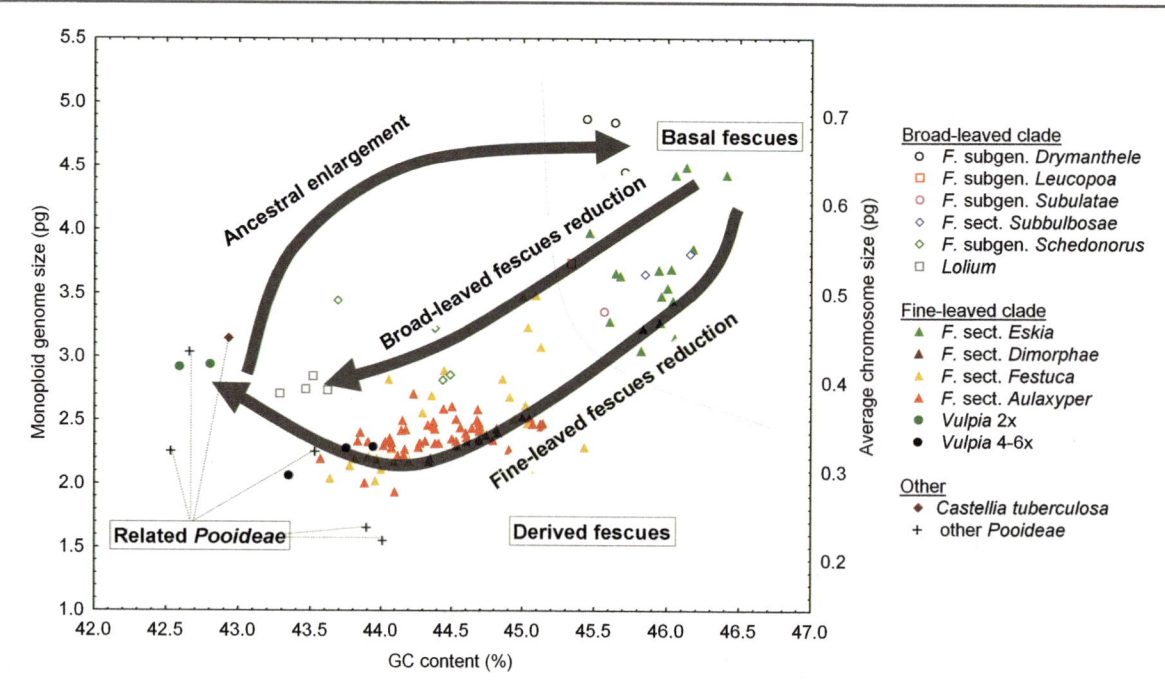

Fig. 14.2 Possible cyclic scenario of genome size and GC content evolution in *Festuca*. The divergence of basal fescues from related *Poeae* was predated by approximately a twofold increase in monoploid genome size (average chromosome size), and the apparent GC content enrichment is hypothesized to have resulted from an amplification of GC-rich repetitive DNA. The subsequent reduction of GC content and genome size running parallel in both main evolutionary lineages, the fine- and broad-leaved fescues, appears to be associated with the diversification of modern species groups (Adapted from Šmarda et al. 2008)

mitochondrial GC contents ranging from 40.6% in the moss *Physcomitrella patens* to 49.0% in the quillwort *Isoetës engelmannii* (Sloan and Taylor 2010).

14.3   Basic Properties of GC Bases in the Genome

The presence of GC base pairs plays a critical role in gene and genome regulation, as well as in determining the physical properties of DNA molecules. Among the most important features of the GC base pair is its higher thermal stability compared to the AT base pair. This is largely due to the existence of stronger stacking interactions between guanine and cytosine bases compared to AT base pairs. Contrary to the common view, the difference in hydrogen bonding (a triple-bond between guanine and cytosine and only a double-bond between adenine and thymine) seems less influential in this respect (Yakovchuk et al. 2006). This is why GC base pairs appear to be preferred in double-stranded regions of high-order DNA and RNA structures that are critical for high temperature-stressed organisms (see Sect. 14.4.3.3).

The stability of GC base pairs also directly affects other physical features of the DNA double helix, i.e., its bendability and curvature. Bendability refers to a local (two or three base region) ability of DNA to bend in correspondence with thermal fluctuations or DNA-protein interactions (Goodsell and Dickerson 1994). Bendability generally increases with increasing GC content. Additionally, it is important for determining the stability and positioning of a nucleosome (Anselmi et al. 1999; Kiyama and Trifonov 2002; Segal et al. 2006), the basic packing unit of eukaryotic chromatin. The positioning of nucleosomes is one of the key regulatory mechanisms of gene transcription via the modulation of the accessibility of regulatory molecules and transcription machinery to the underlying DNA sequence (Bai and Morozov 2010). The presence of GC-rich sequences is believed to facilitate nucleosome formation by increasing DNA flexibility, whereas relatively rigid poly-AT sequences discourage nucleosome assembly (Tillo and Hughes 2009). In addition to their relative bendability, WW (W = A or T) and SS (S = G or C) dinucleotides bend differently in different directions. Hence, they occur with a 10- to 11-bp periodicity in the DNA helix that corresponds with the periodic orientation of the DNA helix in the histone-DNA complex of a nucleosome. WW dinucleotides are favored at sites where the minor groove faces the histone core, whereas SS dinucleotides are favored at sites where the minor groove faces away from the histone core (Segal et al. 2006). The out of phase peaks in the occurrence of these dinucleotides promote optimal DNA bending, with W nucleotides causing a negative base roll and S nucleotides causing a positive base roll (Segal et al. 2006). The conserved nucleosome structure corresponds well with the 10.4-base periodicity of WW and SS dinucleotides that occurs across all eukaryotic genomes (Trifonov and Sussman 1980; Herzel et al. 1999; Bettecken and Trifonov 2009).

In contrast to bendability, DNA curvature refers to the intrinsic tendency of DNA to follow a non-linear pathway over an appreciable length ($\geq$20 bases) as a result of variation in local bends that are in phase with the DNA helix (Goodsell and Dickerson 1994). DNA curvature is calculated over several helical turns and generally increases with the proportion of AT bases. Increased curvature is important for easy, large-scale DNA packing, and thus, DNA packed in heterochromatin is generally GC-poor compared to GC-rich euchromatin, which contains the majority of expressed genes (Vinogradov 2003). The GC content and respective chromatin structure of large DNA regions (or even of whole chromosomes) is also related to their positioning within the nucleus. In general, individual GC-rich and gene-rich chromosomes, as well as GC-rich regions within chromosomes, are preferentially localized within the nuclear interior, while gene poor and AT-rich constitutive-heterochromatin regions are positioned toward the nuclear lamina (Göndör and Ohlsson 2009).

Cytosine is well known as the only base that is naturally methylated in eukaryotic nuclear genomes (though adenine methylation is common in bacteria and is found in plant mitochondria; Vanyushin 2006). The presence of methylcytosines affects the accessibility of DNA molecules to binding by regulatory molecules, and methylation plays a significant role in the regulation of gene expression and retrotransposon defense (Zhang 2008; Law and Jacobsen 2010; see Zhang 2012, this volume). The majority of methylated cytosines in animals and, to a lesser extent, in plants occurs as a C followed by a G (i.e., a CpG dinucleotide). Most CpG dinucleotides are regularly methylated, with the exceptions of CpGs in the transcription starting sites of genes that must remain accessible for the binding of regulatory molecules (Law and Jacobsen 2010). In both plants and animals, there is a clear ~10-bp periodicity in the occurrence of methylated cytosines that has been assumed to be related to the physical structure of methyltransferase enzymes (Law and Jacobsen 2010). However, in a recent study, Chodavarapu et al. (2010) demonstrated that this periodic pattern is due to the preferential targeting of methyltransferases to nucleosome binding sites, and this is supported by a 10.4-bp periodic pattern of methylatable CpG dinucleotides (Bettecken and Trifonov 2009, see above).

The deamination of methylated cytosines (metC) results in thymidine, a change that cannot be recognized by DNA repair mechanisms (in contrast to the deamination of cytosine that leads to uracil; Coulondre et al. 1978; Pfeifer 2006). Hence, methylcytosines represent one of the most important

mutational hotspots (Pfeifer 2006), and a C → T transition is the most frequent spontaneous single nucleotide mutation (e.g., those observed in the genome of *Arabidopsis*; Ossowski et al. 2010; see Sect. 14.4.2.4). In addition to methylcytosine and cytosine deamination, frequent instability is also observed with guanine, which may be oxidized to 7,8-dihydro-8-oxoG (8-oxoG). The 8-oxoG molecule subsequently mispairs with A, and this occasionally leads to GC → AT substitutions (Michaels and Miller 1992). The increased mutability of GC bases is assumed to be a reason why an increased substitution rate is generally observed in synonymous sites of coding (GC-rich) regions compared to non-coding AT-rich regions of DNA (DeRose-Wilson and Gaut 2007).

14.4 Reasons for and Consequences of GC Content Variation Between and Within Genomes

Genomic GC contents may differ between organisms for several reasons (Fig. 14.3). The organisms may (1) have different compositions of structural elements that vary in GC content (e.g., genes, retroelements or other repetitive sequences). An alternative source of GC content variation is single point AT ↔ GC substitutions. These mutations may be (2) enforced with selection for a GC-associated function of specific DNA sequences or whole genomes (e.g., those of orthologous or paralogous genes with different GC contents). Conversely, they may result from (3) a neutral mutational bias appearing as the passive consequence of some genome-wide process where GC content does not primarily matter (e.g., DNA repair or DNA replication). An effect of selection is commonly assumed when preferential mutations differ between nucleotide sequences of different functions (e.g., between redundant third codon positions of exons and sequences of introns), while a neutral mutational bias is assumed to be the case when preferential mutations are homogeneous along larger regions of DNA (or, minimally, whole genes).

Because of the complexity of genomes and the dynamics of mutation, selection, and drift, compositional variation among organisms may thus be viewed as an emergent trait reflecting a suite of internal genomic forces that collectively operate in particular ecological contexts. Within a genome, insights into the relative roles of some key processes may be determined from sequence and genomic data. What drives the variation in genomic GC content between organisms is still hypothesized based largely on studies of model organisms. Noting the relatively long evolutionary time required for mutational variation to substantially affect the composition of a genome, it may be expected that dramatic changes in observed GC contents between closely related

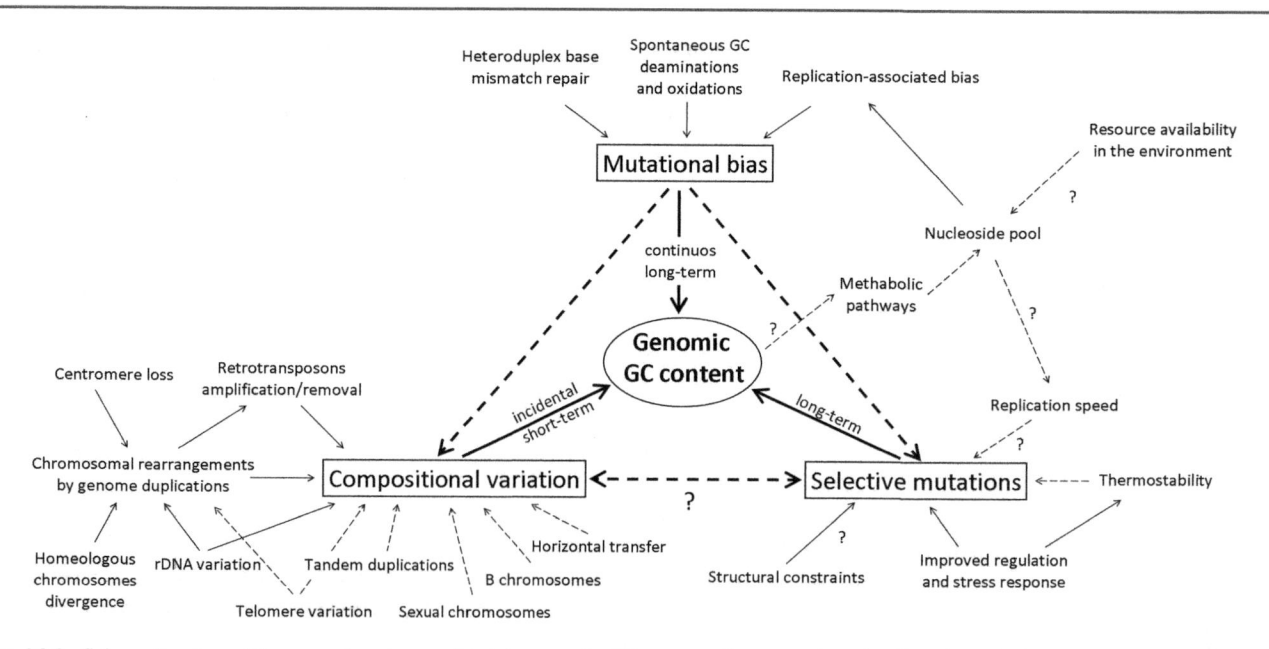

Fig. 14.3 Schematic view of forces and processes shaping genomic GC content. Strong and weak effects/associations are marked with *full* and *interrupted lines*, respectively. If the effects are unclear/hypothesized, they are marked with a *question mark*

Table 14.2 GC content characteristics of four completely sequenced plants

Species	Group	Genome size (1C, Mbp)	Genomic GC content (%)	GC of genes (%)	GC of intergenic regions (%)
Oryza sativa[a]	Monocots (Poaceae)	389	43.6	45.3	42.9
Carica papaya[b]	Dicots (Caricaceae)	372	35.3	37.5	34.9[e]
Vitis vinifera[c]	Dicots (Vitaceae)	487	34.8[e]	36.2	33.0
Chlamydomonas reinhardtii[d]	Unicellular green algae	~127[e]	64	68	~63[e]

[a]International Rice Genome Sequencing Project (2005)
[b]Ming et al. (2008)
[c]Jaillon et al. (2007)
[d]Merchant et al. (2007)
[e]Values that were not given in the original paper but that may be calculated from the provided genomic data

taxa would require extreme compositional variation, e.g., that arising from extensive variation in genome size. The remarkable association of a change in genomic GC content and the GC compositions of some orthologous genome regions between distantly related taxa (e.g., between grasses and *Arabidopsis*) and the relative homogeneity of GC contents within particular families, observed in spite of great variation in genome sizes, indicates that mutational variation is perhaps the prevalent force shaping the GC contents of plants over longer evolutionary time-scales.

14.4.1 Compositional Variation

14.4.1.1 Variation of GC Content Between Genome Components

The most GC-rich sequences within a genome are coding regions of genes (exons). Therefore, a clear difference may

be observed between the GC contents of exons and introns, between genes and intergenic regions and between coding regions and non-coding regions of repetitive DNA. These differences are observed in plant and other genomes independent of their overall genomic GC content (Tables 14.2, 14.3). The main reason for the higher GC content of exons may reflect structural constraints on the stability of secondary structures of mRNA transcripts, which generally increase with the proportion of GC bases in the stem regions of such structures (Biro 2008, also see below). The importance of the stability of mRNA supports the fact that in genes from bacteria to plants and animals, codons are nearly always used in a way that increases the stability of an mRNA transcript compared to the way in which synonymous codons would be used by chance (Seffens and Digby 1999; Gu et al. 2010). The level of GC-richness is also critical for the stability of structures of other essential RNA molecules, such as ribosomal and transfer RNAs,

Table 14.3 GC contents (%) of orthologous genes and average GC contents of transfer and ribosomal RNAs from fully sequenced model angiosperms

Species	Group[a]	Genes[b]	Exons[b]	Introns[b]	tRNA[c]	5S rRNA[d]	45S rDNA[d]
Zea mays	Poaceae (M)	47.2	52.7	42.1	55.0	51.1	58.6
Sorghum bicolor	Poaceae (M)	45.8	53.8	39.1	56.1	50.8	57.7
Oryza sativa	Poaceae (M)	44.6	51.3	37.1	55.5	53.2	57.0
Arabidopsis thaliana	Brassicaceae (D)	39.2	42.5	32.5	55.9	51.7	55.1

[a]M monocots, D dicots
[b]Adapted from Schnable et al. (2009: Supplementary Table 5)
[c]Data from the genomic tRNA database (Chan and Lowe 2009; http://lowelab.ucsc.edu/GtRNAdb/)
[d]Data from the Silva rRNA database (Pruesse et al. 2007; http://www.arb-silva.de/)

which often have the highest GC contents of all genome components (cf. Table 14.3). Furthermore, the contrast between the GC contents of exons and introns and the corresponding secondary structures of a pre-mRNA may be involved in the regulation of pre-mRNA alternative splicing (Zhang et al. 2008).

Because genes generally form a very small portion of any eukaryotic genome, the differences in total genomic GC between eukaryotes are largely driven by the composition and abundance of non-coding DNA. In this respect, the most influential regions may be expected to be retrotransposons because they are the most abundant type of non-coding repetitive DNA in eukaryotic genomes (see Kejnovsky et al. 2012, this volume). Various classes of retrotransposons show a wide range of GC-content composition. For instance, one of the most common low-GC content groups is the short non-autonomous miniature inverted repeat transposable elements (MITEs, with typical GC contents of ~28–34%, Turcotte et al. 2001), accounting for 49% of the total number of repetitive DNA elements of rice chromosome 4 (Feng et al. 2002). On the opposite (high GC) side of GC-content variation are some of the long terminal repeat (LTR) retrotransposons, such as Huck elements, which have an average GC content slightly >60% and are the most abundant repetitive elements found in the genome of maize (Meyers et al. 2001).

14.4.1.2 Mechanisms of Structural Variation in Genomic GC Contents Between Organisms

Changes in the proportion of coding and non-coding sequences is perhaps the major reason for variation in genomic GC contents over short evolutionary time-scales, typically within genera or families. The most dramatic changes in GC content should occur in cases of the amplification or removal of repetitive DNA with either extremely low or high GC contents. For example, the genome of Oryza australiensis nearly doubled during the last 3 million years due to the amplification of three retrotransposons (RIRE1, Kangourou and Wallabi) that account for ~60% of the O. australiensis genome (Piegu et al. 2006). The GC contents of these retrotransposons are 44.6%, 50.0% and 50.9%, respectively (Noma et al. 1997; Piegu et al. 2006), which are high compared to the GC content of the whole genome (43.6%) or even the genes (45.3%) of the related Oryza sativa; thus, the final GC content of O. australiensis increased to ~45.0% (Šmarda et al., unpublished). The amplification of extremely GC-rich Huck elements in maize, forming >10% of its genome, is perhaps the major reason for the high genomic GC content in this plant (Meyers et al. 2001; SanMiguel and Vitte 2009), being among the highest known in grasses and angiosperms as a whole (see Table 14.1, Fig. 14.1). Retrotransposon amplifications and removals are the most important sources of variation in genome size (Bennetzen et al. 2005; see Leitch 2012), and this is why the prevalent amplification or removal of the same type of retrotransposon within a related species can result in a correlation between GC content and genome size, namely, in the case of extremely GC-rich or -poor retrotransposons. This correlation is frequently found in genera that have undergone a dramatic evolution in genome size (Bureš et al. 2007; Šmarda et al. 2008), and the correlation disappears among distantly related taxa with different genome organizations and distinct repetitive DNA compositions, as observed by Barow and Meister (2002) across tracheophytes.

In addition to retrotransposon dynamics, dramatic changes in the proportion of coding regions or GC-rich repetitive DNA may frequently be associated with chromosomal rearrangements following allopolyploid events (see Lysak and Schubert 2012). In wheat, for example, the success of newly formed allopolyploids may depend on the rapid elimination (within two or three generations) of non-coding DNA, which avoids potential homology with parental chromosomes (Feldman et al. 1997; Ozkan et al. 2001; Shaked et al. 2001). This mechanism may facilitate suppression of inadvertent formation of tetravalents in meiosis, which is essential for the stability of newly established polyploids (Feldman et al. 1997; Ozkan et al. 2001). Noting a general tendency for non-coding DNA to be GC-poor, this mechanism may be expected to result in genome downsizing (e.g., Ozkan et al. 2003) and overall GC content enrichment in polyploids. However, when testing this prediction in

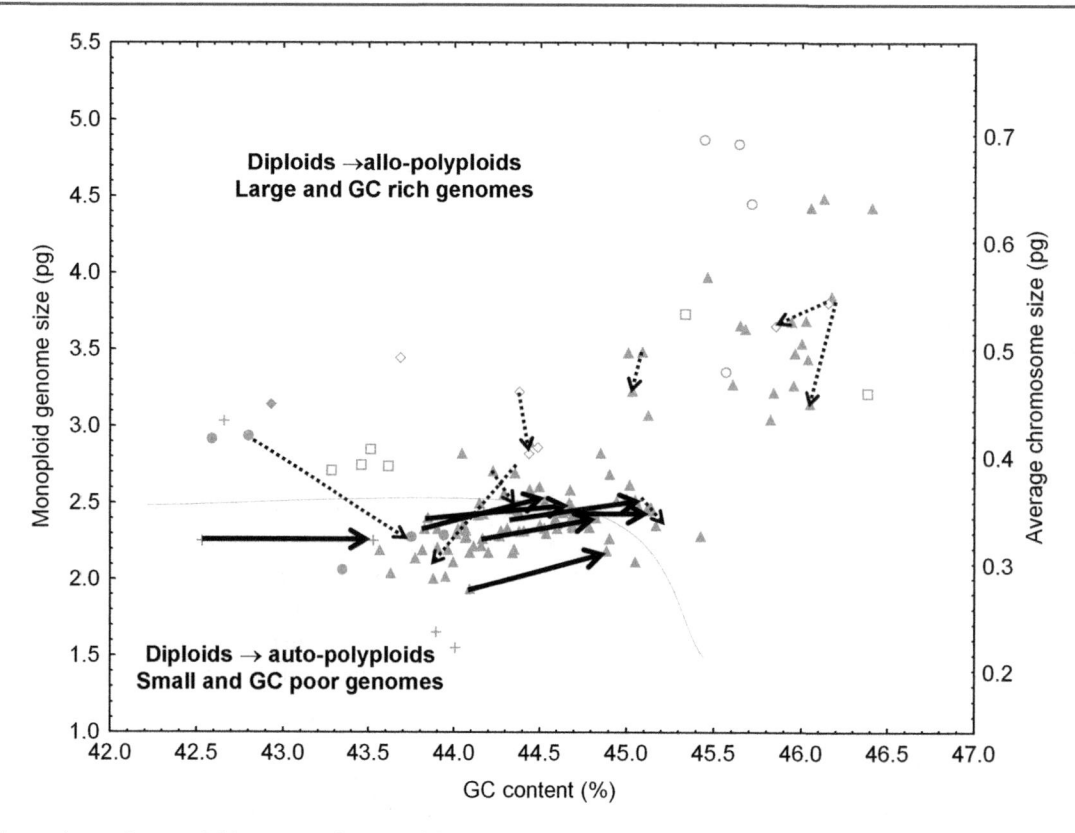

Fig. 14.4 Comparison of monoploid genome size and GC content evolution in closely related diploid-polyploid species pairs or taxa groups in *Festuca*. The *arrows* run from diploid to polyploid. A common genome downsizing with an equivocal effect on GC content is observed in allopolyploids (*interrupted arrows*; see Leitch 2012), while an unusual GC enrichment and slight increase in genome size is observed in autopolyploids (*solid arrows*). The latter seems to be best explained as the combined effect of genome downsizing and the massive proliferation GC-rich repetitive DNA (Adapted from Šmarda et al. 2008)

Festuca polyploids, it was shown that the decreases in genome size and chromosome size of putative allotetraploids were not associated with a consistent change in GC content. Instead, a remarkable increase in the GC content of putative autopolyploids was observed (Fig. 14.4). Because this increase of GC was also coupled with an increase in genome (chromosome) size,[1] it was assumed to be caused by the amplification of GC-rich retrotransposons, which seem to commonly accompany hybridization events in plants (Kawakami et al. 2010; Petit et al. 2010; see also Kejnovsky et al. 2012; Lysak and Schubert 2012).

The loss of redundant centromeres in fused chromosomes is another potential cause of changes in GC content during chromosomal rearrangements in some polyploid genomes. For example, in *Arabidopsis thaliana*, centromere regions account for approximately 17% of its 157 Mb genome (Hosouchi et al. 2002; Bennett et al. 2003) and contain a number of GC-rich repeats, making centromeres (in addition to genes) one of the most GC-rich genomic regions (Zhang

and Zhang 2004). As recently shown for some *Arabidopsis* relatives from Australia, the redundant parental centromeres may be extensively removed during genome restructuring and diploidization of the ancestral allopolyploid genome (Mandáková et al. 2010). Therefore, it should be expected that the observed reductions in chromosome number are associated with a decrease in genomic GC content, if not compensated for by the selective removal of AT-rich repetitive DNA and increased gene survival rates.

A further dynamic, abundant and GC-rich component of polyploid genomes is ribosomal DNA (rDNA). In most eukaryotes, the 5S and 45S (18-5.8-26S) rDNA subunits occur in tandem arrays, each at one to several loci. The number of units may vary from 500 to many thousands, with these copies accounting for 0.015% (*Trillium grandiflorum*) to 1.17% (*Brassica rapa*) of the genomic DNA (Ingle et al. 1975). In spite of their relatively low abundance, their possible impact on variation in genomic GC content is due to their extreme GC-richness, exceeding that of any other genomic sequence (see above and Table 14.3). Variation in rDNA copy numbers may occur between individuals of a species, but the most dramatic changes are documented during allopolyploidy because redundant and silenced rDNA copies (mostly from one of

---

[1] The relatively small increase in genome size may be due to the combined effect of genome downsizing and the massive proliferation of GC-rich retrotransposons.

the parents) may be eliminated from the genome over a relatively short evolutionary time period (Kovarik et al. 2008; Malinska et al. 2010). Similarly to the GC-rich centromeres of *Brassicaceae*, the process of rDNA elimination in allotetraploids should presumably cause a decrease in overall genomic GC content.

Analogous situations to the dynamics of centromeres and rDNA may also be expected for other chromosomal satellite components, such as GC-rich telomeres (Watson and Riha 2010; see also Siomos and Riha 2012, this volume). However, due to their relatively small proportion in the genome, their effect would be rather negligible. A minor effect on genomic GC content may be generally expected in any tandem duplication of genes or gene-rich/-poor regions due to horizontal transfer of genes from parasitic organisms or due to the incorporation of plastid and mitochondrial genome components into the nuclear genome. Repetitive and GC-poor elements are frequently found in sexual chromosomes (Kubát et al. 2008; Kejnovský et al. 2009; see also Janousek et al. 2012), and small differences in GC content may be expected to exist between plant sexes. These differences are indeed occasionally reported, in addition to variation in the genome size of some dioecious species (Marie and Brown 1993). Highly repetitive and AT-rich DNA is also typical of accessory B chromosomes (Puertas 2002; Cheng and Lin 2003), which represent another way for GC content variation to occur between individuals within a species.

14.4.2 Mutational Bias

14.4.2.1 Mutational Bias and/or Selection?

From analyzing genes from prokaryotes and eukaryotes it is clear that genomic GC content (primarily from abundant non-coding DNA in eukaryotes) is a major determinant of the pattern in which codons are used in genes, and conversely, knowledge of codon usage and gene design enables a prediction of the overall composition of non-coding and repetitive DNA segments (Knight et al. 2001; Chen et al. 2004). The close relationship between coding and non-coding DNA is also seen in the overall correlation between the GC contents in exons and their respective introns (Vinogradov 2001). These associations may most easily be explained by the existence of a global mutational bias that does not discriminate between regions of coding, non-coding and repetitive DNA. Mutational bias may couple both with selection and compositional variation that may have consistent or contradictory effects on genomic GC content. Hence, the major effect of mutational bias on genomic GC may be expected when acting over a long timeframe with a low level of selection. Breaks of some common equilibria, such as those between the sequence composition and pool of

available nucleotides or pool of tRNAs, may initiate periods of sequence adaptation, including the mutual action of mutational bias and selection (Vetsigian and Goldenfeld 2009). Hence, it is sometimes difficult to separate the effects of mutational bias and selection. Below, we summarize several known principles and theories that include mutation bias as one of the important sources of GC content change in DNA sequences.

14.4.2.2 Replication-Associated Bias and the Effect of Dominant Deoxyribonucleosides

Eukaryotic cells contain a delicate balance of minute amounts of the four deoxyribonucleoside triphosphates (dNTPs) that are sufficient for DNA replication for only a short time (Ferraro et al. 2010). The relative amounts of each dNTP are important for correct DNA synthesis, and both deficiencies and excesses of some dNTPs may result in increased mutation rates or faulty DNA repair. Experiments have shown, for instance, that an increased concentration of dGTP and dCTP during DNA replication leads to an increased misincorporation rate of C and G into the newly synthesized DNA (Meuth 1989; Martomo and Mathews 2002). This fact has given rise to the hypothesis that the relatively fast decrease of dCTP and dGTP observed during human *HeLa* cell replication (Leeds et al. 1985) may be responsible for the difference in GC content between early and late replicating regions of DNA in vertebrate genomes (Wolfe et al. 1989; Eyre-Walker and Hurst 2001). Indeed, more detailed studies have shown that GC-rich isochores generally replicate earlier compared to low-GC isochores in the human genome (Constantini and Bernardi 2008b).

During replication, there is a difference in mutation rate between the leading and lagging strands of the DNA double helix. While replication of the leading strand proceeds continuously from the $3'$ to the $5'$ end of the template, the synthesis of the lagging strand occurs via a series of Okazaki fragments and is comparatively slower. Hence, the lagging strand remains single-stranded longer, during which time it is much more prone to mutation compared to the more chemically stable double-stranded DNA (Frank and Lobry 1999; Rocha 2004). In the case of (methyl)cytosine deamination, the differences in the mutability between single-stranded and double-stranded DNA may be ~140 times (Frederico et al. 1990). While methylcytosine deamination results directly in a $CG \rightarrow AT$ substitution in the newly synthesized DNA (that cannot be recognized by DNA repair mechanisms), the pure C deamination results in a U-A mispair in the newly synthesized DNA that results in a $GC \rightarrow AT$ substitution only if not repaired prior to the next round of DNA replication. The prevalence of (met)C deaminations on the slowly synthesized lagging strand leads to a strand-specific changes in nucleotide compositions:

C-depleted (T-enriched) lagging strands and A-enriched (G depleted) leading strands, the so-called replication-associated bias.

A different pattern of strand-specific nucleotides is generally observed in genes, which have effective DNA repair mechanisms functioning during transcription (Svejstrup 2002). Transcription-coupled repair (TCR) effectively resolves all mispairings originating during DNA replication by re-synthesizing the sequence on the lagging strand using the leading strand as a template. Thus, all recognizable mutations originating on lagging strands during DNA replication are removed, and only those that occurred on the leading strand remain. This so-called transcription-coupled bias leads to a strand-specific base composition that is different from that found in non-transcribed DNA (Green et al. 2003). Transcription-coupled bias is assumed to be the major reason for the excess of T in the leading strands of DNA, which is found in genes across all eukaryotic organisms (Touchon et al. 2004).

Similar to the example with metC and C deaminations, strand asymmetries may also exist between other bases. Conspicuous variations in replication-associated strand-specific base composition are found in prokaryotic and viral genomes (Mrázek and Karlin 1998). An excess of guanine in leading strands has been observed in the genes of vertebrates and plants (opposite to the trend observed in invertebrates), and excess cytosine is common in the leading strands of GC-rich genes in grasses (Touchon et al. 2004; Tatarinova et al. 2010). This variation in strand-specific biases may account for differences in mechanisms of transcription-coupled repair, but there is increasing opinion that this is a consequence of multiple mechanisms acting on gene sequences, including perhaps some form of selection (Frank and Lobry 1999; Rocha et al. 2006).

A likely explanation for the variation in the skew and direction of replication-related mutational biases may be that these are consequences of the more general processes of co-adaptation of sequence composition and pools of available dNTPs (Vetsigian and Goldenfeld 2009). In contrast to the conventional view of the independent actions of mutation and selection on sequence composition, this view suggests that both are mutually affected by the concentration of available dNTPs, and the availability of dNTPs may be conversely driven by changes in the sequence composition of genes involved in dNTP synthesis (Fig. 14.5). Dominance of a particular dNTP would lead to its increased misincorporation into the DNA, and this misincorporation may be supported by selection because this type of mutation would (perhaps) leads to easier and faster replication. This cyclic scenario could lead to extreme cases of positive selection toward some bases (e.g., GC) until an equilibrium is achieved between the sequence composition and resource pool. In principle, even a small imbalance of nucleotides given by changes in dNTP metabolism (e.g., driven by intrinsic metabolic or extrinsic environment-associated reasons) may initiate a new equilibrium, the constitution of which would be associated with specific patterns of nucleotide substitutions and changes in base composition. This principle may be applied not only to DNA replication but may also be valid for transcription and translation (i.e., for co-adaptation of codon usage and the pool of available dNTPs or co-adaptation of codon usage with the pool of available tRNAs, respectively; Vetsigian and Goldenfeld 2009). However, the effect of transcription and translation on genomic GC composition may be negligible compared that of DNA replication.

The hypothesis of Vetsigian and Goldenfeld (2009) is difficult to test directly because measurements of nucleotide concentrations suffer from some uncertainty (Ferraro et al. 2010) and are largely lacking for germline cells that only enable mutations to be transmitted to the next generation in animals and sexually reproducing plants. An indirect test

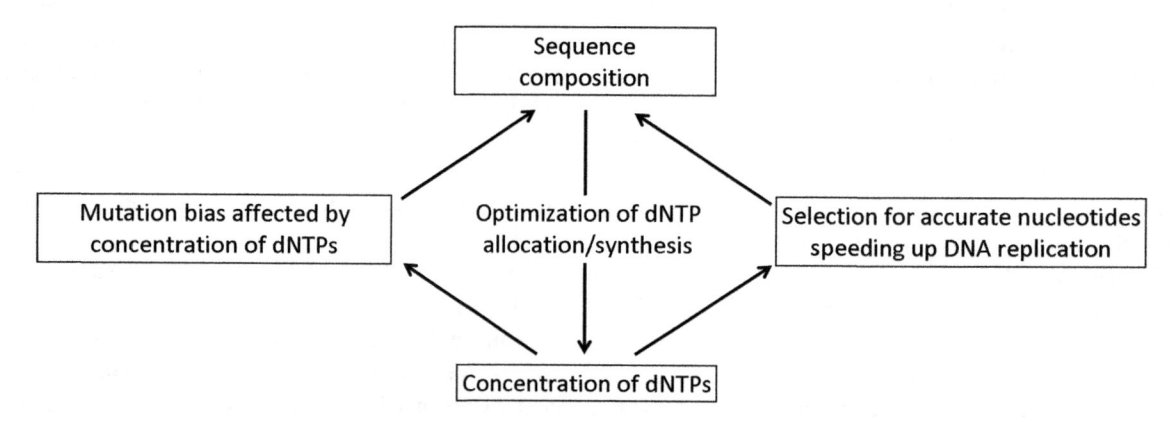

Fig. 14.5 Scheme of co-adaptation between the pool of deoxyribonucleosides (dNTPs) and sequence composition based on the hypothesis by Vetsigian and Goldenfeld (2009)

may provide a comparison of the skew in nucleotide composition (naturally associated with that kind of mutational bias) and speed of DNA transcription. In the case of an imbalance in the proportion of dNTPs, the lagging strand would perhaps be selected to contain the dominant nucleotide in order to not slow DNA replication. Therefore, higher skews in nucleotide composition can be expect to be found in organisms with short-life spans or extremely large genome sizes for which the speed of DNA replication is of critical importance (Bennett 1987; see also Leitch 2012). However, testing these relationships is still challenging.

14.4.2.3 Nucleotide Economy Hypothesis

Due to the common biochemical machinery that exists within cells, any parasitic or symbiotic genomes may be expected to mimic the codon usage and genome composition of the host organism (Lawrence and Ochman 1998; Greenbaum et al. 2009). Contrary to this assumption, Rocha and Danchin (2002) have documented that the genomic GC content of obligatory pathogenic and symbiotic bacteria generally is low compared to other free-living bacteria. A similar relationship has also been found between nuclear genomes and plastomes, which are generally GC-poor compared to nuclear DNA (Liu and Xue 2005; Smith 2009; see Wolf 2012, this volume). Rocha and Danchin (2002) proposed that this might result from competition for metabolic resources, noting that the synthesis of dGTP and dCTP is generally more energetically expensive compared to that of dATP and dTTP (e.g., cytidine nucleotides formed from the transamination of UTP by CTP synthetase require one ATP; among purines, the synthesis of GMP requires one more NAD + than AMP).

The argument by Rocha and Danchin (2002) provides one reason why some disproportion in the relative availability of dNTPs may exist in a cell. Indeed, detailed measurements of the proportion of dNTPs in human cells indicate that dATP and dTTP are much more abundant than dCTP and dGTP throughout the entire cell cycle (Ferraro et al. 2010). This dNTP imbalance may be important for explaining the existence of a continuous mutational bias, and it may lead to a number of testable ecological (selectionist) hypotheses related to resource availability in the environment.

One explanation for the lower GC content in pathogenic and symbiotic bacteria may be that parasites exploiting a host at a lower cost with easily available resources would be selectively favored (Rocha and Danchin 2002). A mutational explanation would indicate that the small genomes of pathogenic and symbiotic bacteria have lost many of their DNA repair systems, and their genomes are thus more prone to mutational biases (Rocha and Danchin 2002) that are assumed to be generally directed more toward the most frequent (lower cost) AT nucleotides. This latter view was recently supported by the detailed experimental study of spontaneous mutation in the bacterium *Salmonella typhimurium* that demonstrated that strains defective in their DNA repair mechanisms show a preference for GC → TA mutations (91% of all observed base pair substitutions; Lind and Andersson 2008).

Based on the nucleotide resource hypothesis, fast growing organisms are predicted to be biased toward easily synthesized AT nucleotides. Indeed, fast growing bacteria usually posses GC contents <50%, though this fact may also be due to the close relationship of fast growth with a small tRNA diversity that is sufficient for the translation of GC-poor but not GC-rich genes (Satapathy et al. 2010). It may be also expected that high GC contents would be avoided in extremely large genomes in order not to burden the transcription machinery with the requirement of expensive GC nucleotides. However, this is certainly not the case for bacteria. In bacteria, genome size is generally positively correlated with genomic GC content (Musto et al. 2006; Mann and Phoebe-Chen 2010), perhaps reflecting other types of structural or functional associations.

The nucleotide economy hypothesis proposed based on the study of bacterial and phage genomes raises some interesting ideas that may also be applicable to explain the variation in the genomic GC content of eukaryotes. Based on this hypothesis, a decrease in genomic GC content (compared to close relatives) may be expected in parasitic and fast growing species, species with extremely large genomes or species living in various resource-limited environments (e.g., nutrients, light or competitors). Only recently GC content was shown to have unimodal relationship with genome size in geophytic plants (Veselý et al. 2012). The positive correlation between both parameters found within small genomed species agree with the trend in bacteria and within some plant genera (Šmarda et al. 2008, Bureš and Šmarda unpublished data), while the negative correlation found in species with large and extreme genome sizes is in agreement with the nucleotide economy hypothesis (although some alternative explanations are also possible; Veselý et al. 2012). The understanding of genomic background for this pattern as well as studies on the ecological significance of GC contents, however, still remain a challenge.

14.4.2.4 Methylcytosine Mutability

The high mutability of metC has been assumed to be responsible for the origin of AT-rich genomes and genomic regions (Bird 1980). Vertebrate genomes are highly depleted in methylatable CpGs, which occur at a frequency approximately four times less than the frequency of alternative non-methylatable GpC dinucleotides. However, studies have shown: (1) that the decrease of CpGs in vertebrates does not correspond with a respective increase in TpG dinucleotides and (2) that the concentration of TpG dinucleotides is fairly similar in animal genomes with either high or low levels of methylatation (i.e., human and *Drosophila melanogaster*, respectively). Hence, the depletion of CpGs in vertebrates likely occurs for a different primary reason (Jabbari and Bernardi 2004). Due to the role of CpG dinucleotides in the formation and positioning of nucleosomes, the reason for CpG depletion may be related to chromatin-mediated regulation of gene expression (Vinogradov 2005; Bettecken and Trifonov 2009; Bai and Morozov 2010).

An effect of metC mutability has also been observed in *Arabidopsis thaliana*, where GC → AT transitions form the majority of spontaneous base substitutions (Ossowski et al. 2010). Here, however, the effect of metC mutability seems less apparent compared to the number of GC → AT transitions that originated from UV-induced mutagenesis (Ossowski et al. 2010). Calculating that 11 of the 71 total GC → AT transitions that accumulated in five different *Arabidopsis* lines after 30 generations are caused with metC mutability, (cf. Ossowski et al. 2010), a 1% decrease in the GC content of the 157-Mb *Arabidopsis* genome would take ~21 million years if exclusively the result of metC deaminations at the current mutation rate. Including reverse AT → GC transitions in the calculation, metC deaminations would likely not be of great significance in plants, even over larger evolutionary timescales. This view is supported by the fact that despite the long history of intensive methylation in plant genomes (Jeltsch 2010), they are not significantly depleted in methylatable CpGs (Schwartz et al. 1962).

14.4.2.5 Heteroduplex Base Mismatch Repair (HBMR) and Recombination Rate

During homologous recombination a heteroduplex DNA is regularly formed which can lead to a base mismatch if the heteroduplex extends across a heterozygous site. This mismatch is sometimes repaired, but this process tends to be biased toward GC pairs (Brown and Jiricny 1988). Therefore, it is possible that the extent of the recombination rate of a particular DNA sequence is responsible for a locally increased variation in GC content, as well as for differences in genomic GC content between organisms with different overall rates of recombination (Eyre-Walker and Hurst 2001). This is well documented within mammalian genomes where GC content of a particular chromosome is positively correlated with its recombination rate (Eyre-Walker 1993; Romiguier et al. 2010), and the local recombination rate on/within chromosomes positively correlates with the local GC content (Fullerton et al. 2001; Meunier and Duret 2004; Duret and Arndt 2008). However, HBMR is unlikely to explain the gross variation in genomic GC content across vertebrates and other animals at all. Here, highest GC contents are in low recombining genomes of caudate amphibians

(Vinogradov 1998) and the highest recombination rates are in GC-poor insects and urochordates (Lattorff and Moritz 2008).

A relationship between recombination rate and GC content has also been observed in *Arabidopsis*, but contrary to mammals, the relationship is negative, with higher recombination rates found in AT-rich regions irrespective of whether they are comprised preferentially of genes or retrotransposons (Drouaud et al. 2006). A different situation may be found in the genomes of grasses (maize and wheat), for which recombination rates generally correlate with gene density and generally increase with increasing distance from the centromere (Akhunov et al. 2003; Li et al. 2007). However, a recent analysis of the maize genome indicates that gene density itself cannot fully explain the distribution of recombination sites (Schnable et al. 2009) and that the recombination rate is epigenetically controlled, with an increased rate associated with the epigenetic marks for open chromatin (Liu et al. 2009). Across diverse plant genomes, the overall recombination rate is closely related to phylogeny and life-form, with conifers exhibiting lower recombination rates compared to angiosperms. Within angiosperms, the highest rates are found in trees, while lower rates are typical of short-lived herbs and shrubs (Jaramillo-Correa et al. 2010). These life-form traits, as well as the core data on genome-wide recombination rates by Jaramillo-Correa et al. (2010), are not in correspondence/correlation with the respective GC contents of plants. As noted by Tatarinova et al. (2010), a greater effect of HBMR may be expected in more heterogeneous genomes that have arisen from outcrossing compared to self-pollinating or asexually reproducing taxa. Indeed, a comparison of the orthologous genes of outcrossing *Raphanus sativus*, *Brassica rapa* and *B. napus* with self-pollinating *Arabidopsis* indicate that the gene orthologs of outcrossing species tend to be slightly more GC-rich in redundant codon sites compared to *Arabidopsis* (Tatarinova et al. 2010). However, a wider sampling is required to verify this hypothesis. Even if valid, the overall impact on total genomic GC content would perhaps be negligible.

Similar to animals, recent data on the factors controlling recombination across plants indicate that HBMR is unlikely to be the primary force driving the gross variation in genomic GC content. Nevertheless, HBMR (mostly discussed in relation to biased gene conversion in the literature) may be still viewed as an important mechanism that participates in the modulation of DNA sequence GC composition in gene-rich and, thus, highly recombining regions. At the same time, the GC-biased nature of HBMR seems to be an important force counteracting the common AT-prone effect of mutational biases acting on GC bases (i.e., deaminations and oxidative damage), maintaining the GC level of genes at a sufficiently high level.

14.4.3 Selection

14.4.3.1 Evidence of Selection

Although the correlation between the GC compositions of coding DNA and non-coding DNA (see Sect. 14.4.2.1) is tight in the case of prokaryotes, significant deviation has been observed in the case of the functional compartmentalization of genomes, namely, that of homeotherm (warm-blooded) vertebrates (Chen et al. 2004). For this group, genomes are organized into several classes of megabase-sized DNA regions (called isochores) with surprisingly homogeneous GC contents (Bernardi et al. 1985; Bernardi 2000a, 2007). In this case of genome organization, the similarity of coding and non-coding DNA was only found at a local scale, i.e., by comparing the GC contents of genes and their flanking regions (Elhaik et al. 2009).

A typical genomic feature of both evolutionary lineages of homeotherms (mammals and birds) is the presence of extremely GC-rich isochores (Bernardi 2007; Constantini et al. 2009). These have developed from the comparatively GC-poor regions of their ancestral genomes, leaving other ancestral genome regions principally unchanged (Bernardi 2007). This GC content increase has occurred in both genes and large gene-flanking regions, and this non-random association is the key determinant of isochore organized genomes (Eyre-Walker and Hurst 2001; Elhaik et al. 2009). GC-rich isochores share several properties that differentiate them from the remaining portion of the genome (GC-poor isochores). Among others, they are early replicating, highly recombinogenic and contain a majority of genes that are highly expressed and have short introns (Bernardi 2007: Fig. 14.2). Evidence that the origins of GC-rich isochores in homeotherm vertebrates are associated with selection on specific gene and genomic functions related to increased body temperature has been summarized by Bernardi (2007).

An isochore structure has also been proposed to exist in plants (Salinas et al. 1988). Indeed, three classes of isochores (centromeric, CG-rich and AT-rich) that also showed some degree of compositional difference were defined in *Arabidopsis* (Zhang and Zhang 2004). However, these isochores are not as well defined as those of vertebrate genomes, and in general, the major determinant of the GC content seems to be the concentration of genes or specific GC/AT-rich repeats. Nevertheless, an analogous situation to vertebrates is observed in the genomes of grasses (Poaceae). Grass genomes are composed of a mixture of GC-standard (compared to *Arabidopsis*) and extremely GC-rich genes (see Sect. 14.4.3.2), making the overall GC content of coding DNA high compared to other angiosperms (Table 14.2). While an intragenomic divergence in the GC content of genes may arise due to their location in isochores differing in GC content in vertebrates, grass genes that differ in their GC richness do not concentrate in certain specific regions and are, for the most part, evenly distributed across the genome (Tatarinova et al. 2010). Consequently, the GC contents of genes do not correlate with the GC contents of their corresponding flanking regions in grasses (Tatarinova et al. 2010), in contrast to homeotherm vertebrates (Elhaik et al. 2009). The distribution pattern of GC-rich grass genes also indicates that they are not a remnant of ancient hybridization events or a product of a genome-wide mutational bias, but that they perhaps originated from the selection for some, still unrecognized or hypothesized, function.

A widely discussed hypothesis is the idea that the evolution and origin of GC-rich genes (or genomes as a whole) is related to selection for increased thermal stability. This is evidenced by: (1) the higher thermal stability of GC base pairs and (2) the compositional similarly between GC-rich genes in thermophilic bacteria, genes in GC-rich isochores of vertebrates and GC-rich genes of grasses. In an analysis of *Oryza sativa*, Tatarinova et al. (2010) also found a possible association of GC-rich genes with improved regulation of complex genomes and stress tolerance.

Genes form only a minor proportion of large eukaryotic genomes. Therefore, the occurrence of a subgroup of GC-rich genes itself cannot be expected to cause dramatic changes in the total genomic GC content. In spite of this simple deduction, large grass genomes distinctly belong to the most GC-rich genomes among seed plants. Hence, their high genomic GC content must be due to the high GC content of their repetitive DNA, as evidenced with high GC content of some grass retrotransposons (e.g., in maize; Meyers et al. 2001; see above) or higher GC of grass microsatellites (Crane 2007). Still, it is unclear whether selection for high GC-richness of some genes (Shi et al. 2007; Guo et al. 2007) is coupled to selection for the composition of non-coding DNA, and vice versa. In addition to selection, it remains a possibility that the correspondence between base compositions of coding and non-coding DNA has also resulted from neutral mutational processes (cf. Chen et al. 2004).

14.4.3.2 Divergence and Structural Heterogeneity in the GC Content of Genes in Grasses

Even the early studies of GC content in grass genomes determined with melting profiles, CsCl buoyant densities and nuclear sequences indicated that the internal structure of grass genomes is extremely variable and can frequently contain large portions of GC-rich compartments that mostly represent coding genes (King and Ingrouille 1987; Salinas et al. 1988; Matassi et al. 1989; Carels et al. 1998). While the genes of dicots usually vary within a relatively narrow GC range and only rarely exceed 65% (as in *Arabidopsis*), grass genes display a wider variation in GC content. This is caused by the presence of extremely GC-rich genes (~70% or

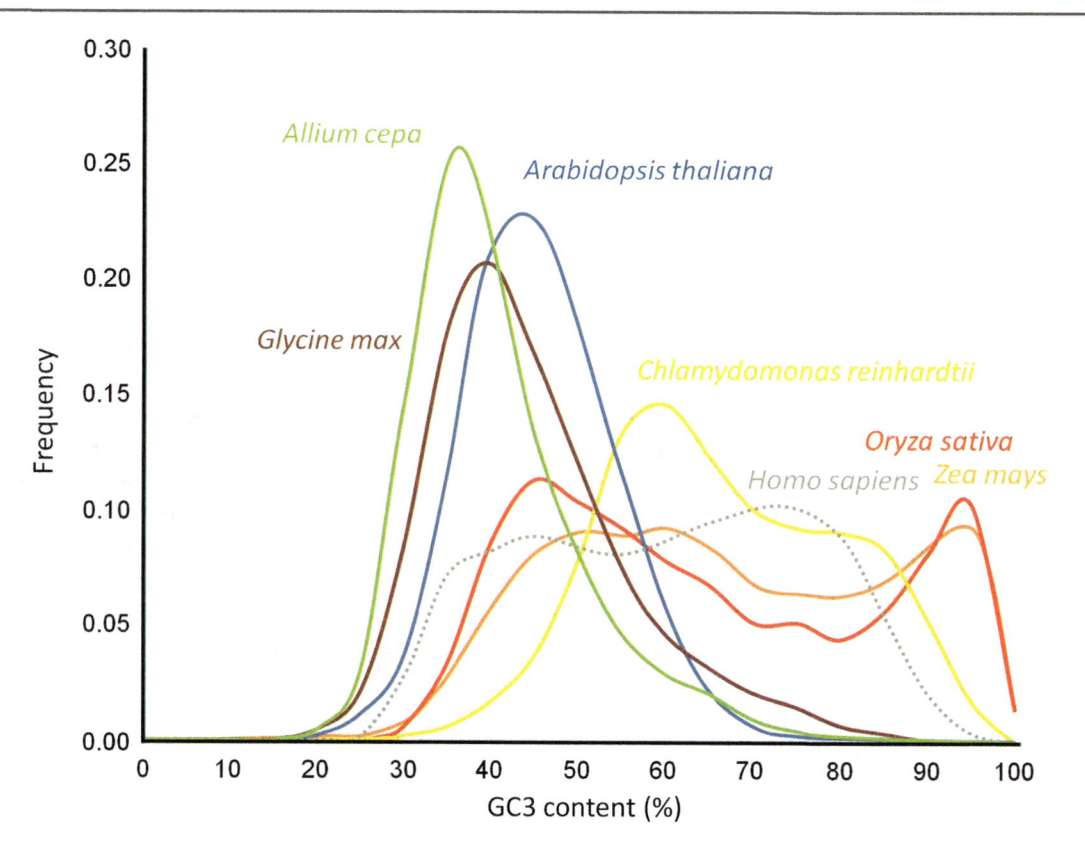

Fig. 14.6 Profile of the distribution of the GC content of the third codon position of genes in some low-GC monocots (*Allium cepa*), low-GC dicots (*Arabidopsis thaliana*, *Glycine max*), high-GC algae (*Chlamydomonas reinhardtii*), the isochore-structured genome of humans and typical GC-rich genomes of grasses (*Oryza sativa, Zea mays*) (Redrawn from Tatarinova et al. 2010)

more), causing a bimodality in GC content among genes (Carels et al. 1998; Carels and Bernardi 2000; Guo et al. 2007). This effect is most pronounced at the third codon position (GC3). In GC-rich genes of rice, the GC3 content regularly exceeds 80% (for approximately one third of rice genes) with most possessing a GC3 content of ~95% (Tatarinova et al. 2010; Fig. 14.6). Although GC-rich genes are more frequent in functional groups of genes responsible for basic metabolic functions and stress responses (Tatarinova et al. 2010), they do not form a unique functional group, and they frequently only represent functional variants of standard genes (i.e., functional paralogs; Zhang et al. 2001; Guo et al. 2007). Hence, high variation or even bimodality in the distribution of the GC content of genes in grasses may be observed despite a 100% protein BLAST similarity (Fig. 14.6).

The GC-rich genes in grasses are generally shorter because they contain either no introns or a limited number of introns (Carels and Bernardi 2000; Guo et al. 2007; Alexandrov et al. 2009; Tatarinova et al. 2010). Perhaps the most peculiar feature of GC-rich grass genes is the existence of a gradient in GC content. In contrast to GC-standard genes of grasses and other plants, in which the GC content is nearly identical along the entire gene length, the 5′ end of GC-rich genes is typically ~25% GC-richer compared

to the 3′ ends of the gene (Yu et al. 2002; Wong et al. 2002; Kuhl et al. 2004; Guo et al. 2007; Fig. 14.7). This increase is observed both in exons and introns, though the increase in GC of introns is generally less dramatic compared to that of exons (Carels and Bernardi 2000; Guo et al. 2007; Shi et al. 2007; Tatarinova et al. 2010). This leads to the speculation that GC-rich grass genes have originated from the combined action of natural selection and some type of transcriptionally coupled mutational bias (Guo et al. 2007; Shi et al. 2007; Liu et al. 2010).

The compositional differences of GC-rich grass genes are also reflected in the substantial codon usage bias and the changes in usage of amino acids. The spectrum of codons used in GC-rich genes is restricted, typically ranging from 25 to 45 codons (of 61 possible), indicating a strong codon usage bias with most amino acids being specified by only 1 or 2 codons (Mukhopadhyay et al. 2007; Liu et al. 2010). Beyond the conspicuous tendency for choosing a G or C for the third redundant codon position and the restricted use of AT-rich synonymous codons, the 5′ ends of GC-rich grass genes are also frequently modified with non-synonymous mutations toward a G or C in the first two positions that lead to a change in amino-acid composition. Specifically, the 5′ starting regions are enriched in alanine (GCN codons) at the expense of serine (UCN codons), which dominates the

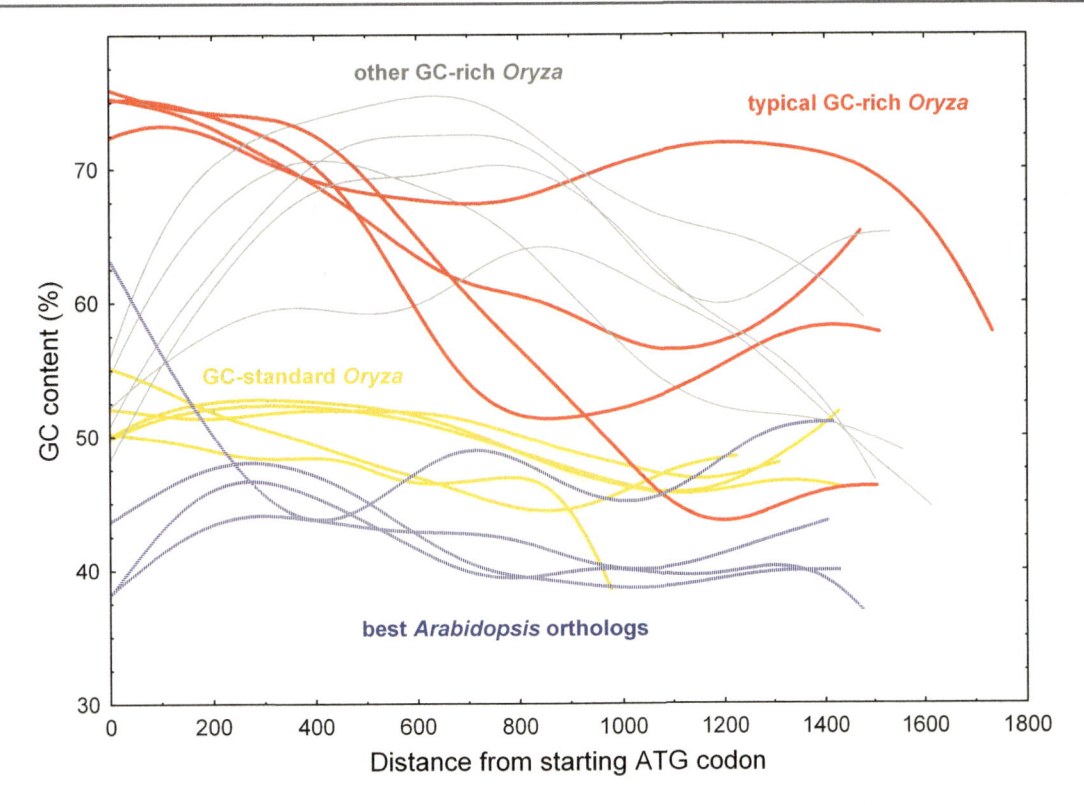

Fig. 14.7 The GC content profiles of CDS sequences of paralogous genes from *Oryza sativa* (paralogous group 3630 encoding a cytochrome P450 protein; Lin et al. 2008) and their best orthologs in *Arabidopsis. Lines* indicate GC contents obtained from distance weighted least square fitting (stiffness = 0.25) of sequence data from the Rice Genome Annotation Project and the Arabidopsis Information Resource (Ouyang et al. 2008; Swarbreck et al. 2008)

starting regions of eudicot plants (Wong et al. 2002). There is also a decreasing trend in the concentration of arginine, and an overall decrease has been observed in the usage of leucine in GC-rich genes (Wong et al. 2002; Guo et al. 2007). This amino acid replacement is generally assumed to increase the thermal stability of the resulting proteins (see Sect. 14.4.3.3 for details) and has been independently observed in the genes from GC-rich isochores in the human genome (Constantini and Bernardi 2008a) and in the genomes of some thermotolerant bacteria (Nishio et al. 2003).

Another interesting property of GC-rich grass genes is a clear GC skew toward the preference of C in the leading strand of DNA that prevails over the entire gene length. In the GC-standard grass genes and genes of other plants, this tendency is usually observed for only approximately the first 50 codons, while in the remaining portion of the sequence, there is a slight tendency to prefer G over C (Tatarinova et al. 2010: Fig. 14.6). This difference goes hand in hand with the clear difference observed between GC-rich and standard grass genes in the concentration of methylatable CpG-dinucleotides. The difference is not only a simple passive consequence of an increased GC content and a GC skew but is seemingly also as a result of positive selection for CpG because the concentrations of other C-methylatable

motifs (e.g., CWG (W = A or T)) do not differ between GC-rich and GC-poor genes (Tatarinova et al. 2010: Figs. 14.8–14.10). Noting the important functions of CpG dinucleotides in plant methylation and nucleosome positioning (see also Richards et al. 2012; Zhang 2012; this volume), this invokes questions as to their specific role in the regulation of gene expression and chromatin structure (see Sect. 14.4.3.4). This view is further supported by the fact that compared to GC-standard paralogs, GC-rich genes have a higher number of regulatory TATA boxes in their promoter regions, and in general, they display a wider tissue-specific variation in expression (Tatarinova et al. 2010). In addition to the existence of a gradient in GC composition, the enrichment in CpG dinucleotides in GC-rich grass genes makes them structurally different from the GC-rich genes in vertebrates, which are usually highly depleted of CpG dinucleotides (see Sect. 14.4.2.4).

Grass-like genes have also been found to a limited extent in some (but not all) other monocots, such as in banana and some Zingiberales (Lescot et al. 2008; Tatarinova et al. 2010). A divergence in GC content between gene paralogs has also been observed in some gymnosperms and ferns (Jansson et al. 1994), and a clear bimodality in the GC contents of genes is also evident in the genome of the GC-rich alga *Chlamydomonas reinhardtii* (Merchant et al. 2007;

Fig. 14.6). A similar distribution in gene GC content has not yet been found in dicots. Although exactly comparable data are still lacking, the phylogenetic patterns in the occurrence of GC-rich genes agree well with the observed ranges of variation in genomic GC contents (i.e., a broad range is found in algae, gymnosperm, ferns, and monocots, but a generally narrower range in dicots (Fig. 14.1)). This supports the view that the presence of GC-rich genes may coincide with selection or neutral processes acting on the global genomic level. Several hypotheses that might explain the incidence of GC-rich genes and/or GC-rich genomes are discussed below.

14.4.3.3 Thermostability Hypothesis

As discussed earlier, the most remarkable difference between GC and AT base pairs is their difference in thermal stability, which is higher in GC base pairs and lower in AT base pairs.

Thermal stability is critical for the secondary structures of structural RNA molecules (rRNA and tRNA). A close positive relationship was observed between the optimum growth temperature and GC contents of the stem regions of ribosomal and transfer RNAs in prokaryotes (Galtier and Lobry 1997; Hurst and Merchant 2001), and between the stem GC content of 18s RNA and body temperature in vertebrates (Varriale et al. 2008). However, the question still remains as to whether such a general relationship with body (or environment) temperature may be found when overall genomic GC content is considered. At least in the case of simple prokaryotic genomes, genomic GC content has indeed been shown to be one of the important factors determining optimum growth temperatures. However, because of the complex determination of thermostability (Barton 2005; Trivedi et al. 2005) and confounding effect of other life-history traits

and functional constraints (Naya et al. 2002; Rocha and Danchin 2002; Mann and Phoebe-Chen 2010), correlations between genomic GC contents and the optimum growth temperatures of prokaryotes are mainly observed when comparing closely related taxa with a relatively similar genome structure and organization, such as among species within families (Musto et al. 2004, 2006). An illustrative example of an approximately 10% increase in genomic GC associated with increase of thermotolerance provides Nishio et al. (2003) between the two closely related species of *Corynebacterium*. Here, the increase of GC was associated with the preferential usage of GC rich synonymous codons, and GC biased incidence of nonsynonymous mutations leading to amino-acid replacements in favor of charged and hydrophobic amino acids (Nishio et al. 2003) that result in an increase in protein hydrophobicity (Sterpone et al. 2010; Vieille and Zeikus 2001). A close positive correlation between the expected protein hydrophobicity and the GC content of the corresponding DNA sequences has been observed universally across prokaryotes and eukaryotes (D'Onofrio et al. 1999; Cruveiller et al. 1999; Jabbari et al. 2003; Constantini and Bernardi 2008a; Bucciarelli et al. 2009), indicating that an increase in GC content may indeed be a common signature for increased thermotolerance. However, the question still remain to what extent the in silico-calculated hydrophobicity reflects the actual in vitro thermotolerance of corresponding proteins. Moreover, it is also not clear if the calculated protein hydrophobicity may simply be a passive consequence of GC enrichment driven by selection on different genomic property and whether a very GC-rich protein can be predicted to be less hydrophobic compared to a paralogous or orthologous GC-poor gene? In the case of true positive selection for protein hydrophobicity, the same proportional increase in the GC content of a protein that results from mutation that is randomly

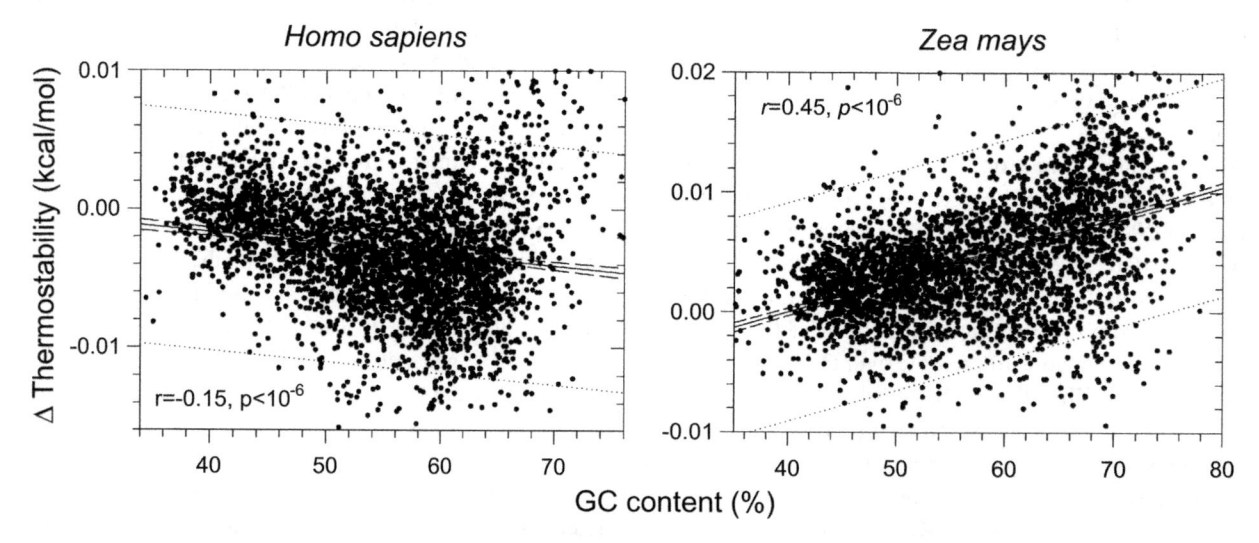

Fig. 14.8 The correlation between the relative thermostability and GC content of gene coding sequences from maize and humans. The relative thermostability is calculated as the deviation between the actual

thermostability of a sequence and the mean thermostability of ten randomly generated sequences with the same amino acid composition and GC content (Adapted from Vinogradov 2003) applied to the original ancestral GC-poor sequence (preserving the proportion of synonymous and nonsynonymous mutations) should clearly result in a lower or equal hydrophobicity (but generally not higher) to the observed protein. However, a similar test on protein hydrophobicities is still lacking to our knowledge.

Comparing several properties of DNA sequences of the human genome in a similar way, Vinogradov (2003) demonstrated that while the overall thermal stability of the DNA calculated from dinucleotide-free melting energies necessarily increases with increasing GC content, this increase is substantially suppressed in GC-rich genes (Fig. 14.8). For many human GC-rich genes, the actual thermal stability is even lower compared to randomly generated amino acid homologs (preserving the same GC content), indicating that they are not likely to be selected directly for higher thermostability. Instead, several correlations were observed between gene GC content and DNA bendability, the ability for a B–Z transition, DNA curvature and nucleosome formation potential (Vinogradov 2003, 2005). Together with the later studies on single nucleotide polymorphism, these results lead to the conclusion that the GC enrichment of vertebrate genes and genomes is a consequence of selection for improved genomic structure and organization (i.e., chromatin conformation and nucleosome positioning; Vinogradov 2003, 2005; Pozzoli et al. 2008; see Sect. 14.4.3.4), and selection only passively affects the calculated thermal stability of DNA sequences.

In contrast to the human genome, thermostability tests with maize genes demonstrate that grass genes may have higher thermal stabilities. In maize, the thermostability of the actual gene sequence regularly exceeds the thermostability of other possible gene variants, and this increase is prominent in the case of GC-rich genes (Vinogradov 2003; Fig. 14.8). This finding indicates that different forces may have shaped the GC contents of grass and vertebrate genes, which may be reflected in the structural difference of GC-rich genes in both groups (e.g., the existence of a GC content gradient and high CpG dinucleotide content in GC-rich grass genes, Sect. 14.4.3.4). Unlike the situation for vertebrates, studying the effect of the overall genome structure and base composition of plants on the ecology and distribution of plant species still remains challenging.

Recent analyses of GC-rich genes and their GC-standard paralogs in rice indicate that the former have increased stability of secondary mRNA structures (Mukhopadhyay et al. 2007). Indeed, the stability of the mRNA secondary structure seems to be critical for effective protein translation. Across all organisms (from bacteria to eukaryotes), the actual synonymous codon usage in expressed genes results in higher mRNA stability than would be the situation when synonymous codons are used by chance (Gu et al. 2010). However, it remains unclear as to what extent the contemporary *in silico* calculations of the stability of secondary

mRNA structures (highly correlated with the overall GC content) relate to the actual longevity of mRNA in the cytosol (Biro 2008). The question also remains open as to what extent changes in local mRNA structure and its stability may play in their targeting by regulatory molecules (Stalder and Mühlemann 2008; Carthew and Sontheimer 2009), which could be at the core of selection instead of simply mRNA thermal stability. The extent to which mRNA stability is conditioned by ecology has only been tested in bacteria. A comparison of the extremophilous *Thermus thermophilus* (GC = 69.4%, optimum temperatute [T opt]= 73°C) with the related mezophilous *Deinococcus radiodurans* (GC = 66.6%, T opt = 30°C) indeed indicated that an increased GC content and high temperature tolerance were associated with increased stability of mRNA (Basak et al. 2010). However, this increase in mRNA stability was not found for the entire length of the mRNA sequence, and certain regions were documented to be selected rather for translation efficiency, which is likewise assumed to be critical for proper genome functioning at higher temperatures (Basak et al. 2010).

14.4.3.4 Improved Regulation and Stress Response

Although increases in the GC content in grass genes and their genomes seem to coincide with selection for higher thermal stability, it is not clear how a genome might profit from the existence of two different, thermally stabile classes of genes. The extreme GC-richness of some grass genes is also associated with other peculiar characteristics (e.g., low numbers of introns, an increased frequency of TATA boxes in promoter regions, an increased CpG richness and a clear GC skew) which indicate that increased thermostability may be only one of a complex set of adaptations that follow some common principle. Below, it is argued that this principle may be enhanced ability to fine-tune gene and genome regulation, enabling an organism to better respond to external stress stimuli.

Important information on the function of GC-rich genes may be found by observing their expression patterns. The short and intron-poor GC-rich genes of rice were observed not to differ from standard genes in their overall level of expression (Mukhopadhyay et al. 2007), but importantly, their expression was more variable depending on the tissue type and ontogenetic stage (Tatarinova et al. 2010). A similar trend for more variable expression of short genes has also been observed in bacteria and humans (Jeffares et al. 2008; Vinogradov 2004). It is assumed that this gene design may result from transcription economy. Expression of short genes is less expensive and faster, enabling the rapid adjustment of an organism to changing environmental conditions or conditions of resource limitation (Jeffares et al. 2008). In addition to the economical reasons and possible responses to resource limitation, a less complex gene structure with limited numbers or the absence of introns is perhaps also less prone to mRNA splicing errors that may occur under extreme stress conditions.

The GC-rich grass genes are usually related to basic metabolic processes (e.g., histone, ribosomal and chlorophyll a-b binding proteins; electron transport and energy pathways; and transcription and signal transduction) and abiotic or biotic stress responses (Tatarinova et al. 2010). Indirect evidence of a key role for GC-rich genes in stress responses also comes from the fact that their promoter regions often contain TATA boxes. In genomes, TATA boxes are important gene regulation sequences, and in humans, TATA-containing genes are often highly regulated by biotic or stress stimuli (Yang et al. 2007). Similar to rice, human TATA-containing genes are also shorter and show highly variable expression patterns (Moshonov et al. 2008; Troukhan et al. 2009), indicating that both humans and rice may share a common regulatory mechanisms that is somehow related to their physical structure and organization.

Another peculiar regulation function of GC-rich grass genes may be associated with their increased content of methylatable CpG dinucleotides. Two detailed methylation studies of rice genomes were recently published by Lu et al. (2010) and Yan et al. (2010). Unfortunately, neither considered the possible differences in methylation patterns between standard and GC-rich genes in detail. In general, the methylation pattern of the rice genome is similar to that of *Arabidopsis* (Cokus et al. 2008; Lister et al. 2008) and contains highly methylated heterochromatic regions (Zhang 2008, 2012, this volume; Yan et al. 2010). Similar to human genes, *Arabidopsis* and rice genes display hypomethylation in the promoter and 5′ regions of genes, keeping them free for binding by regulatory molecules. In rice, but generally not in *Arabidopsis*, the degree of methylation increases toward the 3′ end of the gene (Yan et al. 2010). This trend may correspond to the peculiar GC-skew and overall higher GC content of grass genes. However, compared to the common expectation, these data indicate that the 5′ regions of GC-rich genes are enriched in cytosine, and the methylatable CpG dinucleotides remain perhaps mostly unmethylated. Hence, their function may be assumed to be different from providing proper substrates for methyltransferases and for methylation-driven regulation of genes. One hypothesis is that an increased concentration of CpG dinucleotides and the peculiar GC skew of GC-rich grass genes may act in nucleosome positioning and the chromatin-driven regulation of gene expression.

The effect of GC richness on gene regulation is observed in both the sequences of genes and in non-coding DNA. For example, in a recent experiment, Jia et al. (2010) mounted ~1-kb long, extremely GC-rich non-coding fragments (both naturally occurring and synthetic) in the flanking regions of 20 different eukaryotic genes (either before the 5′ or after the 3′ end). Such prepared DNA fragments were then transformed into chicken and human cell lines. This resulted in a substantial increase in the expression levels of all genes (mostly 5–100-fold), indicating that GC-rich non-coding DNA may serve as a general chromatin opening element and that the chromatin-based gene expression of eukaryotic organisms may be (to some extent) encoded in the primary DNA structure and its base composition (Jia et al. 2010).

In accordance with the above evidence, it may be hypothesized that the reason for the GC enrichment of grass genes and the associated increase in the GC content of non-coding DNA might be to optimize the overall genome organization and gene regulation (e.g., as an advantage for coping with stressful conditions). The expansion of grasses, now dominating approximately one third of the Earth´s vegetative cover (Shantz 1954), began relatively recently in congruence with global climate changes in the Tertiary that opened the closed Tertiary forests (Linder and Rudall 2005; Edwards and Smith 2010). It may be speculated that the evolutionary success of grasses and their worldwide dominance in various harsh environments is related to their peculiar genome organization. It might be also speculated that the presence of GC-rich grass-like genes observed in some other monocot groups and extant gymnosperms and ferns (Jansson et al. 1994) may be related to their ability to adapt to changing environments and survive historical climate oscillations. In addition to the verification of a relationship between the incidence of GC-rich genes and an increased genomic GC content in plants, these ecological and evolutionary hypotheses remain to be tested.

14.5 Measurements of Genomic GC Content

14.5.1 Complete Genome Sequence Data

The large genomes of eukaryotes contain a number of various repetitive sequences, such as retrotransposons or centromeric and telomeric repeats (see Hirsch and Jiang 2012; Siomos and Riha 2012, this volume) which are difficult to correctly assemblage in genomic sequence data (Imelfort and Edwards 2009; Flagel and Blackman 2012). Therefore, these types of repetitive sequences are frequently lacking, even in the most complete sequencing projects. Such repeats may consistently deviate in GC content from the average genomic GC content of a species and may cause a significant discrepancy between the GC content observed in the (nearly) complete sequence projects and GC content estimates from other methods, e.g., between genomic sequence of *Arabidopsis thaliana*, lacking most GC-rich centromeric regions (GC = 34.7%; Arabidopsis Genome Initiative 2000; Bennett et al. 2003), and later analyses with flow cytometry (GC = 39.1%; Barow and Meister 2002).

In addition to the potential incompleteness of genome assemblages, a GC-bias may even appear in raw sequencing data. Such a bias may result, for example, from the difficulties of amplifying extremely GC- or AT-rich fragments by PCR used for library preparation for genome analyzers, which may then be underrepresented in the pool of sequenced fragments (Hillier et al. 2008; Kozarewa et al. 2009).

These difficulties are expected to be eliminated in the near future (Aird et al. 2010), and unassembled shotgun sequencing data has certainly become an important source of accurate and highly informative GC content data available across a large phylogenetic range. In spite of the assumed availability of high quality and complete sequencing data for many taxa in the near future, the testing of various ecological and evolutionary hypotheses concerning the evolution of GC content will require dozens to hundreds of species or individuals and (perhaps for some time) will still also be based on comparatively faster and more economical biochemical methods. Two of the commonly used methods are described below.

14.5.2 DNA Melting

Since the 1960s (Marmur and Doty 1962), the difference in thermal stability between GC and AT base pairs has been frequently used for the quantification of the GC content in genomic DNA. In a typical experiment, fragmented genomic DNA is gradually heated, and the changing proportion of single-stranded (melted) and double-stranded DNA is quantified. The proportions may be measured using various biophysical techniques, such as spectrophotometry, based on the increased absorbance of disassociated single- vs. double-stranded DNA. Typically, an experiment monitors the absorbance as a function of temperature to provide a "melting curve" (Fig. 14.9). The midpoint of the melting curve (i.e., the point at which 50% of the DNA has melted/disassociated) is called the melting temperature (Tm) and is used in subsequent calculations of the exact GC content using the formula by Mandel and Marmur (1968):

$$GC[\%] = 2.439 * (Tm[°C] - 81.5 - 16.6 * \log(Na^+[mol/l])),$$

where the concentration of monovalent ions (typically Na+) is also taken into account (Doktycz 2002).

This simple model provides a good approximation for long DNA sequences because various other counteracting effects are assumed to cancel out over the entire sequence. In fact, the stability of a particular base pair is also affected by the hydrophobic interactions of neighboring bases. This is why a more exact stability of a known DNA sequence is calculated from

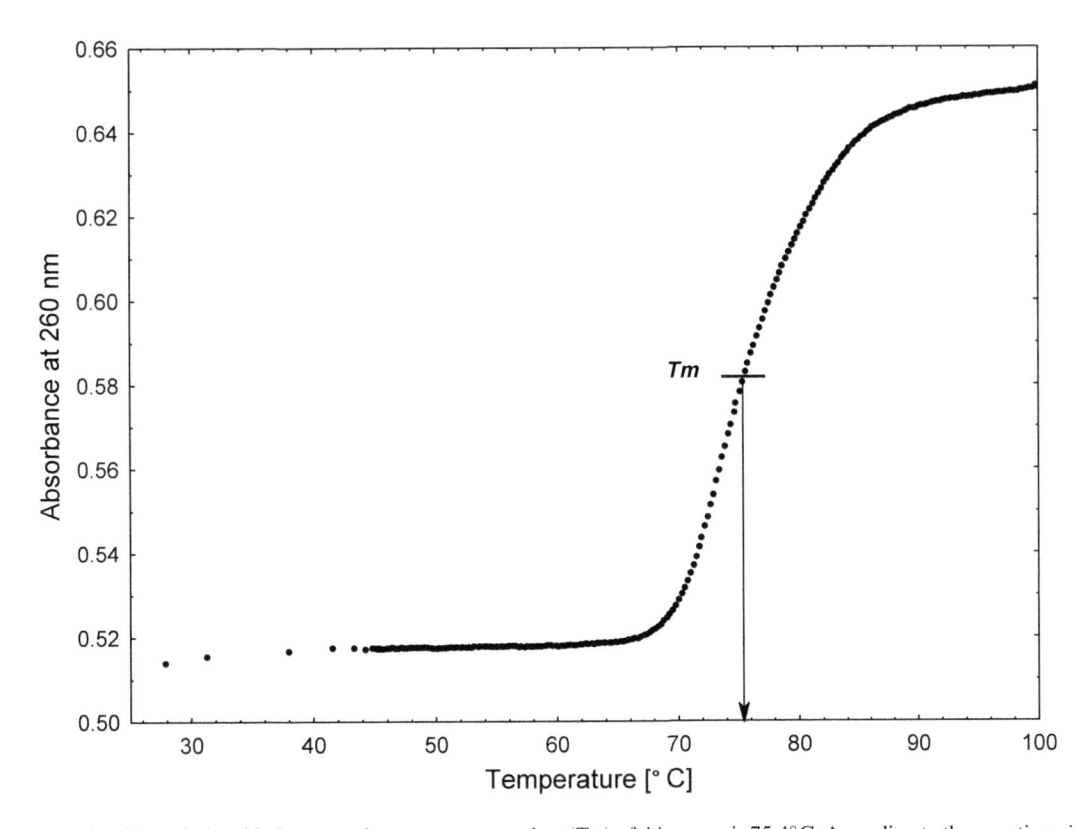

Fig. 14.9 An example of the relationship between the temperature and absorbance of a 0.039 M sodium chloride solution of genomic DNA of *Curculigo capitulata* (Hypoxidaceae) observed during slow (0.5°C/min) solution heating (Šmarda et al. 2012). The melting temperature (Tm) of this curve is 75.4°C. According to the equation given above, the GC content of this sample is 2.439 * (75.4 − 81.5 − 16.6 * log 0.039) = 42.17%

Fig. 14.10 The dependence of sample fluorescence on the binding length (n) of the base-specific dye used and the specific base content. (**a**) Relationship of sample fluorescence and the content of target base pair (here AT) for different binding lengths of a base specific dye. (**b**) Relationship between sample AT content and its dye factor obtained from sample measurement with the AT-specific DAPI dye (binding length = 4) and *Oryza sativa* 'Nipponbare' (AT = 56.4%) as the standard. A dye factor >1 indicates that the AT content of the sample is higher than *Oryza*, while a dye factor <1 indicates that the sample is less AT-rich compared to *Oryza*. The greater the difference of the dye factor from 1, the greater the peak shifts that will be observed between measurements of the sample and standard with intercalary and base specific dyes

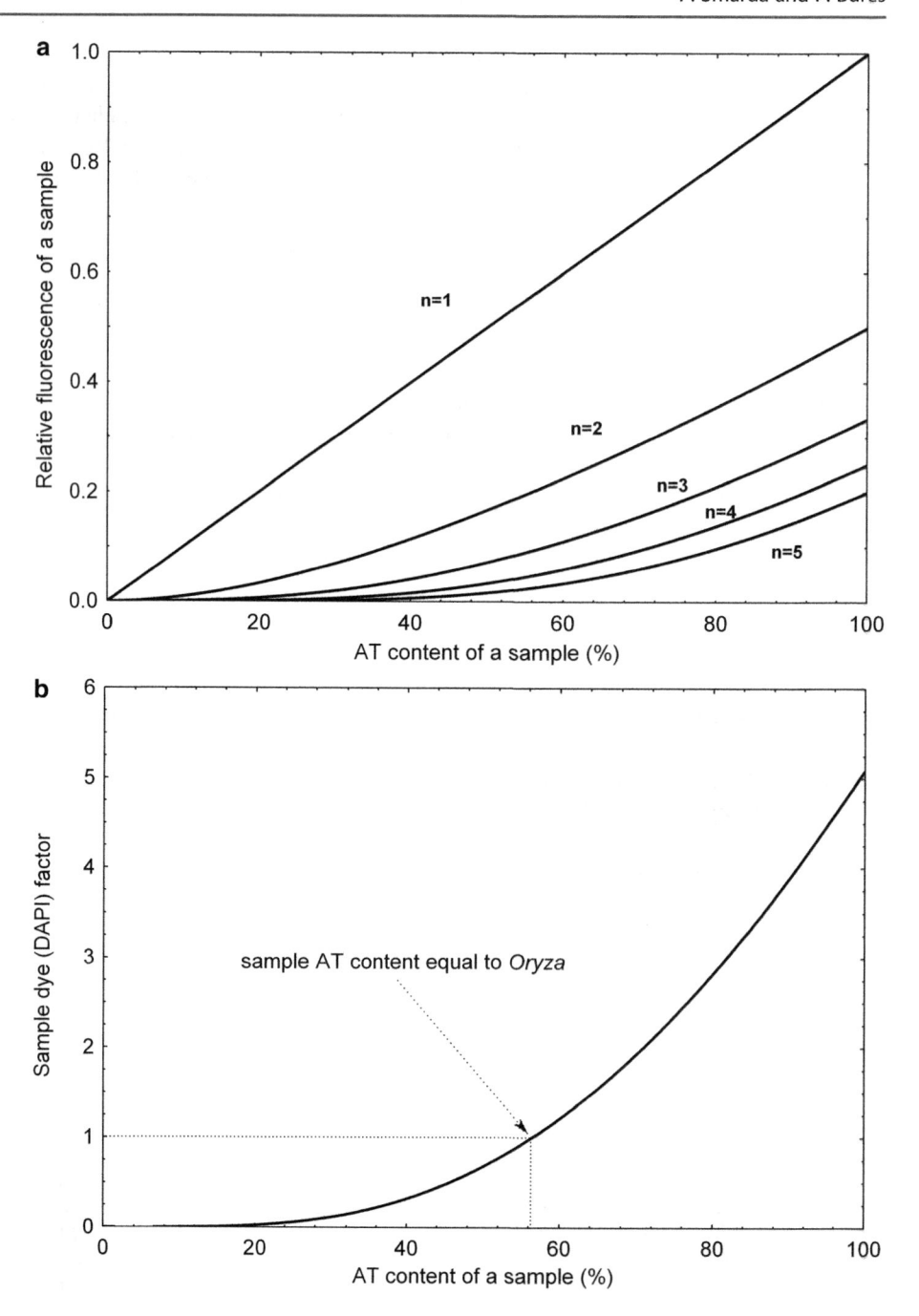

the comparison of the stability of dinucleotide base pairs using more complex models (SantaLucia 1998). Among other interactions, the thermal stability of DNA increases with the presence of methylated pyrimidines, such as methylcytosine (Guldberg et al. 2002). Despite a potential bias resulting from an unknown level of DNA methylation and patterns in base neighboring in melting experiments with unknown genomic DNA sequences, the GC contents from melting experiments (at least in bacteria) (De Ley 1970) correspond well with the data obtained later from complete sequencing projects (Šmarda et al. 2012). Also, in our pilot comparison, estimates of base composition from DNA melting curves corresponded

well with flow cytometric estimates of GC content (Šmarda et al. 2012).

14.5.3 Flow Cytometry

Flow cytometry is the most frequently used method to measure total genomic GC content in plants (Meister and Barow 2007). The method is based on measurements of the emission fluorescence of stained nuclei detected in a hydrodynamically-focused fluid stream (Robinson and Grégori 2007). Conventionally, a small piece of sample tissue and the tissue of a standard (mostly fresh leaves) are cut together in a buffer with a sharp razor blade. The suspension is then filtered, stained with a specific dye, and then the fluorescence of the sample's and the standard's nuclei are measured (Doležel et al. 2007). When using some intercalating dyes, such as propidium iodide or ethidium bromide, this is a powerful method for the measurement of genome size (see Leitch 2012), which is calculated from the comparison of sample and standard fluorescence intensities in a linear fashion (Doležel and Bartoš 2005; Greilhuber et al. 2007).

The base-specific dyes typically require three to five consecutive bases of the same type (AT or GC) to bind to the DNA molecule (i.e., the binding length; Meister and Barow 2007). Because the probability of such a clustering of base pairs is non-linearly related to the GC content of the genome (Godelle et al. 1993; Fig. 14.10), the calculation is much more complicated than in the case of estimating the amount of DNA. Frequently, GC content is calculated by flow cytometry using two different dyes: (1) an intercalating dye, enabling the measurement of a sample/standard ratio for the fluorescences of their total DNA contents and (2) a base-specific dye (such as AT-specific DAPI or GC-specific mithramycin) that enables the estimation of a sample/standard ratio between the AT- or GC-bound portions of their genomic DNA. The ratio between (2) and (1) has been referred to as the dye factor (Barow and Meister 2002; Meister and Barow 2007). When a sample is relatively base-rich compared to the standard, the respective dye factor is >1 (Fig. 14.10). Thus, measurements made for the GC-poor *Juncus effusus* using a GC-rich rice as a standard provide a dye factor <1 for GC-specific dyes and dye factors >1 for AT-specific dyes (Figs. 14.10, 14.11).

Based on the relationship described by Godelle et al. (1993; Fig. 14.10) for the situation of a random distribution of bases in the genome, the AT-content of an unknown sample measured with an AT-specific dye and an intercalary dye may be calculated from the following equation:

$$\frac{(1 - AT_{sample})AT^n_{sample}}{1 - AT^n_{sample}}$$
$$- DF_{sample} \times \frac{(1 - AT_{reference})AT^n_{reference}}{1 - AT^n_{reference}} = 0,$$

where AT is the AT content in a range from 0 to 1, DF_{sample} is the dye factor of the unknown sample, and n is the binding length of the AT-specific dye used. In the case of measurements with GC-specific dyes, AT should be substituted with the GC content.

Solving this equitation requires the use of mathematical approximation. An excel sheet for automatic calculations of GC content data from flow cytometric measurements using AT or GC specific dyes of various binding lengths is available at:

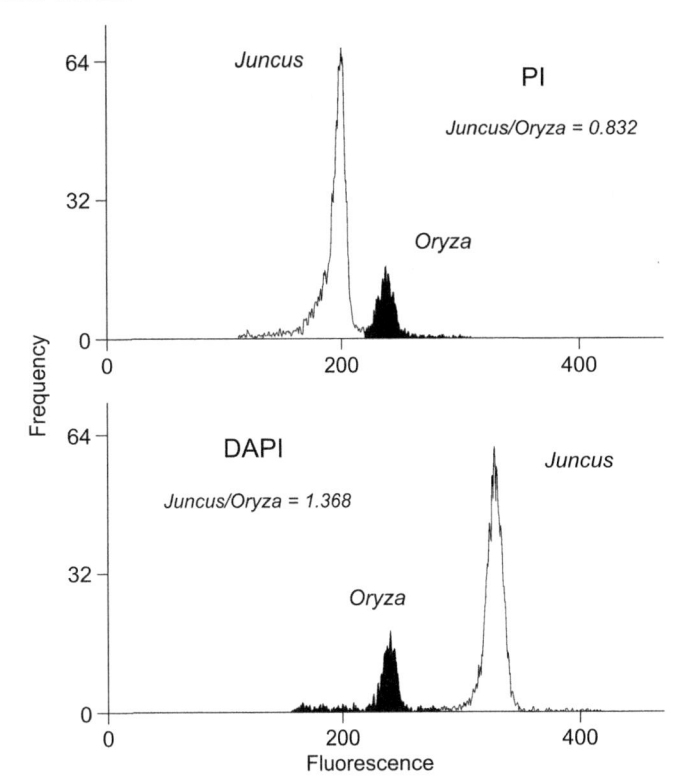

Fig. 14.11 Comparison of nuclei fluorescences in flow cytometry histograms obtained by measuring two species with contrasting GC contents (AT-rich *Juncus effusus* and GC-rich *Oryza sativa* 'Nipponbare') using intercalary (propidium iodide; *PI*) and AT base-specific (4′,6-diamidino-2-phenylindole; *DAPI*) dyes. While *Juncus* clearly has a smaller DNA content then *Oryza* as seen by measurements with the intercalary dye (PI), measurement of its AT fraction made with DAPI indicate that it has an AT-rich genome compared to *Oryza*. Based on the known GC content of rice (43.6%), the GC content of *Juncus* is calculated as 33.7% (see text for details)

http://www.sci.muni.cz/botany/systemgr/download/Festuca/ ATGCFlow.xls (Šmarda et al. 2008). Using the example in Fig. 14.11, the AT/GC content of *J. effusus* may be calculated as outlined below (with a GC content of the standard *O. sativa* equal to 43.6% (i.e., 56.4% AT) and the binding length of DAPI taken to be equal to 4; (Barow and Meister 2002; International Rice Genome Sequencing Project 2005; Šmarda et al. 2012)):

$$\frac{(1 - AT_{Juncus})AT^4_{Juncus}}{1 - AT^4_{Juncus}} - \frac{1.368}{0.832} \times \frac{(1 - 0.564)0.564^4}{1 - 0.564^4} = 0$$
$$AT_{Juncus} = 0.663$$
$$GC_{Juncus}(\%) = 100 \times (1 - AT_{Juncus}) = 33.7\%$$

The main problem with the estimation of GC content by flow cytometry is in the nonrandom distribution of base pairs in measured genomes (e.g., high GC content of coding sequences and AT- or GC-rich repetitive DNA) that may cause deviations from the assumed random-base-distribution model. This may be the case for comparisons of genomes with very distinct genomic structures and organizations, such as the isochore-organized genomes of homeotherm vertebrates (e.g., humans, chicken and mice). Such comparisons might provide biased GC content estimates and it is not generally recommended to measure the GC content of plants using animal (vertebrate) standards.

The above recommendation raises a concern regarding the use of proper standards because most contemporary standards are based on human cells (Doležel and Greilhuber 2010). As noted above, the best plant standard to date is *O. sativa* because a substantial portion of its repetitive DNA is has been assembled using genome sequencing. Some other very complete plant genomes will be finished in the near future (see Flagel and Blackman 2012) helping to overcome the insufficiency of standards for measurements of species with small genomes. However, a challeng will remain for measurements of GC content of species with large genomes. The technical limitations of flow cytometry allow for comparisons of samples that have relatively similar genome sizes (up to a two to threefold difference). Otherwise, the results may be flawed due to the effect of the non-linearity of the detection system of flow cytometry instruments (Doležel and Bartoš 2005; Greilhuber et al. 2007). Therefore, the GC contents for plants with larger genomes are measured using several consecutive standards with increasing genome sizes (e.g. Doležel et al. 1992; Barow and Meister 2002).This cascade-like design accumulates the measurement errors of all previous estimates and hence GC contents of large genomes used to be estimated with relatively high uncertainty. Therefore, some discrepancy may occur in GC content measurements done in the same species at different times or by different laboratories.

Despite these complications, flow cytometry is a reliable method that provides GC content estimates that are highly correlated with estimates from other biochemical methods and sequence data (Meister and Barow 2007: Fig. 8.6; Šmarda et al. 2012).

Acknowledgements The work of authors was supported by the Czech Ministry of Education, Youth and Sports (grants MSM 0021622416 and LC06073) and the Czech Science Foundation (grants GACR206/08/P222 and GACR506/11/0890).

References

Aird D, Chen WS, Ross M, Connolly K, Meldrim J, Russ C, Fisher S, Jaffe D, Nusbaum C, Gnirke A (2010) Analyzing and minimazing bias in Illumina sequencing libraries. Genome Biol 11(Suppl 1):P3

Akhunov ED, Goodyear AW, Geng S, Qi LL, Echalier B, Gill BS, Miftahudin, Gustafson JP, Lazo G, Chao SM et al (2003) The organization and rate of evolution of wheat genomes are correlated with recombination rates along chromosome arms. Genome Res 13:753–763

Alexandrov BV, Freidin S, Troukhan M, Tatarinova T, Zhang H, Swaller T, Lu Y, Bouck J, Flavell R et al (2009) Insights into corn genes derived from large-scale cDNA sequencing. Plant Mol Biol 69:179–194

Anselmi C, Bocchinfuso G, De Santis P, Savino M, Scipioni A (1999) Dual role of DNA intrinsic curvature and flexibility in determining nucleosome stability. J Mol Biol 286:1293–1301

Arabidopsis Genome Initiative (2000) Analysis of the genome sequence of the flowering plant *Arabidopsis thaliana*. Nature 408:796–815

Bai L, Morozov AV (2010) Gene regulation by nucleosome positioning. Trends Genet 26:476–483

Barow M, Meister A (2002) Lack of correlation between AT frequency and genome size in higher plants and the effect of nonrandomness of base sequences on dye binding. Cytometry 47:1–7

Barton LL (2005) Physiological basis for the growth in extreme environments. In: Barton L (ed) Structural and functional relationships in prokaryotes. Springer, Berlin/Heidelberg/New York, pp 348–393

Basak S, Mukhopadhyay P, Gupta SK, Ghosh TC (2010) Genomic adaptation of prokaryotic organisms at high temperature. Bioinformation 4:352–356

Bennett MD (1987) Variation in genomic form in plants and its ecological implications. New Phytol 106(Suppl):177–200

Bennett MD, Leitch IJ, Price HJ, Johnston JS (2003) Comparison of *Caenorhabditis* (100 Mb) and *Drosophila* (175 Mb) using flow cytometry show genome size in *Arabidopsis* to be (157 Mb) and thus 25% larger than the Arabidopsis Genome Initiative estimate of 125 Mb. Ann Bot 91:547–557

Bennetzen JL, Ma J, Devos KM (2005) Mechanisms of recent genome size variation in flowering plants. Ann Bot 95:127–132

Bernardi G (2000a) Isochores and the evolutionary dynamics of vertebrates. Gene 241:3–17

Bernardi G (2000b) The compositional evolution of vertebrate genomes. Gene 259:31–43

Bernardi G (2007) The neoselectionist theory of genome evolution. Proc Natl Acad Sci U S A 104:8385–8390

Bernardi G, Olofsson B, Filipski J, Zerial M, Salinas J, Cuny G, Meunier-Rotival M, Rodier F (1985) The mosaic genome of warm-blooded vertebrates. Science 228:953–958

Bettecken T, Trifonov EN (2009) Repertoires of the nucleosome-positioning dinucleotides. PLoS One 4:e7654

Bird AP (1980) DNA methylation and the frequency of CpG in animal DNA. Nucleic Acids Res 8:1499–1504

Biro JC (2008) Correlation between nucleotide composition and folding energy of coding sequences with special attention to wobble bases. Theor Biol Med Model 5:14

Brown TC, Jiricny J (1988) Different base/base mispairs are corrected with different efficiencies and specifities in monkey kidney-cells. Cell 54:705–711

Bucciarelli G, Di Filippo M, Costagliola D, Alvarez-Valin F, Bernardi G, Bernardi G (2009) Environmental genomics: a tale of two fishes. Mol Biol Evol 26:1235–1243

Bureš P, Wang YF, Horová L, Suda J (2004) Genome size variation in Central European species of *Cirsium* (Compositae) and their natural hybrids. Ann Bot 94:353–363

Bureš P, Šmarda P, Hralová I, Fuentes-Soriano S, Lysák M, Řepka R, Helánová K, Rotreklová O, Procházková J, Úradníček L (2007) Correlation between GC content and genome size in plants. Cytometry 71A:764

Carels N, Bernardi G (2000) Two classes of genes in plants. Genetics 154:1819–1825

Carels N, Hatey P, Jabbari K, Bernardi G (1998) Compositional properties of homologous coding sequences from plants. J Mol Evol 46:45–53

Carthew RW, Sontheimer EJ (2009) Origins and mechanisms of miRNAs and siRNAs. Cell 136:642–655

Chan PP, Lowe TM (2009) GtRNAdb: a database of transfer RNA genes detected in genomic sequence. Nucleic Acids Res 37(Database issue):D93–D97

Chen SL, Lee W, Hottes AK, Shapiro L, McAdams HH (2004) Codon usage between genomes is constrained by genome-wide mutational processes. Proc Natl Acad Sci U S A 101:3480–3485

Cheng YM, Lin BY (2003) Cloning and characterization of maize B chromosome sequences derived from microdissection. Genetics 164:299–310

Chodavarapu RK, Feng S, Bernatavichute YV, Chen PY, Stroud H, Yu Y, Hetzel JA, Kuo F, Kim J, Cokus SJ et al (2010) Relationship between nucleosome positioning and DNA methylation. Nature 466:388–392

Cokus SJ, Feng S, Zhang X, Chen Z, Merriman B, Haudenschild CD, Pradhan S, Nelson SF, Pellegrini M, Jacobsen SE (2008) Shotgun bisulphite sequencing of the Arabidopsis genome reveals DNA methylation patterning. Nature 452:215–219

Constantini M, Bernardi G (2008a) The short-sequence designs of isochores from the human genome. Proc Natl Acad Sci U S A 105:13971–13976

Constantini M, Bernardi G (2008b) Replication timing, chromosomal bands, and isochores. Proc Natl Acad Sci U S A 105:3433–3437

Constantini M, Cammarano R, Bernardi G (2009) The evolution of isochore patterns in vertebrate genomes. BMC Genomics 10:146

Coulondre C, Miller JH, Farabaugh PJ, Gilbert W (1978) Molecular basis of base substitution hotspots in Escherichia coli. Nature 274:775–780

Crane CF (2007) Patterned sequence in the transcriptome of vascular plants. BMC Genomics 8:173

Cruveiller S, Jabbari K, D'Onofrio G, Bernardi G (1999) Different hydrophobicities of orthologous proteins from Xenopus and human. Gene 238:15–21

D'Onofrio G, Jabbari K, Bernardi G, Musto H (1999) The correlation of protein hydropathy with the base composition of coding sequences. Gene 238:3–14

De Ley J (1970) Reexamination of the association between melting point, buoyant density, and chemical base composition of deoxyribonucleic acid. J Bacteriol 101:738–754

De Rose-Wilson LJ, Gaut BS (2007) Transcription-related mutations and GC content drive variation in nucleotide substitution rates across the genomes of Arabidopsis thaliana and Arabidopsis lyrata. BMC Evol Biol 7:66

Doktycz MJ (2002) Nucleic acids: thermal stability and denaturation. In: Encyclopedia of life sciences. Wiley. Wiley Online Library http://onlinelibrary.wiley.com/doi/10.1038/npg.els.0003123

Doležel J, Bartoš J (2005) Plant DNA flow cytometry and the estimation of nuclear genome size. Ann Bot 95:99–110

Doležel J, Greilhuber J (2010) Nuclear genome size: are we getting closer? Cytometry 77A:635–642

Doležel J, Sgorbati S, Lucretti S (1992) Comparison of three DNA fluorochromes for flow cytometric estimation of nuclear DNA content in plants. Physiol Plantarum 85:625–631

Doležel J, Greilhuber J, Suda J (2007) Estimation of nuclear DNA content in plants using flow cytometry. Nat Protoc 2:2233–2244

Drouaud J, Camilleri C, Bourguignon PY, Canaguier A, Bérard A, Vezon D, Giancola S, Brunel D, Colot V, Prum B et al (2006) Variation in crossing-over rates across chromosome 4 of Arabidopsis thaliana reveals the presence of meiotic recombination "hot spots". Genome Res 16:106–114

Duret L, Arndt PF (2008) The impact of recombination on nucleotide substitution in the human genome. PLoS Genet 4:e1000071

Edwards EJ, Smith SA (2010) Phylogenetic analyses reveal the shady history of C4 grasses. Proc Natl Acad Sci U S A 107:2532–2537

Elhaik E, Landan G, Graur D (2009) Can GC content at third-codon positions be used as a proxy for isochore composition? Mol Biol Evol 26:1829–1833

Eyre-Walker A (1993) Recombination and mammalian genome evolution. Proc R Soc Lond B Biol Sci 252:237–243

Eyre-Walker A, Hurst LD (2001) The evolution of isochores. Nat Rev Genet 2:549–555

Feldman M, Liu B, Segal G, Abbo S, Levy AA, Vega JM (1997) Rapid elimination of low-copy DNA sequences in polyploid wheat, a possible mechanism for differentiation of homeologous chromosomes. Genetics 147:1381–1387

Feng Q, Zhang Y, Hao P, Wang S, Fu G, Huang Y, Li Y, Zhu J, Liu Y, Hu X et al (2002) Sequence and analysis of the rice chromosome 4. Nature 420:316–320

Ferraro P, Franzolin E, Pontarin G, Reichard P, Bianchi V (2010) Quantification of cellular deoxynucleoside triphosphates. Nucleic Acids Res 38:e85

Flagel L, Blackman B (2012) The first ten years of plant genome sequencing and prospects for the next decades. In: Wendel JF (ed) Plant genome diversity, vol 1, Plant genomes, their residents, and their evolutionary dynamics. Springer, Wien/New York

Frank AC, Lobry JR (1999) Asymmetric substitution patterns: a review of possible underlying mutational or selective mechanisms. Gene 238:65–77

Frederico LA, Kunkel TA, Shaw BR (1990) A sensitive genetic assay for the detection of cytosine deamination—determination of rate constants and the activation energy. Biochemistry 29:2532–2537

Fullerton SM, Carvalho AB, Clark AG (2001) Local rates of recombination are positively correlated with GC content in the human genome. Mol Biol Evol 18:1139–1142

Galtier N, Lobry JR (1997) Relationships between genomic G + C content, RNA secondary structures, and optimal growth temperature in prokaryotes. J Mol Evol 44:632–636

Godelle B, Cartier D, Marie D, Brown SC, Siljak-Yakovlev S (1993) Heterochromatin study demonstrating the non-linearity of fluorometry useful for calculating genomic base composition. Cytometry 14:618–626

Göndör A, Ohlsson R (2009) Chromosome crosstalk in three dimensions. Nature 461:212–217

Goodsell DS, Dickerson RE (1994) Bending and curvature calculations in B-DNA. Nucleic Acids Res 22:5497–5503

Green P, Ewing B, Mille W, Thomas PJ, NISC Comparative Sequencing Program, Green ED (2003) Transcription-associated mutational asymmetry in mammalian evolution. Nat Genet 33:514–517

Greenbaum BD, Levine AJ, Bhanot G, Rabadan R (2009) Patterns of evolution and host gene mimicry in influenza and other RNA viruses. PLoS Pathog 4:e1000079

Greilhuber J, Temsch EM, Loureiro JCM (2007) Nuclear DNA content measurement. In: Doležel J, Greilhuber J, Suda J (eds) Flow cytometry with plant cells. Analysis of genes, chromosomes and genomes. Wiley-VCH, Weinheim, pp 67–101

Gu W, Zhou T, Wilke C (2010) A universal trend of reduced mRNA stability near the translation-initiation site in prokaryotes and eukaryotes. PLoS Comput Biol 6:e1000664

Guldberg P, Worm J, Grønbæk K (2002) Profiling DNA methylation by melting analysis. Methods 27:121–127

Guo X, Bao J, Fan L (2007) Evidence of selectively driven codon usage in rice: implications for GC content evolution of Gramineae genes. FEBS Lett 581:1015–1021

Herzel H, Weiss O, Trifonov EN (1999) 10-11 bp periodicities in complete genomes reflect protein structure and DNA folding. Bioinformatics 15:187–193

Hillier LW, Marth GT, Quinlan AR, Dooling D, Fewell G, Barnett D, Fox P, Glasscock JI, Hickenbotham M, Huang W et al (2008) Whole-genome sequencing and variant discovery in C. elegans. Nat Methods 5:183–188

Hirsch C, Jiang J (2012) Centromeres: sequence, structure, and biology. In: Wendel JF (ed) Plant genome diversity, vol 1, Plant genomes, their residents, and their evolutionary dynamics. Springer, Wien/New York

Hosouchi T, Kumekawa N, Tsuruoka H, Kotani H (2002) Physical map-based size of centromeric regions of Arabidopsis thaliana chromosome 1, 2, and 3. DNA Res 9:117–121

Hurst LD, Merchant AR (2001) High guanine-cytosine content is not an adaptation to high temperature: a comparative analysis among prokaryotes. Proc R Soc Lond B Biol Sci 268:493–497

Imelfort M, Edwards D (2009) De novo sequencing of plant genomes using second-generation technologies. Brief Bioinform 10:609–618

Ingle J, Timmis JN, Sinclair J (1975) The relationship between satellite deoxyribonucleic acid, ribosomal ribonucleic acid gene redundancy, and genome size in plants. Plant Physiol 55:496–501

International Rice Genome Sequencing Project (2005) The map based sequence of the rice genome. Nature 436:793–800

Jabbari K, Bernardi G (2004) Cytosine methylation and CpG, TpG (CpA) and TpA frequencies. Gene 333:143–149

Jabbari K, Cruveiller S, Clay O, Bernardi G (2003) The correlation between GC3 and hydropathy in human genes. Gene 317:137–140

Jaillon CO, Aury JM, Noel B, Policriti A, Clepet C, Casagrande A, Choisne N, Aubourg S, Vitulo N, Jubin C et al (2007) The grapevine genome sequence suggests ancestral hexaploidization in major angiosperm phyla. Nature 449:463–467

Janousek B, Hobza R, Vyskot B (2012) Chromosomes and sex differentiation. In: Leitch IJ, Dolezel J, Greilhuber J (eds) Plant genome diversity, vol 2, Physical structure and evolution of plant genomes. Springer, Wien/New York

Jansson S, Meyer-Gauen G, Cerff R, Martin W (1994) Nucleotide distribution in gymnosperm nuclear sequences suggests a model for GC-content change in land-plant nuclear genomes. J Mol Evol 39:34–46

Jaramillo-Correa JP, Verdú M, González-Martínez SC (2010) The contribution of recombination to heterozygosity differs among plant evolutionary lineages and life-forms. BMC Evol Biol 10:22

Jeffares DC, Penkett CJ, Bähler J (2008) Rapidly regulated genes are introns poor. Trends Genet 24:375–378

Jeltsch A (2010) Phylogeny of methylomes. Science 328:837–838

Jia Q, Wu HT, Zhou XJ, Gao J, Zhao W, Aziz JD, Wei JS, Hou LH, Wu SY, Zhang Y et al (2010) A "GC-rich" method for mammalian gene expression: a dominant role of non-coding DNA GC content in regulation of mammalian gene expression. Sci China Life Sci 53:94–100

Kawakami T, Strakosh SC, Zhen Y, Ungerer MC (2010) Different scales of Ty1/copia-like retrotransposon proliferation in the genomes of three diploid hybrid sunflower species. Heredity 104:341–350

Kejnovsky E, Hawkins J, Feschotte C (2012) Plant transposable elements: biology and evolution. In: Wendel JF (ed) Plant genome diversity, vol 1, Plant genomes, their residents, and their evolutionary dynamics. Springer, Wien/New York

Kejnovsky E, Hobza R, Cermak T, Kubat Z, Vyskot B (2009) The role of repetitive DNA in structure and evolution of sex chromosomes in plants. Heredity 102:533–541

King GJ, Ingrouille MJ (1987) Genome heterogenity and classification of the Poaceae. New Phytol 107:633–644

Kiyama R, Trifonov EN (2002) What positions nucleosomes?—a model. FEBS Lett 523:7–11

Knight RD, Freeland SJ, Landweber LF (2001) A simple model based on mutation and selection explains trends in codon and amino-acid usage and GC composition within and across genomes. Genome Biol 2:research0010.1–0010.13

Kovarik A, Dadejova M, Lim YK, Chase MW, Clarkson JJ, Knapp S, Leitch AR (2008) Evolution of rDNA in Nicotiana allopolyploids: a potential link between rDNA homogenization and epigenetics. Ann Bot 101:815–823

Kozarewa I, Ning Z, Quail MA, Sanders MJ, Berriman M, Turner DJ (2009) Amplification free Illumina sequencing-library preparation facilitates improved mapping and assembly of GC-biased genomes. Nat Methods 6:291–295

Kubát Z, Hobza R, Vyskot B, Kejnovský E (2008) Microsatellite accumulation on the Y chromosome in Silene latifolia. Genome 51:350–356

Kuhl JC, Cheung F, Yuan Q, Martin W, Zewdie Y, McCallum J, Catanach A, Rutherford P, Sink KC, Jenderek M et al (2004) A unique set of 11,008 onion expressed sequence tags reveals expressed sequence and genomic differences between the monocot orders Asparagales and Poales. Plant Cell 16:114–125

Lang D, Zimmer AD, Rensing SA, Reski R (2008) Exploring plant biodiversity: the Physcomitrella genome and beyond. Trends Plant Sci 13:542–549

Lattorff HMG, Moritz RFA (2008) Recombination rate and AT-content show opposite correlations in mammalian and other animal genomes. J Evol Biol 35:146–149

Law JA, Jacobsen SE (2010) Establishing, maintaining and modifying DNA methylation patterns in plants and animals. Nat Rev Genet 11:204–220

Lawrence JG, Ochman H (1998) Molecular archaeology of the Escherichia coli genome. Proc Natl Acad Sci U S A 95:9413–9417

Leeds JM, Slabaugh MB, Mathews CK (1985) DNA precursor pools and ribonucleotide reductase activity: distribution between the nucleus and cytoplasm of mammalian cells. Mol Cell Biol 5:3443–3450

Leitch IJ (2012) Genome size diversity and evolution in land plants. In: Leitch IJ, Dolezel J, Greilhuber J (eds) Plant genome diversity, vol 2, Physical structure and evolution of plant genomes. Springer, Wien/New York

Lescot M, Piffanelli P, Ciampi AY, Ruiz M, Blanc G, Leebens-Mack J, da Silva FR, Santos CMR, D'Hont A, Garsmeur O et al (2008) Insights into the Musa genome: syntenic relationships to rice and between Musa species. BMC Genomics 9:58

Li J, Hsia AP, Schnable PS (2007) Recent advances in plant recombination. Curr Opin Plant Biol 10:131–135

Lin H, Ouyang S, Egan A, Nobuta K, Haas BJ, Zhu W, Gu X, Silva JC, Meyers BC, Buell RC (2008) Characterization of paralogous protein families in rice. BMC Plant Biol 8:18

Lind P, Andersson DI (2008) Whole genome mutational bias in bacteria. Proc Natl Acad Sci U S A 105:17878–17883

Linder PH, Rudall PJ (2005) Evolutionary history of Poales. Annu Rev Ecol Evol Syst 36:107–124

Lister R, O'Malley RG, Tonti-Filippini J, Gregory BD, Berry CC, Millar AH, Ecker JR (2008) Highly integrated single-base resolution maps of the epigenome in Arabidopsis. Cell 133:523–536

Liu QP, Xue QZ (2005) Comparative studies on codon usage pattern of chloroplasts and their host nuclear genes in four plant species. J Genet 84:55–62

Liu S, Yeh CT, Ji T, Ying K, Wu H, Tang HM, Fu Y, Nettleton D, Schnable PS (2009) Mu transposon insertion sites and meiotic recombination events co-localize with epigenetic marks for open chromatin across the maize genome. PLoS Genet 5:e1000733

Liu H, He R, Zhang H, Huang Y, Tian M, Zhang J (2010) Analysis of synonymous codon usage in Zea mays. Mol Biol Rep 37:677–684

Lu T, Lu G, Fan D, Zhu C, Li W, Zhao Q, Feng Q, Zhao Y, Guo Y, Li W et al (2010) Function annotation of the rice transcriptome at single-nucleotide resolution by RNA-seq. Genome Res 20:1238–1249

Lysak M, Schubert I (2012) Mechanisms of chromosome rearrangements. In: Leitch IJ, Dolezel J, Greilhuber J (eds) Plant genome diversity, vol 2, Physical structure and evolution of plant genomes. Springer, Wien/New York

Malinska H, Tate JA, Matyasek R, Leitch AR, Soltis DE, Soltis PS (2010) Similar patterns of rDNA evolution in synthetic and recently formed natural populations of *Tragopogon* (Asteraceae) allotetraploids. BMC Evol Biol 10:291

Mandáková T, Joly S, Krzywinski M, Mummenhoff K, Lysak MA (2010) Fast diploidization in close mesopolyploid relatives of *Arabidopsis*. Plant Cell 22:2277–2290

Mandel M, Marmur J (1968) Use of ultraviolet absorbance-temperature profile for determining the guanine plus cytosine content of DNA. Methods Enzymol 12B:195–206

Mann S, Phoebe-Chen YP (2010) Bacterial genomic G + C composition-eliciting environmental adaptation. Genomics 95:7–15

Marie D, Brown SC (1993) A cytometric exercise in plant DNA histograms, with 2C values for 70 species. Biol Cell 78:41–51

Marmur J, Doty P (1962) Determination of the base composition of deoxyribonucleic acid from its thermal denaturation temperature. J Mol Biol 5:109–118

Martomo SA, Mathews CK (2002) Effects of biological DNA precursor pool asymmetry upon accuracy of DNA replication in vitro. Mutat Res-Fund Mol M 499:197–211

Matassi G, Montero LM, Salinas J, Bernardi G (1989) The isochore organization and compositional distribution of homologous coding sequences in the nuclear genomes of plants. Nucleic Acids Res 17:5273–5290

Meister A, Barow M (2007) DNA base composition of plant genomes. In: Doležel J, Greilhuber J, Suda J (eds) Flow cytometry with plant cells. Analysis of genes, chromosomes and genomes. Wiley-VCH, Weinheim, pp 177–215

Merchant SS, Prochnik SE, Vallon O, Harris EH, Karpowicz SJ, Witman GB, Terry A, Salamov A, Fritz-Laylin LK, Maréchal-Drouard L et al (2007) The *Chlamydomonas* genome reveals the evolution of key animal and plant functions. Science 318:245–251

Meunier J, Duret L (2004) Recombination drives the evolution of GC-content in the human genome. Mol Biol Evol 21:984–990

Meuth M (1989) The molecular basis of mutations induced by deoxyribonucleoside triphosphate pool imbalances in mammalian cells. Exp Cell Res 181:305–316

Meyers BC, Tingey SV, Morgante M (2001) Abundance, distribution, and transcriptional activity of repetitive elements in the maize genome. Genome Res 11:1660–1676

Michaels ML, Miller JH (1992) The GO system protects organisms from the mutagenic effect of the spontaneous lesion 8-hydroxyguanine (7,8-dihydro-8-oxoguanine). J Bacteriol 174:6321–6325

Ming R, Hou S, Feng Y, Yu Q, Dionne-Laporte A, Saw JH, Senin P, Wang W, Ly BV, Lewis KLT et al (2008) The draft genome of the transgenic tropical fruit tree papaya (*Carica papaya* Linnaeus). Nature 452:991–997

Moshonov S, Elfakess R, Golan-Mashiach M, Sinvani H, Dikstein R (2008) Links between core promoter and basic gene features influence gene expression. BMC Genomics 9:92

Mrazek J, Karlin S (1998) Strand compositional asymmetry in bacterial and large viral genomes. Proc Natl Acad Sci U S A 95:3720–3725

Mukhopadhyay P, Basak S, Ghosh TC (2007) Nature of selective constraints on synonymous codon usage of rice differs in GC-poor and GC-rich genes. Gene 400:71–81

Musto H, Naya H, Zavala A, Romero H, Alvarez-Valín F, Bernardi G (2004) Correlations between genomic GC levels and optimal growth temperatures in prokaryotes. FEBS Lett 573:73–77

Musto H, Naya H, Zavala A, Romero H, Alvarez-Valín F, Bernardi G (2006) Genomic GC level, optimal growth temperature, and genome size in prokaryotes. Biochem Biophys Res Commun 347:1–3

Nakabachi A, Yamashita A, Toh H, Ishikawa H, Dunbar HE, Moran NA, Hattori M (2006) The 160 -kilobase genome of the bacterial endosymbiont *Carsonella*. Science 314:267

Naya H, Romero H, Zavala A, Alvarez B, Musto H (2002) Aerobiosis increases the genomic guanine plus cytosine content (GC%) in prokaryotes. J Mol Evol 55:260–264

Nishio Y, Nakamura Y, Kawarabayasi Y, Usuda Y, Kimura E, Sugimoto S, Matsui K, Yamagishi A, Kikuchi H, Ikeo K et al (2003) Comparative complete genome sequence analysis of the amino acid replacements responsible for thermostability of *Corynebacterium efficiens*. Genome Res 13:1572–1579

Noma K, Nakajima R, Ohtsubo H, Ohtsubo E (1997) RIRE1, a retrotransposon from wild rice *Oryza australiensis*. Genes Genet Syst 72:131–140

Ossowski S, Schneeberger K, Lucas-Lledó JI, Warthmann N, Clark RM, Shaw RG, Weigel D, Lynch M (2010) The rate and molecular spectrum of spontaneous mutations in *Arabidopsis thaliana*. Science 327:92–94

Ouyang S, Zhu W, Hamilton J, Lin H, Campbell M, Childs K, Thibaud-Nissen F, Malek RL, Lee Y, Zheng L et al (2008) The TIGR Rice Genome Annotation Resource: improvements and new features. Nucleic Acids Res 35:D883–D887

Ozkan H, Levy AA, Feldman M (2001) Allopolyploidy-induced rapid genome evolution in the wheat (*Aegilops-Triticum*) group. Plant Cell 13:1735–1747

Ozkan H, Tuna M, Arumuganathan K (2003) Nonadditive changes in genome size during allopolyploidisation in the wheat (*Aegilops-Triticum*) group. J Hered 94:260–264

Petit M, Guidat C, Daniel J, Denis E, Montoriol E, Bui QT, Lim KY, Kovarik A, Leitch AR, Grandbastien MA et al (2010) Mobilization of retrotransposons in synthetic allotetraploid tobacco. New Phytol 186:135–147

Pfeifer GP (2006) Mutagenesis at methylated CpG sequences. In: Doerfler W, Böhm P (eds) DNA methylation: basic mechanisms, vol 301, Current topics in microbiology and immunology. Springer, Berlin/Heidelberg/New York, pp 259–281

Piegu B, Guyot R, Picault N, Roulin A, Saniyal A, Kim H, Collura K, Brar DS, Jackson S, Wing RA et al (2006) Doubling genome size without polyploidization: dynamics of retrotransposition-driven genomic expansion in *Oryza australensis*, a wild relative of rice. Genome Res 16:1262–1269

Pozzoli U, Menozzi G, Fumagalli M, Cereda M, Comi GP, Cagliani R, Bresolin N, Sironi M (2008) Both selective and neutral processes drive GC content evolution in the human genome. BMC Evol Biol 8:99

Prochnik SE, Umen J, Nedelcu AM, Hallmann A, Miller SM, Nishii I, Ferris P, Kuo A, Mitros T, Fritz-Laylin LK et al (2010) Genomic analysis of organismal complexity in the multicellular green alga *Volvox carteri*. Science 329:223–226

Pruesse E, Quast C, Knittel K, Fuchs B, Ludwig W, Peplies J, Glöckner FO (2007) SILVA: a comprehensive online resource for quality checked and aligned ribosomal RNA sequence data compatible with ARB. Nucleic Acids Res 35:7188–7196

Puertas MJ (2002) Nature and evolution of B chromosomes in plants: a non-coding but information-rich part of plant genomes. Cytogenet Genome Res 96:198–205

Richards C, Verhoeven KJF, Bossdorf O (2012) Evolutionary significance of epigenentic variation. In: Wendel JF (ed) Plant genome diversity, vol 1, Plant genomes, their residents, and their evolutionary dynamics. Springer, Wien/New York

Robinson JP, Grégori G (2007) Principles of flow cytometry. In: Doležel J, Greilhuber J, Suda J (eds) Flow cytometry with plant cells. Analysis of genes, chromosomes and genomes. Wiley-VCH, Weinheim, pp 19–40

Rocha EPC (2004) The replication-related organization of the bacterial chromosome. Microbiology 150:1609–1627

Rocha EPC, Danchin A (2002) Competition for scarce resources might bias bacterial genome composition. Trends Genet 18:291–294

Rocha EPC, Touchon M, Feil EJ (2006) Similar compositional biases are caused by very different mutational effects. Genome Res 16:1537–1547

Romiguier J, Ranwez V, Douzery EJP, Galtier N (2010) Contrasting GC-content dynamics across 33 mammalian genomes: relationship with life-history traits and chromosome sizes. Genome Res 20:1001–1009

Salinas J, Matassi G, Montero LM, Bernardi G (1988) Compositional compartmentalization and compositional patterns in the nuclear genomes of plants. Nucleic Acids Res 16:4269–4285

Sanford RA, Cole JR, Tiedje JM (2002) Characterization and description of *Anaeromyxobacter dehalogenans* gen. nov., sp. nov., an aryl-halorespiring facultative anaerobic myxobacterium. Appl Environ Microbiol 68:893–900

SanMiguel P, Vitte C (2009) The LTR-retrotransposons of maize. In: Bennetzen J, Hake S (eds) Handbook of maize genetics and genomics. Springer, New York, pp 307–327

SantaLucia J Jr (1998) A unified view of polymer, dumbbell, and oligonucleotide DNA nearest-neighbor thermodynamics. Proc Natl Acad Sci U S A 95:1460–1465

Satapathy SS, Dutta M, Ray SK (2010) Variable correlation of genome GC% with transfer RNA number as well as with transfer RNA diversity among bacterial groups: a-Proteobacteria and Tenericutes exhibit strong positive correlation. Microbiol Res 165:232–242

Schnable PS, Ware D, Fulton RS, Stein JC, Wei F, Pasternak S, Liang C, Zhang J, Fulton L, Graves TA et al (2009) The B73 maize genome: complexity, diversity, and dynamics. Science 326:1112–1115

Schwartz MN, Trautner TA, Kornberg A (1962) Enzymatic synthesis of deoxyribonucleic acid: XI. Further studies on nearest neighbor base sequences in deoxyribonucleic acids. J Biol Chem 237:1961–1967

Seffens W, Digby D (1999) mRNAs have greater negative folding free energies than shuffled or codon choice randomized sequences. Nucleic Acids Res 27:1578–1584

Segal E, Fondufe-Mittendorf Y, Chen L, Thåström AC, Field Y, Moore IK, Wang JPZ, Widom J (2006) A genomic code for nucleosome positioning. Nature 442:772–778

Serenkov GP (1962) Nucleic acids in the evolution of algae [in Russiian]. Izv Akad Nauk SSSR Biol 1962:857–868

Shaked H, Kashkush K, Ozkan H, Feldman M, Levy AA (2001) Sequence elimination and cytosine methylation are rapid and reproducible responses of the genome to wide hybridization and allopolyploidy in wheat. Plant Cell 13:1749–1759

Shantz HL (1954) The place of grasslands in the Earths cover of vegetation. Ecology 35:143–145

Shapiro HS (1976) Distribution of purines and pyrimidines in deoxyribonucleic acids. In: Fasman GD (ed) Handbook of biochemistry and molecular biology, 3rd edn. CRC Press, Cleveland, pp 241–281

Shi XL, Wang XY, Li Z, Zhu Q, Yang J, Ge S, Luo J (2007) Evidence that natural selection is the primary cause of the guanine-cytosine content variation in rice genes. J Integr Plant Biol 49:1393–1399

Siomos M, Riha K (2012) Telomeres and their biology. In: Wendel JF (ed) Plant genome diversity, vol 1, Plant genomes, their residents, and their evolutionary dynamics. Springer, Wien/New York

Sloan DB, Taylor DR (2010) Testing for selection on synonymous sites in plant mitochondrial DNA: the role of codon bias and RNA editing. J Mol Evol 70:479–491

Šmarda P, Bureš P, Horová L (2010) The evolution of base composition in monocots. In: Proceedings of the international workshop on structural and functional diversity of the eukaryotic genome, Brno. Muni Press, Brno, p 69

Šmarda P, Bureš P, Horová L, Foggi B, Rossi G (2008) Genome size and GC content evolution of *Festuca*: ancestral expansion and subsequent reduction. Ann Bot 101:421–433

Šmarda P, Bureš P, Šmerda J, Horová L (2012) Measurements of genomic GC content in plant genomes with flow cytometry: a test for reliability. New Phytol 193:513–521

Smith DR (2009) Unparalleled GC content in the plastid DNA of *Selaginella*. Plant Mol Biol 71:627–639

Stalder L, Mühlemann O (2008) The meaning of nonsense. Trends Cell Biol 18:315–321

Sterpone F, Bertonati C, Briganti G, Melchionna S (2010) Water around thermophilic proteins: the role of charged and apolar atoms. J Phys Condens Matter 22:284113

Svejstrup JQ (2002) Mechanisms of transcription-coupled DNA repair. Nat Rev Mol Cell Biol 3:21–29

Swarbreck D, Wilks C, Lamesch P, Berardini TZ, Garcia-Hernandez M, Foerster H, Li D, Meyer T, Muller R, Ploetz L et al (2008) The *Arabidopsis* Information Resource (TAIR): gene structure and function annotation. Nucleic Acids Res 36:D1009–D1014

Tatarinova TV, Alexandrov NN, Bouck JB, Feldmann KA (2010) GC3 biology in corn, rice, sorghum and other grasses. BMC Genomics 11:308

Tillo D, Hughes TR (2009) G plus C content dominates intrinsic nucleosome occupancy. BMC Bioinformatics 10:442

Touchon M, Arneodo A, d'Aubenton-Carafa Y, Thermes C (2004) Transcription-coupled and splicing-coupled strand asymmetries in eukaryotic genomes. Nucleic Acids Res 32:4969–4978

Trifonov EN, Sussman JL (1980) The pitch of chromatin DNA is reflected in its nucleotide sequence. Proc Natl Acad Sci U S A 77:3816–3820

Trivedi S, Rao SR, Gehlot HS (2005) Nucleic acid stability in thermophilic prokaryotes: a review. J Cell Mol Biol 4:61–69

Troukhan M, Tatarinova T, Bouck J, Flawell R, Alexandrov N (2009) Genome-wide discovery of cis-elements in promoter sequences using gene expression data. OMICS 13:139–151

Turcotte K, Srinivasan S, Bureau T (2001) Survey of transposable elements from rice genomic sequences. Plant J 25:169–179

Vanyushin B (2006) DNA methylation in plants. In: Doerfler W, Böhm P (eds) DNA methylation: basic mechanisms, vol 2. Springer, Berlin/Heidelberg/New York, pp 67–122

Varriale A, Torelli G, Bernardi G (2008) Compositional properties and thermal adaptation of 18S rRNA in vertebrates. RNA 14:1492–1500

Veselý P, Bureš P, Šmarda P, Pavlíček T (2012) Genome size and DNA base composition of geophytes: the mirror of phenology and ecology? Ann Bot 109:65–75

Vetsigian K, Goldenfeld N (2009) Genome rhetoric and the emergence of compositional bias. Proc Natl Acad Sci U S A 106:215–220

Vieille C, Zeikus GJ (2001) Hyperthermophilic enzymes: sources, uses, and molecular mechanisms for thermostability. Microbiol Mol Biol Rev 65:1–43

Vinogradov AE (1998) Genome size and GC-percent in vertebrates as determined by flow cytometry: the triangular relationship. Cytometry 31:100–109

Vinogradov AE (2001) Within-intron correlation with base composition of adjacent exons in different genomes. Gene 276:143–151

Vinogradov AE (2003) DNA helix: the importance of being GC-rich. Nucleic Acids Res 31:1838–1844

Vinogradov AE (2004) Compactness of human housekeeping genes: selection for economy or genomic design? Trends Genet 20:248–253

Vinogradov AE (2005) Noncoding DNA, isochores and gene expression: nucleosome formation potential. Nucleic Acids Res 33:559–563

Watson JM, Riha K (2010) Comparative biology of telomeres: where plants stand. FEBS Lett 584:3752–3759

Wolf P (2012) Plastid genome diversity. In: Wendel JF (ed) Plant genome diversity, vol 1, Plant genomes, their residents, and their evolutionary dynamics. Springer, Wien/New York

Wolfe KH, Sharp PM, Li W-H (1989) Mutation rates differ among regions of the mammalian genome. Nature 337:283–285

Wong GKS, Wang J, Tao L, Tan J, Zhang JG, Passey DA, Yu J (2002) Compositional gradients in Gramineae genes. Genome Res 12:851–856

Yakovchuk P, Protozanova E, Frank-Kamenetskii MD (2006) Base-stacking and base-pairing contributions into thermal stability of the DNA double helix. Nucleic Acids Res 34:564–574

Yan H, Kikuchi S, Neumann P, Zhang W, Wu Y, Chen F, Jiang J (2010) Genome-wide mapping of cytosine methylation revealed dynamic DNA methylation patterns associated with genes and centromeres in rice. Plant J 63:353–365

Yang C, Bolotin E, Jiang T, Sladek FM, Martinez M (2007) Prevalence of the initiator over the TATA box in human and yeast genes and identification of DNA motifs enriched in human TATA-less core promoters. Gene 389:52–65

Yu J, Hu S, Wang J, Wong GKS, Li S, Liu B, Deng Y, Dai L, Zhou Y, Zhang X et al (2002) A draft sequence of the rice genome (*Oryza sativa* L. ssp. *indica*). Science 296:79–92

Zhang X (2008) The epigenetic landscape of plants. Science 320:489–492

Zhang X (2012) Chromatin modifications. In: Wendel JF (ed) Plant genome diversity, vol 1, Plant genomes, their residents, and their evolutionary dynamics. Springer, Wien/New York

Zhang R, Zhang CT (2004) Isochore structures in the genome of the plant *Arabidopsis thaliana*. J Mol Evol 59:227–238

Zhang L, Kosakovsky Pond S, Gaut BS (2001) A survey of the molecular evolutionary dynamics of twenty-five multigene families from four grass taxa. J Mol Evol 52:144–156

Zhang C, Li WH, Krainer AR, Zhang MQ (2008) RNA landscape of evolution of optimal exon and intron discrimination. Proc Natl Acad Sci U S A 105:5797–5802

Chromatin Modifications in Plants

15

Xiaoyu Zhang

Contents

15.1   Introduction

In eukaryotic cells, nuclear DNA is wrapped around the core histone octamers to form nucleosomes, which are further packaged into the chromatin. This process is thought to have been evolutionarily advantageous as a packaging mechanism to allow eukaryotic cells to accommodate longer DNA molecules. However, the discovery that both DNA and histones can be covalently modified led to the realization that chromatin structure may not be uniform across the genome, and that the local structural properties of the chromatin may have profound effects on the biological processes that occur in the chromatin environment. Two major types of modifications have been described, including the addition of a methyl group to the C5 position of the cytosine ring (referred to as cytosine methylation or DNA methylation), and the post-translational modifications of various residues in core histones (i.e., H2A, H2B, H3 and H4). These modifications are referred to as "epigenetic" as they can affect gene expression in a mitotically and meiotically stable manner without changing the underlying DNA sequences.

Chromatin modifications in plants are highly complex. Unlike in mammalian genomes where DNA methylation is largely limited in the 5′-CG-3′ context, plant genomes contain DNA methylation in all sequence contexts, which is established and maintained by three distinct DNA methylation pathways. In addition, dozens of residues in all four core histones can be modified by a variety of modifications, including methylation, acetylation, phosphorylation and ubiquitination. The establishment, maintenance and removal of DNA methylation and histone modifications are described in Sect. 15.2.

Early experiments to characterize chromatin modifications were often performed at individual loci. The results have provided important clues about the mechanisms and functions of chromatin modifications. However, the design of such experiments and the interpretation of the results were sometimes confounded by the fact that chromatin modifications are highly locus-specific, and that opposite

X. Zhang (✉)

Department of Plant Biology, University of Georgia, 4409 Miller Plant Sciences Building, Athens, GA 30602-7271, USA
e-mail: xiaoyu@plantbio.uga.edu

J.F. Wendel et al. (eds.), *Plant Genome Diversity Volume 1*,
DOI 10.1007/978-3-7091-1130-7_15, © Springer-Verlag Wien 2012

changes may take place at different loci in the same mutant background. For example, in the DNA methyltransferase mutant *met1*, the *FWA* gene is hypomethylated and overexpressed, causing a late flowering phenotype. However, the *SUPERMAN* gene becomes hypermethylated and repressed in the same mutant, resulting in the homeotic transformation of floral organs. Recent advances in genomic technologies, such as the availability of high-density whole-genome tiling microarrays and high-throughput sequencing platforms, have opened a new area of chromatin modification research, referred to as "epigenomics". It is now possible to determine the genome-wide distribution profiles of chromatin modifications at very high resolution. The results not only provide novel insight into the mechanisms and functions of chromatin modifications, but also broaden our understanding of the structure and organization of the genome as a whole. The methods used in epigenomic studies as well as the major findings are described in Sects. 15.3 and 15.4, respectively.

Results from chromatin modification profiling studies also provide the necessary data to address questions regarding the functions of these modifications. It has become clear that genomic regions with different functions, such as actively transcribed or developmentally repressed genes, silenced transposons, replication origins and chromatin cohesion sites, are preferentially associated with distinct combinations of chromatin modifications. These observations have allowed us to formulate hypotheses using a "guilt-by-association" approach, which can then be tested by genetic and biochemical methods. As described in Sect. 15.5, a growing body of evidence suggests that chromatin modifications play critically important roles in regulating many biological processes.

Most of the findings described in this chapter were obtained from the model dicot plant *Arabidopsis thaliana*, largely as a result of the availability of the entire genome sequence, various high-throughput genomic tools, a large collection of T-DNA insertion mutants and an enormous amount of transcriptional profiling data. Because of these advantages, a number of epigenomic approaches were first developed in *Arabidopsis* and then applied to other organisms, including humans,

15.2 Chromatin Modification Pathways

15.2.1 DNA Methylation

15.2.1.1 Plant DNA Methyltransferases

In mammalian genomes, DNA methylation is catalyzed by the DNA methyltransferases DNMT1, DNMT3a and DNMT3b, and occurs predominantly at CG sites (Bird 2002; Goll and Bestor 2005). In flowering plants, three

distinct DNA methylation pathways are involved in DNA methylation in all three sequence contexts, namely CG, CHG and CHH (H = A, C and T). The DNMT1 homologue METHYLTRANSFERASE 1 (MET1) is required for the maintenance of DNA methylation at CG sites (Finnegan et al. 1996; Ronemus et al. 1996; Kankel et al. 2003; Saze et al. 2003), the DNMT3a/3b homologues DOMAIN REARRANGED METHYLTRANSFERASE 1 and 2 (DRM1 and DRM2) are responsible for the maintenance of DNA methylation at CHH sites as well as the de novo methylation in all sequence contexts (Cao and Jacobsen 2002a), and the plant-specific CHROMOMETHYLASE 3 (CMT3) functions in DNA methylation at CHG sites (Bartee et al. 2001; Lindroth et al. 2001). In addition to the differences in sequence preferences, MET1, DRM1/2 and CMT3 require different signals and interact with different sets of protein partners. The mechanisms and functions of these DNA methyltransferases in de novo and maintenance DNA methylation are described below.

15.2.1.2 de novo DNA Methylation

In mammals, DNA methylation at most genomic sites is erased and reestablished in each generation (Surani et al. 2007; Sasaki and Matsui 2008; Hemberger et al. 2009). This process does not appear to take place in plants, where de novo DNA methylation primarily occurs when "foreign DNA" invades the genome, such as new transposons or transgenes. Loss of DRM1/2 in *Arabidopsis* eliminates all de novo methylation activity, indicating that DRM1/2 are responsible for de novo methylation in all sequence contexts (Cao and Jacobsen 2002a). Early genetic studies showed that the signals that direct DNA methylation by DRM1/2 might come from the siRNA pathway, as the loss of siRNA pathway components also abolished de novo DNA methylation (Dalmay et al. 2000; Chan et al. 2004; Zilberman et al. 2003; Kanno et al. 2005; Pontier et al. 2005). This process is termed "RNA-directed DNA methylation (RdDM)", which has been the topic of several recent reviews (Henderson and Jacobsen 2007; Matzke et al. 2009; Law and Jacobsen 2010; Lahmy et al. 2010; Zhang and Zhu 2011). Briefly, the single-stranded RNA generated by the plant-specific RNA polymerase IV (Pol IV) serves as the template for RNA-dependent RNA polymerase 2 (RDR2), and the resulting double-stranded RNA is processed by DICER-LIKE 3 (DCL3) into mostly 24nt siRNAs. These siRNAs are then loaded into ARGONAUTE 4 (AGO4), which interacts with DRM2. AGO4 also interacts with two subunits of another plant-specific RNA polymerase called Pol V, and targets de novo methylation at siRNA-generating loci.

15.2.1.3 Maintenance of DNA Methylation

DNA methylation at CG and CHG sites is considered "symmetric" as it occurs on both cytosines on the two opposite DNA strands. Following DNA replication, these sites become hemi-methylated, and the MET1 and CMT3 pathways function to maintain DNA methylation by converting hemi-methylated sites back to a fully methylated state. MET1 interacts with VARIANT IN METHYLATION (VIM), which recognizes hemi-methylated CG sites through its Set-or-Ring-associated (SRA) domain (Bostick et al. 2007; Woo et al. 2007, 2008; Hashimoto et al. 2008; Kraft et al. 2008). It therefore appears that the only signal required for the maintenance of CG-methylation following DNA replication is the existence of CG-methylation prior to replication. This notion was supported in an elegant genetic study, which showed that once CG methylation was lost in a *met1* gamete, it could not be reestablished by introducing a functional copy of MET1 in the *met1*/+ zygote (Saze et al. 2003).

The maintenance of methylation at CHG sites is similar to that at CG sites in that it is facilitated by the prior existence of CHG methylation. In addition, the finding that CHG methylation is compromised in mutants with decreased levels of H3 lysine 9 dimethylation (H3K9me2) suggests that the maintenance of CHG methylation also requires the H3K9me2 pathway (Jackson et al. 2002). The mechanistic basis for this requirement was later elucidated by biochemical studies. CMT3 binds H3K9me2 through its chromodomain, and the H3K9me2 methyltransferase KRYPTONITE (KYP) binds hemi-methylated CHG sites through its SRA domain (Lindroth et al. 2004; Johnson et al. 2007). These observations indicate that the coordinated actions of CMT3 and KYP may lead to a self-reinforcing loop for the stable maintenance of CHG methylation and H3K9me2 at the same sites. Further support for this model comes from the finding that CHG methylation and H3K9me2 tightly co-localized with each other at a genome-wide level. The maintenance of CHG methylation and H3K9me2 thus represent another example where the only signal required for the maintenance of a certain chromatin modification pattern is the prior existence of this pattern.

Unlike CG and CHG methylation, CHH methylation is considered "asymmetric" because only one of the two parental DNA strands is methylated at a particular site. This poses a problem for the maintenance of CHH methylation, since methylation is absent from one of the two double helices after DNA replication. What signal directs DRM1/2 to the appropriate sites? In addition to the defects in de novo methylation described above, siRNA pathway mutants also display severely decreased levels of CHH methylation at endogenous sites (Hamilton et al. 2002; Chan et al. 2004, 2006; Herr et al. 2005; Kanno et al. 2005; Onodera et al. 2005; Pontes et al. 2006). This observation suggests that the RdDM process is responsible for not only de novo methylation in all sequence contexts, but also the maintenance of CHH methylation.

15.2.1.4 Removal of DNA Methylation

The direct removal of the methyl group from a methyl-cytosine is thermodynamically unfavorable as it involves the breakage of a carbon-carbon bond, and the enzymes that can catalyze this reaction have not been identified. However, a genetic screen for *Arabidopsis* mutants in which a deregulated transgene became hypermethylated and silenced turned up *Repressor Of Silencing 1* (*ROS1*), which belongs to a family of glycosylase genes in *Arabidopsis* that also include *DEMETER* (*DME*), *DEMETER-LIKE 1* (*DML1*) and *DML2* (Choi et al. 2002; Gong et al. 2002). Both ROS1 and DME are capable of releasing methyl-cytosines from DNA through a base excision mechanism, and presumably demethylation is achieved through the incorporation of unmethylated cytosines during the subsequent DNA repair process. Importantly, hundreds of regions become hypermethylated in the *ros1 dml1 dml2* triple mutant, indicating that the ROS1 family of glycosylases normally function to remove deleterious DNA methylation from the genome (Penterman et al. 2007; Lister et al. 2008). The most dramatic case of DNA demethylation takes place in the developing endosperm, where the entire maternal genome is hypomethylated in a DME-dependent manner (Hsieh et al. 2009).

15.2.2 Histone Modifications

Four types of post-translational histone modifications have been detected by mass spectrometry studies (Johnson et al. 2004; Zhang et al. 2007a). These modifications include methylation, acetylation, phosphorylation and ubiquitination. In addition, in the case of methylation, a lysine residue can be mono-, di- or trimethylated, and an arginine residue can be mono- or di-methylated (either symmetrically or asymmetrically). The combinatorial patterns of histone modifications are therefore highly complex. The pathways involved in histone modifications are described below.

15.2.2.1 Lysine Methylation

Histone lysine methylation is catalyzed by a group of proteins containing the SET histone methyltransferase (HMTase) domain, referred to as "SET domain genes" (SDGs). Four lysines are methylated in *Arabidopsis* and all are located on H3, namely H3K4, H3K9, H3K27 and H3K36 (Johnson et al. 2004; Zhang et al. 2007a). Phylogenetic analysis of plant SDGs (roughly 40 in *Arabidopsis*) defined five major classes based on sequence homology (Baumbusch et al. 2001; Springer et al. 2003). Genetic and biochemical studies also showed that these five classes have distinct substrate specificities and enzymatic activities.

Class I SDGs are homologous to the animal H3K27 trimethyltransferase ENHANCER OF ZESTE [E(z)] and include three members in *Arabidopsis*: SDG1/CURLY LEAF (CLF),

SDG5/MEDEA (MEA) and SDG10/SWINGER (SWN) (Pien and Grossniklaus 2007). Loss of Class I SDG activity leads to decreased levels of di- and tri-methylation of H3K27 (H3K27me2/3) as well as the overexpression of many developmentally regulated genes, indicating that Class I SDGs function in H3K27me2/3 and transcriptional repression. Several protein complexes containing Class I SDGs have been isolated and found to play distinct roles during development (for review, see Pien and Grossniklaus 2007), including the MEA-FIS2-FIE-MSI1 complex that functions in gametophyte development, the EMF2-CLF-FIE-MSI1 complex involved in repressing precocious transition from vegetative to reproductive growth, and the VRN2-FIE-CLF-MSI1 complex responsible for promoting the transition to flowering after prolonged exposure to cold temperature (called "vernalization").

Class II SDGs are homologous to the animal H3K36 methyltransferase ABSENT, SMALL OR HOMEOTIC 1 (ASH1). The loss of the Class II protein SDG8 (also referred to as EFS or ASHH2) results in significant decrease in H3K36me2/3 and a number of developmental abnormalities, indicating that SDG8 functions primarily in H3K36me2/3 (Kim et al. 2005; Xu et al. 2008; Zhao et al. 2005). Interestingly, SDG8 is also capable of methylating H3K4 in vitro, although the global level of H3K4 methylation does not appear to be affected in sdg8 (Ko et al. 2010). Two other Class II SDGs have also been found to be required for normal development: the transition to flowering is delayed in sdg26/ashh1, and the sdg4/ashr3 mutant displays pollen tube growth defects (Thorstensen et al. 2008; Xu et al. 2008). The substrate specificities of SDG26 and SDG4 have yet to be determined.

Class III SDGs are homologous to the animal H3K4 methyltransferase TRITHORAX (TRX). Two Class II proteins, SDG2/ATXR3 and SDG27/ATX1, have been shown to methylate H3K4 in vitro (Alvarez-Venegas et al. 2003; Guo et al. 2010). Consistent with their role in H3K4me2/3, loss of SDG2 and SDG27 results in decreased levels of H3K4me2/3 in vivo. The SDG30/ATX2 and SDG25/ATXR7 proteins in Class III are also required for H3K4me2/3 in a locus-specific manner (Berr et al. 2009; Saleh et al. 2008).

Class IV SDGs do not have animal homologues, and their counterparts in yeast (SET3 and SET4 in S. cerevisiae) have not been shown to be enzymatically active. The two members in this class, SDG15/ATXR5 and SDG34/ATXR6, appear to be partially redundant in function. Both ATXR5 and ATXR6 interact with PROLIFERATING CELL NUCLEAR ANTIGEN (PCNA) and specifically catalyze the mono-methylation of H3K27 (Raynaud et al. 2006; Jacob et al. 2009). The atxr5 atx6 double mutant produces smaller leaves, and the overexpression of either ATXR5 or ATXR6 leads to male sterility. The loss of H3K27me1 in atxr5 atxr6 is also associated with DNA replication defects (described below).

Class V SDGs are homologous to the animal H3K9 methyltransferase SUPPRESSOR OF VARIEGATION 3-9 [Su(var)3-9]. The founding member of this class, SDG33/SUVH4/KYP, functions in H3K9me2 and heterochromatic silencing (Jackson et al. 2002; Malagnac et al. 2002). SDG9/SUVH5 and SDG23/SUVH6 are closely related to SUVH4, and both possess H3K9 methyltransferase activity in vitro, as does a more distantly related member in this class, SDG3/SUVH2 (Ebbs et al. 2005; Ebbs and Bender 2006; Naumann et al. 2005). The severe decrease of H3K9me2 in Class V mutants leads to the reactivation of previously silenced transposons, but not obvious developmental phenotypes. It therefore appears that H3K9me2 is functionally specialized in transposon silencing in Arabidopsis.

15.2.2.2 Lysine Demethylation

Unlike in the case of methyl-cytosines in DNA, methyl groups can be readily removed from methylated lysine residues in histones by two types of demethylases: the Lysine-specific demethylase1 (LSD1) type of demethylases contain the Polyamine oxidase (PAO) domain and remove methyl groups by amine oxidation, and the Jumonji C (JmjC) domain-containing demethylases remove methyl groups by hydroxylation (Shi et al. 2004; Tsukada et al. 2006). These two types of enzymes also act on different substrates and require different cofactors. The LSD1-type demethylases act on di- and mono-methylated (but not tri-methylated) H3K4 and H3K9, and require flavin adenine dinucleotide (FAD). In contrast, the JmjC-type demethylases act on mono-, di- and tri-methylated H3K4, H3K9, H3K27 and H3K36, and require Fe(II) and α-ketoglutarate (αKG) as cofactors.

The Arabidopsis genome harbors four LSD1 type demethylases: FLOWERING LOCUS D (FLD), LSD1-LIKE 1 (LDL1), LDL2 and LDL3. Interestingly, these four genes are represented by four ancient lineages that have diverged prior to the diversification of monocots and dicots, and each lineage has been maintained at low copy number in most plant species, indicating that they may play divergent yet conserved roles in lysine demethylation (Zhou and Ma 2008). Increased levels of H3K4me2 have been reported at specific loci in the fld, ldl1 (also referred to as kdm1c) and ldl1 ldl2 mutants, indicating that plant LSD1-type demethylases function in H3K4 demethylation (Jiang et al. 2007; Liu et al. 2007). Whether these enzymes are also involved in H3K9 demethylation has yet to be determined.

JmjC-type demethylases are much greater in number in plants, totaling 21 in Arabidopsis and 20 in rice (Lu et al. 2008). Comparative phylogenetic analysis of animal and plant JmjC-type demethylases showed that some animal genes might be missing from plant genomes. In addition, substantial differences in substrate specificities may exist between some closely related JmjC-type demethylases in plants and animals (see below). Five groups of JmjC-type demethylases have been defined in plants, including the KDM5/JARID1 group, the KDM4/JHDM3 group, the

KDM3/JHDM2 group, the JMJD6 group and the JmjC domain-only group. The activity of plant JmjC-type demethylases is currently an active area of research, and the specificities of several demethylases have been determined. For example, JMJ14 in the KDM5/JARID1 group demethylates H3K4, and REF6 and Increase in Bonsai Methylation 1 (IBM1/JMJ25) in the KDM4/JHDM3 group demethylates H3K27 and H3K9, respectively (Saze et al. 2008; Lu et al. 2010, 2011). The enzymes responsible for demethylation of H3K36 have not been identified in plants.

15.2.2.3 Arginine Methylation

In addition to lysines, several arginine residues in core histones can also be mono- or di-methylated, although a comprehensive list of methylated arginines is not yet available. Nine Protein Arginine Methyltransferases (PRMTs) have been identified in *Arabidopsis* (Niu et al. 2007, 2008). AtPRMT4a and AtPRMT4b are capable of di-methylating H3 arginine 2 (H3R2), H3R17 and H3R26, and AtPRMT5/SKB1 and AtPRMT10 can dimethylate H4R3. Developmental abnormalities have been reported for the *atprmt5*, *atprmt10* and *atprmt4a atprmt4b* mutants, indicating that histone arginine methylation plays important roles in regulating gene expression and plant development.

15.2.2.4 Lysine Acetylation and Deacetylation

Acetylation and deacetylation of histone lysine residues are catalyzed by histone acetyltransferases (HATs) and histone deacetylases (HDACs), respectively. Twelve HATs have been identified in *Arabidopsis* and grouped into three families, including the GNAT/MYST family (five members; HAG1–HAG5), the CBP family (five members; HAC1, 2, 4, 5 and 12), and the TAFII250 family (two members; HAF1 and HAF2) (Pandey et al. 2002). The sixteen HDACs in *Arabidopsis* have also been grouped into three families: the RPD3/HDA1 family (ten members; referred to as HDAs), the HD2 family (four members; HDT1–4) and the SIR2 family (two members, SRT1 and SRT2) (Pandey et al. 2002). Altered levels of H3 and H4 lysine acetylation (H3ac and H4ac, respectively) have been reported in several HATs and HDACs mutants, which is often accompanied by developmental phenotypes and defects in transposon silencing. The specificities of several HATs have been determined using in vitro assays. For example, HAG1 acetylates H3K14, HAG2 acetylates H4K12, and HAG4/HAM1 and HAG5/HAM2 both acetylate H4K5 (Earley et al. 2007). However, other HATs (such as HAC1, HAC5 and HAC12) appear more promiscuous and can act on multiple lysines.

15.2.2.5 Lysine Ubiquitination and Deubiquitination

Ubiquitination is critically important in protein degradation pathways where proteins to be degraded are marked by long ubiquitin chains. In contrast, only a single ubiquitin is added to lysine residues in histones ("mono-ubiquitination"). Histone lysine ubiquitination is thought to disrupt histone-DNA methylation, thereby creating a more open chromatin environment. Consistent with this view, histone lysine ubiquitination is associated with active transcription in animal genomes.

Lysine 143 on histone H2B (H2BK143) is the only ubiquitinated residue in *Arabidopsis* core histones, and H2BK143ubq depends on the E3 ligases HISTONE MONOUBIQUITINATION1 (HUB1) and HUB2 (Cao et al. 2008). Both *hub1* and *hub2* mutants display decreased levels of H2BK143ubq and several developmental phenotypes, including early flowering, reduced plant size and pale green leaves. Deubiquitination of H2BK143 is catalyzed by several ubiquitin-specific proteases/deubiquitinases in *Arabidopsis*, including UBP26/SUP32 and OTLD1 (Sridhar et al. 2007; Krichevsky et al. 2011). Consistent with the role of histone lysine ubiquitination in transcriptional activation, the loss of H2BK143ubq from the *FLC* gene is associated with the down-regulation of *FLC* expression (Schmitz et al. 2009). In addition, the effective removal of ubiquitin from H2BK143ubq is required for transcriptional repression, as the *ubp26/sup32* and *otld1* mutants are both defective in down-regulating *FLC* expression (Krichevsky et al. 2011).

15.2.2.6 Serine and Threonine Phosphorylation

A number of serine and threonine residues on histone H2A, H2B and H3 are phosphorylated in *Arabidopsis*. Histone phosphorylation has been linked to cell cycle progression and DNA damage repair in animals. As described below, H3 phosphorylation in *Arabidopsis* also appears to be tightly regulated in a cell-cycle-dependent manner. Two types of kinases are involved in H3 phosphorylation: the Aurora kinases phosphorylate serines 10 and 28 on histone H3 (H3S10p and H3S28p, respectively), and AtHaspin phosphorylates threonines 3 and 11 on histone H3 (H3T3p and H3T11p, respectively) (Demidov et al. 2005; Kawabe et al. 2005; Ashtiyani et al. 2011; Kurihara et al. 2011). The kinases involved in H2A and H2B phosphorylation or the phosphatases responsible for removing phosphates from histones have yet to be identified in plants.

15.3 Methods for Genome-Wide Analysis of Chromatin Modifications

The combination of biochemical methods previously used for locus-specific studies and high-throughput genomic approaches has provided us with many powerful tools to study the genome-wide distribution patterns of chromatin modifications at very high resolution. Some recently developed methods are described here.

15.3.1 DNA Methylation

Several biochemical methods have been used to distinguish methylated and unmethylated DNA. These methods take advantage of the differences between cytosines and methyl-cytosines in the susceptibilities to methylation-sensitive or methylation-specific enzyme digestion, the binding affinities by several families of proteins or antibodies, and the resistance to bisulfite conversion. The combination of these biochemical methods with high-throughput genomic approaches has made it possible to detect DNA methylation in a genome-wide manner.

15.3.1.1 Methods Involving Enzymatic Digestion of DNA

Eukaryotic DNA methyltransferases evolved from prokaryotic DNA methyltransferases that are components of the restriction modification system. In this system, site-specific methyltransferases catalyze the methylation of cellular DNA at restriction sites, which blocks digestion by corresponding restriction enzymes. This process provides a defense mechanism against "invading DNA" (such as that from bacteriophages), which is unmethylated and therefore digested upon entering the cell. For example, the restriction enzyme *Hpa*II digests DNA at 5′-CCGG-3′ sites, but its activity is blocked by the methylation of either cytosine at these sites catalyzed by the *Hpa*II methyltransferase. In addition, many methylation-sensitive restriction enzymes have isoschizomers that digest DNA at the same sites, but are either fully or partially insensitive to methylation. For example, *Msp*I also digests DNA at 5′-CCGG-3′ sites, but it is insensitive to the methylation status of the inner C. The different digestion efficiencies of methylation-sensitive and methylation-insensitive isoschizomer pairs can therefore be used to probe the methylation status at specific sites. *Hpa*II and *Msp*I are a commonly used combination, particularly in mammalian systems where DNA methylation predominantly occurs at CG sites. Other isoschizomer pairs with different recognition sites can be used in plants to detect DNA methylation in additional sequence contexts. For example, the methylation-sensitive *Sau*3AI and the methylation-insensitive *Mbo*I both cut DNA at 5′-GATC-3′ sites, which can be used to detect CG, CHG and CHH methylation depending on the 3′ flanking sequence of the 5′-GATC-3′ sites. Methylation-sensitive digestion can be used in combination with Southern blot or Amplification Fragment Length Polymorphism (AFLP) analyses to detect DNA methylation at a single or dozens of sites, respectively. Moreover, digested DNA samples can also be analyzed by microarray hybridization to simultaneously detect DNA methylation at numerous restriction sites (Tran et al. 2005a, b). The major limitation of this method is that each enzyme can only be used at the corresponding restriction sites.

A second type of enzyme is exemplified by the methylation-specific McrBC. In contrast to methylation-sensitive restriction enzymes, McrBC only digests methylated DNA. One advantage of McrBC over typical restriction enzymes is the weaker sequence specificity: McrBC cuts between two Pu^mC sites (Pu: A or G; ^mC: methylated C) that can be up to several kilobases apart. The methylation status in a specific genomic region can therefore be measured by methods such as qPCR amplification of digested and undigested DNA. McrBC digestion can also be coupled with high-throughput methods for large-scale detection of DNA methylation: unmethylated DNA can be recovered as longer fragments after McrBC digestion and identified by microarray hybridization or high-throughput sequencing (Lippman et al. 2004; Vaughn et al. 2007; Li et al. 2008; Wang et al. 2009; He et al. 2010).

15.3.1.2 Methods Involving Affinity Purification or Immunoprecipitation

Several eukaryotic proteins have been found to preferentially bind methylated DNA, such as those containing the methyl-CpG-binding domain (MBD) or the SET and RING associated (SRA) domains (Cross et al. 1994; Johnson et al. 2007). These proteins are involved in the maintenance of DNA methylation or in mediating the down-stream functions of DNA methylation. However, they can also be used to isolate methylated DNA by affinity purification because they exhibit higher affinities toward methylated DNA. The most commonly used protein is the MBD domain of the human protein MeCP2. Recombinant MeCP2 protein binds both methylated and unmethylated DNA under low salt concentration, but it only binds DNA methylated at CG sites under high salt concentration (Cross et al. 1994). It should be noted, however, that MeCP2 cannot be used to isolate DNA that only contains CHG and CHH methylation.

Another effective method to isolate methylated DNA from the total genomic DNA pool is methyl-C immunoprecipitation (mCIP; or methylated DNA immunoprecipitation, mDIP). This method takes advantage of a commercially available mouse monoclonal antibody, which binds methylated cytosine in single-stranded DNA (presumably when the methyl-cytosine ring is exposed) (Keshet et al. 2006). One advantage of mCIP/mDIP over affinity purification is that the anti-methyl-C antibody recognizes methylated cytosines in all three sequence contexts.

Methylated DNA isolated by affinity purification or immunoprecipitation can be analyzed by microarray hybridization or high-throughput sequencing (Zhang et al. 2006; Penterman et al. 2007; Zilberman et al. 2007; Cokus et al. 2008; Miura et al. 2009). These methods have been successfully applied to produce the first whole-genome DNA methylation maps. The resolution of these methods is roughly a few hundred basepairs, which largely depends on the fragment size of input DNA as well as the computational analysis algorithms. Unlike the digestion-based approaches described above, affinity purification and immunoprecipitation do not require the presence of specific restriction sites. These methods are highly effective in the identification of methylated regions, although it is usually not possible to determine which particular cytosines are methylated within a region.

15.3.1.3 Methods Involving Bisulfite Treatment

Treating genomic DNA with sodium bisulfite converts cytosines to uracils, but methyl-cytosines resist this conversion. Following PCR amplification of bisulfite-treated DNA, unmethylated cytosines appear as thymines whereas methyl-cytosines remain as cytosines (Frommer et al. 1992; Henderson et al. 2010). This method, called bisulfite genomic sequencing, has been the gold standard for detecting DNA methylation for the past two decades. Unlike the methods described above, bisulfite genomic sequencing can unveil the methylation status of every cytosine within a genomic region. Bisulfite-treated DNA can also be analyzed by hybridization to microarrays with probes corresponding to unconverted sequences, as methylated DNA is unaffected by bisulfite treatment and therefore retains the ability to hybridize to the microarray (Reinders et al. 2008).

The combination of bisulfite conversion and deep sequencing using high-throughput methods has led to a major breakthrough in methylation studies, as it allows the detection of DNA methylation at single-base resolution in the entire genome (Cokus et al. 2008; Lister et al. 2008; Hsieh et al. 2009; Feng et al. 2010; Zemach et al. 2010a, b). Compared to other approaches described above, this "BS-seq" method offers several significant advantages. First, BS-seq is highly sensitive and can be used to detect very light methylation as well as subtle changes (e.g., in different tissues or mutant backgrounds). Second, it reveals the sequence contexts of DNA methylation. This is particularly informative in plant genomes where CG, CHG and CHH methylation is catalyzed by MET1, CMT3 and DRM2, respectively. BS-seq results therefore provide important information to identify the specific DNA methylation pathway(s) that function(s) at any given site. Third, the deep coverage of BS-seq makes it possible to quantitatively measure the percentage of DNA fragments in the input DNA pool that are methylated at a particular cytosine. This effectively measures the cell-to-cell variations in DNA methylation in the starting biological material and therefore offers important information regarding cell-specific methylation. The continuous decrease in sequencing costs and the maturation of analysis methods should make BS-seq an increasingly attractive method for DNA methylation detection.

15.3.2 Nucleosome Positioning and Occupancy

The position of a nucleosome is defined by the ~147 bp DNA wrapped around the histone octamer (referred to as "nucleosomal DNA" here), and nucleosome occupancy refers to the frequency at which a nucleosome is positioned at a given locus. It is possible to enrich for nucleosomal DNA by immunoprecipitation from cross-linked chromatin fragments (e.g., using an antibody against one of the histones). However, the most effective method utilizes an endo-exonuclease from *Staphylococcus aureus* called the Micrococcal Nuclease ("MNase"; also called S7 Nuclease). MNase digests both single-stranded and double-stranded nucleic acids, but nucleosomal DNA wrapped around histones is less susceptible to MNase digestion. Under appropriate conditions, it is possible to preferentially digest inter-nucleosomal DNA (but not nucleosomal DNA), thus releasing single nucleosomes from the chromatin. Nucleosomal DNA can then be recovered and analyzed by high-throughput sequencing (Chodavarapu et al. 2010). In this case, each nucleosomal DNA fragment defines the position of a nucleosome, and the frequency at which this fragment is recovered from the sequencing results provides a quantitative measurement of nucleosome occupancy at this position. It should be noted that the stability of nucleosomes may be variable, making some nucleosomal DNA more susceptible to MNase digestion. Therefore, MNase digestion conditions should be carefully controlled to avoid experimental artifacts caused by over-digestion.

15.3.3 Histone Modifications

The principle approach to study the distribution of histone modifications is by chromatin immuniprecipitation (ChIP). Chromatin can be cross-linked using formaldehyde to stabilize the histone-DNA interaction, and fragmented by sonication to a few hundred basepairs in length. Alternatively, MNase digestion can be performed to produce single or oligonucleosomes. Although MNase digestion can potentially lead to higher resolution (e.g., single-nucleosome level), it may be complicated by the variations in nucleosome stability, as mentioned above. Specific antibodies against modified histones are used in ChIP experiments to isolate chromatin fragments containing the corresponding modifications. The associated DNA can then be recovered and analyzed by microarray hybridization ("ChIP-chip") or high-throughput sequencing ("ChIP-seq") to determine the genomic localization of modified histones (Lippman et al. 2004; Turck et al. 2007; Zhang et al. 2007b, 2009; Bernatavichute et al. 2008; Oh et al. 2008; Charron et al. 2009; Elling and Deng 2009; Zhou et al. 2010; He et al. 2010; Ding et al. 2011a; Ha et al. 2011; Roudier et al. 2011).

ChIP results can be normalized to input DNA to determine the absolute level of modified histones. One can also normalize ChIP results against nucleosomal DNA (e.g., isolated using antibodies against histones regardless of the modifications), which minimizes the effect of nucleosome occupancy and provides a measurement of the abundance of modified histones relative to total histones at a give position.

15.4 Genome-Wide Distribution of Chromatin Modifications

15.4.1 DNA Methylation

15.4.1.1 *DNA Methylation in* Arabidopsis Thaliana

Genome-wide BS-seq results show that, in *Arabidopsis*, approximately 24%, 6.7% and 1.7% of CG, CHG and CHH sites are methylated, respectively (Cokus et al. 2008; Lister et al. 2008). Interestingly, the methylation frequency at CG sites appears to be bimodal. That is, a particular CG site is either always methylated or always unmethylated. In contrast, methylation frequency at CHH sites is consistently low (~10%), and it frequency varies significantly at CHG sites (from 20% to 100%). Considering the fact that the methylation frequency at any given site in fact represents the percentage of cells in the starting material containing methylation at this site, these results seem to indicate that CG methylation is relatively stably maintained throughout plant development, whereas a considerable amount of cell-specific methylation takes place at CHG and CHH sites. As described above, DRM2-mediated CHH methylation depends on the siRNA pathway, and significant differences have been reported between the siRNA populations that accumulate in different *Arabidopsis* tissues (Lu et al. 2005). It is therefore possible that the RdDM pathway may function at different target sites during development.

All three types of DNA methylation are highly enriched in transposons and other repeats, whereas non-repetitive intergenic regions generally contain very low levels of methylation (Zhang et al. 2006; Zilberman et al. 2007; Cokus et al. 2008; Lister et al. 2008). At endogenous genes, the promoters and 3′ ends are usually unmethylated. Interestingly, roughly one third of genes are methylated within transcribed regions with a pronounced bias toward the 3′ half (referred to as "body methylation") (Tran et al. 2005a; Zhang et al. 2006; Zilberman et al. 2007; Cokus et al. 2008; Lister et al. 2008). Gene body methylation is MET1-dependent and occurs almost exclusively at CG sites. Unlike at repetitive sequences where DNA methylation is associated with transcriptional repression, most genes containing body methylation are expressed at moderate to high levels. The origin and function of gene body methylation remain largely unknown (see more discussions below).

15.4.1.2 Evolution of DNA Methylation in Plants

The availability of the genome sequences of a growing number of plant species has made it possible to address questions regarding the evolution of DNA methylation patterns in plants. To this end, the genome-wide distribution of DNA methylation has been determined by BS-seq for several plant species across the phylogenetic spectrum, including an additional dicot (*Populus trichocarpa*), a monocot (*Oryza sativa*), early land plants (*Physcomitrella patens* and *Selaginella moellendorffii*), and green algae (*Chlamydomonas reinhardtii, Chlorella* sp. NC64A and *Volvox carteri*) (Feng et al. 2010; Zemach et al. 2010b).

All three flowering plants share a similar DNA methylation pattern as described above for *Arabidopsis*, including the dense CG, CHG and CHH methylation at repetitive sequences, the depletion of DNA methylation at promoters, and the presence of CG methylation at moderately expressed endogenous genes. *Physcomitrella patens* and *S. moellendorffii* share a similar DNA methylation pattern that is distinct from that in flowering plants. That is, although repetitive sequences are also heavily methylated at CG, CHG and CHH sites in these two species, little methylation occurs within genes. More diverse patterns of genic methylation were observed in green algae: *Chlorella* sp. NC64A genes are uniformly methylated at CG sites with a sharp decrease at promoters, whereas the genic methylation level is significantly lower in *V. carteri*, and only slightly higher than in intergenic regions in *C. reinhardtii*. These results indicate that although DNA methylation at both repetitive sequences and genes is ancient, several major changes in the distribution and possibly the functions of DNA methylation have taken place during plant evolution.

15.4.2 Nucleosome Positioning

The nucleosome hypothesis, that DNA is packaged around histones to form the basic units of the chromatin, was first proposed in the 1970s. However, the distribution pattern of nucleosomes in plants remained largely unknown until recently, when a genome-wide and detailed nucleosome map was generated in *Arabidopsis* by MNase digestion and high-throughput sequencing (Chodavarapu et al. 2010). The average distance between regularly spaced nucleosomes was found to be ~30 bp. However, nucleosomes are not evenly distributed across a chromosome. Consistent with the higher level of compaction of the hetrochromatin, pericentromeric hetrochromatin contains denser and more regularly spaced nucleosomes than the euchromatin.

Nucleosome occupancy is low at the promoters and 3′ ends of highly transcribed genes, perhaps to make DNA more accessible to protein complexes involved in transcriptional initiation and termination. Consistent with this notion, genes that are poorly transcribed in specific tissues assayed (but may be highly transcribed in other tissues) have significantly higher nucleosome occupancy at both ends, including a well-positioned nucleosome centered on the transcription start site (TSS). This observation also suggests that the nucleosome occupancy at the promoters and 3' ends of genes can be dynamically regulated. Interestingly, immediately downstream of the TSS, nucleosomes become very well positioned with the first nucleosome centered at approximately +40 bp and the second at +200 bp, and so on. Another interesting finding is that introns and exons are associated with different densities of nucleosomes. That is, exons have a significantly higher density of nucleosomes than introns. Moreover, both exon-intron and intron-exon boundaries (i.e., donor and acceptor splicing sites) are marked by well-positioned nucleosomes. Such a differential distribution does not appear to be affected by transcription, as it is similar in both highly and poorly transcribed genes. The functional significance of marking introns and exons with different densities of nucleosomes is not yet clear. However, the movement of RNA Polymerase II (Pol II) appears to be slower when it passes well-positioned nucleosomes (referred to as "stalling"). It is therefore possible that Pol II stalling in exons caused by the higher nucleosome density might prevent exon skipping and promote accurate splicing of upstream introns.

The observation that numerous nucleosomes are well positioned in the genome raises an important question: how is nucleosome position determined? DNA bends sharply in the nucleosome, and some dinucleotides (such as AA, AT and TT) facilitate directional bends. Analyses of nucleosomal DNA sequences in *Arabidopsis* revealed strong 10 bp periodicities in AA, GC and CG dinucleotides in the minor groove of the double helix facing the histone core (Widom 2001; Segal et al. 2006). Similar results have also been found in fungi and animals (Albert et al. 2007; Mavrich et al. 2008a, b; Valouev et al. 2008). The importance of DNA sequence in nucleosome positioning is highlighted by the fact that computational predictions of nucleosome positions based on dinucleotide periodicities can produce virtually the same results as MNase-seq. Collectively, these findings suggest that the positions of nucleosomes are profoundly affected by the DNA sequences.

15.4.3 Histone Modifications

The number of possible combinations of different histone modifications is so large that every nucleosome in the *Arabidopsis* genome can have its own unique combination. In reality, however, this does not appear to be the case. Instead, a distinct type of histone modification profile is associated with each of the four major components of the genome classified based on their transcriptional activity, namely intergenic regions, actively transcribed genes, repressed genes and transposons (Lippman et al. 2004; Turck et al. 2007; Zhang et al. 2007b, 2009; Bernatavichute et al. 2008; Oh et al. 2008; Charron et al. 2009; Elling and Deng 2009; Wang et al. 2009; He et al. 2010; Zhou et al. 2010; Ding et al. 2011a; Ha et al. 2011; Lafos et al. 2011; Roudier et al. 2011). Intergenic regions have little transcription activity and are virtually devoid of histone modifications. The profiles of the remaining three components are described below.

15.4.3.1 Actively Transcribed Genes

Actively transcribed genes are associated with H3K4me1, H3K4me2, H3K4me3, H3K36me1, H3K36me2, H3K36me3, H3K9ac, H3K14ac, H3K23ac, H4K5ac, H4K8ac, H4K12ac and H4K16ac (Lippman et al. 2004; Oh et al. 2008; Charron et al. 2009; Wang et al. 2009; Zhang et al. 2009; Ding et al. 2011a; Ha et al. 2011; Roudier et al. 2011) (Fig. 15.1). Importantly, none of these modifications is uniformly distributed across actively transcribed genes, and each modification is enriched only in a specific region. Specifically, H3K4me3 and all seven types of histone acetylation are enriched at the 5' ends of transcribed regions, covering roughly 1–1.2 kb downstream of the TSS, with the highest level found at ~400 bp downstream of the TSS. The distribution of H3K4me2 is similar to that of H3K4me3, with H3K4me2 being slightly downstream of H3K4me3 by roughly one nucleosome. The remainder of the transcribed region is associated with H3K4me1. H3K36me3 is distributed in regions also associated with either H3K4me2 or H3K4me1, and H3K36me2 is located near the 3' ends of transcribed regions. H3K36me1 is the most downstream modification and covers the 3' ends of genes as well as a few hundred base pairs downstream of the polyadenylation site. It is interesting to consider that histone modifications essentially partition actively transcribed genes into three regions: (1) the first 1–1.4 kb is associated with early transcriptional elongation activity and enriched for H3K4me2, H3K4me3 and all seven types of acetylation; (2) the 3' end is associated with transcriptional termination and polyadenylation and is enriched for H3K36me1 and H3K36me2; (3) the region between (1) and (2) is associated with transcriptional elongation and is enriched for H3K4me1 and H3K36me3.

Although the modifications described above are found almost exclusively at active genes, a given active gene may or may not contain all the modifications. Interestingly, the assortment of histone modifications at a particular gene appears to be determined by two simple factors: the length of this gene, and a hierarchy among histone modifications (H3K4me3 >H3K4me2 > H3K4me1; H3K36me3 > H3K36me2 > H3K36me1). For example, short genes (e.g. 1 kb in length) would only contain H3K4me2, H3K4me3 and acetylation, genes of moderate length would

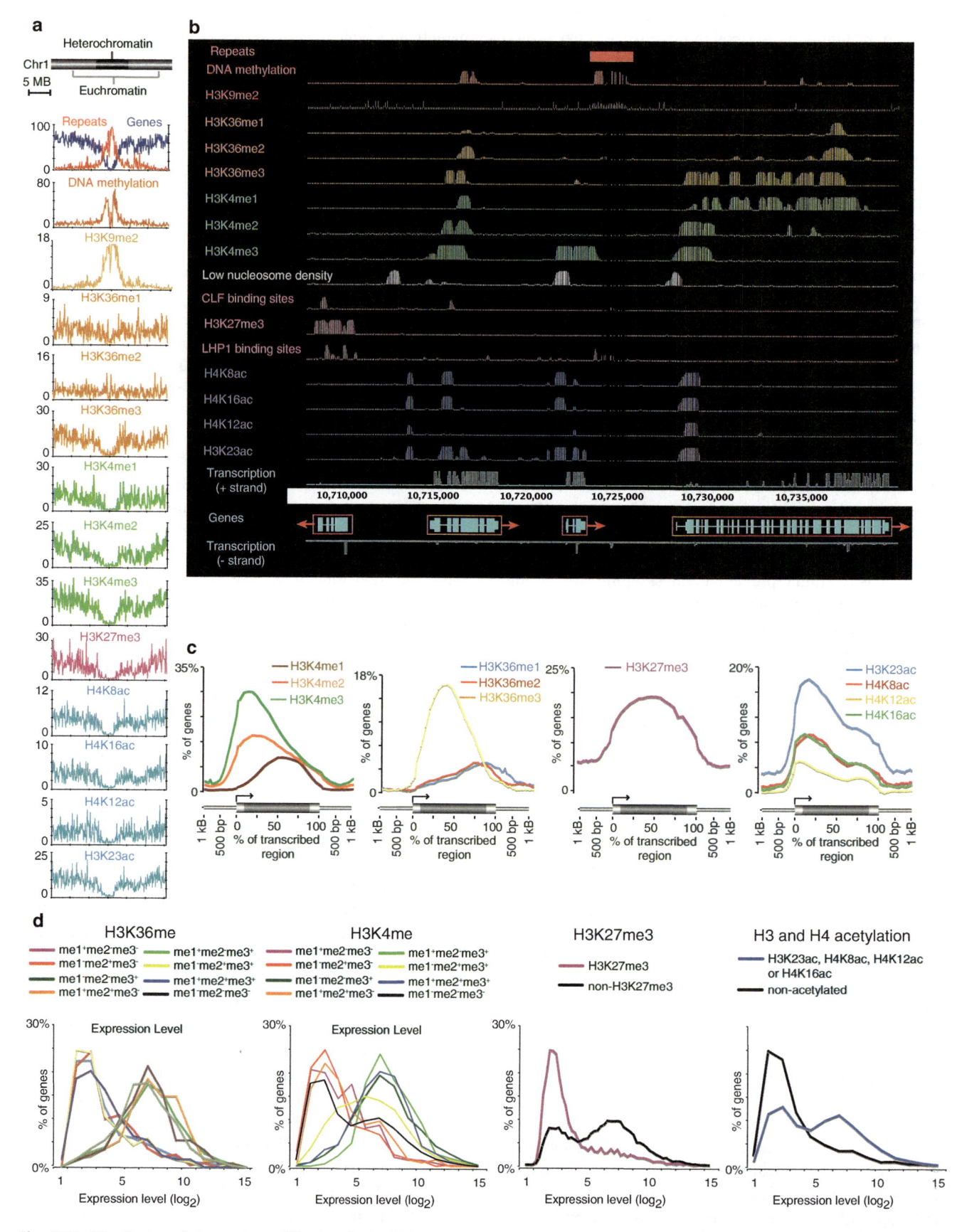

Fig. 15.1 Distribution of chromatin modifications in *Arabidopsis*. (**a**) Chromosomal distribution of chromatin modifications (Chr. 1 shown as an example). *x*-axis: chromosomal positions; *y*-axis: the length (kb) of genomic regions associated with each modification. A schematic representation of Chr. 1 is shown on *top*. (**b**) Detailed distribution patterns of chromatin modifications in a representative euchromatic region on

additionally contain H3K4me1 and H3K36me3, and long genes would contain all the modifications described above. Although the mechanistic basis for this hierarchy is not entirely clear, these observations suggest that the proper transcription of short, moderate and long genes may require different combinations of chromatin modifications.

15.4.3.2 Developmentally Repressed Genes

Developmentally repressed genes are hypoacetylated and generally do not contain H3K4me or H3K36me. Instead, most repressed genes are associated with H3K27me3 (Turck et al. 2007; Zhang et al. 2007b; Oh et al. 2008; Wang et al. 2009; Lafos et al. 2011; Roudier et al. 2011) (Fig. 15.1). In contrast to the histone modifications in active genes that are distributed in a 5′ to 3′ gradient, H3K27me3 is relatively uniformly distributed in repressed genes, covering the entire transcribed region but rarely extending up- or downstream to neighboring genes.

Both repressed genes and silenced transposons are transcriptionally inactive. However, a distinction should be made between these two types of loci. Whereas transposons are typically constitutively silenced, repressed genes are usually expressed elsewhere or at another time. Therefore, for repressed genes, the repressive modification profile (i.e. H3K27me3) must be converted to an active one (described above) during transcriptional activation and vice versa during transcriptional repression. Indeed, a recently comparative analysis of H3K27me3 in different *Arabidopsis* tissues showed that hundreds of genes gained or lost H3K27me3 during differentiation when they were repressed or activated, respectively (Lafos et al. 2011). Multi-protein complexes responsible for simultaneously removing H3K27me3 and adding H3K4me3 during transcriptional activation and for adding H3K27me3 while removing H3K4me3 during transcriptional repression have been identified in animals. It is tempting to speculate that similar protein complexes may also be involved in regulating histone modifications during transcriptional activation and repression in plants, but this has yet to be demonstrated.

15.4.3.3 Transposons and Other Repeats

The activities of transposons have been a major driving force of evolution by creating enormous amounts of genetic diversity. However, transposition also generates deleterious mutations and chromosomal rearrangements (e.g., by illegitimate recombination). To minimize these undesirable effects, it may be in the best interest of an organism to control transposon activities in the short term by transcriptional and post-transcriptional silencing. Genome-wide profiling studies have shown that transposons generally do not contain any of the histone modifications seen in active or repressed genes. Rather, transposons are marked by H3K9me2, heavy DNA methylation in CG, CHG and CHH contexts as well as dense clusters of siRNAs (Lippman et al. 2004; Lu et al. 2005; Zhang et al. 2006, 2007c; Zilberman et al. 2007; Bernatavichute et al. 2008; Cokus et al. 2008; Lister et al. 2008; Wang et al. 2009; Zhou et al. 2010; Roudier et al. 2011) (Fig. 15.1). As detailed above, CG methylation is effectively maintained by the MET1 pathway, the self-reinforcing interaction between the CMT3-mediated CHG methylation and KYP-mediated H3K9me2 pathways ensures the colonization of CHG methylation and H3K9me2, and CHH methylation by DRM2 persists as a result of the RdDM pathway. Consequently, these chromatin modifications are highly stable and function to constitutively repress transposon activities.

An interesting observation from chromatin profiling studies is that the modifications associated with transposons rarely extend beyond the elements themselves. Apparently, any spreading of constitutive silencing modifications into nearby genes may lead to undesirable consequences. But how are the boundaries specified and maintained? It is possible that insulator-like sites with specific DNA sequences or chromatin modifications might border most transposons. However, such sites have not been identified. Alternatively, the chromatin modification boundaries may be maintained dynamically by opposing modification pathways (Tamaru 2010). A recent study suggests that the latter scenario may be the case for CHG methylation and H3K9me2 (Saze et al. 2008). The *Bonsai* gene in *Arabidopsis* contains a LINE-type transposon in its 3′ UTR. In wild-type plants, the LINE element is heavily methylated in CG, CHG and CHH sites and contains H3K9me2. These modifications are limited to the LINE element itself and they do not appear to interfere with the normal expression of *Bonsai*. However, when the H3K9me2 demethylase *IBM1* is mutated, both CHG methylation and H3K9me2 spread beyond the LINE element into the *Bonsai* gene, causing the silencing of *Bonsai* and a number of developmental abnormalities. Importantly, thousands of other genes become ectopically associated with

Fig. 15.1 (Continued) Chr.1. The enrichment of each modification is shown as HHM posterior probability values (*bottom* = 0, *top* = 1). Genes are indicated by *red open boxes* (*arrows*: direction of transcription; *filled boxes*: exons; lines: introns). (**c**) Gene-level distribution of histone modifications. Each gene (*thick horizontal bar*) is divided into 20 intervals (5% each interval), and the 1-kb regions upstream and downstream of each gene (*thin horizontal bars*) are divided into 50-bp intervals. The percentage of genes with the corresponding modification in each interval is graphed (*y*-axis). (**d**) Genes with different expression level and patterns are associated with different types and combinations of histone methylation. *x*-axis: gene expression level (log$_2$ scale). *y*-axis: the percentage of genes with corresponding histone modification and expression values. Because mono-, di- and tri-methylation data are available for H3K4 and H3K36, the eight possible combinations for each of these two residues are analyzed separately

H3K9me2 and hypermethylated at CHG sites in the *ibm1* mutant background (Miura et al. 2009). These results strongly suggest that the H3K9me2 demethylase pathway plays a major role in determining the boundaries of H3K9me2 and CHG methylation.

15.5 Functions of Chromatin Modifications

15.5.1 Role of Chromatin Modifications in Regulating Transcription and Plant Development

The prevalence and the distinct distribution patterns of different types of modifications at both transposons and endogenous genes indicate that chromatin modifications may play important roles in regulating transcription. Our current understanding of the transcriptional regulatory functions of chromatin modifications relies largely on two types of evidence: the presence of particular types of chromatin modifications at specific sites in the genome, and the effects on transcription when chromatin modifications are compromised in mutants as revealed by genetic studies. The mechanistic details of how chromatin modifications might interact with the transcription apparatus have yet to be determined in plants. However, results from genetic studies have provided a wealth of information to address the transcription regulatory roles of chromatin modifications.

15.5.1.1 DNA Methylation and H3K9me2

The best-characterized function of DNA methylation is transposon silencing. Transcriptional reactivation of individual transposons was first observed in early genetic studies (Kakutani et al. 2004), and genome-wide transcriptional profiling studies have identified numerous cases of transposon reactivation in mutants with decreased levels of DNA methylation such as *met1* and the *drm1 drm2 cmt3* triple mutant (Zhang et al. 2006; Zilberman et al. 2007; Lister et al. 2008). In addition, some exogenous transposons introduced into wild-type *Arabidopsis* by transgenic methods are transcriptionally and transpositionally active, but they are progressively silenced over a few generations. Importantly, this silencing process does not occur in mutants that are defective in DNA methylation (Hirochika et al. 2000; Perez-Hormaeche et al. 2008). These results indicate that DNA methylation is critically important in both maintaining the silent state of endogenous transposons and the de novo silencing of new, "invading" transposons.

DNA methylation is also required for normal plant development. In addition to transposon reactivation, loss of CG methylation in *met1* causes several developmental abnormalities, such as delayed transition to flowering and floral organ transformation (Finnegan et al. 1996; Saze et al.

2003). Although the *drm1 drm2* and *cmt3* mutants appear phenotypically normal, gross loss of non-CG methylation (but with little change in CG-methylation) in the *drm1 drm2 cmt3* triple mutant leads to several developmental defects, such as twisted leaves, short stature and partial sterility (Cao and Jacobsen 2002b). The *met1 cmt3* double mutant and the *met1 drm1 drm2* triple mutant exhibit even more severe defects, and the simultaneous loss of all DNA methyltransferases in the *met1 drm1 drm2 cmt3* quadruple mutant leads to embryo lethality (Zhang and Jacobsen 2006). It is therefore likely that DNA methylation may be required for the proper expression of many endogenous genes throughout development.

Several genes regulated by DNA methylation haven been studied in detail, For example, the *FWA* gene is methylated in its promoter and silenced in all tissues except the endosperm in wild-type plants, but it becomes hypomethylated and ectopically expressed in the *ddm1* and *met1* mutants, resulting in a late flowering phenotype (Ronemus et al. 1996; Soppe et al. 2000; Kankel et al. 2003). Similarly, the *SDC* gene is normally methylated in the promoter and silenced, but it loses promoter methylation and becomes overexpressed in the *drm1 drm2 cmt3* mutant, causing the twisted leaf phenotype (Henderson and Jacobsen 2008). A common feature shared by *FWA* and *SDC* is that DNA methylation is localized to transposon-like repeats in their promoter regions. It therefore appears that the function of DNA methylation in transposon silencing has been co-opted during evolution to also affect endogenous genes located in close proximity of transposons. Many other genes regulated by CG and non-CG methylation have also been identified by transcriptional profiling studies in *met1* and *drm1 drm2 cmt3* (Zhang et al. 2006; Zilberman et al. 2007; Lister et al. 2008).

The function of gene body methylation remains an open question. It is unlikely that gene body methylation grossly represses transcription, as most genes with body methylation are transcribed at moderate to high levels. It has been proposed that body methylation might function to repress ectopic initiation within transcribed regions, but the loss of CG methylation in the *met1* mutant does not appear to cause rampant ectopic initiation in gene bodies (Tran et al. 2005a; Lister et al. 2008). It has also been proposed that body CG methylation might interfere with Pol II elongation and thereby fine-tune transcription. Consistent with this idea, the most highly expressed genes appear to be devoid of body methylation, and a moderate increase in transcript level has been observed for body-methylated genes in *met1* (Zilberman et al. 2007).

Consistent with the mechanistic interaction between the H3K9me2 and CHG methylation pathways, decreased level of H3K9me2 leads to very similar defects as those seen in *cmt3*. The moderate loss of H3K9me2 in the *kyp/suvh4* mutant or a more severe loss in the *kyp/suvh4 suvh5 suvh6* triple mutant causes transposon reactivation, but both mutants are phenotypically normal. In addition, when combined with other mutations, *kyp/suvh4* causes the same effects as *cmt3*. For example, the *drm1 drm2 kyp/suvh4* triple mutant exhibits the same phenotypic defects as *drm1 drm2 cmt3* as well as *SDC* overexpression.

15.5.1.2 H3K27me3

The observation that most developmentally repressed genes are associated with H3K27me3 suggests that this modification may play important roles in transcriptional repression. The moderate loss of H3K27me3 in the *clf* mutant causes pleiotropic phenotypes, and the severe loss of H3K27me3 leads to a much more dramatic phenotype: the *clf swn* mutant remains as immortal callus-like embryonic tissue and appears unable to differentiate. The transcriptional profile of *clf swn* is drastically different from that of wild-type plants, and most genes repressed by H3K27me3 are ectopically overexpressed in the mutant (X.Z., unpublished results).

It should be noted that, although H3K27me3 is critically important in maintaining its target genes in a repressed state, it does not appear to be required for the initial repression process. The best example comes from studies of vernalization, which refers to the promotion of flowering in some plant species in response to a prolonged exposure to cold temperatures. In *Arabidopsis*, a key gene in vernalization is the flowering repressor *FLC*. Before vernalization, *FLC* is highly expressed and represses flowering. During vernalization, the expression level of *FLC* decreases while the H3K27me3 level increases over several weeks, until *FLC* becomes transcriptionally repressed. Following vernalization, although the environmental signal that triggers *FLC* repression (i.e. cold temperature) is removed, *FLC* remains repressed. Importantly, *FLC* repression still occurs during vernalization in the *vrn2* mutant, but it does not remain repressed following vernalization. Instead, *FLC* expression recovers to a level comparable to that before vernalization (Gendall et al. 2001). It is therefore likely that, although H3K27me3 plays a broad and important role in maintaining transcriptional repression, the repression process may occur independently at individual genes through the actions of genetic factors.

15.5.1.3 H3K4 and H3K36 Methylation

Most actively transcribed genes are associated with H3K4me3 at the 5′ end, and longer genes are also marked by a 5′ to 3′ gradient of H3K4me3/H3K4me2/H3K4me1 and H3K36me3/H3K36me2/H3K36me1 (Lippman et al. 2004; Oh et al. 2008; Zhang et al. 2009; Roudier et al. 2011). Several loss-of-function H3K4 or H3K36 HMTase mutants have been found to display many developmental defects. These mutants include *atx1/sdg27*, *atxr3/sdg2*, *atxr7/sdg25*, *ashh1/sdg26*, *ashh2/sdg8* and *ashr3/sdg4* (Alvarez-Venegas et al. 2003; Kim et al. 2005; Zhao et al. 2005; Dong et al. 2008; Thorstensen et al. 2008; Xu et al. 2008; Berr et al. 2009, 2010a, b; Cazzonelli et al. 2009, 2010; Grini et al. 2009; Tamada et al. 2009; Guo et al. 2010; Palma et al. 2010). Consistent with the notion that these modifications may be important for active transcription, many genes have been found to be down-regulated by transcriptional profiling studies of these mutants (Alvarez-Venegas et al. 2006; Xu et al. 2008; Guo et al. 2010). Interestingly, loss of H3K4me and H3K36me is not always associated with transcriptional down-regulation. For example, many genes are unaffected in the *atxr3/sdg2* mutant, although their H3K4me3 levels are severely decreased. In addition, the H3K36me3 level is universally reduced in the *ashh2/sdg8* mutant, but the expression of the vast majority of genes remain unchanged. The interpretation of these results is complicated by the genetic redundancy of H3K4 and H3K36 HMTases (i.e. H3K4me and H3K36me levels are reduced but not eliminated in these mutants) (Alvarez-Venegas et al. 2006; Xu et al. 2008; Guo et al. 2010). Nevertheless, these results suggest that some genes appear to be more sensitive to changes in chromatin modifications than others. The mechanistic basis for this difference remains unclear.

Two lines of evidence suggest that, in addition to maintaining active transcription, H3K4me3 may also be required for the transcriptional activation process. First, the H3K4me3 HMTase ATX1/SDG27 is required for the recruitment of TATA-box Binding Protein and pre-initiation Pol II, which occurs prior to and is the prerequisite of active transcription (Ding et al. 2011b). Second, the *atx1/sdg27* mutant is more susceptible to drought stress, and the defect is in part due to the failure to rapidly activate the *NCED3* gene upon dehydration treatment (Ding et al. 2011a).

15.5.2 Role of Chromatin Modifications in Other Chromatin-Based Processes

In addition to transcriptional regulation, chromatin modifications also play important roles in other biological processes that take place in the chromatin environment, including DNA replication, chromatin condensation, DNA damage repair and recombination.

15.5.2.1 DNA Replication

Faithful duplication of the chromosomes during cell division is critically important for the maintenance of genome integrity and the inheritance of genetic materials by the two daughter cells. A recent analysis of the timing of DNA replication of *Arabidopsis* chromosome 4 identified two phases of replication: early replication occurs preferentially in euchromatic arms, whereas the heterochromatic regions are mostly replicated in a later phase (Lee et al. 2010). This observation indicates that chromatin modifications may play import roles in specifying the origin and timing of DNA replication.

In *S. cerevisiae*, the origins of replication are defined by specific DNA sequences (Bell and Stillman 1992; Wyrick et al. 2001). In contrast, chromatin modification profiles might play a more important role in specifying the replication origins in plants. Results from genome-wide mapping of replication origins in *Arabidopsis* showed that the midpoint of most origin regions are preferentially located within the 5′ half of genes, where they are associated with H3K4me2, H3K4me3, H4K5ac and the histone H2A variant H2A.Z (Costas et al. 2011). DNA methylation, H3K4me1 and H3K9me2 are generally depleted at replication origins (Costas et al. 2011). Importantly, the large subunit of the origin recognition complex, ORC1, binds histone H3 containing H3K4me3 through its PHD finger domain (de la Paz Sanchez and Gutierrez 2009). These results indicate that H3K4me3 may function to define replication origins by directly recruiting the origin recognition complex.

Chromatin modifications are also critically important for the correct timing of DNA replication. A breakthrough finding came recently during the analysis of the *Arabidopsis* H3K27me1 HMTases ATXR5 and ATXR6. H3K27me1 is enriched in the heterochromatin, and the decrease of H3K27me1 in the *atxr5 atxr6* mutant leads to transposon activation without perturbing DNA methylation or H3K9me2 (Jacob et al. 2009, 2010). The observation that ATXR5 and ATXR6 interact with the replication fork protein PCNA suggests that H3K27me1 may be involved in DNA replication (Raynaud et al. 2006). Indeed, the primary defect in the *atxr5 atxr6* mutant appears to be the over-replication of heterochromatic regions (Jacob et al. 2010). That is, the replication origins in heterochromatin "fire" more than once during a single S phase, resulting in multiple copies of the DNA near heterochromatic replication origins. These results indicate that H3K27me1 functions in the heterochromatin to ensure that only one round of replication is initiated from each origin.

15.5.2.2 Chromatin Cohesion and Condensation

The assembly of DNA into nucleosomes and chromatin is thought to have evolved as a packaging mechanism to allow eukaryotic organisms to accommodate longer DNA. It is therefore not surprising that some chromatin modifications have been implicated in chromatin condensation and segregation during cell divisions. The best characterized modification that affect chromatin cohesion and condensation is the phosphorylation of several serine and threonine residues on histone H3.

In *Arabidopsis*, the Aurora kinases are responsible for the phosphorylation of serines 10 and 28 on histone H3 (H3S10p and H3S28p, respectively) (Demidov et al. 2005; Kawabe

et al. 2005). H3S10p and H3S28p accumulate in the pericentromeric regions at metaphase in mitosis and at metaphase II in meiosis. In contrast, these modifications are seen throughout the chromatin at metaphase I in meiosis (Kaszas and Cande 2000; Manzanero et al. 2000; Gernand et al. 2003). It is therefore possible that H3S10p and H3S28p may play multiple roles: the cohesion of pericentromeric regions during mitosis and meiosis II, and the cohesion of sister chromatids during meiosis I. Inhibition of Aurora kinase activity by the inhibitor Hesperadin in tobacco cells does not affect chromatid condensation, but frequently leads to lagging chromosomes at anaphase, indicating that H3S10p and H3S28p may be required for the dissociation of chromatid cohesion (Kurihara et al. 2006).

A second type of kinase, AtHaspin, catalyzes the phosphorylation of threonines 3 and 11 on histone H3 in *Arabidopsis* (H3T3p and H3T11p, respectively) (Ashtiyani et al. 2011; Kurihara et al. 2011). H3T3p and H3T11p are present over the entire condensed chromosomes in both mitosis and meiosis (Houben et al. 2005). Importantly, the reduced levels of H3T3p and H3T11p in *AtHaspin* RNAi plants is associated with reduced chromatin condensation and a number of developmental phenotypes (Ashtiyani et al. 2011).

Phosphorylation is not restricted to canonical histone H3. For example, the centromere-specific histone H3 variant CENH3 in maize is also phosphorylated at serine 50 ("phCENH3-Ser50") (Zhang et al. 2005). The occurrence of phCENH3-Ser50 is first apparent at prophase, peaks at metaphase, and decreases at anaphase. The sharp drop of phCENH3-Ser50 between metaphase and anaphase suggests that this modification may play a role in the spindle checkpoint.

15.6   Concluding Remarks

The mechanisms and functions of plant chromatin modifications have been an active area of research in the past two decades. The collective results of numerous genetic, genomic and biochemical studies have significantly broadened our understanding of the enzymes involved in chromatin modifications and how these modifications are distributed in the genome. However, we know relatively little about how these enzymes are targeted to specific genes during development and, once established, how chromatin modifications may affect the myriad of biological processes taking place in the chromatin environment. Detailed mechanistic studies are underway to address these questions, which should lead to many new and exciting findings in the near future.

From a technical perspective, the availability of high-throughput tools has made it possible to profile genome-wide chromatin modification patterns in great detail. However, a growing body of evidence suggests that the mechanisms and functions of chromatin modifications may be tissue- or cell-type-specific. For example, significant differences in chromatin modifications have been reported in different *Arabidopsis* tissues, the expression of chromatin modifying enzymes is often developmentally regulated, the same enzyme can be incorporated into different protein complexes in different tissues, and mutations in chromatin modification pathways often leads to specific developmental defects. Our ability to understand the developmental regulation and the functions of chromatin modifications is limited by the lack of effective methods to isolate large quantities of specific cells, particularly those that are of low abundance but nevertheless critically important in plant development and reproduction, such as germ cells, early embroys and meristem cells. A number of methods are being developed to tackle this problem, including laser capture dissection (LCM), fluorescent activated cell sorting (FACS), and a recently developed method to tag the nuclear membranes of specific cells (Deal and Henikoff 2010). The combination of these methods with high-throughput approaches will undoubtedly further our understanding of the dynamic structure and functions of plant genomes.

Acknowledgement The author is grateful to Drs. Jonathan Wendel and Christina Richards for their critical reading of the manuscript. Research in X.Z.'s laboratory is suppoted by the National Science Foundation Grant 0960425.

References

Albert I, Mavrich TN, Tomsho LP, Qi J, Zanton SJ, Schuster SC, Pugh BF (2007) Translational and rotational settings of H2A.Z nucleosomes across the Saccharomyces cerevisiae genome. Nature 446:572–576

Alvarez-Venegas R, Pien S, Sadder M, Witmer X, Grossniklaus U, Avramova Z (2003) ATX-1, an *Arabidopsis* homolog of trithorax, activates flower homeotic genes. Curr Biol 13:627–637

Alvarez-Venegas R, Sadder M, Hlavacka A, Baluska F, Xia Y, Lu G, Firsov A, Sarath G, Moriyama H, Dubrovsky JG, Avramova Z (2006) The *Arabidopsis* homolog of trithorax, ATX1, binds phosphatidylinositol 5-phosphate, and the two regulate a common set of target genes. Proc Natl Acad Sci U S A 103:6049–6054

Ashtiyani RK, Moghaddam AM, Schubert V, Rutten T, Fuchs J, Demidov D, Blattner FR, Houben A (2011) AtHaspin phosphorylates histone H3 at threonine 3 during mitosis and contributes to embryonic patterning in *Arabidopsis*. Plant J. doi:10.1111/j.1365-313X.2011.04699.x [Epub ahead of print]

Bartee L, Malagnac F, Bender J (2001) *Arabidopsis* cmt3 chromomethylase mutations block non-CG methylation and silencing of an endogenous gene. Genes Dev 15:1753–1758

Baumbusch LO, Thorstensen T, Krauss V, Fischer A, Naumann K, Assalkhou R, Schulz I, Reuter G, Aalen RB (2001) The *Arabidopsis thaliana* genome contains at least 29 active genes encoding SET domain proteins that can be assigned to four evolutionarily conserved classes. Nucleic Acids Res 29:4319–4333

Bell SP, Stillman B (1992) ATP-dependent recognition of eukaryotic origins of DNA replication by a multiprotein complex. Nature 357:128–134

Bernatavichute YV, Zhang X, Cokus S, Pellegrini M, Jacobsen SE (2008) Genome-wide association of histone H3 lysine nine methylation with CHG DNA methylation in *Arabidopsis thaliana*. PLoS One 3:e3156

Berr A, McCallum EJ, Alioua A, Heintz D, Heitz T, Shen WH (2010a) *Arabidopsis* histone methyltransferase SET DOMAIN GROUP8 mediates induction of the jasmonate/ethylene pathway genes in plant defense response to necrotrophic fungi. Plant Physiol 154:1403–1414

Berr A, McCallum EJ, Menard R, Meyer D, Fuchs J, Dong A, Shen WH (2010b) *Arabidopsis* SET DOMAIN GROUP2 is required for H3K4 trimethylation and is crucial for both sporophyte and gametophyte development. Plant Cell 22:3232–3248

Berr A, Xu L, Gao J, Cognat V, Steinmetz A, Dong A, Shen WH (2009) SET DOMAIN GROUP25 encodes a histone methyltransferase and is involved in FLOWERING LOCUS C activation and repression of flowering. Plant Physiol 151:1476–1485

Bird A (2002) DNA methylation patterns and epigenetic memory. Genes Dev 16:6–21

Bostick M, Kim JK, Esteve PO, Clark A, Pradhan S, Jacobsen SE (2007) UHRF1 plays a role in maintaining DNA methylation in mammalian cells. Science 317:1760–1764

Cao X, Jacobsen SE (2002a) Locus-specific control of asymmetric and CpNpG methylation by the DRM and CMT3 methyltransferase genes. Proc Natl Acad Sci U S A 99(Suppl 4):16491–16498

Cao X, Jacobsen SE (2002b) Role of the *Arabidopsis* DRM methyltransferases in de novo DNA methylation and gene silencing. Curr Biol 12:1138–1144

Cao Y, Dai Y, Cui S, Ma L (2008) Histone H2B monoubiquitination in the chromatin of FLOWERING LOCUS C regulates flowering time in *Arabidopsis*. Plant Cell 20:2586–2602

Cazzonelli CI, Cuttriss AJ, Cossetto SB, Pye W, Crisp P, Whelan J, Finnegan EJ, Turnbull C, Pogson BJ (2009) Regulation of carotenoid composition and shoot branching in *Arabidopsis* by a chromatin modifying histone methyltransferase, SDG8. Plant Cell 21:39–53

Cazzonelli CI, Roberts AC, Carmody ME, Pogson BJ (2010) Transcriptional control of SET DOMAIN GROUP 8 and CAROTENOID ISOMERASE during *Arabidopsis* development. Mol Plant 3:174–191

Chan SW, Zilberman D, Xie Z, Johansen LK, Carrington JC, Jacobsen SE (2004) RNA silencing genes control de novo DNA methylation. Science 303:1336

Chan SW-L, Henderson IR, Zhang X, Shah G, Chien JS-C, Jacobsen SE (2006) RNAi, DRD1 and istone methylation actively target developmentally important non-CG DNA methylation in *Arabidopsis*. PLoS Genet 2:e83

Charron JB, He H, Elling AA, Deng XW (2009) Dynamic landscapes of four histone modifications during deetiolation in *Arabidopsis*. Plant Cell 21:3732–3748

Chodavarapu RK, Feng S, Bernatavichute YV, Chen PY, Stroud H, Yu Y, Hetzel JA, Kuo F, Kim J, Cokus SJ, Casero D, Bernal M, Huijser P, Clark AT, Kramer U, Merchant SS, Zhang X, Jacobsen SE, Pellegrini M (2010) Relationship between nucleosome positioning and DNA methylation. Nature 466:388–392

Choi Y, Gehring M, Johnson L, Hannon M, Harada JJ, Goldberg RB, Jacobsen SE, Fischer RL (2002) DEMETER, a DNA glycosylase domain protein, is required for endosperm gene imprinting and seed viability in arabidopsis. Cell 110:33–42

Cokus SJ, Feng S, Zhang X, Chen Z, Merriman B, Haudenschild CD, Pradhan S, Nelson SF, Pellegrini M, Jacobsen SE (2008) Shotgun bisulphite sequencing of the *Arabidopsis* genome reveals DNA methylation patterning. Nature 452:215–219

Costas C, de la Paz Sanchez M, Stroud H, Yu Y, Oliveros JC, Feng S, Benguria A, Lopez-Vidriero I, Zhang X, Solano R, Jacobsen SE,

Gutierrez C (2011) Genome-wide mapping of *Arabidopsis thaliana* origins of DNA replication and their associated epigenetic marks. Nat Struct Mol Biol 18:395–400

Cross SH, Charlton JA, Nan X, Bird AP (1994) Purification of CpG islands using a methylated DNA binding column. Nat Genet 6:236–244

Dalmay T, Hamilton A, Rudd S, Angell S, Baulcombe DC (2000) An RNA-dependent RNA polymerase gene in *Arabidopsis* is required for posttranscriptional gene silencing mediated by a transgene but not by a virus. Cell 101:543–553

de la Paz Sanchez M, Gutierrez C (2009) Arabidopsis ORC1 is a PHD-containing H3K4me3 effector that regulates transcription. Proc Natl Acad Sci U S A 106:2065–2070

Deal RB, Henikoff S (2010) A simple method for gene expression and chromatin profiling of individual cell types within a tissue. Dev Cell 18:1030–1040

Demidov D, Van Damme D, Geelen D, Blattner FR, Houben A (2005) Identification and dynamics of two classes of aurora-like kinases in *Arabidopsis* and other plants. Plant Cell 17:836–848

Ding Y, Avramova Z, Fromm M (2011a) The Arabidopsis trithorax-like factor ATX1 functions in dehydration stress responses via ABA-dependent and ABA-independent pathways. Plant J 66:735–744

Ding Y, Avramova Z, Fromm M (2011b) Two distinct roles of ARABIDOPSIS HOMOLOG OF TRITHORAX1 (ATX1) at promoters and within transcribed regions of ATX1-regulated genes. Plant Cell 23:350–363

Dong G, Ma DP, Li J (2008) The histone methyltransferase SDG8 regulates shoot branching in *Arabidopsis*. Biochem Biophys Res Commun 373:659–664

Earley KW, Shook MS, Brower-Toland B, Hicks L, Pikaard CS (2007) In vitro specificities of Arabidopsis co-activator histone acetyltransferases: implications for histone hyperacetylation in gene activation. Plant J 52:615–626

Ebbs ML, Bartee L, Bender J (2005) H3 lysine 9 methylation is maintained on a transcribed inverted repeat by combined action of SUVH6 and SUVH4 methyltransferases. Mol Cell Biol 25:10507–10515

Ebbs ML, Bender J (2006) Locus-specific control of DNA methylation by the *Arabidopsis* SUVH5 histone methyltransferase. Plant Cell 18:1166–1176

Elling AA, Deng XW (2009) Next-generation sequencing reveals complex relationships between the epigenome and transcriptome in maize. Plant Signal Behav 4:760–762

Feng S, Cokus SJ, Zhang X, Chen PY, Bostick M, Goll MG, Hetzel J, Jain J, Strauss SH, Halpern ME, Ukomadu C, Sadler KC, Pradhan S, Pellegrini M, Jacobsen SE (2010) Conservation and divergence of methylation patterning in plants and animals. Proc Natl Acad Sci U S A 107:8689–8694

Finnegan EJ, Peacock WJ, Dennis ES (1996) Reduced DNA methylation in *Arabidopsis thaliana* results in abnormal plant development. Proc Natl Acad Sci U S A 93:8449–8454

Frommer M, McDonald LE, Millar DS, Collis CM, Watt F, Grigg GW, Molloy PL, Paul CL (1992) A genomic sequencing protocol that yields a positive display of 5- methylcytosine residues in individual DNA strands. Proc Natl Acad Sci U S A 89:1827–1831

Gendall AR, Levy YY, Wilson A, Dean C (2001) The VERNALIZA-TION 2 gene mediates the epigenetic regulation of vernalization in *Arabidopsis*. Cell 107:525–535

Gernand D, Demidov D, Houben A (2003) The temporal and spatial pattern of histone H3 phosphorylation at serine 28 and serine 10 is similar in plants but differs between mono- and polycentric chromosomes. Cytogenet Genome Res 101:172–176

Goll MG, Bestor TH (2005) Eukaryotic cytosine methyltransferases. Annu Rev Biochem 74:481–514

Gong Z, Morales-Ruiz T, Ariza RR, Roldan-Arjona T, David L, Zhu JK (2002) ROS1, a repressor of transcriptional gene silencing in *Arabidopsis*, encodes a DNA glycosylase/lyase. Cell 111:803–814

Grini PE, Thorstensen T, Alm V, Vizcay-Barrena G, Windju SS, Jorstad TS, Wilson ZA, Aalen RB (2009) The ASH1 HOMOLOG 2 (ASHH2) histone H3 methyltransferase is required for ovule and anther development in *Arabidopsis*. PLoS One 4:e7817

Guo L, Yu Y, Law JA, Zhang X (2010) SET DOMAIN GROUP2 is the major histone H3 lysine [corrected] 4 trimethyltransferase in Arabidopsis. Proc Natl Acad Sci U S A 107:18557–18562

Ha M, Ng DW, Li WH, Chen ZJ (2011) Coordinated histone modifications are associated with gene expression variation within and between species. Genome Res 21:590–598

Hamilton A, Voinnet O, Chappell L, Baulcombe D (2002) Two classes of short interfering RNA in RNA silencing. EMBO J 21:4671–4679

Hashimoto H, Horton JR, Zhang X, Bostick M, Jacobsen SE, Cheng X (2008) The SRA domain of UHRF1 flips 5-methylcytosine out of the DNA helix. Nature 455:826–829

He G, Zhu X, Elling AA, Chen L, Wang X, Guo L, Liang M, He H, Zhang H, Chen F, Qi Y, Chen R, Deng XW (2010) Global epigenetic and transcriptional trends among two rice subspecies and their reciprocal hybrids. Plant Cell 22:17–33

Hemberger M, Dean W, Reik W (2009) Epigenetic dynamics of stem cells and cell lineage commitment: digging Waddington's canal. Nat Rev Mol Cell Biol 10:526–537

Henderson IR, Chan SR, Cao X, Johnson L, Jacobsen SE (2010) Accurate sodium bisulfite sequencing in plants. Epigenetics 5:47–49

Henderson IR, Jacobsen SE (2007) Epigenetic inheritance in plants. Nature 447:418–424

Henderson IR, Jacobsen SE (2008) Tandem repeats upstream of the Arabidopsis endogene SDC recruit non-CG DNA methylation and initiate siRNA spreading. Genes Dev 22(12):1597–1606

Herr AJ, Jensen MB, Dalmay T, Baulcombe DC (2005) RNA Polymerase IV Directs silencing of endogenous DNA. Science 308:118–120

Hirochika H, Okamoto H, Kakutani T (2000) Silencing of retrotransposons in arabidopsis and reactivation by the ddm1 mutation. Plant Cell 12(3):357–369

Houben A, Demidov D, Rutten T, Scheidtmann KH (2005) Novel phosphorylation of histone H3 at threonine 11 that temporally correlates with condensation of mitotic and meiotic chromosomes in plant cells. Cytogenet Genome Res 109:148–155

Hsieh TF, Ibarra CA, Silva P, Zemach A, Eshed-Williams L, Fischer RL, Zilberman D (2009) Genome-wide demethylation of *Arabidopsis* endosperm. Science 324:1451–1454

Jackson JP, Lindroth AM, Cao X, Jacobsen SE (2002) Control of CpNpG DNA methylation by the KRYPTONITE histone H3 methyltransferase. Nature 416:556–560

Jacob Y, Feng S, LeBlanc CA, Bernatavichute YV, Stroud H, Cokus S, Johnson LM, Pellegrini M, Jacobsen SE, Michaels SD (2009) ATXR5 and ATXR6 are H3K27 monomethyltransferases required for chromatin structure and gene silencing. Nat Struct Mol Biol 16:763–768

Jacob Y, Stroud H, Leblanc C, Feng S, Zhuo L, Caro E, Hassel C, Gutierrez C, Michaels SD, Jacobsen SE (2010) Regulation of heterochromatic DNA replication by histone H3 lysine 27 methyltransferases. Nature 466:987–991

Jiang D, Yang W, He Y, Amasino RM (2007) *Arabidopsis* relatives of the human lysine-specific Demethylase1 repress the expression of FWA and FLOWERING LOCUS C and thus promote the floral transition. Plant Cell 19:2975–2987

Johnson L, Mollah S, Garcia BA, Muratore TL, Shabanowitz J, Hunt DF, Jacobsen SE (2004) Mass spectrometry analysis of *Arabidopsis* histone H3 reveals distinct combinations of post-translational modifications. Nucleic Acids Res 32:6511–6518

Johnson LM, Bostick M, Zhang X, Kraft E, Henderson I, Callis J, Jacobsen SE (2007) The SRA methyl-cytosine-binding domain links DNA and histone methylation. Curr Biol 17:379–384

Kakutani T, Kato M, Kinoshita T, Miura A (2004) Control of development and transposon movement by DNA methylation in *Arabidopsis thaliana*. Cold Spring Harb Symp Quant Biol 69:139–143

Kankel MW, Ramsey DE, Stokes TL, Flowers SK, Haag JR, Jeddeloh JA, Riddle NC, Verbsky ML, Richards EJ (2003) *Arabidopsis* MET1 cytosine methyltransferase mutants. Genetics 163:1109–1122

Kanno T, Huettel B, Mette MF, Aufsatz W, Jaligot E, Daxinger L, Kreil DP, Matzke M, Matzke AJ (2005) Atypical RNA polymerase subunits required for RNA-directed DNA methylation. Nat Genet 37:761–765

Kaszas E, Cande WZ (2000) Phosphorylation of histone H3 is correlated with changes in the maintenance of sister chromatid cohesion during meiosis in maize, rather than the condensation of the chromatin. J Cell Sci 113:3217–3226

Kawabe A, Matsunaga S, Nakagawa K, Kurihara D, Yoneda A, Hasezawa S, Uchiyama S, Fukui K (2005) Characterization of plant Aurora kinases during mitosis. Plant Mol Biol 58:1–13

Keshet I, Schlesinger Y, Farkash S, Rand E, Hecht M, Segal E, Pikarski E, Young RA, Niveleau A, Cedar H, Simon I (2006) Evidence for an instructive mechanism of de novo methylation in cancer cells. Nat Genet 38:149–153

Kim SY, He Y, Jacob Y, Noh YS, Michaels S, Amasino R (2005) Establishment of the vernalization-responsive, winter-annual habit in *Arabidopsis* requires a putative histone H3 methyl transferase. Plant Cell 17:3301–3310

Ko JH, Mitina I, Tamada Y, Hyun Y, Choi Y, Amasino RM, Noh B, Noh YS (2010) Growth habit determination by the balance of histone methylation activities in *Arabidopsis*. EMBO J 29:3208–3215

Kraft E, Bostick M, Jacobsen SE, Callis J (2008) ORTH/VIM proteins that regulate DNA methylation are functional ubiquitin E3 ligases. Plant J 56:704–715

Krichevsky A, Zaltsman A, Lacroix B, Citovsky V (2011) Involvement of KDM1C histone demethylase-OTLD1 otubain-like histone deubiquitinase complexes in plant gene repression. Proc Natl Acad Sci U S A 108:11157–11162

Kurihara D, Matsunaga S, Kawabe A, Fujimoto S, Noda M, Uchiyama S, Fukui K (2006) Aurora kinase is required for chromosome segregation in tobacco BY-2 cells. Plant J 48:572–580

Kurihara D, Matsunaga S, Omura T, Higashiyama T, Fukui K (2011) Identification and characterization of plant Haspin kinase as a histone H3 threonine kinase. BMC Plant Biol 11:73

Lafos M, Kroll P, Hohenstatt ML, Thorpe FL, Clarenz O, Schubert D (2011) Dynamic regulation of H3K27 trimethylation during *Arabidopsis* differentiation. PLoS Genet 7:e1002040

Lahmy S, Bies-Etheve N, Lagrange T (2010) Plant-specific multisubunit RNA polymerase in gene silencing. Epigenetics 5:4–8

Law JA, Jacobsen SE (2010) Establishing, maintaining and modifying DNA methylation patterns in plants and animals. Nat Rev Genet 11:204–220

Lee TJ, Pascuzzi PE, Settlage SB, Shultz RW, Tanurdzic M, Rabinowicz PD, Menges M, Zheng P, Main D, Murray JA, Sosinski B, Allen GC, Martienssen RA, Hanley-Bowdoin L, Vaughn MW, Thompson WF (2010) *Arabidopsis thaliana* chromosome 4 replicates in two phases that correlate with chromatin state. PLoS Genet 6:e1000982

Li X, Wang X, He K, Ma Y, Su N, He H, Stolc V, Tongprasit W, Jin W, Jiang J, Terzaghi W, Li S, Deng XW (2008) High-resolution mapping of epigenetic modifications of the rice genome uncovers interplay between DNA methylation, histone methylation, and gene expression. Plant Cell 20:259–276

Lindroth AM, Cao X, Jackson JP, Zilberman D, McCallum CM, Henikoff S, Jacobsen SE (2001) Requirement of CHROMOMETHYLASE3 for maintenance of CpXpG methylation. Science 292:2077–2080

Lindroth AM, Shultis D, Jasencakova Z, Fuchs J, Johnson L, Schubert D, Patnaik D, Pradhan S, Goodrich J, Schubert I, Jenuwein T, Khorasanizadeh S, Jacobsen SE (2004) Dual histone H3 methylation marks at lysines 9 and 27 required for interaction with CHROMOMETHYLASE3. EMBO J 23:4286–4296

Lippman Z, Gendrel AV, Black M, Vaughn MW, Dedhia N, McCombie WR, Lavine K, Mittal V, May B, Kasschau KD, Carrington JC, Doerge RW, Colot V, Martienssen R (2004) Role of transposable elements in heterochromatin and epigenetic control. Nature 430:471–476

Lister R, O'Malley RC, Tonti-Filippini J, Gregory BD, Berry CC, Millar AH, Ecker JR (2008) Highly integrated single-base resolution maps of the epigenome in Arabidopsis. Cell 133:523–536

Liu F, Quesada V, Crevillen P, Baurle I, Swiezewski S, Dean C (2007) The *Arabidopsis* RNA-binding protein FCA requires a lysine-specific demethylase 1 homolog to downregulate FLC. Mol Cell 28:398–407

Lu C, Tej SS, Luo S, Haudenschild CD, Meyers BC, Green PJ (2005) Elucidation of the small RNA component of the transcriptome. Science 309:1567–1569

Lu F, Cui X, Zhang S, Jenuwein T, Cao X (2011) *Arabidopsis* REF6 is a histone H3 lysine 27 demethylase. Nat Genet 43:715–719

Lu F, Cui X, Zhang S, Liu C, Cao X (2010) JMJ14 is an H3K4 demethylase regulating flowering time in *Arabidopsis*. Cell Res 20:387–390

Lu F, Li G, Cui X, Liu C, Wang XJ, Cao X (2008) Comparative analysis of JmjC domain-containing proteins reveals the potential histone demethylases in *Arabidopsis* and rice. J Integr Plant Biol 50:886–896

Malagnac F, Bartee L, Bender J (2002) An *Arabidopsis* SET domain protein required for maintenance but not establishment of DNA methylation. EMBO J 21:6842–6852

Manzanero S, Arana P, Puertas MJ, Houben A (2000) The chromosomal distribution of phosphorylated histone H3 differs between plants and animals at meiosis. Chromosoma 109:308–317

Matzke M, Kanno T, Daxinger L, Huettel B, Matzke AJ (2009) RNA-mediated chromatin-based silencing in plants. Curr Opin Cell Biol 21:367–376

Mavrich TN, Ioshikhes IP, Venters BJ, Jiang C, Tomsho LP, Qi J, Schuster SC, Albert I, Pugh BF (2008a) A barrier nucleosome model for statistical positioning of nucleosomes throughout the yeast genome. Genome Res 18:1073–1083

Mavrich TN, Jiang C, Ioshikhes IP, Li X, Venters BJ, Zanton SJ, Tomsho LP, Qi J, Glaser RL, Schuster SC, Gilmour DS, Albert I, Pugh BF (2008b) Nucleosome organization in the *Drosophila* genome. Nature 453:358–362

Miura A, Nakamura M, Inagaki S, Kobayashi A, Saze H, Kakutani T (2009) An *Arabidopsis* jmjC domain protein protects transcribed genes from DNA methylation at CHG sites. EMBO J 28:1078–1086

Naumann K, Fischer A, Hofmann I, Krauss V, Phalke S, Irmler K, Hause G, Aurich AC, Dorn R, Jenuwein T, Reuter G (2005) Pivotal role of AtSUVH2 in heterochromatic histone methylation and gene silencing in *Arabidopsis*. EMBO J 24:1418–1429

Niu L, Lu F, Pei Y, Liu C, Cao X (2007) Regulation of flowering time by the protein arginine methyltransferase AtPRMT10. EMBO Rep 8:1190–1195

Niu L, Zhang Y, Pei Y, Liu C, Cao X (2008) Redundant requirement for a pair of PROTEIN ARGININE METHYLTRANSFERASE4 homologs for the proper regulation of *Arabidopsis* flowering time. Plant Physiol 148:490–503

Oh S, Park S, van Nocker S (2008) Genic and global functions for Paf1C in chromatin modification and gene expression in *Arabidopsis*. PLoS Genet 4:e1000077

Onodera Y, Haag JR, Ream T, Nunes PC, Pontes O, Pikaard CS (2005) Plant nuclear RNA polymerase IV mediates siRNA and DNA methylation-dependent heterochromatin formation. Cell 120:613–622

Palma K, Thorgrimsen S, Malinovsky FG, Fiil BK, Nielsen HB, Brodersen P, Hofius D, Petersen M, Mundy J (2010) Autoimmunity in *Arabidopsis* acd11 is mediated by epigenetic regulation of an immune receptor. PLoS Pathog 6:e1001137

Pandey R, Muller A, Napoli CA, Selinger DA, Pikaard CS, Richards EJ, Bender J, Mount DW, Jorgensen RA (2002) Analysis of histone acetyltransferase and histone deacetylase families of *Arabidopsis thaliana* suggests functional diversification of chromatin modification among multicellular eukaryotes. Nucleic Acids Res 30:5036–5055

Penterman J, Zilberman D, Huh JH, Ballinger T, Henikoff S, Fischer RL (2007) DNA demethylation in the *Arabidopsis* genome. Proc Natl Acad Sci U S A 104:6752–6757

Perez-Hormaeche J, Potet F, Beauclair L, Le Masson I, Courtial B, Bouche N, Lucas H (2008) Invasion of the *Arabidopsis* genome by the tobacco retrotransposon Tnt1 is controlled by reversible transcriptional gene silencing. Plant Physiol 147:1264–1278

Pien S, Grossniklaus U (2007) Polycomb group and trithorax group proteins in *Arabidopsis*. Biochim Biophys Acta 1769:375–382

Pontes O, Li CF, Nunes PC, Haag J, Ream T, Vitins A, Jacobsen SE, Pikaard CS (2006) The *Arabidopsis* chromatin-modifying nuclear siRNA pathway involves a nucleolar RNA processing center. Cell 126:79–92

Pontier D, Yahubyan G, Vega D, Bulski A, Saez-Vasquez J, Hakimi MA, Lerbs-Mache S, Colot V, Lagrange T (2005) Reinforcement of silencing at transposons and highly repeated sequences requires the concerted action of two distinct RNA polymerases IV in *Arabidopsis*. Genes Dev 19:2030–2040

Raynaud C, Sozzani R, Glab N, Domenichini S, Perennes C, Cella R, Kondorosi E, Bergounioux C (2006) Two cell-cycle regulated SET-domain proteins interact with proliferating cell nuclear antigen (PCNA) in *Arabidopsis*. Plant J 47:395–407

Reinders J, Delucinge Vivier C, Theiler G, Chollet D, Descombes P, Paszkowski J (2008) Genome-wide, high-resolution DNA methylation profiling using bisulfite-mediated cytosine conversion. Genome Res 18:469–476

Ronemus MJ, Galbiati M, Ticknor C, Chen J, Dellaporta SL (1996) Demethylation-induced developmental pleiotropy in *Arabidopsis*. Science 273:654–657

Roudier F, Ahmed I, Berard C, Sarazin A, Mary-Huard T, Cortijo S, Bouyer D, Caillieux E, Duvernois-Berthet E, Al-Shikhley L, Giraut L, Despres B, Drevensek S, Barneche F, Derozier S, Brunaud V, Aubourg S, Schnittger A, Bowler C, Martin-Magniette ML, Robin S, Caboche M, Colot V (2011) Integrative epigenomic mapping defines four main chromatin states in *Arabidopsis*. EMBO J 30:1928–1938

Saleh A, Alvarez-Venegas R, Yilmaz M, Le O, Hou G, Sadder M, Al-Abdallat A, Xia Y, Lu G, Ladunga I, Avramova Z (2008) The highly similar *Arabidopsis* homologs of trithorax ATX1 and ATX2 encode proteins with divergent biochemical functions. Plant Cell 20:568–579

Sasaki H, Matsui Y (2008) Epigenetic events in mammalian germ-cell development: reprogramming and beyond. Nat Rev Genet 9:129–140

Saze H, Scheid OM, Paszkowski J (2003) Maintenance of CpG methylation is essential for epigenetic inheritance during plant gametogenesis. Nat Genet 34:65–69

Saze H, Shiraishi A, Miura A, Kakutani T (2008) Control of genic DNA methylation by a jmjC domain-containing protein in *Arabidopsis thaliana*. Science 319:462–465

Schmitz RJ, Tamada Y, Doyle MR, Zhang X, Amasino RM (2009) Histone H2B deubiquitination is required for transcriptional activation of FLOWERING LOCUS C and for proper control of flowering in *Arabidopsis*. Plant Physiol 149:1196–1204

Segal E, Fondufe-Mittendorf Y, Chen L, Thastrom A, Field Y, Moore IK, Wang JP, Widom J (2006) A genomic code for nucleosome positioning. Nature 442:772–778

Shi Y, Lan F, Matson C, Mulligan P, Whetstine JR, Cole PA, Casero RA (2004) Histone demethylation mediated by the nuclear amine oxidase homolog LSD1. Cell 119:941–953

Soppe WJ, Jacobsen SE, Alonso-Blanco C, Jackson JP, Kakutani T, Koornneef M, Peeters AJ (2000) The late flowering phenotype of fwa mutants is caused by gain-of-function epigenetic alleles of a homeodomain gene. Mol Cell 6:791–802

Springer NM, Napoli CA, Selinger DA, Pandey R, Cone KC, Chandler VL, Kaeppler HF, Kaeppler SM (2003) Comparative analysis of SET domain proteins in maize and *Arabidopsis* reveals multiple duplications preceding the divergence of monocots and dicots. Plant Physiol 132:907–925

Sridhar VV, Kapoor A, Zhang K, Zhu J, Zhou T, Hasegawa PM, Bressan RA, Zhu JK (2007) Control of DNA methylation and heterochromatic silencing by histone H2B deubiquitination. Nature 447:735–738

Surani MA, Hayashi K, Hajkova P (2007) Genetic and epigenetic regulators of pluripotency. Cell 128:747–762

Tamada Y, Yun JY, Woo SC, Amasino RM (2009) ARABIDOPSIS TRITHORAX-RELATED7 is required for methylation of lysine 4 of histone H3 and for transcriptional activation of FLOWERING LOCUS C. Plant Cell 21:3257–3269

Tamaru H (2010) Confining euchromatin/heterochromatin territory: jumonji crosses the line. Genes Dev 24:1465–1478

Thorstensen T, Grini PE, Mercy IS, Alm V, Erdal S, Aasland R, Aalen RB (2008) The *Arabidopsis* SET-domain protein ASHR3 is involved in stamen development and interacts with the bHLH transcription factor ABORTED MICROSPORES (AMS). Plant Mol Biol 66:47–59

Tran RK, Henikoff JG, Zilberman D, Ditt RF, Jacobsen SE, Henikoff S (2005a) DNA methylation profiling identifies CG methylation clusters in *Arabidopsis* genes. Curr Biol 15:154–159

Tran RK, Zilberman D, de Bustos C, Ditt RF, Henikoff JG, Lindroth AM, Delrow J, Boyle T, Kwong S, Bryson TD, Jacobsen SE, Henikoff S (2005b) Chromatin and siRNA pathways cooperate to maintain DNA methylation of small transposable elements in *Arabidopsis*. Genome Biol 6:R90

Tsukada Y, Fang J, Erdjument-Bromage H, Warren ME, Borchers CH, Tempst P, Zhang Y (2006) Histone demethylation by a family of JmjC domain-containing proteins. Nature 439:811–816

Turck F, Roudier F, Farrona S, Martin-Magniette ML, Guillaume E, Buisine N, Gagnot S, Martienssen RA, Coupland G, Colot V (2007) *Arabidopsis* TFL2/LHP1 specifically associates with genes marked by trimethylation of histone H3 lysine 27. PLoS Genet 3:e86

Valouev A, Ichikawa J, Tonthat T, Stuart J, Ranade S, Peckham H, Zeng K, Malek JA, Costa G, McKernan K, Sidow A, Fire A, Johnson SM (2008) A high-resolution, nucleosome position map of C. elegans reveals a lack of universal sequence-dictated positioning. Genome Res 18:1051–1063

Vaughn MW, Tanurd Ic M, Lippman Z, Jiang H, Carrasquillo R, Rabinowicz PD, Dedhia N, McCombie WR, Agier N, Bulski A, Colot V, Doerge RW, Martienssen RA (2007) Epigenetic natural variation in *Arabidopsis thaliana*. PLoS Biol 5:e174

Wang X, Elling AA, Li X, Li N, Peng Z, He G, Sun H, Qi Y, Liu XS, Deng XW (2009) Genome-wide and organ-specific landscapes of epigenetic modifications and their relationships to mRNA and small RNA transcriptomes in maize. Plant Cell 21:1053–1069

Widom J (2001) Role of DNA sequence in nucleosome stability and dynamics. Q Rev Biophys 34:269–324

Woo HR, Dittmer TA, Richards EJ (2008) Three SRA-domain methylcytosine-binding proteins cooperate to maintain global CpG methylation and epigenetic silencing in *Arabidopsis*. PLoS Genet 4:e1000156

Woo HR, Pontes O, Pikaard CS, Richards EJ (2007) VIM1, a methylcytosine-binding protein required for centromeric heterochromatinization. Genes Dev 21:267–277

Wyrick JJ, Aparicio JG, Chen T, Barnett JD, Jennings EG, Young RA, Bell SP, Aparicio OM (2001) Genome-wide distribution of ORC and MCM proteins in *S. cerevisiae*: high-resolution mapping of replication origins. Science 294:2357–2360

Xu L, Zhao Z, Dong A, Soubigou-Taconnat L, Renou JP, Steinmetz A, Shen WH (2008) Di- and tri- but not monomethylation on histone H3 lysine 36 marks active transcription of genes involved in flowering time regulation and other processes in *Arabidopsis thaliana*. Mol Cell Biol 28:1348–1360

Zemach A, Kim MY, Silva P, Rodrigues JA, Dotson B, Brooks MD, Zilberman D (2010a) Local DNA hypomethylation activates genes in rice endosperm. Proc Natl Acad Sci U S A 107:18729–18734

Zemach A, McDaniel IE, Silva P, Zilberman D (2010b) Genome-wide evolutionary analysis of eukaryotic DNA methylation. Science 328:916–919

Zhang H, Zhu JK (2011) RNA-directed DNA methylation. Curr Opin Plant Biol 14:142–147

Zhang K, Sridhar VV, Zhu J, Kapoor A, Zhu JK (2007a) Distinctive core histone post-translational modification patterns in Arabidopsis thaliana. PLoS One 2:e1210

Zhang X, Bernatavichute YV, Cokus S, Pellegrini M, Jacobsen SE (2009) Genome-wide analysis of mono-, di- and trimethylation of histone H3 lysine 4 in *Arabidopsis thaliana*. Genome Biol 10:R62

Zhang X, Clarenz O, Cokus S, Bernatavichute YV, Pellegrini M, Goodrich J, Jacobsen SE (2007b) Whole-genome analysis of histone H3 lysine 27 trimethylation in *Arabidopsis*. PLoS Biol 5:e129

Zhang X, Henderson IR, Lu C, Green PJ, Jacobsen SE (2007c) Role of RNA polymerase IV in plant small RNA metabolism. Proc Natl Acad Sci U S A 104:4536–4541

Zhang X, Jacobsen SE (2006) Genetic analyses of DNA methyltransferases in Arabidopsis thaliana. Cold Spring Harb Symp Quant Biol 71:439–447

Zhang X, Li X, Marshall JB, Zhong CX, Dawe RK (2005) Phosphoserines on maize CENTROMERIC HISTONE H3 and histone H3 demarcate the centromere and pericentromere during chromosome segregation. Plant Cell 17:572–583

Zhang X, Yazaki J, Sundaresan A, Cokus S, Chan SW, Chen H, Henderson IR, Shinn P, Pellegrini M, Jacobsen SE, Ecker JR (2006) Genome-wide high-Resolution mapping and functional analysis of DNA methylation in *Arabidopsis*. Cell 126:1189–1201

Zhao Z, Yu Y, Meyer D, Wu C, Shen WH (2005) Prevention of early flowering by expression of FLOWERING LOCUS C requires methylation of histone H3 K36. Nat Cell Biol 7:1256–1260

Zhou J, Wang X, He K, Charron JB, Elling AA, Deng XW (2010) Genome-wide profiling of histone H3 lysine 9 acetylation and dimethylation in *Arabidopsis* reveals correlation between multiple histone marks and gene expression. Plant Mol Biol 72:585–595

Zhou X, Ma H (2008) Evolutionary history of histone demethylase families: distinct evolutionary patterns suggest functional divergence. BMC Evol Biol 8:294

Zilberman D, Cao X, Jacobsen SE (2003) ARGONAUTE4 control of locus-specific siRNA accumulation and DNA and histone methylation. Science 299:716–719

Zilberman D, Gehring M, Tran RK, Ballinger T, Henikoff S (2007) Genome-wide analysis of *Arabidopsis thaliana* DNA methylation uncovers an interdependence between methylation and transcription. Nat Genet 39:61–69

Evolutionary Significance of Epigenetic Variation 16

Christina L. Richards, Koen J.F. Verhoeven, and Oliver Bossdorf

Contents

16.1 Introduction

Several chapters in this volume demonstrate how epigenetic work at the molecular level over the last few decades has revolutionized our understanding of genome function and developmental biology. However, epigenetic processes not only further our understanding of variation and regulation at the genomic and cellular levels, they also challenge our understanding of heritable phenotypic variation at the level of whole organisms and even the process of evolution by natural selection (Jablonka and Lamb 1989, 1995; Danchin et al. 2011). Although many of the epigenetic mechanisms involved in differential gene expression are reset each generation, some epigenetic marks are faithfully transmitted across generations (Jablonka and Raz 2009; Verhoeven et al. 2010a). In addition, we now know that natural variation exists not only at the DNA sequence level but also the epigenetic level (e.g., Vaughn et al. 2007; Herrera and Bazaga 2010). This may be particularly common in plants, and several studies suggest that epigenetic variation alone can cause significant heritable variation in phenotypic traits (e.g., Cubas et al. 1999; Johannes et al. 2009; Scoville et al. 2011). Because of these observations, there is currently increasing interest in understanding the role of epigenetic processes in ecology and evolution (e.g., Richards 2006, 2011; Bossdorf et al. 2008; Johannes et al. 2008; Richards et al. 2010a).

In spite of the speculation about the potential evolutionary implications of epigenetic processes, most previous work has involved agricultural crops and model species such as *Arabidopsis thaliana*, frequently under artificial conditions, and we therefore still know little about the importance of epigenetic processes in natural populations (Richards 2008, 2011; Richards et al. 2010b). With this in mind, we review some of the salient examples of known heritable phenotypic effects of epigenetic mechanisms as well as how epigenetic variation can be created. We address the important issue of disentangling genetic and epigenetic components of heritable phenotypic variation with particular emphasis on how some epigenetic effects may be determined by genotype, while

C.L. Richards (✉)
Department of Integrative Biology, University of South Florida, 4202 E Fowler Ave, Tampa, FL 33620, USA
e-mail: clr@usf.edu; christinalrichards@gmail.com

J.F. Wendel et al. (eds.), *Plant Genome Diversity Volume 1*,
DOI 10.1007/978-3-7091-1130-7_16, © Springer-Verlag Wien 2012

others act independently of genotype. We discuss what little we know about how these patterns are manifest in natural populations and what we need to do to discover how these mechanisms work in the real world. Finally, we conclude with some ideas about how these concepts may enhance our understanding of evolutionary processes.

16.2 Phenotypic Effects of Epigenetic Variation

The logical first step to understanding the importance of epigenetic effects is characterizing the phenotypic response to epigenetic variation. There are few known simple and obvious phenotypic effects that result from changes in epigenetic marks at single genes. However, through manipulation of methylation levels and isolation of methylation mutants, researchers have made substantial progress in demonstrating important phenotypic effects that result from changes at only the epigenetic level.

16.2.1 Single Gene to Phenotype Epigenetic Effects

One of the most celebrated studies of single-gene epigenetic effects on the phenotype and epigenetic inheritance is an elegant example of the change from bilateral to radial floral symmetry in *Linaria vulgaris* (Fig. 16.1). Cubas et al. (1999) found that the radial symmetry phenotype was associated with methylation changes at the gene *Lcyc* which controls dorsoventral asymmetry. In the variety of *Linaria* with radial flowers, *Lcyc* is extensively methylated and transcriptionally silent. The epigenetic modification is heritable and co-segregates with the phenotype. Occasionally, the phenotype reverts during somatic development, which correlates with demethylation of *Lcyc*, restoration of gene expression and the development of normal bilateral flowers.

Although the *Linaria* example is satisfying, there are not many other simple examples of the phenotypic effects of epigenetic alteration of one or few genes, especially changes that are stably inherited for generations (see further discussion in Paszkowski and Grossniklaus 2011). There are several examples of epigenetic effects that are not heritable in plant developmental biology: for instance, studies have shown that epigenetic silencing (caused by cold treatment) of the floral repressor Flowering Locus C (FLC, Sung and Amasino 2004; Shindo et al. 2006), followed by increased methylation of histone 3 lysine 9 (H3K9) and histone 3 lysine 27 (H3K27), triggers flowering in many *A. thaliana* accessions that overwinter as a rosette. However, this response is necessarily non-heritable, so that the next generation of seedlings will also overwinter before flowering.

Fig. 16.1 A clear example of phenotypic effects of natural epigenetic variation, and epigenetic inheritance. Cubas et al. (1999) showed that the naturally occurring change from normal bilateral (*right*) to radial symmetry (*left*) of *Linaria vulgaris* was associated with methylation and silencing of the gene *Lcyc* (Photos from Palevitz 1999)

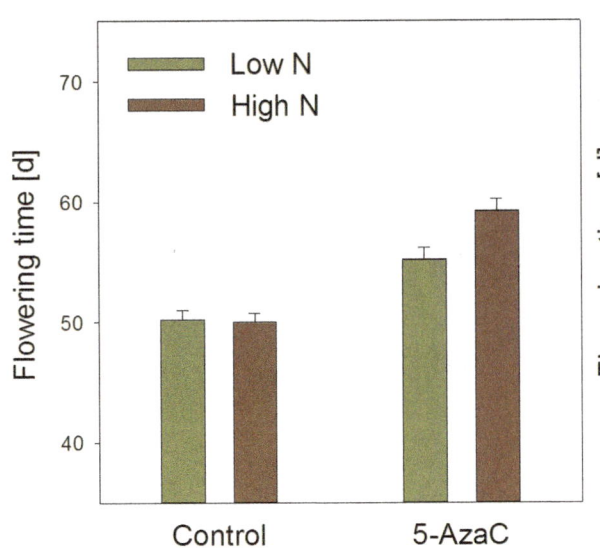 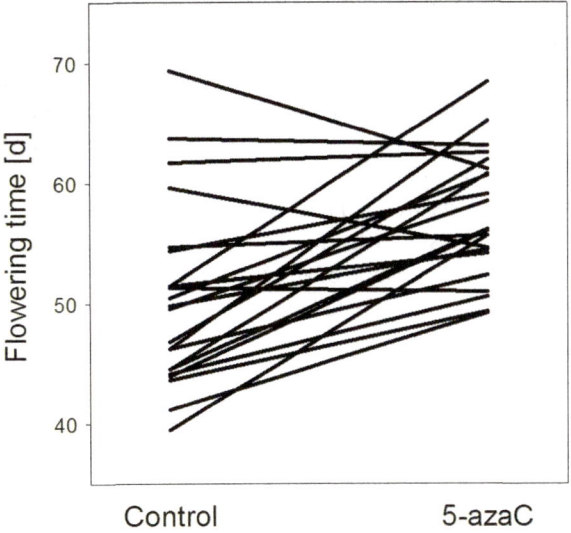

Fig. 16.2 Experimental demethylation through 5-azacytidine (5-azaC) significantly alters the phenotype of *A. thaliana*. *Left panel*: the phenotypic plasticity to nutrient addition of control vs. 5-azaC-treated plants. *Right panel*: variation in flowering time responses to 5-azaC among 22 different *A. thaliana* ecotypes (Modified from Bossdorf et al. 2010)

16.2.2 Experimental Manipulations of DNA Methylation

One approach to understanding the phenotypic effects of DNA methylation has been to manipulate genome wide levels of methylation through the application of the demethylating agent 5-azacytidine (5-azaC), a chemical that is incorporated into DNA during replication and thereby inhibits the enzyme methyltransferase, causing partial demethylation of the DNA (e.g., Burn et al. 1993; Fieldes and Amyot 1999a; Tatra et al. 2000). This creates different epigenetic variants of the same genotypes and therefore allows researchers to demonstrate phenotypic effects of epigenetic changes. Several studies have demonstrated that the effects of demethylation on ecologically important traits can be significant. For example, Burn et al. (1993) showed that treatment of *A. thaliana* and *Thlaspi arvense* with 5-azaC significantly altered plant flowering time, while Fieldes and colleagues (Fieldes 1994; Fieldes and Amyot 1999b; Fieldes et al. 2005) found that 5-azaC affected the growth, fitness and phenology of *Linum usitatissimum*. More recently, Bossdorf et al. (2010) found that experimental alteration of DNA methylation not only altered the growth, fitness and phenology of *A. thaliana*, but also the phenotypic plasticity of these traits in response to nutrient addition (Fig. 16.2). Moreover, there were significant differences among the 22 studied *A. thaliana* genotypes in the degree to which trait means and plasticities were affected by 5-azaC. Demethylation also altered overall patterns of among-line variability, which indicates that epigenetic changes can not only affect the short-term environmental

responses (phenotypic plasticity) of plants, but also the evolutionary potential of important traits and their plasticities (Bossdorf et al. 2010).

16.2.3 Methylation Mutants

Another approach to isolating the phenotypic effects of methylation has been the isolation of mutants that have non-functional or reduced function in the methylation machinery. Using southern blot analysis, Vongs et al. (1993) isolated three hypomethylation mutants from approximately 2,000 ethylmethanesulfonate (EMS) mutagenized plants from the Columbia genotype (Col-0) of *A. thaliana*. The three lines were referred to as *ddm* mutants in reference to their *d*ecrease in *D*NA *m*ethylation. A large portion of the change in methylation was found in the repeat regions of the genome, and two of the mutants (*ddm1-1* and *ddm 1-2*) had methylation levels reduced to only 25–30% of wild type, while the third was reduced to 83% of wild type. The genes involved in these mutations were later characterized as chromatin remodeling proteins (Jeddeloh et al. 1999). However, the original screening of the homozygous mutants did not show any obvious difference in phenotype from wild type (Vongs et al. 1993). Backcrosses to wild type demonstrated that hypomethylation was gradually lost through segregation, but hypomethylated fragments were slow to be re-methylated.

Screening of another 5,000 EMS mutagenized plants revealed two more mutants with decreased methylation, this time in the METHYLTRANSFERASE1 (MET1) gene (Kankel et al. 2003). These were initially referred to as *ddm2-1* and *ddm2-2*, but were renamed *met1-1* and *met1-2* when the mutations were identified in the MET1 gene. In this case, methylation levels were reduced by 50% in *met1-2* and plants displayed normal development and morphology. In contrast, *met1-1* showed delayed flowering time, which was associated specifically with the demethylation of the floral repressor FWA and creation of an FWA epiallele. As with the *ddm1* mutants, hypomethylated segments of the genome from *met1* mutant could be inherited in wild type backcrosses.

16.2.4 Epigenetic Recombinant Inbred Lines (epiRILs)

Although the first set of *A. thaliana ddm1* mutants did not initially demonstrate any obvious phenotypic effects of reduced methylation, epigenetic recombinant inbred lines (epiRILs) developed from backcrosses of these and the *met1* mutants to Col-0 wild type have proven to be powerful tools for detecting effects of variation in DNA methylation on quantitative traits (Johannes et al. 2009; Reinders et al. 2009). Johannes et al. (2009) created 505 epiRILs by crossing the *ddm1-2* to wild type Col-0 and backcrossing a single F1 female with Col-0. From this first backcross, 509 offspring with the *DDM1/DDM1* wild type genotype were chosen to initiate lines of single seed descent for six self-fertilizing generations. Despite the fact that these lines are nearly isogenic, the authors found increased variance and significant among-line variation in plant height and flowering time across the 505 lines that survived the inbreeding process, and they ascribed the phenotypic variation to the differences in DNA methylation patterns. Subsequent work with these epiRILs has confirmed that there is consistent and significant heritable variation in many other ecologically important traits among these lines, e.g. fitness traits (Roux et al. 2011) and drought responses (Zhang and Bossdorf, unpublished data; Fig. 16.3).

In a similar recombination study, Reinders et al. (2009) created 68 epiRILs by crossing the *met1-3* to wild type Col-0 and backcrossing a single F1 female with Col-0. From this first backcross, F2 offspring with *MET1/MET1* wild type genotype were selfed, which provided for 100 F3 lines to initiate lines of single seed descent for four additional generations. At F7, seed from nine random plants were bulked to provide the F8 used for phenotyping and methylation analysis. As in the Johannes et al. (2009) epiRIL study, this population of epiRILs showed significant heritable among-line variation in growth and response to stress. In addition, the population was characterized by a bimodal distribution of flowering time, which the authors related to

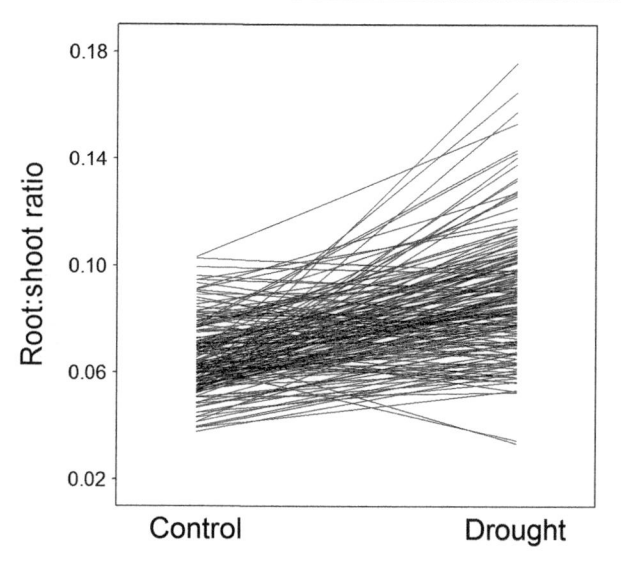

Fig. 16.3 Variation in drought responses of 160 epiRILs (Zhang and Bossdorf, unpublished data)

the fact that the floral repressor FWA is methylated in wild type and demethylated in *met1-3*. The FWA locus in 11 late flowering lines each had a "*met*-like epiallele".

Both epiRIL studies elegantly demonstrate the potentially strong effects of epigenetic changes on plant phenotypes. However, they also show that the stability of phenotypes based on epi-alleles can be difficult to predict and is locus-dependent (Reinders and Paszkowski 2009). Both studies show only partial inheritance of methylation changes, and partial remethylation, especially of transposable element (TE) loci whose active remethylation is ensured by an RNA-dependent DNA methylation machinery that recognizes TE-derived transcripts and subsequently silences the TE locus (Teixeira et al. 2009). Johannes et al. (2009) tested the stability of 11 loci (including FWA) that were methylated in wild type and hypomethylated in *ddm1-2* as well as three loci that were not methylated in either parental line. They examined the methylation of these 14 loci across 22 epiRILs representing the two ends of the flowering time spectrum (Fig. 16.4). The three loci that were unmethylated in both parents were stably inherited as unmethylated in the offspring. Five of the eleven differentially methylated sequences segregated close to the expected Mendelian fashion; for the other six loci (including FWA) however, there was almost complete reversion to the methylated state. This indicates that there are mechanisms to correct for hypomethylation in the long term, and that these mechanisms are able to methylate de novo in later generations, without using methylation inherited from either parent (Reinders and Paszkowski 2009; Teixeira et al. 2009). Potentially, such mechanisms could contribute to population level variation and ultimately evolution.

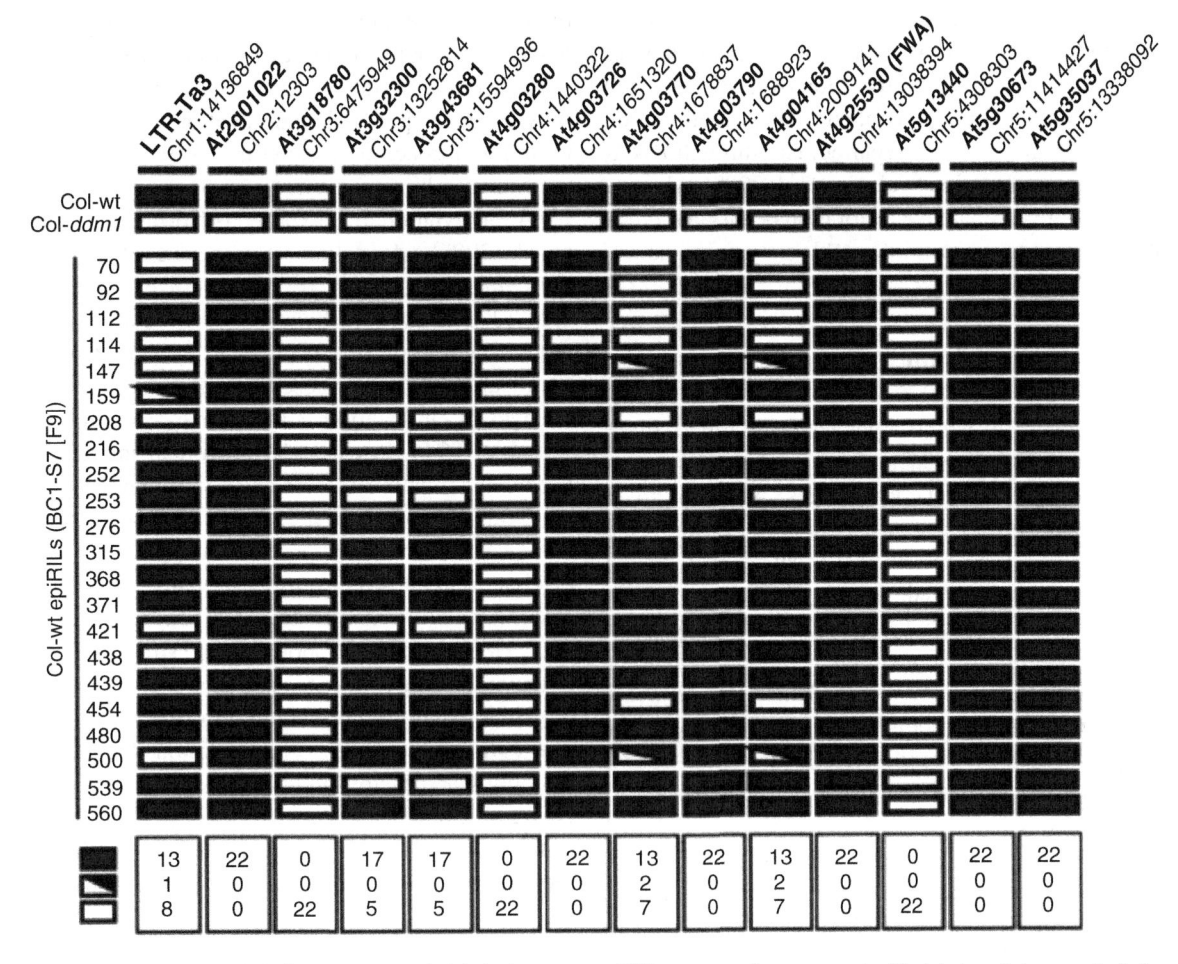

Fig. 16.4 Segregation of methylation patterns of 14 loci across 22 Col-0 × ddm1-2 A. thaliana epiRILs. Name and position of loci are listed across the *top* with *horizontal bars* indicating closely linked sequences. *Black rectangles* represent high (*wild type*) methylation. *White rectangles* represent *ddm1*-induced hypomethylation. *Partial rectangles* represent intermediate levels of methylation. Segregation is summarized across the 22 epiRILs for each sequence at the *bottom* (Reprinted from Johannes et al. 2009)

16.3 Creation of Heritable Epigenetic Variation

Having established that epigenetic variation can have significant phenotypic effects, the next question is how heritable epigenetic variation is created. From what is known, there seem to be three main mechanisms: (1) natural epimutations, (2) environmental induction, and (3) genomic events such as hybridization and polyploidization.

16.3.1 Epimutations

One source of natural epialleles (that are not triggered by defective enzymes in the DNA methylation machinery, as in the case of the epiRILs) are natural epimutations created by imperfect maintenance of DNA methylation patterns and other epigenetic marks through mitosis and meiosis (Genereux

et al. 2005). For instance, methylation polymorphisms may develop by a failure of maintenance methyltransferase enzymes to copy methylation patterns to daugther strands during DNA replication (Vaughn et al. 2007).

While the source of most epialleles is largely unknown, the studies with the *A. thaliana* epiRILs discussed earlier indicate increases in epigenetic variation even after the function of the methylation machinery has been restored (i.e. in *DDM1/DDM1* or *MET1/MET1* homozygous epiRILs). In addition to suppressed DNA demethylation activities, this increase in variation is manifest through the redistribution across the genome of other silencing marks like H3K9 and H3K27 histone methylation (Lippman et al. 2004; Reinders et al. 2009), which could dramatically alter phenotypes.

Existing empirical evidence suggests that epimutations may be more frequent but also more labile than mutations of DNA sequence. For instance, in a study of apomictic dandelions, Verhoeven et al. (2010a) showed that even in a constant environment, some DNA methylation differences developed between individual plants (7.5% of the polymorphic DNA methylation loci) and that most of these changes were inherited across generations. Moreover, it has been shown that in some sequence contexts, DNA methylation errors are quickly and actively repaired, e.g. through RNA-directed DNA re-methylation (Teixeira et al. 2009), whereas in other DNA contexts no active restoration occurs and DNA methylation changes may turn into stable polymorphisms.

16.3.2 Environmental Induction

Another mechanism by which epigenetic variation can be created is environmental induction. Exposure to environmental stress triggers epigenetic changes, which regulate the transcriptomic stress response (Chinnusamy and Zhu 2009), and some of these stress-induced changes may persist even after the stress is relieved, resulting in a stress 'memory' that can be stable throughout the lifetime of an organism or even across generations. For instance, humans that were prenatally exposed to famine during the Dutch Hunger Winter (1944–1945) still show modified patterns of DNA methylation at *IGF2* after six decades (Heijmans et al. 2008). When gestating female rats were transiently exposed to endocrine disruptors, their male offspring showed reduced spermatogenic activity even four generations later, and this phenotypic effect was correlated to altered DNA methylation in the germ line (Anway et al. 2005).

Environmental induction of epigenetic variation that is stably inherited across generations seems to be particularly important in plant biology. One explanation for this might be that in plants, unlike in mammals, the germ line is separated from somatic tissue at a much later developmental stage (Jablonka and Lamb 1989; Wessler 1996). As a result,

there might be more opportunities for plants to transmit epigenetic modifications to offspring that were acquired in somatic tissue during the plant's life, including environmentally-induced epigenetic modifications. Several studies have used multi-generation experiments to show that parental exposure to biotic or abiotic stresses resulted in modified DNA methylation in unexposed offspring in tobacco (Boyko et al. 2007), dandelion (Verhoeven et al. 2010a) and *A. thaliana* (Boyko et al. 2010). For example, Verhoeven et al. (2010a) used methylation sensitive amplified fragment length polymorphism (MS-AFLP) to show that plants with identical genotypes exposed to different stresses had more changes in methylation sensitive markers compared to control (Fig. 16.5). In particular, chemical induction of herbivore and pathogen defenses triggered considerable methylation variation throughout the genome. Although some of these methylation differences reverted back to the original in the next generation of plants grown in a common environment, Verhoeven et al. (2010a) provide some of the first evidence that the majority of the stress-induced changes in methylation are inherited in the next generation. However, the authors did not present any information on the potential phenotypic effects of these methylation changes, and cannot rule out the possibility that the stress treatment induced changes at the DNA sequence level (e.g. through mutation or the activity of transposable elements). Further studies are required of these plants with stress-induced changes to see if the changes may be adaptive.

In *A. thaliana*, parental plants exposed to abiotic stresses produced offspring that not only showed modified transcriptomes and genomic DNA methylation, but sometimes also altered responses to stress exposure. For instance, parental exposure to salt stress resulted in offspring with a greater salt tolerance, and this transgenerational response

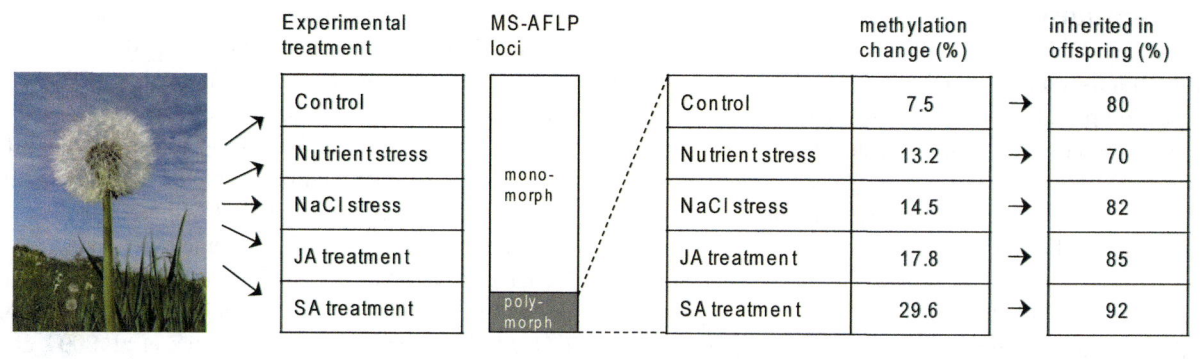

Experimental treatment	MS-AFLP loci		methylation change (%)		inherited in offspring (%)
Control			Control	7.5 →	80
Nutrient stress	mono-morph		Nutrient stress	13.2 →	70
NaCl stress			NaCl stress	14.5 →	82
JA treatment			JA treatment	17.8 →	85
SA treatment	poly-morph		SA treatment	29.6 →	92

Fig. 16.5 Induction of DNA methylation changes by ecological stresses and their heritability in asexual dandelions (*Taraxacum officinale*). A single apomictic dandelion genotype was exposed to different experimental environments. From a total of 359 MS-AFLP marker loci, 20 showed polymorphism within the experiment. Although these 'susceptible' loci showed some background level of methylation change also in the control group, the rate of methylation change that was observed within the subset of susceptible loci increased significantly due to stress treatment, particularly due to treatment with jasmonic (JA) or salicylic (SA) acid. JA and SA are plant hormones involved in herbivore and pathogen defenses. Most of the induced methylation changes were inherited in apomictic offspring that were not exposed to stress but raised in a common control environment (Verhoeven et al. 2010a)

appeared to depend on DNA methylation because the enhanced salt tolerance in the offspring disappeared when plants were treated with the demethylating agent 5-azaC (Boyko et al. 2010). This study suggests that one function of stress-induced epigenetic changes could be to prime offspring for enhanced stress resistance, presumably by heritably modulating the expression of stress-related genes. However, such a function implies that epigenetic changes are gene-specific, and we know that at least in some cases stress-induced epigenetic changes can also be untargeted or random. In dandelion, for instance, replicated individuals showed only limited consistency in their stress-induced DNA methylation changes (at anonymous MS-AFLP marker loci, Verhoeven et al. 2010a).

16.3.3 Polyploidization and Hybridization

Novel epigenetic variation can also be generated through genomic events such as hybridization and polyploidization (Liu and Wendel 2003; Adams and Wendel 2005; Grant-Downton and Dickinson 2006). The merging and doubling of genomes creates significant challenges for complex organisms, and epigenetic modifications may play an important role in reconciling the regulatory incompatibilities that are created through such genomic events (Chen 2007), for instance through epigenetic regulation or silencing of duplicated genes and re-activation of epigenetically silenced transposons (Liu and Wendel 2003; Wang et al. 2004; Lukens et al. 2006; Chen 2007; Paun et al. 2007; Hegarty and Hiscock 2008). Full genome duplication means complete functional redundancy of all genes, allowing for duplicates to take on new function or to be completely silenced and preserved for possible future function, thereby providing a sort of latent evolutionary potential (Adams et al. 2003; Rodin and Riggs 2003; Rapp and Wendel 2005; Ainouche et al. 2009; Slotkin et al. 2012, this volume). Although the epigenetic consequences of genome doubling have been documented, for instance in *A. thaliana* (Mittelsten Scheid et al. 1996, 2003), *Gossypium hirsutum* (Keyte et al. 2006) and dandelions (Verhoeven et al. 2010b), the hybridization of different genomes appears to have even more dramatic effects on epigenetic variation (Salmon et al. 2005; Parisod et al. 2009; Chelaifa et al. 2010).

A well-studied example of plant hybridization and polyploidization, and its epigenetic consequences, are the recent studies on *Spartina* (salt marsh grass) by Ainouche and coworkers. Within the last 150 years, natural hybridization between *S. alterniflora* and *S. maritima* produced the hybrid *S. ×townsendii*, which through polyploidization produced the highly invasive and phenotypically plastic allopolyploid *S. anglica* (Salmon et al. 2005; Ainouche et al. 2009). The *Spartina* system provides a rare opportunity to contrast genetic and epigenetic effects of hybridization and polyploidization. Ainouche and coworkers investigated genetic and epigenetic differences, as well as transposable element activity and gene expression differences between hybrids, allopolyploids and parental species (Salmon et al. 2005; Parisod et al. 2009; Chelaifa et al. 2010). They found that both hybridization and polyploidization were associated with substantial DNA methylation changes, particularly around transposon loci. However, these changes were more dramatic during the hybridization event, and therefore, if epigenetic changes contributed to the invasive success of *S. anglica*, then the preceeding hybridization event may have been an important step in the evolution of this invader.

16.4 Relationships Between Genetic and Epigenetic Variation

One of the main challenges in understanding the importance of epigenetics in evolution is the relationship between genetic and epigenetic effects, and the degree to which phenotypic variation can be explained by epigenetic effects independent of genetic effects. Richards (2006) speculated that while some epigenetic effects will be entirely determined by genotype, others may be only "facilitated" by specific genotypes, or may be completely independent. Disentangling these different possibilities in reality, however, is complicated not only because we know little about these interactions, but also because so far the genetics of most complex traits is not well understood.

The network of genes involved in the flowering time pathway in *A. thaliana* is one of the best-studied regulatory pathways that control a complex life history trait in plants (Mouradov et al. 2002; Simpson and Dean 2002), and is therefore a useful starting point for unraveling genetic-epigenetic relationships. This is particularly true because flowering at the appropriate time requires the integration of many environmental signals to initiate a developmental switch in plants. The flowering time response requires the coordination of hundreds of genes organized into four main pathways responsive to specific environmental stimuli: the autonomous, gibberellin, photoperiod, and vernalization pathways (Mouradov et al. 2002; Simpson and Dean 2002). Within these pathways, mutant and transgenic studies have revealed genes that respond directly to specific environmental factors (e.g., PHYA detects far red light, PHYB, D and E detect red light; Schmitt et al. 1999; Simpson and Dean 2002), whereas quantitative genetic studies have shown dramatic phenotypic effects of natural polymorphisms in some of these genes (Caicedo et al. 2004; Olsen et al. 2004; Stinchcombe et al. 2004; Korves et al. 2007). The two best-known genes involved in the vernalization pathway are FRIGIDA (FRI) and the floral repressor FLC (Johanson et al. 2000; Weinig et al. 2002; Olsen et al. 2004). FRI represses flowering by activating the expression of FLC. Candidate gene association studies, while not definitive, have suggested that common allelic variation at FRI and FLC are associated with flowering time diversity in natural field conditions (Caicedo et al. 2004; Stinchcombe et al. 2004). In particular, an active FRI-FLC pathway results in late flowering, whereas an inactive FRI-FLC pathway results in early flowering.

While these studies of the flowering time response in *A. thaliana* have been generally focused on candidate genes and DNA sequence based mechanisms, recent studies indicate that epigenetic processes play an important role in this regulatory network, in particular in the incorporation of environmental cues into genomic responses.

16.4.1 Epigenotype Dependent on or Inseparable from Genotype

The vernalization pathway in *A. thaliana* controls epigenetic silencing of the floral repressor FLC, and is an important component of the regulation of flowering time (Sung and Amasino 2004). Moreover, Shindo et al. (2006) recently explored how this epigenetic component of FLC regulation differed among different *A. thaliana* genotypes, and they found that there was significant variation among genotypes in the critical length of cold treatment that was required for FLC repression (when exposed to shorter cold treatments, FLC would become reactivated and flowering time would be repressed). Molecular comparisons between a genotype that rapidly responded to vernalization (Edi-0) to one that required much longer vernalization (Lov-1) suggested that the critical length of vernalization period required was related to (1) different levels of initial histone trimethylation at H3K27 in the FLC promoter and (2) different rates of acquiring the silencing marks at H3K27.

While in this case epigenetic variation is clearly associated with phenotypic variation in an important plant trait, different genotypes are involved, so the epigenetic differences could be a consequence of genetic instead of epigenetic differences. In fact, Shindo et al. (2006) hypothesize that sequence polymorphisms in the FLC gene itself could be partially responsible for the observed differences in accumulation of trimethylation at H3K27.

Although the epigenetic patterns of FLC seem to be genotype specific, the mechanism of inheritance for this epigenetic effect is unknown. Is the epigenetic pattern re-established de novo in each generation, or after its initial establishment does it persist regardless of genotype? To discriminate between these possibilities, it would be informative to do controlled crosses between genotypes with either quick or slow vernalization responses to see how the

trimethylation of H3K27 segregates. Another approach to understanding the epigenetic programming of FLC would be to explore the patterns in the mutant lines. Crosses with the *ddm1* mutant show that in general expression profiles look like that of *ddm1* (Lippman et al. 2004), but no reports have specifically diagnosed the epigenetic status of FLC in these crosses. Targeted studies of the epigenetic silencing of FLC in the epi-RILs may be helpful in understanding the relative contribution of genetic and epigenetic regulation of this ecologically important gene.

16.4.2 Epigenotype Can Act Independently of Genotype

In the case of silencing of FLC through accumulation of trimethylation of H3K27, variation in the epigenetic response is associated with different genotypes. But are there also cases where epigenetic effects are independent of genetic effects? Evidence that this is possible comes from studies of epimutations such as the *Linaria* study of Cubas et al. (1999; Fig. 16.1), because in the absence of genetic variation, phenotypic effects can be unambiguously ascribed to the epigenetic alterations. However, even in such studies one cannot rule out the possibility that certain genotypes are more prone to epigenetic changes and that there may be some interdependence of genotype and epigenotype (i.e., "facilitated" epigenetic effects, sensu Richards 2006).

Essentially, all of the approaches described in Sect. 16.2 that have been used to examine phenotypic effects of epigenetic variation, i.e. natural or artificial epimutations, or epigenetic recombinant inbred lines, have successfully demonstrated that there *can* be epigenetic effects independent of genotype, but many of these methods were artificial. The question is still open as to the extent that independent epigenetic effects exist and play a role in natural populations (see Sect. 16.5 below).

16.4.3 Genotypic Variation Created by Epigenetic Effects: Transposable Elements and Heat Shock Proteins

In addition to the possibility of "facilitated" epigenetic effects (Richards 2006), there are other ways for genetic and epigenetic mechanisms to interact. One important phenomenon in this context is the possibility of genetic changes caused by transposons. Most insertions of transposable elements (TEs) are to be likely neutral or deleterious (reviewed in Brookfield 2005; Slotkin et al. 2012, this volume), but some may be adaptive in providing novel regulatory function or by promoting adaptive alternative splicing or transcription factor binding sites (Kidwell and Lisch 2000; Feschotte 2008; Slotkin et al. 2012, this volume).

Transposable elements are concentrated in the hetero-chromatin of plants where they typically are epigenetically silenced by histone and DNA sequence methylation (Lippman et al. 2004; Slotkin and Martienssen 2007; Vaughn et al. 2007). Lippman et al. (2004) showed that in the *A. thaliana* Col-0 wild type, both DNA and H3K9 methylation are significantly correlated with the location of transposable elements and repeats. In *A. thaliana ddm1* mutant lines, both DNA and H3K9 methylation are reduced, specifically in the genomic areas where transposable elements are located, and several classes of transposons, which are not active in the wild type, have a high frequency of transposition (Kakutani 2002; Lippman et al. 2004; Tsukahara et al. 2009).

When transposable elements reinsert in the genome, they not only change DNA sequence, but they can also affect gene expression (Lippman et al. 2004; Feschotte 2008) and have dramatic effects on phenotype through the production of epialleles (Wessler 1996; Slotkin and Martienssen 2007; Tsukahara et al. 2009; Slotkin et al. 2012, this volume). Many studies have speculated that environmental stress could result in phenotypic variation that results from changes in epigenetic silencing of transposons, which could ultimately cause transposition and change in the genetic make-up of an individual (Kidwell and Lisch 2000; Rapp and Wendel 2005; Slotkin and Martienssen 2007; Jablonka and Raz 2009; Mirouze and Paszkowski 2011). While this type of transposon activity in response to environmental stress has not yet been demonstrated in plants (Colot V 2010, personal communication, see Slotkin and Martienssen 2007 for examples in *Drosophila* and *Schizosaccaromyces pombe*), increased transposon activity in response to *genomic* stress, including hybridization and polyploidization, is well-documented (McClintock 1984; Wessler 1996; Liu and Wendel 2003; Rapp and Wendel 2005; Mirouze and Paszkowski 2011). In a recent example, Parisod et al. (2009) used methyl-sensitive transposon display (MSTD) to investigate if increased transposon activity resulted from hybridization of *Spartina* species. They found that hybridization resulted in major methylation changes, especially near TE insertions in the maternal *S. alterniflora* genome, but in the three families of transposable elements that they examined, they did not find evidence of transposition. In general, the adaptive significance of such (possibly epigenetically induced) transposable element activity remains unclear.

One emerging story about the epigenetic silencing of TEs involves the activity of heat shock proteins (Hsp). Hsp90 is a required component of many regulatory complexes and as such Hsp90 activity affects many different pathways. Studies have found that the Hsp90 is a *phenotypic capacitor* and that its normal activity buffers phenotypes against environmental variation by suppressing the expression of genetic variation in several developmental pathways (Rutherford and Lindquist 1998; Queitsch et al. 2002). When Hsp90 is altered, a variety of developmental abnormalities develop which can become fixed even when Hsp90 activity is restored. In addition, the response of a given genotype appears to be dependent on the nature of the previously silent genetic variation (Queitsch et al. 2002).

The concept of phenotypic capacitors contributing to phenotypic robustness (also known as canalization) dates back to Waddington's experiments with *Drosophila* over 50 years ago (Waddington 1942, 1953). Waddington found that after 12 generations of selecting a phenotype elicited by exposure to 40 C for 4 h in *Drosophila*, some offspring would elicit the phenotype without the 40 C treatment. He later described this phenomenon as genetic assimilation. Without any understanding of the mechanistic basis for the inheritance of the trait, Waddington argued that the coding of the trait had become assimilated by the genotype (Waddington 1953). It is only recently that this buffering has been linked to an epigenetic effect (Sollars et al. 2003; Specchia et al. 2010). In a mutant screen of *Drosophila*, Sollars et al. (2003) found that Hsp90 is involved in chromatin regulation. The authors argue that instead of just relying on cryptic genetic variation that is revealed when Hsp90 is reduced as the sole source of heritable phenotypic variation, adaptive differences may arise epigenetically and allow for a more rapid response to selection. This source of phenotypic variation may be less stable than strictly DNA sequence based variation, and it is most likely that a combination of genetic and epigenetic mechanisms interact to produce adaptation.

In addition to affecting expression through alteration of chromatin structure, recent studies have connected the activity of Hsp90 with the activity of Piwi-interacting RNA. These small RNAs are involved in the silencing of TEs and mutations in Hsp90 show increased transcription for all TEs tested (Specchia et al. 2010). Similar to previous studies of Hsp90 where the response depends on the genetic background, Specchia et al. (2010) found that different genotypes may induce different transposon insertions. Combined, these studies suggest that Hsp90 activity (and perhaps other phenotypic capacitors) can contribute to phenotypic variation through both restructuring of chromatin or the indirect activation of TEs. In both cases, gene expression may be altered and the effects can be inherited. We know of no studies that have tested these effects in an ecological context, but the possible effects for evolutionary biology are intriguing.

16.5 Patterns of Natural Epigenetic Variation

Epigenetic inheritance in natural contexts is currently little explored, although some of the best-known heritable epialleles occur naturally in wild populations (e.g., *Linaria vulgaris*, Fig. 16.1). It could well be that many epialleles have more subtle effects and are not as easily discovered, but they may nevertheless significantly contribute to quantitative variation in ecologically relevant traits and therefore, ultimately, to adaptation.

To date, analyses of natural epigenetic variation have either used high-resolution genetic information to understand variation in specific traits (usually on model species without explicit links to populations or environments), or using low-resolution genetic information such as MS-AFLPs to address population-level questions. We expect progress to be made from more intimate merging of these two approaches, which should allow linking of epigenetic polymorphisms with known functional effects on gene expression and/or traits to performance and fitness under ecologically relevant conditions and in natural populations. Still, some important insights already have been obtained from analyses of natural epigenetic variation.

16.5.1 High Levels of Heritable Epigenetic Variation Exist Within and Between Natural Populations

Early studies of natural variation in DNA methylation in *A. thaliana* revealed that methylation patterns are similar between plants from the same inbred accession (demonstrating faithful inheritance of DNA methylation marks) but different between plants from different accessions (Cervera et al. 2002). Subsequent genomic analyses confirmed that there are high levels of DNA methylation polymorphism between accessions. The extent of polymorphism differs considerably between genomic regions, and polymorphisms are particularly abundant within genes (Vaughn et al. 2007). This probably has to do with the mechanisms that generate and maintain DNA methylation. For instance, the methylation of transposable elements is often guided by small RNAs and can be actively restored de novo, which leads to high levels of TE methylation but limited methylation polymorphism among different accessions. In contrast, many *A. thaliana* genes have DNA methylation that is heritable but fairly unstable and that apparently is not restored de novo after loss of methylation marks. Over time, this seems to lead to high levels of genic methylation polymorphism between different accessions that don't seem to reflect changes in gene expression or isolation by distance patterns (Vaughn et al. 2007).

High levels of DNA methylation variation within and between populations are also detected in several other plant species including cotton (Keyte et al. 2006), wild barley (Li et al. 2008), mangrove trees (Lira-Medeiros et al. 2010) and Mediterranean violets (Herrera and Bazaga 2010). Several such studies in non-model species have used MS-AFLP analysis to quantify DNA methylation polymorphisms at anonymous marker loci throughout the genome. In these studies, population genetic measures of diversity within or among populations were often significantly higher for methylation-sensitive markers than for normal (methylation-insensitive) genetic markers, i.e. there was often both greater diversity within populations and greater population differentiation at the epigenetic level than at the genetic level.

16.5.2 Natural Epigenetic Variation May Be Partly Autonomous

In principle, only epigenetic variation that is not under complete control of DNA sequence variation has the potential to explain phenotypic variation beyond that already explained by DNA sequence variation. It is clear that heritable, natural epigenetic variation is sometimes not autonomous, but under genetic control (e.g. FLC as discussed above). This genetic control can not only be direct, but also indirect through small RNAs, as among-accession differences in DNA methylation can be controlled by differences in accession-specific small interfering RNAs (Zhai et al. 2008), which presumably reflect genetic differences in siRNA-generating loci (such as TEs and repetitive sequences). Nevertheless, in the natural population studies described above, correlations between genetic variation and DNA methylation variation were often surprisingly weak. In *A. thaliana*, for instance, a matrix of pairwise similarities between individuals based on methylation polymorphisms was uncorrelated to a similarity matrix based on genetic polymorphisms at genomewide AFLP markers (Cervera et al. 2002), and patterns of gene-level methylation between *A. thaliana* accessions did not reflect genetic relatedness of the accessions (Vaughn et al. 2007).

Lack of correlation between genetic and epigenetic variation suggests that epigenetic variation may not be under strict genetic control. However, lack of correlation does not necessarily prove independence. For instance, in MS-AFLP versus AFLP comparisons, it is possible that MS-AFLP polymorphisms are under trans-acting genetic control (by loci that are not themselves part of the AFLP dataset). The data from natural populations are certainly consistent with partial autonomy of DNA methylation variation, but a more conservative conclusion that can currently be drawn is that many DNA methylation polymorphisms in natural populations are not under cis-acting genetic control (Cervera et al. 2002).

16.5.3 Epigenetic Variation Can Be Correlated to Adaptive Population Differentiation

If heritable epigenetic variation plays a role in adaptation, then local differences in ecological habitat characteristics may select for different epialleles in different populations. Just as with selection on genetic polymorphism, this will result in population-level associations between heritable epigenetic polymorphisms and ecological habitat characteristics. So far, such associations are unexplored for epialleles with well-characterized functional effects on gene expression or traits. However, some interesting population epigenetic observations have recently been made using methylation-sensitive markers on natural populations of *Dactylorhiza* orchid species (Paun et al. 2010), the mangrove *Laguncularia racemosa* (Lira-Medeiros et al. 2010) and the Mediterranean violet *Viola cazorlensis* (Herrera and Bazaga 2010, 2011). Each of these studies found correlations between epigenetic diversity and different habitats, but all three studies were performed on field collected material and did not control for environmentally induced epigenetic effects so the potential for these epigenetic effects to be involved in adaptation is unclear (Richards et al. 2010b).

For example, the *Viola cazorlensis* populations are genetically differentiated, and individual AFLP marker loci can be identified that show higher-than-expected population differentiation compared to genomic background levels of differentiation. Such 'outlier' loci are usually attributed to divergent natural selection between populations, which maintains population-specific alleles at genes closely linked to these loci under divergent selection, whereas the rest of the genome is more homogenized between populations. It turns out that these outlier genetic polymorphisms in *V. cazorlensis* are statistically correlated both to flower morphology and also to DNA methylation variation in plants collected from their natural habitat (Herrera and Bazaga 2008, 2010). Thus, adaptive genetic divergence may be associated with epigenetic differentiation between the populations. However, since Herrera and Bazaga (2010) were unable to grow the plants in common garden, many of these epigenetic differences could also be environmentally induced. Alternatively, this epigenetic differentiation could be a downstream consequence of selectively maintained genetic variation. In another study, Herrera and Bazaga (2011) found both DNA sequence and methylation polymorphisms in *V. cazorlensis* were correlated specifically with herbivory damage. Structural equation models suggested that genotype contributed directly to herbivory damage and epigenotype, but could not discriminate the relationship between epigenotype and herbivory damage. The two best models equally predict a consequential and causal role between epigenetic variation and herbivory

suggesting that there could be a combination of effects that are induced by herbivory and affect the likelihood of herbivory. Another possibility is that random epigenetic mutations arise and build up rapidly within isolated populations, potentially resulting in (neutral) epigenetic differences between populations that correlate with genetic differentiation of the populations.

In summary, the use of anonymous MS-AFLP markers, which has dominated the work in more ecologically-oriented epigenetics research so far, has provided promising first insights from natural populations, in the wild and in non-model species. The high levels of natural epigenetic variation, its limited correlation with genetic variation and its association with adaptive population differentiation are all consistent with (but not conclusive evidence for) a role for epigenetic inheritance in adaptation and evolution. It is possible to investigate whether patterns of natural epigenetic variation in the field associate with phenotypes or environmental factors. However, such questions are currently unexplored and require common environment manipulations because of the environmentally labile nature of epigenetic effects (Richards et al. 2010b).

Ultimately, however, marker-level data provide only limited information. To gain a deeper understanding of the causes and consequences of the observed natural epigenetic variation, the next step should be to merge ecological approaches (linking epigenetic variation to fitness, ecological environments and natural populations), molecular approaches (gene-level information on sequence, epigenetic and activity status), and common garden experiments. Scoville et al. (2011) have made progress in combining these approaches by identifying a target gene that may be epigenetically modified and contribute to epigenetic inheritance of trichome density in *Mimulus guttatus*. The genetic basis of trichome production has been extensively studied in the model plants *A. thaliana* and *Antirrhinum majus,* and Scoville et al. were able to select homologs in *M. guttatus* of genes known to be involved in trichome development. They examined the relationship between expression of candidate genes and inheritance of damage-induced trichome production in high and low trichome parental lines and four recombinant inbred lines (RILs) exposed to damage and control conditions. Their findings indicate that down-regulation of *MgMYBML8* is correlated with the inheritance of increased trichome density in one of the parents and three of the four RILs. However, their study does not explore epigenetic mechanisms that may be involved in regulating this expression, which merits further work (Richards and Wendel 2011). It is this challenge of grafting detailed epigenomic tools onto an ecological genetics approach that will eventually provide a deeper understanding of natural epigenetic variation.

16.6 Impact of Epigenetics on Our Understanding of Evolutionary Concepts

The evolutionary relevance of epigenetic inheritance has been much discussed in recent years (Richards 2006; Bossdorf et al. 2008; Johannes et al. 2008; Jablonka and Raz 2009; Richards et al. 2010a, b). Mechanisms that generate heritable variation are a driving force behind all evolution. The heritable modulation of gene activity through epigenetic inheritance could represent a variation-generating mechanism, in addition to mutation (that creates novel gene polymorphisms) and the joint processes of recombination and segregation (that create novel gene combinations). Because heritable epigenetic variation can be induced by environmental conditions, it has been argued that epigenetic inheritance can be a mechanism for 'soft inheritance', where an environmentally-induced phenotype is transmitted to offspring generations (Richards 2006). This resembles Lamarckism and the inheritance of acquired characters, a concept that was dismissed long ago in the modern evolutionary synthesis but that some researchers argue deserves re-evaluation (Jablonka and Lamb 1989, 1995; Gissis and Jablonka 2011).

16.6.1 Selection on Heritable Epigenetic Variants

In a strict sense, the evolutionary implications of epigenetic inheritance are due to phenotypic effects of epiallelic variants that are (at least partially) independent of DNA sequence variants and that are exposed to natural selection, thereby affecting population responses to selection in ways that cannot be explained by DNA sequence variation alone. Independence or partial independence of epigenetic variation from genetic variation causes partial de-coupling of phenotypic change and genotypic change. As discussed above, this can arise, for instance, due to imperfect maintenance of heritable epigenetic marks, or by environment-induced epigenetic modifications that are subsequently stably transmitted to offspring generations. In addition, epiallelic variants arise more frequently and are more inducible and reversible than DNA sequence mutations. These features can affect micro-evolution in several ways, as described below.

16.6.1.1 Epigenetic Inheritance Facilitates Exploration of the Adaptive Landscape

A dynamic and reversible epigenetic code can add adaptive flexibility to the more stable and hard-wired genetic code. In principle, through random epigenetic variation a single genotype can heritably vary in different phenotypic directions, permitting the exploration of novel niches without abandoning the old one. Pal and Miklos (1999) modeled the consequences of epigenetic inheritance for adaptation, and found that the ability to generate heritable epigenetic variation can speed up the process of reaching a fitness peak in the adaptive landscape. It may also facilitate peak shifts, or the transition from one fit genotypic state to another fit genotypic state despite reduced fitness of intermediate genotypic states. This is because under randomly generated epigenetic variation, some individuals of a reduced-fitness intermediate *genotype* may still possess a high-fitness and heritable *phenotype* due to the added epigenetic contribution to the phenotype, thus effectively flattening the fitness landscape. After approaching a novel fitness peak, processes of canalization and genetic assimilation may take over and genetically stabilize the novel phenotype (Pal and Miklos 1999; see also Sect. 16.6.2.2).

High epigenetic mutation rates could particularly enhance the adaptive possibilities of asexual or low-diversity taxa, or in rapidly changing environments, when the rate of genetic change can be a limiting factor for generating novel variation (Jablonka and Lamb 1989). A concrete example of when epigenetic effects may be a particularly important source of phenotypic variation is in the expansion of invasive species. Understanding local adaptation in plant invasions has been challenging given the likelihood of reduced genetic variation following a population bottleneck, which is assumed to severely constrain the evolutionary potential of a given population or species. Although many invasive species benefit from alternative sources of increased DNA sequence variation through multiple introductions (Durka et al. 2005; Lavergne and Molofsky 2007; Rosenthal et al. 2008; Gammon and Kesseli 2009) or hybridization (Daehler and Strong 1997; Pysek et al. 2003; Bímová et al. 2004; Mandák et al. 2004; Bailey et al. 2009), several invasives appear to do well even with low levels of sequence based variation (Hollingsworth and Bailey 2000; Dlugosch and Parker 2008a, b; Richards et al. 2008; Loomis and Fishman 2009). Despite the observation of decreased sequence based variance in these studies, only one reported a substantial decline in phenotypic variance (Simberloff et al. 2000). Rapp and Wendel (2005) argue that even with reduced genetic variation, epigenetic effects could expand the phenotypic possibilities that may result from epigenetic modifications induced by genomic stresses, such as those caused by extreme environmental selection or ecological change which are often experienced by populations that go through a genetic bottleneck. This possibility is particularly relevant given that past hybridization or polyploidization may have produced latent evolutionary potential through the preservation of epigenetically silenced duplicate genes that are poised to take on new function and contribute to phenotypic variation (Adams et al. 2003; Rodin and Riggs 2003; Rapp and Wendel 2005; Ainouche et al. 2009).

The chance sampling of genotypes involved in the invasion process, combined with non-DNA sequence based sources of phenotypic variation, can lead to divergence in phenotypes of these populations even in the absence of abundant sequence based variation and differential selection (Rapp and Wendel 2005; Keller and Taylor 2008; Prentis et al. 2008). For example, Richards et al. (2008) found that *Fallopia* populations have invaded a diversity of habitats on Long Island, NY and have persistent phenotypic variation with almost no sequence variation (Figure 16.6a, b). However, using MS-AFLP, Richards and colleagues showed that these populations harbor five times as many polymorphic epigenetic loci as DNA sequence loci (Fig. 16.6c, d; Richards, Schrey and Pigliucci unpublished).

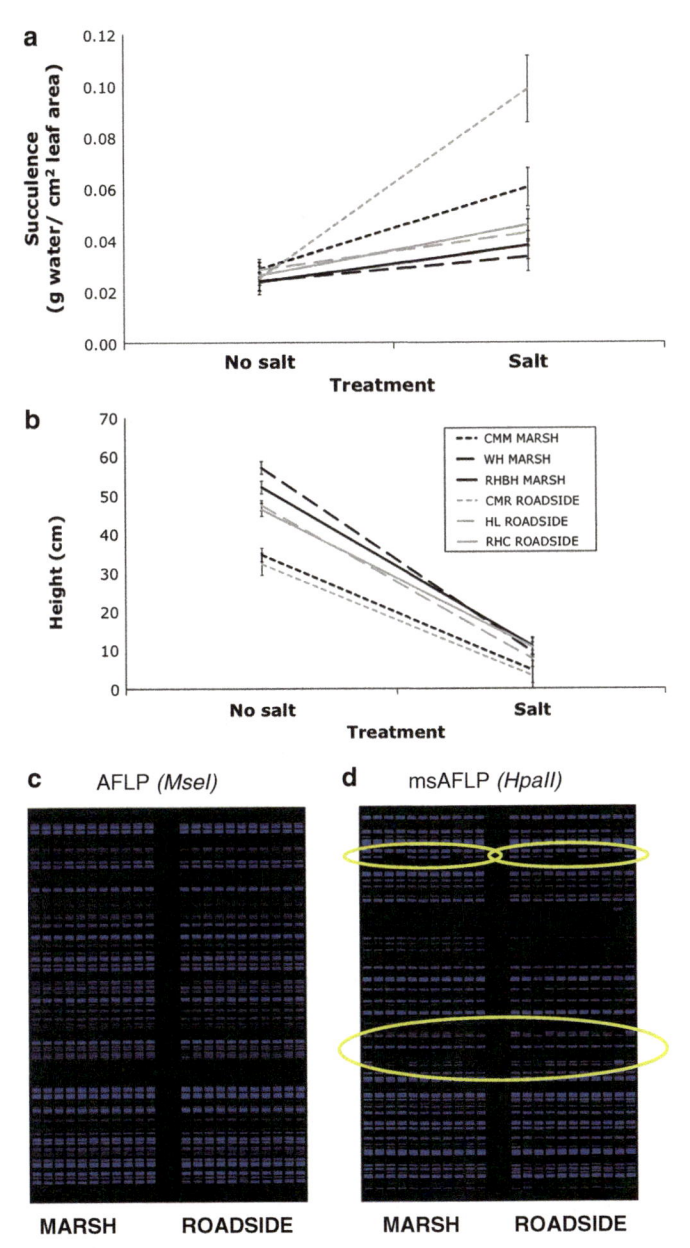

Fig. 16.6 *Fallopia japonica* and *F. ×bohemica* invading Long Island habitats. Shown are reaction norms (means ± 1 SE) for (**a**) succulence and (**b**) height of *Fallopia spp.* in response to two salt treatments for three road side and three salt marsh populations (From Richards et al. 2008). Also shown are (**c**) AFLP markers indicating no polymorphism across one of the roadside and one of the marsh populations, and (**d**) polymorphism within and among the same to populations for MS-AFLP (Epigenetic loci; Richards, Schrey and Pigliucci unpublished)

16.6.1.2 Epigenetic Inheritance Adds a Trans-generational Component to Phenotypic Plasticity

While undirected modifications will contribute to random epigenetic variation and can result in increased offspring trait variances, directed (targeted) modifications that occur in the same way in different individuals can result in shifted offspring trait means. In the latter case, epigenetic inheritance can be an underlying mechanism for trans-generational phenotypic plasticity (or 'maternal effects'; Richards et al. 2010a; Richards 2011; Scoville et al. 2011). Such trans-generational effects are commonly reported in the ecological literature, and may persist for more than one offspring generation. Particularly in cases where trans-generational effects involve highly specific stress responses (such as increased leaf trichome density in offspring of herbivory-exposed parental plants; Holeski 2007; Scoville et al. 2011) or persist for multiple generations (Whittle et al. 2009), authors have speculated that epigenetic inheritance could be an underlying mechanism. Thus, in contrast to genetic mutations, beneficial heritable modifications can be triggered simultaneously in multiple individuals in the population, thereby speeding up the population response to a changed environment (Jablonka and Raz 2009). The microevolutionary consequences of environment-induced parental effects can be diverse (Fox and Mousseau 1998) and they can impact both the rate and direction of evolutionary change in response to selection (Kirkpatrick and Lande 1989). By 'preparing' offspring for specific environmental conditions, parents can increase their offspring performance and fitness, as demonstrated both in the lab (Agrawal et al. 1999) and under natural field conditions (Galloway and Etterson 2007).

16.6.2 Selection on Genetic Variants Mediated by Epigenetic Change

In addition to selective effects of heritable epigenetic variants, epigenetic inheritance mechanisms can add to the evolutionary potential that is based on selective effects of genetic variants.

16.6.2.1 Epigenetic Control Over Transposable Elements Affects Rates of Genetic Change

Release of epigenetic TE silencing, which can occur in response to stressful conditions, can trigger sudden bursts of novel genetic variation due to transposition and associated processes. This adds raw material for natural selection during times of stress, when novel variation can be advantageous (McClintock 1984; Wessler 1996; Rapp and Wendel 2005), but also allows for novel genome function especially by way of regulatory genes and added transcription factor binding sites (Feschotte 2008). It is important to note that novel epigenetic variation (not just novel genetic variation) can be generated in this way as well, as release and subsequent re-silencing of TEs has the potential to generate novel epiallelic variation at functional genes that are physically close to TE insertion sites (Slotkin and Martienssen 2007; Paszkowski and Grossniklaus 2011). Several authors have pointed out potential macro-evolutionary consequences of stress-induced release of TE activity and silencing. Long periods of more or less stable TE silencing that are interrupted with moments of unleashed TE activity could account for a 'punctuated equilibrium' model of evolutionary change, when bursts of TE-induced genetic variation permit brief periods of rapid evolutionary change. Sudden TE-induced genome restructuring can displace a population far from its adaptive peak in the fitness landscape, facilitating the transition to other fitness peaks (Zeh et al. 2009). TE-induced genome restructuring might also result in genetic incompatibilities between subpopulations, leading to rapid reproductive isolation and speciation (Rebollo et al. 2010).

16.6.2.2 Epigenetic Change Can Initiate Heritable Change That Guides Subsequent Selection

An epigenetic code that is more dynamic and flexible than the genetic code can play an important role in initiating evolutionary change that is subsequently taken over by the more stable genetic code. Epigenetic modifications could 'hold' a potentially advantageous phenotype for multiple generations, allowing time for more stable genetic variants to stabilize the phenotype. This sequence of events has been proposed as a mechanism for genetic assimilation, the process by which an initially plastic phenotype that is stimulus-induced becomes heritably fixed and stimulus-independent (Waddington 1953). More generally, West-Eberhard (2005) argued that evolutionary change is commonly initiated by developmental plasticity (for instance in response to environmental stimuli) and is subsequently followed by selection of genetic variants that stabilize or otherwise accommodate the changed phenotype. According to this view, genes may be followers rather than initiators of evolutionary change.

Rapid accumulation of epigenetic modifications can also be a first step in some macro-evolutionary processes. For instance, different populations may rapidly build up epigenetic differences, either due to epigenetic drift or environment-specific induction of epigenetic modifications, and such epigenetic differences can be a first steps towards reproductive isolation (Jablonka and Lamb 1998) because epigenetic incompatibilities can be responsible for compromised fitness of hybrids (Vrana et al. 2000, in Jablonka and Raz 2009). Once the process of reproductive isolation has started, it facilitates the evolution of genetic incompatibilities and speciation.

16.7 The Future of Ecological and Evolutionary Epigenetics

While molecular studies of epigenetic phenomena over the last few decades have revolutionized our understanding of genome function and developmental biology, these processes also challenge our understanding of heritable phenotypic variation at the level of whole organisms and even the process of evolution by natural selection (Jablonka and Lamb 1989; 1995; Danchin et al. 2011). In this chapter, we have discussed how epigenetic variation is associated with variation in phenotypic traits, that natural variation exists at the epigenetic level, that epigenetic variation can be induced by environmental stress and that some epigenetic marks are faithfully transmitted across generations. However, since previous work emphasizes agricultural crops and model species, frequently under artificial conditions, we still know little about the importance of epigenetic processes in natural populations (Richards 2008; Richards et al. 2010b). Fortunately, several researchers have begun to explore questions related to ecology and evolution in natural populations. The future of ecological and evolutionary epigenetics holds many more studies like those of Verhoeven et al. (2010a, b), Hererra and Bazaga (2010, 2011) and Scoville et al. (2011) which begin to address the question of epigenetic responses to the environment and ultimately epigenetic contributions to adaptation. This task will continue to be challenging because of the labile nature of epigenetic effects, the complicated relationship between epigenetic effects and DNA sequence variation and the tools required to tie functionality to epigenetic changes.

Because epigenetic variation is to some extent environmentally labile and reversible and many of the developmental processes that underlie response to different environments involve epigenetic changes, patterns of epigenetic differentiation among individuals that are measured in different environments will most likely include a reversible component (e.g., for FLC) that results in phenotypic plasticity and a non-reversible or relatively stable component due to heritable epigenetic differentiation (e.g., *Lcyc* in *Linaria vulgaris*). As we have discussed recently (Richards et al. 2010b), analyses of epigenetic variation are similar to analyses of phenotypic variation, and common garden experiments are necessary to firmly establish inheritance of epigenetic effects and separate between plastic and heritable components of variation. As in classic analyses of phenotypic variation, demonstrating adaptation requires assessing response to reciprocal transplant studies in the field or studies in a controlled environment. In the case of epigenetic effects, a common environment approach will be even more critical to rule out the possibility that any association of epialleles is not merely a transient and environmentally induced association.

Another issue that will be important to consider in future studies is the level of independence of epigenetic variation. We are particularly interested in whether epigenetic variation that is unrelated to DNA sequence variation has the potential to explain phenotypic variation beyond that already explained by DNA sequence. Verhoeven et al. (2010a) have provided a first glimpse into the inheritance of induced methylation patterns for a single genotype in response to multiple environmental stresses. Expanding this approach to include (1) multiple genotypes and (2) associated phenotypes will be critical information to evaluate the importance of induced and heritable epigenetic changes.

Providing the association between epigenetic changes and phenotypic changes will be particularly challenging. A major limitation to our current understanding of epigenetics in natural systems is that marker-level data provide only limited information, and cannot typically be associated with phenotype. To gain a deeper understanding of the causes and consequences of the observed natural epigenetic variation, we must find a way to explore the phenotypic effects of specific epigenetic activity in natural settings. Scoville et al. (2011) have made some progress in this respect by targeting known genes that underlie inheritance of environmentally induced trichome formation in *M. guttatus*. The increasing application of next generation sequencing combined with epigenetic specific approaches (i.e. bi-sulfite sequence conversion or tiling microarrays) could be incorporated into ecological experimental design and should help in the process (Richards and Wendel 2011).

While there are few ecological epigenetics studies making steps in the right direction, we can continue in this quest by learning from the extensive studies in the field of ecological genetics, which has long been interested in deciphering organismal response to the environment and how natural selection leads to adaptation. Applying the new tools and understanding of epigenetics and genome function in general to a robust ecological design will be powerful for assessing the importance of these effects in the real world.

Acknowledgements The authors would like to thank Jonathan Wendel for the invitation to write this chapter as well as JF Wendel and X Zhang for feedback on previous versions of the manuscript. This work was supported by University of South Florida (CR), the Netherlands Organisation for Scientific Research (KJFV), and the Swiss National Science Foundation (OB).

References

Adams KL, Wendel JF (2005) Polyploidy and genome evolution in plants. Curr Opin Plant Biol 8:135–141

Adams KR, Cronn R, Percifield R, Wendel JF (2003) Genes duplicated by polyploidy show unequal contributions to the transcriptome and organ-specific reciprocal silencing. Proc Natl Acad Sci USA 100:4649–4654

Agrawal AA, Laforsch C, Tollrian R (1999) Transgenerational induction of defences in animals and plants. Nature 401:60–63

Ainouche ML, Fortune PM, Salmon A, Parisod C, Grandbastien MA, Fukunaga K, Ricou M, Misset MT (2009) Hybridization, polyploidy and invasion: lessons from *Spartina* (Poaceae). Biol Invas 11:1159–1173

Anway MD, Cupp AS, Uzumcu M, Skinner MK (2005) Epigenetic transgenerational actions of endocrine disruptors and mate fertility. Science 308:1466–1469

Bailey JP, Bímová K, Mandák B (2009) Asexual spread versus sexual reproduction and evolution in Japanese Knotweed s.l. sets the stage for the "Battle of the Clones". Biol Invas 11:1189–1203

Bímová K, Mandák B, Kasparová I (2004) How does *Reynoutria* invasion fit the various theories of invisibility? J Veg Sci 15:495–504

Bossdorf O, Richards CL, Pigliucci M (2008) Epigenetics for ecologists. Ecol Lett 11:106–115

Bossdorf O, Arcurri D, Richards CL, Pigliucci M (2010) Experimental alteration of DNA methylation affects the phenotypic plasticity of ecologically relevant traits in *Arabidopsis thaliana*. Evol Ecol 24:541–553

Boyko A, Kathiria P, Zemp FJ, Yao YL, Pogribny I, Kovalchuk I (2007) Transgenerational changes in the genome stability and methylation in pathogen-infected plants (virus-induced plant genome instability). Nucleic Acids Res 35:1714–1725

Boyko A, Blevins T, Yao YL, Golubov A, Bilichak A, Ilnytskyy Y, Hollander J, Meins F, Kovalchuk I (2010) Transgenerational adaptation of *Arabidopsis* to stress requires DNA methylation and the function of dicer-like proteins. PLoS One 5:e9514

Brookfield JFY (2005) The ecology of the genome—mobile DNA elements and their hosts. Nat Rev Genet 6:128–136

Burn JE, Bagnall DJ, Metzger JD, Dennis ES, Peacock WJ (1993) DNA methylation, vernalization, and the initiation of flowering. Proc Natl Acad Sci U S A 90:287–291

Caicedo AL, Stinchcombe J, Schmitt J, Purugganan MD (2004) Epistatic interaction between the *Arabidopsis* FRI and FLC flowering time genes establishes a latitudinal cline in a life history trait. Proc Natl Acad Sci U S A 101:15670–15675

Cervera MT, Ruiz-Garcia L, Martinez-Zapater JM (2002) Analysis of DNA methylation in *Arabidopsis thaliana* based on methylation-sensitive AFLP markers. Molec Genet Genomics 268:543–552

Chelaifa H, Monnier A, Ainouche ML (2010) Transcriptomic changes following recent natural hybridization and allopolyploidy in the salt marsh species *Spartina* × *townsendii* and *Spartina anglica* (Poaceae). New Phytol 186:161–174

Chen ZJ (2007) Genetic and epigenetic mechanisms for gene expression and phenotypic variation in plant polyploids. Annu Rev Plant Biol 58:377–406

Chinnusamy V, Zhu JK (2009) Epigenetic regulation of stress responses in plants. Curr Opin Plant Biol 12:133–139

Cubas P, Vincent C, Coen E (1999) An epigenetic mutation responsible for natural variation in floral symmetry. Nature 401:157–161

Daehler CC, Strong DR (1997) Reduced herbivore resistance in introduced smooth cordgrass (*Spartina alterniflora*) after a century of herbivore-free growth. Oecologia 110:99–108

Danchin É, Charmantier A, Champagne FA, Mesoudi A, Pujol B, Blanchet S (2011) Beyond DNA: integrating inclusive inheritance into an extended theory of evolution. Nat Rev Genet 12:475–486

Dlugosch KM, Parker IM (2008a) Founding events in species invasions: genetic variation, adaptive evolution, and the role of multiple introductions. Molec Ecol 17:431–449

Dlugosch KM, Parker IM (2008b) Invading populations of an ornamental shrub show rapid life history evolution despite genetic bottlenecks. Ecol Lett 11:701–709

Durka W, Bossdorf O, Prati D, Auge H (2005) Molecular evidence for multiple introductions of invasive garlic mustard (*Alliaria petiolata*, Brassicaceae) to North America. Molec Ecol 14:1697–1706

Feschotte C (2008) Transposable elements and the evolution of regulatory networks. Nat Rev Genet 9:397–405

Fieldes MA (1994) Heritable effects of 5-azacytidine treatments on the growth and development of flax (*Linum usitatissimum*) genotrophs and genotypes. Genome 37:1–11

Fieldes MA, Amyot LM (1999a) Evaluating the potential of using 5-Azacytidine as an epimutagen. Can J Bot 77:1617–1622

Fieldes MA, Amyot LM (1999b) Epigenetic control of the early flowering in flax lines induced by 5-Azacytidine applied to germinating seeds. J Hered 90:199–206

Fieldes MA, Schaeffer SM, Krech MJ, Brown JCL (2005) DNA hypomethylation in 5-azacytidine-induced early-flowering lines of flax. Theor Appl Genet 111:136–149

Fox CW, Mousseau TA (1998) Maternal effects as adaptations for transgenerational phenotypic plasticity in insects. In: Mousseau TA, Fox CW (eds) Maternal effects as adaptations. Oxford University Press, New York, pp 159–177

Galloway LF, Etterson JR (2007) Transgenerational plasticity is adaptive in the wild. Science 318:1134–1136

Gammon MA, Kesseli R (2009) Haplotypes of *Fallopia* introduced into the US. Biol Invas. doi:10.1007/s10530-009-9459-7

Genereux DP, Miner BE, Bergstrom CT, Laird CD (2005) A population-epigenetic model to infer site-specific methylation rates from double-stranded DNA methylation patterns. Proc Natl Acad Sci U S A 102:5802–5807

Gissis SB, Jablonka E (2011) Transformations of Lamarckism. MIT Press, Cambridge, MA

Grant-Downton RT, Dickinson HG (2006) Epigenetics and its implications for plant biology 2. The 'epigenetic epiphany': epigenetics, evolution and beyond. Ann Bot 97:11–27

Hegarty MJ, Hiscock SJ (2008) Genomic clues to the evolutionary success of polyploid plants. Curr Biol 18:R435–R444

Heijmans BT, Tobi EW, Stein AD, Putter H, Blauw GJ, Susser ES, Slagboom PE, Lumey LH (2008) Persistent epigenetic differences associated with prenatal exposure to famine in humans. Proc Natl Acad Sci U S A 105:17046–17049

Herrera CM, Bazaga P (2008) Population-genomic approach reveals adaptive floral divergence in discrete populations of a hawk moth-pollinated violet. Molec Ecol 17:5378–5390

Herrera CM, Bazaga P (2010) Epigenetic differentiation and relationship to adaptive genetic divergence in discrete populations of the violet *Viola cazorlensis*. New Phytol 187:867–876

Herrera CM, Bazaga P (2011) Untangling individual variation in natural populations: ecological, genetic and epigenetic correlates of long-term inequality in herbivory. Molec Ecol 20:1675–1688

Holeski LM (2007) Within and between generation phenotypic plasticity in trichome density of *Mimulus guttatus*. J Evol Biol 20:2092–2100

Hollingsworth ML, Bailey JP (2000) Evidence for massive clonal growth in the invasive weed *Fallopia japonica* (Japanese knotweed). Bot J Linn Soc 133:463–472

Jablonka E, Lamb MJ (1989) The inheritance of acquired epigenetic variations. J Theor Biol 139:69–83

Jablonka E, Lamb MJ (1995) Epigenetic inheritance and evolution: the Lamarckian dimension. Oxford University Press, Oxford

Jablonka E, Lamb MJ (1998) Epigenetic inheritance in evolution. J Evol Biol 11:159–183

Jablonka E, Raz G (2009) Transgenerational epigenetic inheritance: prevalence, mechanisms, and implications for the study of heredity and evolution. Q Rev Biol 84:131–176

Jeddeloh JA, Stokes TL, Richards EJ (1999) Maintenance of genomic methylation requires a SWI2/SNF2-like protein. Nat Genet 22:94–97

Johannes F, Colot V, Jansen RC (2008) Epigenome dynamics: a quantitative genetics perspective. Nat Rev Genet 9:883–890

Johannes F, Porcher E, Teixeira FK, Saliba-Colombani V, Simon M, Agier N, Bulski A, Albuisson J, Heredia F, Audigier P, Bouchez D, Dillmann C, Guerche P, Hospital F, Colot V (2009) Assessing the impact of transgenerational epigenetic variation on complex traits. PLoS Genet 5:e1000530

Johanson U, West J, Lister C, Michaels S, Amasino R, Dean C (2000) Molecular analysis of FRIGIDA, a major determinant of natural variation in *Arabidopsis* flowering time. Science 290:344–347

Kakutani T (2002) Epi-alleles in plants: inheritance of epigenetic information over generations. Plant Cell Physiol 43:1106–1111

Kankel MW, Ramsey DE, Stokes TL, Flowers SK, Haag JR, Jeddeloh JA, Riddle NC, Verbsky ML, Richards EJ (2003) *Arabidopsis* MET1 cytosine methyltransferase mutants. Genetics 163:1109–1122

Keller SR, Taylor DR (2008) History, chance and adaptation during biological invasion: separating stochastic phenotypic evolution from response to selection. Ecol Lett 11:852–866

Keyte AL, Percifield R, Liu B, Wendel JF (2006) Infraspecific DNA methylation polymorphism in cotton (*Gossypium hirsutum* L.). J Hered 97:444–450

Kidwell MG, Lisch DR (2000) Transposable elements and host genome evolution. Trends Ecol Evol 15:95–99

Kirkpatrick M, Lande R (1989) The evolution of maternal characters. Evolution 43:485–503

Korves T, Schmid KJ, Caicedo AL, Mays C, Stinchcombe JR, Purugganan MD, Schmitt J (2007) Fitness effects associated with the major flowering time gene FRIGIDA in *Arabidopsis thaliana* in the field. Am Nat 169:E141–E157

Lavergne S, Molofsky J (2007) Increased genetic variation and evolutionary potential drive the success of an invasive grass. Proc Natl Acad Sci U S A 104:3883–3888

Li YD, Shan XH, Liu XM, Hu LJ, Guo WL, Liu B (2008) Utility of the methylation-sensitive amplified polymorphism (MSAP) marker for detection of DNA methylation polymorphism and epigenetic population structure in a wild barley species (*Hordeum brevisubulatum*). Ecol Res 23:927–930

Lippman Z, Gendrel AV, Black M, Vaughn MW, Dedhia N, McCombie WR, Lavine K, Mittal V, May B, Kasschau KD, Carrington JC, Doerge RW, Colot V, Martienssen R (2004) Role of transposable elements in heterochromatin and epigenetic control. Nature 430:471–476

Lira-Medeiros CF, Parisod C, Fernandes RA, Mata CS, Cardoso MA, Ferreira PCG (2010) Epigenetic variation in mangrove plants occurring in contrasting natural environment. PLoS One 5:e10326

Liu B, Wendel JF (2003) Epigenetic phenomena and the evolution of plant allopolyploids. Molec Phylogenet Evol 29:365–379

Loomis ES, Fishman L (2009) A continent-wide clone: Population genetic variation of the invasive plant *Hieracium aurantiacum* (Orange Hawkweed; Asteraceae) in North America. Int J Plant Sci 170:759–765

Lukens LN, Pires JC, Leon E, Vogelzang R, Oslach L, Osborn T (2006) Patterns of sequence loss and cytosine methylation within a population of newly resynthesized *Brassica napus* allopolyploids. Plant Physiol 140:336–348

Mandák B, Pysek P, Bímová K (2004) History of the invasion and distribution of *Reynoutria* taxa in the Czech Republic: a hybrid spreading faster than its parents. Preslia 76:15–64

McClintock B (1984) The significance of responses of the genome to challenge. Science 226:792–801

Mirouze M, Paszkowski J (2011) Epigenetic contribution to stress adaptation in plants. Curr Opin Plant Biol 14:267–274

Mittelsten Scheid O, Jakovleva L, Afsar K, Maluszynska J, Paszkowski J (1996) A change of ploidy can modify epigenetic silencing. Proc Natl Acad Sci U S A 93:7114–7119

Mittelsten Scheid O, Afsar K, Paszkowski J (2003) Formation of stable epialleles and their paramutation-like interaction in tetraploid *Arabidopsis thaliana*. Nat Genet 34:450–454

Mouradov A, Cremer F, Coupland G (2002) Control of flowering time: interacting pathways as a basis for diversity. Plant Cell 14:S111–S130

Olsen KM, Haldorsdottir S, Stinchcombe J, Weinig C, Schmitt J, Purugganan MD (2004) Linkage disequilibrium mapping of *Arabidopsis* CRY2 flowering time alleles. Genetics 157:1361–1369

Pal C, Miklos I (1999) Epigenetic inheritance, genetic assimilation and speciation. J Theor Biol 200:19–37

Palevitz BA (1999) Helical science. Scientist 13:31

Parisod C, Salmon A, Zerjal T, Tenaillon M, Grandbastien MA, Ainouche ML (2009) Rapid structural and epigenetic reorganization near transposable elements in hybrid and allopolyploid genomes in *Spartina*. New Phytol 184:1003–1015

Paszkowski J, Grossniklaus U (2011) Selected aspects of transgenerational epigenetic inheritance and resetting in plants. Curr Opin Plant Biol 14:195–203

Paun O, Fay MF, Soltis DE, Chase MW (2007) Genetic and epigenetic alterations after hybridization and genome doubling. Taxon 56:649–656

Paun O, Bateman RM, Fay MF, Hedrén M, Civeyrel L, Chase MW (2010) Stable epigenetic effects impact adaptation in allopolyploid orchids (*Dactylorhiza*: Orchidaceae). Molec Biol Evol 27:2465–2473

Prentis PJ, Wilson JRU, Dormontt EE, Richardson DM, Lowe AJ (2008) Adaptive evolution in invasive species. Trends Plant Sci 13:288–294

Pysek P, Brock JH, Bímová K, Mandák B, Jarosík V, Koukolíková I, Pergl J, Stepánek J (2003) Vegetative regeneration in invasive *Reynoutria* (Polygonaceae) taxa: the determinant of invisibility at the genotype level. Am J Bot 90:1487–1495

Queitsch C, Sangster TA, Lindquist S (2002) Hsp90 as a capacitor of phenotypic variation. Nature 417:618–624

Rapp RA, Wendel JF (2005) Epigenetics and plant evolution. New Phytol 168:81–91

Rebollo R, Horard B, Hubert B, Vieira C (2010) Jumping genes and epigenetics: towards new species. Gene 454:1–7

Reinders J, Paszkowski J (2009) Unlocking the *Arabidopsis* epigenome. Epigenetics 4:557–563

Reinders J, Wulff BBH, Mirouze M, Marí-Ordóñez A, Dapp M, Rozhon W, Bucher E, Theiler G, Paszkowski J (2009) Compromised stability of DNA methylation and transposon immobilization in mosaic *Arabidopsis* epigenomes. Gene Dev 23:939–950

Richards EJ (2006) Inherited epigenetic variation—revisiting soft inheritance. Nat Rev Genet 7:395–401

Richards EJ (2008) Population epigenetics. Cu Op Genet Dev 18:221–226

Richards EJ (2011) Natural epigenetic variation in plant species: a view from the field. CurrOp Plant Biol 14:204–209

Richards CL, Wendel JF (2011) The hairy problem of epigenetics in evolution. New Phytol 191:7–9

Richards CL, Walls R, Bailey JP, Parameswaran R, George T, Pigliucci M (2008) Plasticity in salt tolerance traits allows for invasion of salt marshes by Japanese knotweed s.l. (*Fallopia japonica* and *F. × bohemica*, Polygonaceae). Am J Bot 95:931–942

Richards CL, Bossdorf O, Pigliucci M (2010a) What role does heritable epigenetic variation play in phenotypic evolution? Bioscience 60:232–237

Richards CL, Bossdorf O, Verhoeven KJF (2010b) Understanding natural epigenetic variation. New Phytol 187:562–564

Rodin SN, Riggs AD (2003) Epigenetic silencing may aid evolution by gene duplication. J Molec Evol 56:718–729

Rosenthal DM, Ramakrishnan AP, Cruzan MB (2008) Evidence for multiple sources of invasion and intraspecific hybridization in *Brachypodium sylvaticum* (Hudson) Beauv. in North America. Molec Ecol 17:4657–4669

Roux F, Colomé-Tatché M, Edelist C, Wardenaar R, Guerche P, Hospital F, Colot V, Jansen RC, Johannes F (2011) Genome-wide epigenetic perturbation jump-starts patterns of heritable variation found in nature. Genetics 188(4):1015–7, in press

Rutherford SL, Lindquist S (1998) Hsp90 as a capacitor for morphological evolution. Nature 396:336–342

Salmon A, Ainouche ML, Wendel JF (2005) Genetic and epigenetic consequences of recent hybridization and polyploidy in *Spartina* (Poaceae). Molec Ecol 14:1163–1175

Schmitt J, Dudley SA, Pigliucci M (1999) Manipulative approaches to testing adaptive plasticity: phytochrome-mediated shade-avoidance responses in plants. Am Natur 154:S43–S54

Scoville AG, Barnett LB, Bodbyl-Roels S, Kelly JK, Hileman LC (2011) Differential regulation of a MYB transcription factor predicts transgenerational epigenetic inheritance of trichome density in *Mimulus guttatus*. New Phytol 191:251–263

Shindo C, Lister C, Crevillen P, Nordborg M, Dean C (2006) Variation in the epigenetic silencing of FLC contributes to natural variation in *Arabidopsis* vernalization response. Gene Dev 20:3079–3083

Simberloff D, Dayan T, Jones C, Ogura G (2000) Character displacement and release in the small Indian mongoose, *Herpestes javanicus*. Ecology 81:2086–2099

Simpson GG, Dean C (2002) *Arabidopsis*, the Rosetta stone of flowering time? Science 296:285–289

Slotkin RK, Martienssen R (2007) Transposable elements and the epigenetic regulation of the genome. Nat Rev Genet 8:272–285

Slotkin RK, Nuthikattu S, Jiang N (2012) The evolutionary impact of transposable elements on gene and genome evolution. In: Wendel JF (ed) Plant genome diversity, vol 1, Plant genomes, their residents, and their evolutionary dynamics. Springer, Wien/New York, pp

Sollars V, Lu X, Xiao L, Wang X, Garfinkel MD, Ruden DM (2003) Evidence for an epigenetic mechanism by which Hsp90 acts as a capacitor for morphological evolution. Nat Genet 33:70–74

Specchia V, Piacentini L, Tritto P, Fanti L, D'Alessandro R, Palumbo G, Pimpinelli S, Bozzetti MP (2010) Hsp90 prevents phenotypic variation by suppressing the mutagenic activity of transposons. Nature 463:662–665

Stinchcombe JR, Weinig C, Ungerer M, Olsen KM, Mays C, Halldorsdottir S, Purugganan MD, Schmitt J (2004) A latitudinal cline in flowering time in *Arabidopsis thaliana* modulated by the flowering time gene FRIGIDA. Proc Natl Acad Sci USA 101:4712–4717

Sung SB, Amasino RM (2004) Vernalization and epigenetics: how plants remember winter. Curr Opin Plant Biol 7:4–10

Tatra GS, Miranda J, Chinnappa CC, Reid DM (2000) Effect of light quality and 5-azacytidine on genomic methylation and stem elongation in two ecotypes of *Stellaria longipes*. Physiol Plant 109:313–321

Teixeira FK, Heredia F, Sarazin A, Roudier F, Boccara M, Ciaudo C, Cruaud C, Poulain J, Berdasco M, Fraga MF et al (2009) A role for RNAi in the selective correction of DNA methylation defects. Science 323:1600–1604

Tsukahara S, Kobayashi A, Kawabe A, Mathieu O, Miura A, Kakutani T (2009) Bursts of retrotransposition reproduced in *Arabidopsis*. Nature 461:423–434

Vaughn MW, Tanurdzic M, Lippman Z, Jiang H, Carrasquillo R, Rabinowicz PD, Dedhia N, McCombie WR, Agier N, Bulski A, Colot V, Doerge RW, Martienssen RA (2007) Epigenetic natural variation in *Arabidopsis thaliana*. PLoS Biol 5:1617–1629

Verhoeven KJF, Jansen JJ, van Dijk PJ, Biere A (2010a) Stress-induced DNA methylation changes and their heritability in asexual dandelions. New Phytol 185:1108–1118

Verhoeven KJF, van Dijk PJ, Biere A (2010b) Changes in genomic methylation patterns during the formation of triploid asexual dandelion lineages. Molec Ecol 19:315–324

Vongs A, Kakutani T, Martienssen RA, Richards EJ (1993) *Arabidopsis thaliana* DNA methylation mutants. Science 260:1926–1928

Vrana PB, Fossella JA, Matteson P, del Rio T, O'Neill MJ, Tilghman SM (2000) Genetic and epigenetic incompatibilities underlie hybrid dysgenesis in *Peromyscus*. Nat Genet 25:120–124

Waddington CH (1942) Canalization of development and the inheritance of acquired characteristics. Nature 150:563–565

Waddington CH (1953) Genetic assimilation of an acquired character. Evolution 7:118–126

Wang JL, Tian L, Madlung A, Lee HS, Chen M, Lee JJ, Watson B, Kagochi T, Comai L, Chen ZJ (2004) Stochastic and epigenetic changes of gene expression in *Arabidopsis* polyploids. Genetics 167:1961–1973

Weinig C, Ungerer M, Dorn LA, Kane NC, Halldorsdottir S, Mackay TFC, Purugganan MD, Schmitt J (2002) Novel loci control variation in reproductive timing in *Arabidopsis thaliana* in natural environments. Genetics 162:1875–1881

Wessler SR (1996) Turned on by stress. Plant retrotransposons. Curr Biol 6:959–961

West-Eberhard MJ (2005) Developmental plasticity and the origin of species differences. Proc Natl Acad Sci U S A 102:6543–6549

Whittle CA, Otto SP, Johnston MO, Krochko JE (2009) Adaptive epigenetic memory of ancestral temperature regime in *Arabidopsis thaliana*. Botany-Botanique 87:650–657

Zeh DW, Zeh JA, Ishida Y (2009) Transposable elements and an epigenetic basis for punctuated equilibria. Bioessays 31:715–726

Zhai JX, Liu J, Liu B, Li PC, Meyers BC, Chen XM, Cao XF (2008) Small RNA-directed epigenetic natural variation in *Arabidopsis thaliana*. PLoS Genet 4:e1000056

Index

J.F. Wendel et al. (eds.), *Plant Genome Diversity Volume 1*,
DOI 10.1007/978-3-7091-1130-7, © Springer-Verlag Wien 2012